WORLD HEALTH ORGANIZATION

INTERNATIONAL AGENCY FOR RESEARCH ON CANCER

IARC MONOGRAPHS

ON THE

EVALUATION OF CARCINOGENIC

RISKS TO HUMANS

Non-Ionizing Radiation, Part 1: Static and Extremely Low-Frequency (ELF) Electric and Magnetic Fields

VOLUME 80

This publication represents the views and expert opinions
of an IARC Working Group on the
Evaluation of Carcinogenic Risks to Humans,
which met in Lyon,

19–26 June 2001

2002

IARC MONOGRAPHS

In 1969, the International Agency for Research on Cancer (IARC) initiated a programme on the evaluation of the carcinogenic risk of chemicals to humans involving the production of critically evaluated monographs on individual chemicals. The programme was subsequently expanded to include evaluations of carcinogenic risks associated with exposures to complex mixtures, life-style factors and biological and physical agents, as well as those in specific occupations.

The objective of the programme is to elaborate and publish in the form of monographs critical reviews of data on carcinogenicity for agents to which humans are known to be exposed and on specific exposure situations; to evaluate these data in terms of human risk with the help of international working groups of experts in chemical carcinogenesis and related fields; and to indicate where additional research efforts are needed.

The lists of IARC evaluations are regularly updated and are available on Internet: http://monographs.iarc.fr/

This project was supported by Cooperative Agreement 5 UO1 CA33193 awarded by the United States National Cancer Institute, Department of Health and Human Services, and was funded in part by the European Commission, Directorate-General EMPL (Employment, and Social Affairs), Health, Safety and Hygiene at Work Unit. Additional support has been provided since 1993 by the United States National Institute of Environmental Health Sciences.

IARC Library Cataloguing in Publication Data

Non-ionizing radiation, Part 1, Static and extremely low-frequency (ELF)
    electric and magnetic fields/IARC Working Group on the Evaluation of
    Carcinogenic Risks to Humans (2002 : Lyon, France)

(IARC monographs on the evaluation of carcinogenic risks to humans ; 80)

1. Carcinogens – congresses   2. Neoplasms, non-ionizing radiation-induced,
part 1 – congresses   I. IARC Working Group on the Evaluation of
Carcinogenic Risks to Humans   II. Series

ISBN 978-9-2832128-05   (NLM Classification: W1)
ISSN 1017-1606

PRINTED IN FRANCE

CONTENTS

NOTE TO THE READER

The term 'carcinogenic risk' in the *IARC Monographs* series is taken to mean the probability that exposure to an agent will lead to cancer in humans.

Inclusion of an agent in the *Monographs* does not imply that it is a carcinogen, only that the published data have been examined. Equally, the fact that an agent has not yet been evaluated in a monograph does not mean that it is not carcinogenic.

The evaluations of carcinogenic risk are made by international working groups of independent scientists and are qualitative in nature. No recommendation is given for regulation or legislation.

Anyone who is aware of published data that may alter the evaluation of the carcinogenic risk of an agent to humans is encouraged to make this information available to the Unit of Carcinogen Identification and Evaluation, International Agency for Research on Cancer, 150 cours Albert Thomas, 69372 Lyon Cedex 08, France, in order that the agent may be considered for re-evaluation by a future Working Group.

Although every effort is made to prepare the monographs as accurately as possible, mistakes may occur. Readers are requested to communicate any errors to the Unit of Carcinogen Identification and Evaluation, so that corrections can be reported in future volumes.

IARC WORKING GROUP ON THE EVALUATION OF CARCINOGENIC RISKS TO HUMANS: NON-IONIZING RADIATION, PART 1, STATIC AND EXTREMELY LOW-FREQUENCY (ELF) ELECTRIC AND MAGNETIC FIELDS

Lyon, 19–26 June 2001

LIST OF PARTICIPANTS

Members[1]

L.E. Anderson, Pacific Northwest National Laboratory, Bioelectromagnetic Program, Battelle, PO Box 999, MS P7-51, Richland, WA 99352, USA

W.H. Bailey, E^{x}ponent, 420 Lexington Avenue, Suite 408, New York, NY 10170, USA

C.F. Blackman, Environmental Carcinogenesis Division (MD-68), US Environmental Protection Agency, Research Triangle Park, NC 27711-2055, USA

N.E. Day, Strangeways Research Laboratory, University of Cambridge, Wort's Causeway, Cambridge CB1 8RN, United Kingdom (*Chairman*)

V. DelPizzo, CA EMF Program, 5300 Twin Springs Road, Reno, NV 89510, USA

P. Guénel, INSERM Unit 88, National Hospital Saint-Maurice, 14, rue du Val d'Osne, 94415 Saint-Maurice, France

K. Hansson Mild, Working Life Institute, PO Box 7654, 907 13 Umeå, Sweden

E. Hatch, Department of Epidemiology and Biostatistics, Boston University School of Public Health, 715 Albany Street T3E, Boston, MA 02118-2526, USA

J. Juutilainen, Department of Environmental Sciences, University of Kuopio, PO Box 1627, 70211 Kuopio, Finland

L.I. Kheifets[2], Environment Division, Electric Power Research Institute, 3412 Hillview Avenue, PO Box 10412, Palo Alto, CA 94303, USA

A.R. Liboff, Department of Physics, Oakland University, Rochester, MI 48309, USA

---

[1] Unable to attend: A.L. Brown, Department of Pathology and Laboratory Medicine, University of Wisconsin-Madison, 1300 University Avenue, 6152 Medical Sciences Center, Madison, WI 53706, USA; J. Michaelis, Institute for Medical Statistics and Documentation, Johannes Gutenberg University, Langenbeckstrasse 1, 55101 Mainz, Germany

[2] Present address: Occupational and Environmental Health, Protection of the Human Environment, World Health Organization, 1211 Geneva 27, Switzerland

D.L. McCormick, IIT Research Institute, 10 West 35th Street, Chicago, IL 60616-3799, USA

M. Mevissen, Department of Veterinary Pharmacology, University of Bern, Länggass-Strasse 124, 3012 Bern, Switzerland

J. Miyakoshi, Department of Radiation Genetics, Graduate School of Medicine, Kyoto University, Yoshida-Konoe-cho, Sakyo-ku, Kyoto 606-8501, Japan

J.H. Olsen, Danish Cancer Society, Institute of Cancer Epidemiology, Strandboule-varden 49, 2100 Copenhagen, Denmark (*Vice-Chairman*)

C.J. Portier, Environmental Toxicology Program, National Institute of Environmental Health Sciences, PO Box 12233, MD A3-02, Research Triangle Park, NC 27709, USA

R.D. Saunders, National Radiological Protection Board, Chilton, Didcot, Oxon, OX11 0RQ, United Kingdom

J. Schüz, Institute for Medical Statistics and Documentation, Johannes Gutenberg University, Langenbeckstrasse 1, 55101 Mainz, Germany

J.A.J. Stolwijk, Yale University School of Medicine, Department of Epidemiology and Public Health, 333 Cedar Street, New Haven, CT 06510, USA

M.A. Stuchly, Department of Electrical and Computer Engineering, Engineering Office Wing, Room 439, University of Victoria, PO Box 3055 STN CSC, Victoria, BC V8W 3P6, Canada

B. Veyret, Laboratory PIOM, EPHE, ENSCPB, University of Bordeaux 1, 33607 Pessac, France

Representatives/Observers

United States Environmental Protection Agency
Represented by N.N. Hankin, Radiation Center for Science and Risk Analysis, Office of Radiation and Indoor Air, United States Environmental Protection Agency, Washington, DC 20460, USA

World Health Organization
Represented by M.H. Repacholi, Occupational and Environmental Health, Protection of the Human Environment, World Health Organization, 1211 Geneva 27, Switzerland

National Grid Company plc
Represented by J. Swanson, Kelvin Avenue, Leatherhead, Surrey KT22 7ST, United Kingdom

European Ramazzini Foundation
Represented by M. Soffritti, Castello di Bentivoglio, Via Saliceto 3, 40010 Bentivoglio (Bologna), Italy

IARC Secretariat

R.A. Baan, Unit of Carcinogen Identification and Evaluation (*Responsible Officer*)

M. Bird, Visiting Scientist in the Unit of Carcinogen Identification and Evaluation

P. Buffler, Unit of Environmental Cancer Epidemiology

E. Cardis, Unit of Radiation and Cancer

J. Cheney

R. Gallagher[1]

Y. Grosse, Unit of Carcinogen Identification and Evaluation

S. Kaplan (*Editor*, Bern, Switzerland)

N. Napalkov[2]

C. Partensky, Unit of Carcinogen Identification and Evaluation

M. Pearce, Unit of Radiation and Cancer

M. Plummer, Unit of Field and Intervention Studies

J. Rice, Unit of Carcinogen Identification and Evaluation (*Head of Programme*)

B.W. Stewart[3]

K. Straif, Unit of Carcinogen Identification and Evaluation

E. Suonio, Unit of Carcinogen Identification and Evaluation

Technical assistance

B. Kajo

M. Lézère

A. Meneghel

J. Mitchell

E. Perez

---

[1] Present address: Cancer Control Research Program, BC Cancer Agency, 600 West Tenth Avenue, Vancouver BC, Canada V5Z 4E6

[2] Present address: Director Emeritus, Petrov Institute of Oncology, Pesochny-2, 197758 St Petersburg, Russian Federation

[3] Present address: Cancer Control Program, South East Health, Locked Bag 88, Randwick NSW 2031, Australia

PREAMBLE

IARC MONOGRAPHS PROGRAMME ON THE EVALUATION OF CARCINOGENIC RISKS TO HUMANS

PREAMBLE

1. BACKGROUND

In 1969, the International Agency for Research on Cancer (IARC) initiated a programme to evaluate the carcinogenic risk of chemicals to humans and to produce monographs on individual chemicals. The *Monographs* programme has since been expanded to include consideration of exposures to complex mixtures of chemicals (which occur, for example, in some occupations and as a result of human habits) and of exposures to other agents, such as radiation and viruses. With Supplement 6 (IARC, 1987a), the title of the series was modified from *IARC Monographs on the Evaluation of the Carcinogenic Risk of Chemicals to Humans* to *IARC Monographs on the Evaluation of Carcinogenic Risks to Humans*, in order to reflect the widened scope of the programme.

The criteria established in 1971 to evaluate carcinogenic risk to humans were adopted by the working groups whose deliberations resulted in the first 16 volumes of the *IARC Monographs series*. Those criteria were subsequently updated by further ad-hoc working groups (IARC, 1977, 1978, 1979, 1982, 1983, 1987b, 1988, 1991a; Vainio *et al.*, 1992).

2. OBJECTIVE AND SCOPE

The objective of the programme is to prepare, with the help of international working groups of experts, and to publish in the form of monographs, critical reviews and evaluations of evidence on the carcinogenicity of a wide range of human exposures. The *Monographs* may also indicate where additional research efforts are needed.

The *Monographs* represent the first step in carcinogenic risk assessment, which involves examination of all relevant information in order to assess the strength of the available evidence that certain exposures could alter the incidence of cancer in humans. The second step is quantitative risk estimation. Detailed, quantitative evaluations of epidemiological data may be made in the *Monographs*, but without extrapolation beyond the range of the data available. Quantitative extrapolation from experimental data to the human situation is not undertaken.

The term 'carcinogen' is used in these monographs to denote an exposure that is capable of increasing the incidence of malignant neoplasms; the induction of benign neoplasms may in some circumstances (see p. 19) contribute to the judgement that the exposure is carcinogenic. The terms 'neoplasm' and 'tumour' are used interchangeably.

Some epidemiological and experimental studies indicate that different agents may act at different stages in the carcinogenic process, and several mechanisms may be involved. The aim of the *Monographs* has been, from their inception, to evaluate evidence of carcinogenicity at any stage in the carcinogenesis process, independently of the underlying mechanisms. Information on mechanisms may, however, be used in making the overall evaluation (IARC, 1991a; Vainio *et al.*, 1992; see also pp. 25–27).

The *Monographs* may assist national and international authorities in making risk assessments and in formulating decisions concerning any necessary preventive measures. The evaluations of IARC working groups are scientific, qualitative judgements about the evidence for or against carcinogenicity provided by the available data. These evaluations represent only one part of the body of information on which regulatory measures may be based. Other components of regulatory decisions vary from one situation to another and from country to country, responding to different socioeconomic and national priorities. **Therefore, no recommendation is given with regard to regulation or legislation, which are the responsibility of individual governments and/or other international organizations.**

The *IARC Monographs* are recognized as an authoritative source of information on the carcinogenicity of a wide range of human exposures. A survey of users in 1988 indicated that the *Monographs* are consulted by various agencies in 57 countries. About 2500 copies of each volume are printed, for distribution to governments, regulatory bodies and interested scientists. The Monographs are also available from IARC*Press* in Lyon and via the Distribution and Sales Service of the World Health Organization in Geneva.

3.   SELECTION OF TOPICS FOR MONOGRAPHS

Topics are selected on the basis of two main criteria: (a) there is evidence of human exposure, and (b) there is some evidence or suspicion of carcinogenicity. The term 'agent' is used to include individual chemical compounds, groups of related chemical compounds, physical agents (such as radiation) and biological factors (such as viruses). Exposures to mixtures of agents may occur in occupational exposures and as a result of personal and cultural habits (like smoking and dietary practices). Chemical analogues and compounds with biological or physical characteristics similar to those of suspected carcinogens may also be considered, even in the absence of data on a possible carcinogenic effect in humans or experimental animals.

The scientific literature is surveyed for published data relevant to an assessment of carcinogenicity. The IARC information bulletins on agents being tested for carcinogenicity (IARC, 1973–1996) and directories of on-going research in cancer epidemiology (IARC, 1976–1996) often indicate exposures that may be scheduled for future meetings. Ad-hoc working groups convened by IARC in 1984, 1989, 1991, 1993 and 1998 gave recommendations as to which agents should be evaluated in the IARC Monographs series (IARC, 1984, 1989, 1991b, 1993, 1998a,b).

As significant new data on subjects on which monographs have already been prepared become available, re-evaluations are made at subsequent meetings, and revised monographs are published.

4.   DATA FOR MONOGRAPHS

The *Monographs* do not necessarily cite all the literature concerning the subject of an evaluation. Only those data considered by the Working Group to be relevant to making the evaluation are included.

With regard to biological and epidemiological data, only reports that have been published or accepted for publication in the openly available scientific literature are reviewed by the working groups. In certain instances, government agency reports that have undergone peer review and are widely available are considered. Exceptions may be made on an ad-hoc basis to include unpublished reports that are in their final form and publicly available, if their inclusion is considered pertinent to making a final evaluation (see pp. 25–27). In the sections on chemical and physical properties, on analysis, on production and use and on occurrence, unpublished sources of information may be used.

5.   THE WORKING GROUP

Reviews and evaluations are formulated by a working group of experts. The tasks of the group are: (i) to ascertain that all appropriate data have been collected; (ii) to select the data relevant for the evaluation on the basis of scientific merit; (iii) to prepare accurate summaries of the data to enable the reader to follow the reasoning of the Working Group; (iv) to evaluate the results of epidemiological and experimental studies on cancer; (v) to evaluate data relevant to the understanding of mechanism of action; and (vi) to make an overall evaluation of the carcinogenicity of the exposure to humans.

Working Group participants who contributed to the considerations and evaluations within a particular volume are listed, with their addresses, at the beginning of each publication. Each participant who is a member of a working group serves as an individual scientist and not as a representative of any organization, government or industry. In addition, nominees of national and international agencies and industrial associations may be invited as observers.

6.   WORKING PROCEDURES

Approximately one year in advance of a meeting of a working group, the topics of the monographs are announced and participants are selected by IARC staff in consultation with other experts. Subsequently, relevant biological and epidemiological data are collected by the Carcinogen Identification and Evaluation Unit of IARC from recognized sources of information on carcinogenesis, including data storage and retrieval systems such as MEDLINE and TOXLINE.

For chemicals and some complex mixtures, the major collection of data and the preparation of first drafts of the sections on chemical and physical properties, on analysis, on production and use and on occurrence are carried out under a separate contract funded by the United States National Cancer Institute. Representatives from industrial associations may assist in the preparation of sections on production and use. Information on production and trade is obtained from governmental and trade publications and, in some cases, by direct contact with industries. Separate production data on some agents may not be available because their publication could disclose confidential information. Information on uses may be obtained from published sources but is often complemented by direct contact with manufacturers. Efforts are made to supplement this information with data from other national and international sources.

Six months before the meeting, the material obtained is sent to meeting participants, or is used by IARC staff, to prepare sections for the first drafts of monographs. The first drafts are compiled by IARC staff and sent before the meeting to all participants of the Working Group for review.

The Working Group meets in Lyon for seven to eight days to discuss and finalize the texts of the monographs and to formulate the evaluations. After the meeting, the master copy of each monograph is verified by consulting the original literature, edited and prepared for publication. The aim is to publish monographs within six months of the Working Group meeting.

The available studies are summarized by the Working Group, with particular regard to the qualitative aspects discussed below. In general, numerical findings are indicated as they appear in the original report; units are converted when necessary for easier comparison. The Working Group may conduct additional analyses of the published data and use them in their assessment of the evidence; the results of such supplementary analyses are given in square brackets. When an important aspect of a study, directly impinging on its interpretation, should be brought to the attention of the reader, a comment is given in square brackets.

7.   EXPOSURE DATA

Sections that indicate the extent of past and present human exposure, the sources of exposure, the people most likely to be exposed and the factors that contribute to the exposure are included at the beginning of each monograph.

Most monographs on individual chemicals, groups of chemicals or complex mixtures include sections on chemical and physical data, on analysis, on production and use and on occurrence. In monographs on, for example, physical agents, occupational exposures and cultural habits, other sections may be included, such as: historical perspectives, description of an industry or habit, chemistry of the complex mixture or taxonomy. Monographs on biological agents have sections on structure and biology, methods of detection, epidemiology of infection and clinical disease other than cancer.

For chemical exposures, the Chemical Abstracts Services Registry Number, the latest Chemical Abstracts Primary Name and the IUPAC Systematic Name are recorded; other synonyms are given, but the list is not necessarily comprehensive. For biological agents, taxonomy and structure are described, and the degree of variability is given, when applicable.

Information on chemical and physical properties and, in particular, data relevant to identification, occurrence and biological activity are included. For biological agents, mode of replication, life cycle, target cells, persistence and latency and host response are given. A description of technical products of chemicals includes trade names, relevant specifications and available information on composition and impurities. Some of the trade names given may be those of mixtures in which the agent being evaluated is only one of the ingredients.

The purpose of the section on analysis or detection is to give the reader an overview of current methods, with emphasis on those widely used for regulatory purposes. Methods for monitoring human exposure are also given, when available. No critical evaluation or recommendation of any of the methods is meant or implied. The IARC published a series of volumes, *Environmental Carcinogens: Methods of Analysis and Exposure Measurement* (IARC, 1978–93), that describe validated methods for analysing a wide variety of chemicals and mixtures. For biological agents, methods of detection and exposure assessment are described, including their sensitivity, specificity and reproducibility.

The dates of first synthesis and of first commercial production of a chemical or mixture are provided; for agents which do not occur naturally, this information may allow a reasonable estimate to be made of the date before which no human exposure to the agent could have occurred. The dates of first reported occurrence of an exposure are also provided. In addition, methods of synthesis used in past and present commercial production and different methods of production which may give rise to different impurities are described.

Data on production, international trade and uses are obtained for representative regions, which usually include Europe, Japan and the United States of America. It should not, however, be inferred that those areas or nations are necessarily the sole or major sources or users of the agent. Some identified uses may not be current or major applications, and the coverage is not necessarily comprehensive. In the case of drugs, mention of their therapeutic uses does not necessarily represent current practice, nor does it imply judgement as to their therapeutic efficacy.

Information on the occurrence of an agent or mixture in the environment is obtained from data derived from the monitoring and surveillance of levels in occupational environments, air, water, soil, foods and animal and human tissues. When available, data on the generation, persistence and bioaccumulation of the agent are also included. In the case of mixtures, industries, occupations or processes, information is given about all agents present. For processes, industries and occupations, a historical description is also given, noting variations in chemical composition, physical properties and levels of occupational exposure with time and place. For biological agents, the epidemiology of infection is described.

Statements concerning regulations and guidelines (e.g., pesticide registrations, maximal levels permitted in foods, occupational exposure limits) are included for some countries as indications of potential exposures, but they may not reflect the most recent situation, since such limits are continuously reviewed and modified. The absence of information on regulatory status for a country should not be taken to imply that that country does not have regulations with regard to the exposure. For biological agents, legislation and control, including vaccines and therapy, are described.

8.   STUDIES OF CANCER IN HUMANS

(a)   Types of studies considered

Three types of epidemiological studies of cancer contribute to the assessment of carcinogenicity in humans—cohort studies, case–control studies and correlation (or ecological) studies. Rarely, results from randomized trials may be available. Case series and case reports of cancer in humans may also be reviewed.

Cohort and case–control studies relate the exposures under study to the occurrence of cancer in individuals and provide an estimate of relative risk (ratio of incidence or mortality in those exposed to incidence or mortality in those not exposed) as the main measure of association.

In correlation studies, the units of investigation are usually whole populations (e.g. in particular geographical areas or at particular times), and cancer frequency is related to a summary measure of the exposure of the population to the agent, mixture or exposure circumstance under study. Because individual exposure is not documented, however, a causal relationship is less easy to infer from correlation studies than from cohort and case–control studies. Case reports generally arise from a suspicion, based on clinical experience, that the concurrence of two events—that is, a particular exposure and occurrence of a cancer—has happened rather more frequently than would be expected by chance. Case reports usually lack complete ascertainment of cases in any population, definition or enumeration of the population at risk and estimation of the expected number of cases in the absence of exposure. The uncertainties surrounding interpretation of case reports and correlation studies make them inadequate, except in rare instances, to form the sole basis for inferring a causal relationship. When taken together with case–control and cohort studies, however, relevant case reports or correlation studies may add materially to the judgement that a causal relationship is present.

Epidemiological studies of benign neoplasms, presumed preneoplastic lesions and other end-points thought to be relevant to cancer are also reviewed by working groups. They may, in some instances, strengthen inferences drawn from studies of cancer itself.

(b)   Quality of studies considered

The Monographs are not intended to summarize all published studies. Those that are judged to be inadequate or irrelevant to the evaluation are generally omitted. They may be mentioned briefly, particularly when the information is considered to be a useful supplement to that in other reports or when they provide the only data available. Their inclusion does not imply acceptance of the adequacy of the study design or of the analysis and interpretation of the results, and limitations are clearly outlined in square brackets at the end of the study description.

It is necessary to take into account the possible roles of bias, confounding and chance in the interpretation of epidemiological studies. By 'bias' is meant the operation of factors in study design or execution that lead erroneously to a stronger or weaker association than in fact exists between disease and an agent, mixture or exposure circumstance. By 'confounding' is meant a situation in which the relationship with disease is made to appear stronger or weaker than it truly is as a result of an association between the apparent causal factor and another factor that is associated with either an increase or decrease in the incidence of the disease. In evaluating the extent to which these factors have been minimized in an individual study, working groups consider a number of aspects of design and analysis as described in the report of the study. Most of these considerations apply equally to case–control, cohort and correlation studies. Lack of clarity of any of these aspects in the reporting of a study can decrease its credibility and the weight given to it in the final evaluation of the exposure.

Firstly, the study population, disease (or diseases) and exposure should have been well defined by the authors. Cases of disease in the study population should have been identified in a way that was independent of the exposure of interest, and exposure should have been assessed in a way that was not related to disease status.

Secondly, the authors should have taken account in the study design and analysis of other variables that can influence the risk of disease and may have been related to the exposure of interest. Potential confounding by such variables should have been dealt with either in the design of the study, such as by matching, or in the analysis, by statistical adjustment. In cohort studies, comparisons with local rates of disease may be more appropriate than those with national rates. Internal comparisons of disease frequency among individuals at different levels of exposure should also have been made in the study.

Thirdly, the authors should have reported the basic data on which the conclusions are founded, even if sophisticated statistical analyses were employed. At the very least, they should have given the numbers of exposed and unexposed cases and controls in a case–control study and the numbers of cases observed and expected in a cohort study. Further tabulations by time since exposure began and other temporal factors are also important. In a cohort study, data on all cancer sites and all causes of death should have been given, to reveal the possibility of reporting bias. In a case–control study, the effects of investigated factors other than the exposure of interest should have been reported.

Finally, the statistical methods used to obtain estimates of relative risk, absolute rates of cancer, confidence intervals and significance tests, and to adjust for confounding should have been clearly stated by the authors. The methods used should preferably have been the generally accepted techniques that have been refined since the mid-1970s. These methods have been reviewed for case–control studies (Breslow & Day, 1980) and for cohort studies (Breslow & Day, 1987).

(c)   *Inferences about mechanism of action*

Detailed analyses of both relative and absolute risks in relation to temporal variables, such as age at first exposure, time since first exposure, duration of exposure, cumulative exposure and time since exposure ceased, are reviewed and summarized when available. The analysis of temporal relationships can be useful in formulating models of carcinogenesis. In particular, such analyses may suggest whether a carcinogen acts early or late in the process of carcinogenesis, although at best they allow only indirect inferences about the mechanism of action. Special attention is given to measurements of biological markers of carcinogen exposure or action, such as DNA or protein adducts, as well as markers of early steps in the carcinogenic process, such as proto-oncogene mutation, when these are incorporated into epidemiological studies focused on cancer incidence or mortality. Such measurements may allow inferences to be made about putative mechanisms of action (IARC, 1991a; Vainio *et al.*, 1992).

(d)   *Criteria for causality*

After the individual epidemiological studies of cancer have been summarized and the quality assessed, a judgement is made concerning the strength of evidence that the agent, mixture or exposure circumstance in question is carcinogenic for humans. In making its judgement, the Working Group considers several criteria for causality. A strong association (a large relative risk) is more likely to indicate causality than a weak association, although it is recognized that relative risks of small magnitude do not imply lack of causality and may be important if the disease is common. Associations that are replicated in several studies of the same design or using different epidemiological approaches or under different circumstances of exposure are more likely to represent a causal relationship than isolated observations from single studies. If there are inconsistent results among investigations, possible reasons are sought (such as differences in amount of exposure), and results of studies judged to be of high quality are given more weight than those of studies judged to be methodologically less sound. When suspicion of carcinogenicity arises largely from a single study, these data are not combined with those from later studies in any subsequent reassessment of the strength of the evidence.

If the risk of the disease in question increases with the amount of exposure, this is considered to be a strong indication of causality, although absence of a graded response is not necessarily evidence against a causal relationship. Demonstration of a decline in risk after cessation of or reduction in exposure in individuals or in whole populations also supports a causal interpretation of the findings.

Although a carcinogen may act upon more than one target, the specificity of an association (an increased occurrence of cancer at one anatomical site or of one morphological type) adds plausibility to a causal relationship, particularly when excess cancer occurrence is limited to one morphological type within the same organ.

Although rarely available, results from randomized trials showing different rates among exposed and unexposed individuals provide particularly strong evidence for causality.

When several epidemiological studies show little or no indication of an association between an exposure and cancer, the judgement may be made that, in the aggregate, they show evidence of lack of carcinogenicity. Such a judgement requires first of all that the studies giving rise to it meet, to a sufficient degree, the standards of design and analysis described above. Specifically, the possibility that bias, confounding or misclassification of exposure or outcome could explain the observed results should be considered and excluded with reasonable certainty. In addition, all studies that are judged to be methodologically sound should be consistent with a relative risk of unity for any observed level of exposure and, when considered together, should provide a pooled estimate of relative risk which is at or near unity and has a narrow confidence interval, due to sufficient population size. Moreover, no individual study nor the pooled results of all the studies should show any consistent tendency for the relative risk of cancer to increase with increasing level of exposure. It is important to note that evidence of lack of carcinogenicity obtained in this way from several epidemiological studies can apply only to the type(s) of cancer studied and to dose levels and intervals between first exposure and observation of disease that are the same as or less than those observed in all the studies. Experience with human cancer indicates that, in some cases, the period from first exposure to the development of clinical cancer is seldom less than 20 years; latent periods substantially shorter than 30 years cannot provide evidence for lack of carcinogenicity.

9. STUDIES OF CANCER IN EXPERIMENTAL ANIMALS

All known human carcinogens that have been studied adequately in experimental animals have produced positive results in one or more animal species (Wilbourn *et al.*, 1986; Tomatis *et al.*, 1989). For several agents (aflatoxins, 4-aminobiphenyl, azathioprine, betel quid with tobacco, bischloromethyl ether and chloromethyl methyl ether (technical grade), chlorambucil, chlornaphazine, ciclosporin, coal-tar pitches, coal-tars, combined oral contraceptives, cyclophosphamide, diethylstilboestrol, melphalan, 8-methoxypsoralen plus ultraviolet A radiation, mustard gas, myleran, 2-naphthylamine, nonsteroidal estrogens, estrogen replacement therapy/steroidal estrogens, solar radiation, thiotepa and vinyl chloride), carcinogenicity in experimental animals was established or highly suspected before epidemiological studies confirmed their carcinogenicity in humans (Vainio *et al.*, 1995). Although this association cannot establish that all agents and mixtures that cause cancer in experimental animals also cause cancer in humans, nevertheless, **in the absence of adequate data on humans, it is biologically plausible and prudent to regard agents and mixtures for which there is *sufficient evidence* (see p. 24) of carcinogenicity in experimental animals as if they presented a carcinogenic risk to humans**. The possibility that a given agent may cause cancer through a species-specific mechanism which does not operate in humans (see p. 27) should also be taken into consideration.

The nature and extent of impurities or contaminants present in the chemical or mixture being evaluated are given when available. Animal strain, sex, numbers per group, age at start of treatment and survival are reported.

Other types of studies summarized include: experiments in which the agent or mixture was administered in conjunction with known carcinogens or factors that modify carcinogenic effects; studies in which the end-point was not cancer but a defined precancerous lesion; and experiments on the carcinogenicity of known metabolites and derivatives.

For experimental studies of mixtures, consideration is given to the possibility of changes in the physicochemical properties of the test substance during collection, storage, extraction, concentration and delivery. Chemical and toxicological interactions of the components of mixtures may result in nonlinear dose–response relationships.

An assessment is made as to the relevance to human exposure of samples tested in experimental animals, which may involve consideration of: (i) physical and chemical characteristics, (ii) constituent substances that indicate the presence of a class of substances, (iii) the results of tests for genetic and related effects, including studies on DNA adduct formation, proto-oncogene mutation and expression and suppressor gene inactivation. The relevance of results obtained, for example, with animal viruses analogous to the virus being evaluated in the monograph must also be considered. They may provide biological and mechanistic information relevant to the understanding of the process of carcinogenesis in humans and may strengthen the plausibility of a conclusion that the biological agent under evaluation is carcinogenic in humans.

(a)   Qualitative aspects

An assessment of carcinogenicity involves several considerations of qualitative importance, including (i) the experimental conditions under which the test was per-formed, including route and schedule of exposure, species, strain, sex, age, duration of follow-up; (ii) the consistency of the results, for example, across species and target organ(s); (iii) the spectrum of neoplastic response, from preneoplastic lesions and benign tumours to malignant neoplasms; and (iv) the possible role of modifying factors.

As mentioned earlier (p. 11), the *Monographs* are not intended to summarize all published studies. Those studies in experimental animals that are inadequate (e.g., too short a duration, too few animals, poor survival; see below) or are judged irrelevant to the evaluation are generally omitted. Guidelines for conducting adequate long-term carcinogenicity experiments have been outlined (e.g. Montesano *et al.*, 1986).

Considerations of importance to the Working Group in the interpretation and eva-luation of a particular study include: (i) how clearly the agent was defined and, in the case of mixtures, how adequately the sample characterization was reported; (ii) whether the dose was adequately monitored, particularly in inhalation experiments; (iii) whether the doses and duration of treatment were appropriate and whether the survival of treated animals was similar to that of controls; (iv) whether there were adequate numbers of animals per group; (v) whether animals of each sex were used; (vi) whether animals were allocated randomly to groups; (vii) whether the duration of observation was adequate; and (viii) whether the data were adequately reported. If available, recent data on the incidence of specific tumours in historical controls, as well as in concurrent controls, should be taken into account in the evaluation of tumour response.

When benign tumours occur together with and originate from the same cell type in an organ or tissue as malignant tumours in a particular study and appear to represent a stage in the progression to malignancy, it may be valid to combine them in assessing tumour incidence (Huff *et al.*, 1989). The occurrence of lesions presumed to be preneoplastic may in certain instances aid in assessing the biological plausibility of any neoplastic response observed. If an agent or mixture induces only benign neoplasms that appear to be end-points that do not readily progress to malignancy, it should nevertheless be suspected of being a carcinogen and requires further investigation.

(b)   Quantitative aspects

The probability that tumours will occur may depend on the species, sex, strain and age of the animal, the dose of the carcinogen and the route and length of exposure. Evidence of an increased incidence of neoplasms with increased level of exposure strengthens the inference of a causal association between the exposure and the development of neoplasms.

The form of the dose–response relationship can vary widely, depending on the particular agent under study and the target organ. Both DNA damage and increased cell division are important aspects of carcinogenesis, and cell proliferation is a strong determinant of dose–response relationships for some carcinogens (Cohen & Ellwein, 1990). Since many chemicals require metabolic activation before being converted into their reactive intermediates, both metabolic and pharmacokinetic aspects are important in determining the dose–response pattern. Saturation of steps such as absorption, activation, inactivation and elimination may produce nonlinearity in the dose–response relationship, as could saturation of processes such as DNA repair (Hoel *et al.*, 1983; Gart *et al.*, 1986).

(c)   Statistical analysis of long-term experiments in animals

Factors considered by the Working Group include the adequacy of the information given for each treatment group: (i) the number of animals studied and the number examined histologically, (ii) the number of animals with a given tumour type and (iii) length of survival. The statistical methods used should be clearly stated and should be the generally accepted techniques refined for this purpose (Peto *et al.*, 1980; Gart *et al.*, 1986). When there is no difference in survival between control and treatment groups, the Working Group usually compares the proportions of animals developing each tumour type in each of the groups. Otherwise, consideration is given as to whether or not appropriate adjustments have been made for differences in survival. These adjustments can include: comparisons of the proportions of tumour-bearing animals among the effective number of animals (alive at the time the first tumour is discovered), in the case where most differences in survival occur before tumours appear; life-table methods, when tumours are visible or when they may be considered 'fatal' because mortality rapidly follows tumour development; and the Mantel-Haenszel test or logistic regression, when occult tumours do not affect the animals' risk of dying but are 'incidental' findings at autopsy.

In practice, classifying tumours as fatal or incidental may be difficult. Several survival-adjusted methods have been developed that do not require this distinction (Gart *et al.*, 1986), although they have not been fully evaluated.

10.   OTHER DATA RELEVANT TO AN EVALUATION OF CARCINOGENICITY AND ITS MECHANISMS

In coming to an overall evaluation of carcinogenicity in humans (see pp. 25–27), the Working Group also considers related data. The nature of the information selected for the summary depends on the agent being considered.

For chemicals and complex mixtures of chemicals such as those in some occupational situations or involving cultural habits (e.g. tobacco smoking), the other data considered to be relevant are divided into those on absorption, distribution, metabolism and excretion; toxic effects; reproductive and developmental effects; and genetic and related effects.

Concise information is given on absorption, distribution (including placental transfer) and excretion in both humans and experimental animals. Kinetic factors that may affect the dose–response relationship, such as saturation of uptake, protein binding, metabolic activation, detoxification and DNA repair processes, are mentioned. Studies that indicate the metabolic fate of the agent in humans and in experimental animals are summarized briefly, and comparisons of data on humans and on animals are made when possible. Comparative information on the relationship between exposure and the dose that reaches the target site may be of particular importance for extrapolation between species. Data are given on acute and chronic toxic effects (other than cancer), such as organ toxicity, increased cell proliferation, immunotoxicity and endocrine effects. The presence and toxicological significance of cellular receptors is described. Effects on reproduction, teratogenicity, fetotoxicity and embryotoxicity are also summarized briefly.

Tests of genetic and related effects are described in view of the relevance of gene mutation and chromosomal damage to carcinogenesis (Vainio *et al.*, 1992; McGregor *et al.*, 1999). The adequacy of the reporting of sample characterization is considered and, where necessary, commented upon; with regard to complex mixtures, such comments are similar to those described for animal carcinogenicity tests on p. 18. The available data are interpreted critically by phylogenetic group according to the end-points detected, which may include DNA damage, gene mutation, sister chromatid exchange, micronucleus formation, chromosomal aberrations, aneuploidy and cell transformation. The concentrations employed are given, and mention is made of whether use of an exogenous metabolic system *in vitro* affected the test result. These data are given as listings of test systems, data and references. The Genetic and Related Effects data presented in the *Monographs* are also available in the form of Graphic Activity Profiles (GAP) prepared in collaboration with the United States Environmental Protection Agency (EPA) (see also

Waters *et al.*, 1987) using software for personal computers that are Microsoft Windows®
compatible. The EPA/IARC GAP software and database may be downloaded free of
charge from *www.epa.gov/gapdb*.

Positive results in tests using prokaryotes, lower eukaryotes, plants, insects and
cultured mammalian cells suggest that genetic and related effects could occur in
mammals. Results from such tests may also give information about the types of genetic
effect produced and about the involvement of metabolic activation. Some end-points
described are clearly genetic in nature (e.g., gene mutations and chromosomal aberrations), while others are to a greater or lesser degree associated with genetic effects (e.g.
unscheduled DNA synthesis). In-vitro tests for tumour-promoting activity and for cell
transformation may be sensitive to changes that are not necessarily the result of genetic
alterations but that may have specific relevance to the process of carcinogenesis. A
critical appraisal of these tests has been published (Montesano *et al.*, 1986).

Genetic or other activity manifest in experimental mammals and humans is regarded
as being of greater relevance than that in other organisms. The demonstration that an
agent or mixture can induce gene and chromosomal mutations in whole mammals indicates that it may have carcinogenic activity, although this activity may not be detectably
expressed in any or all species. Relative potency in tests for mutagenicity and related
effects is not a reliable indicator of carcinogenic potency. Negative results in tests for
mutagenicity in selected tissues from animals treated *in vivo* provide less weight, partly
because they do not exclude the possibility of an effect in tissues other than those
examined. Moreover, negative results in short-term tests with genetic end-points cannot
be considered to provide evidence to rule out carcinogenicity of agents or mixtures that
act through other mechanisms (e.g. receptor-mediated effects, cellular toxicity with rege-
nerative proliferation, peroxisome proliferation) (Vainio *et al.*, 1992). Factors that may
lead to misleading results in short-term tests have been discussed in detail elsewhere
(Montesano *et al.*, 1986).

When available, data relevant to mechanisms of carcinogenesis that do not involve
structural changes at the level of the gene are also described.

The adequacy of epidemiological studies of reproductive outcome and genetic and
related effects in humans is evaluated by the same criteria as are applied to epidemiological studies of cancer.

Structure–activity relationships that may be relevant to an evaluation of the carcinogenicity of an agent are also described.

For biological agents—viruses, bacteria and parasites—other data relevant to
carcinogenicity include descriptions of the pathology of infection, molecular biology
(integration and expression of viruses, and any genetic alterations seen in human
tumours) and other observations, which might include cellular and tissue responses to
infection, immune response and the presence of tumour markers.

11. SUMMARY OF DATA REPORTED

In this section, the relevant epidemiological and experimental data are summarized. Only reports, other than in abstract form, that meet the criteria outlined on p. 11 are considered for evaluating carcinogenicity. Inadequate studies are generally not summarized: such studies are usually identified by a square-bracketed comment in the preceding text.

(*a*)   *Exposure*

Human exposure to chemicals and complex mixtures is summarized on the basis of elements such as production, use, occurrence in the environment and determinations in human tissues and body fluids. Quantitative data are given when available. Exposure to biological agents is described in terms of transmission and prevalence of infection.

(*b*)   *Carcinogenicity in humans*

Results of epidemiological studies that are considered to be pertinent to an assessment of human carcinogenicity are summarized. When relevant, case reports and correlation studies are also summarized.

(*c*)   *Carcinogenicity in experimental animals*

Data relevant to an evaluation of carcinogenicity in animals are summarized. For each animal species and route of administration, it is stated whether an increased incidence of neoplasms or preneoplastic lesions was observed, and the tumour sites are indicated. If the agent or mixture produced tumours after prenatal exposure or in single-dose experiments, this is also indicated. Negative findings are also summarized. Dose-response and other quantitative data may be given when available.

(*d*)   *Other data relevant to an evaluation of carcinogenicity and its mechanisms*

Data on biological effects in humans that are of particular relevance are summarized. These may include toxicological, kinetic and metabolic considerations and evidence of DNA binding, persistence of DNA lesions or genetic damage in exposed humans. Toxicological information, such as that on cytotoxicity and regeneration, receptor binding and hormonal and immunological effects, and data on kinetics and metabolism in experimental animals are given when considered relevant to the possible mechanism of the carcinogenic action of the agent. The results of tests for genetic and related effects are summarized for whole mammals, cultured mammalian cells and nonmammalian systems.

When available, comparisons of such data for humans and for animals, and particularly animals that have developed cancer, are described.

Structure–activity relationships are mentioned when relevant.

For the agent, mixture or exposure circumstance being evaluated, the available data on end-points or other phenomena relevant to mechanisms of carcinogenesis from studies in humans, experimental animals and tissue and cell test systems are summarized within one or more of the following descriptive dimensions:

(i) Evidence of genotoxicity (structural changes at the level of the gene): for example, structure–activity considerations, adduct formation, mutagenicity (effect on specific genes), chromosomal mutation/aneuploidy

(ii) Evidence of effects on the expression of relevant genes (functional changes at the intracellular level): for example, alterations to the structure or quantity of the product of a proto-oncogene or tumour-suppressor gene, alterations to metabolic activation/inactivation/DNA repair

(iii) Evidence of relevant effects on cell behaviour (morphological or behavioural changes at the cellular or tissue level): for example, induction of mitogenesis, compensatory cell proliferation, preneoplasia and hyperplasia, survival of premalignant or malignant cells (immortalization, immunosuppression), effects on metastatic potential

(iv) Evidence from dose and time relationships of carcinogenic effects and interactions between agents: for example, early/late stage, as inferred from epidemiological studies; initiation/promotion/progression/malignant conversion, as defined in animal carcinogenicity experiments; toxicokinetics

These dimensions are not mutually exclusive, and an agent may fall within more than one of them. Thus, for example, the action of an agent on the expression of relevant genes could be summarized under both the first and second dimensions, even if it were known with reasonable certainty that those effects resulted from genotoxicity.

12. EVALUATION

Evaluations of the strength of the evidence for carcinogenicity arising from human and experimental animal data are made, using standard terms.

It is recognized that the criteria for these evaluations, described below, cannot encompass all of the factors that may be relevant to an evaluation of carcinogenicity. In considering all of the relevant scientific data, the Working Group may assign the agent, mixture or exposure circumstance to a higher or lower category than a strict interpretation of these criteria would indicate.

(a) Degrees of evidence for carcinogenicity in humans and in experimental animals and supporting evidence

These categories refer only to the strength of the evidence that an exposure is carcinogenic and not to the extent of its carcinogenic activity (potency) nor to the mechanisms involved. A classification may change as new information becomes available.

An evaluation of degree of evidence, whether for a single agent or a mixture, is limited to the materials tested, as defined physically, chemically or biologically. When the agents evaluated are considered by the Working Group to be sufficiently closely related, they may be grouped together for the purpose of a single evaluation of degree of evidence.

(i) Carcinogenicity in humans

The applicability of an evaluation of the carcinogenicity of a mixture, process, occupation or industry on the basis of evidence from epidemiological studies depends on the variability over time and place of the mixtures, processes, occupations and industries. The Working Group seeks to identify the specific exposure, process or activity which is considered most likely to be responsible for any excess risk. The evaluation is focused as narrowly as the available data on exposure and other aspects permit.

The evidence relevant to carcinogenicity from studies in humans is classified into one of the following categories:

Sufficient evidence of carcinogenicity: The Working Group considers that a causal relationship has been established between exposure to the agent, mixture or exposure circumstance and human cancer. That is, a positive relationship has been observed between the exposure and cancer in studies in which chance, bias and confounding could be ruled out with reasonable confidence.

Limited evidence of carcinogenicity: A positive association has been observed between exposure to the agent, mixture or exposure circumstance and cancer for which a causal interpretation is considered by the Working Group to be credible, but chance, bias or confounding could not be ruled out with reasonable confidence.

Inadequate evidence of carcinogenicity: The available studies are of insufficient quality, consistency or statistical power to permit a conclusion regarding the presence or absence of a causal association between exposure and cancer, or no data on cancer in humans are available.

Evidence suggesting lack of carcinogenicity: There are several adequate studies covering the full range of levels of exposure that human beings are known to encounter, which are mutually consistent in not showing a positive association between exposure to the agent, mixture or exposure circumstance and any studied cancer at any observed level of exposure. A conclusion of 'evidence suggesting lack of carcinogenicity' is inevitably limited to the cancer sites, conditions and levels of exposure and length of observation covered by the available studies. In addition, the possibility of a very small risk at the levels of exposure studied can never be excluded.

In some instances, the above categories may be used to classify the degree of evidence related to carcinogenicity in specific organs or tissues.

(ii)   *Carcinogenicity in experimental animals*

The evidence relevant to carcinogenicity in experimental animals is classified into one of the following categories:

Sufficient evidence of carcinogenicity: The Working Group considers that a causal relationship has been established between the agent or mixture and an increased incidence of malignant neoplasms or of an appropriate combination of benign and malignant neoplasms in (a) two or more species of animals or (b) in two or more independent studies in one species carried out at different times or in different laboratories or under different protocols.

Exceptionally, a single study in one species might be considered to provide sufficient evidence of carcinogenicity when malignant neoplasms occur to an unusual degree with regard to incidence, site, type of tumour or age at onset.

Limited evidence of carcinogenicity: The data suggest a carcinogenic effect but are limited for making a definitive evaluation because, e.g. (a) the evidence of carcinogenicity is restricted to a single experiment; or (b) there are unresolved questions regarding the adequacy of the design, conduct or interpretation of the study; or (c) the agent or mixture increases the incidence only of benign neoplasms or lesions of uncertain neoplastic potential, or of certain neoplasms which may occur spontaneously in high incidences in certain strains.

Inadequate evidence of carcinogenicity: The studies cannot be interpreted as showing either the presence or absence of a carcinogenic effect because of major qualitative or quantitative limitations, or no data on cancer in experimental animals are available.

Evidence suggesting lack of carcinogenicity: Adequate studies involving at least two species are available which show that, within the limits of the tests used, the agent or mixture is not carcinogenic. A conclusion of evidence suggesting lack of carcinogenicity is inevitably limited to the species, tumour sites and levels of exposure studied.

(*b*)   *Other data relevant to the evaluation of carcinogenicity and its mechanisms*

Other evidence judged to be relevant to an evaluation of carcinogenicity and of sufficient importance to affect the overall evaluation is then described. This may include data on preneoplastic lesions, tumour pathology, genetic and related effects, structure–activity relationships, metabolism and pharmacokinetics, physicochemical parameters and analogous biological agents.

Data relevant to mechanisms of the carcinogenic action are also evaluated. The strength of the evidence that any carcinogenic effect observed is due to a particular mechanism is assessed, using terms such as weak, moderate or strong. Then, the Working Group assesses if that particular mechanism is likely to be operative in humans. The strongest indications that a particular mechanism operates in humans come from data on humans or biological specimens obtained from exposed humans. The data may be considered to be especially relevant if they show that the agent in question has caused changes in exposed humans that are on the causal pathway to carcinogenesis. Such data may, however, never become available, because it is at least conceivable that certain compounds may be kept from human use solely on the basis of evidence of their toxicity and/or carcinogenicity in experimental systems.

For complex exposures, including occupational and industrial exposures, the chemical composition and the potential contribution of carcinogens known to be present are considered by the Working Group in its overall evaluation of human carcinogenicity. The Working Group also determines the extent to which the materials tested in experimental systems are related to those to which humans are exposed.

(*c*)   *Overall evaluation*

Finally, the body of evidence is considered as a whole, in order to reach an overall evaluation of the carcinogenicity to humans of an agent, mixture or circumstance of exposure.

An evaluation may be made for a group of chemical compounds that have been evaluated by the Working Group. In addition, when supporting data indicate that other, related compounds for which there is no direct evidence of capacity to induce cancer in humans or in animals may also be carcinogenic, a statement describing the rationale for this conclusion is added to the evaluation narrative; an additional evaluation may be made for this broader group of compounds if the strength of the evidence warrants it.

The agent, mixture or exposure circumstance is described according to the wording of one of the following categories, and the designated group is given. The categorization of an agent, mixture or exposure circumstance is a matter of scientific judgement, reflecting the strength of the evidence derived from studies in humans and in experimental animals and from other relevant data.

Group 1 —The agent (mixture) is carcinogenic to humans.
The exposure circumstance entails exposures that are carcinogenic to humans.

This category is used when there is *sufficient evidence* of carcinogenicity in humans. Exceptionally, an agent (mixture) may be placed in this category when evidence of carcinogenicity in humans is less than sufficient but there is *sufficient evidence* of carcinogenicity in experimental animals and strong evidence in exposed humans that the agent (mixture) acts through a relevant mechanism of carcinogenicity.

Group 2

This category includes agents, mixtures and exposure circumstances for which, at one extreme, the degree of evidence of carcinogenicity in humans is almost sufficient, as well as those for which, at the other extreme, there are no human data but for which there is evidence of carcinogenicity in experimental animals. Agents, mixtures and exposure circumstances are assigned to either group 2A (probably carcinogenic to humans) or group 2B (possibly carcinogenic to humans) on the basis of epidemiological and experimental evidence of carcinogenicity and other relevant data.

Group 2A—The agent (mixture) is probably carcinogenic to humans.
The exposure circumstance entails exposures that are probably carcinogenic to
humans.

This category is used when there is *limited evidence* of carcinogenicity in humans and *sufficient evidence* of carcinogenicity in experimental animals. In some cases, an agent (mixture) may be classified in this category when there is *inadequate evidence* of carcinogenicity in humans, *sufficient evidence* of carcinogenicity in experimental animals and strong evidence that the carcinogenesis is mediated by a mechanism that also operates in humans. Exceptionally, an agent, mixture or exposure circumstance may be classified in this category solely on the basis of *limited evidence* of carcinogenicity in humans.

Group 2B—The agent (mixture) is possibly carcinogenic to humans.
The exposure circumstance entails exposures that are possibly carcinogenic to
humans.

This category is used for agents, mixtures and exposure circumstances for which there is *limited evidence* of carcinogenicity in humans and less than *sufficient evidence* of carcinogenicity in experimental animals. It may also be used when there is *inadequate evidence* of carcinogenicity in humans but there is *sufficient evidence* of carcinogenicity in experimental animals. In some instances, an agent, mixture or exposure circumstance for which there is *inadequate evidence* of carcinogenicity in humans but *limited evidence* of carcinogenicity in experimental animals together with supporting evidence from other relevant data may be placed in this group.

Group 3—The agent (mixture or exposure circumstance) is not classifiable as to its
carcinogenicity to humans.

This category is used most commonly for agents, mixtures and exposure circumstances for which the *evidence of carcinogenicity* is *inadequate* in humans and *inadequate* or *limited* in experimental animals.

Exceptionally, agents (mixtures) for which the *evidence of carcinogenicity* is *inadequate* in humans but *sufficient* in experimental animals may be placed in this category when there is strong evidence that the mechanism of carcinogenicity in experimental animals does not operate in humans.

Agents, mixtures and exposure circumstances that do not fall into any other group are also placed in this category.

Group 4—The agent (mixture) is probably not carcinogenic to humans.

This category is used for agents or mixtures for which there is *evidence suggesting lack of carcinogenicity* in humans and in experimental animals. In some instances, agents or mixtures for which there is *inadequate evidence* of carcinogenicity in humans but *evidence suggesting lack of carcinogenicity* in experimental animals, consistently and strongly supported by a broad range of other relevant data, may be classified in this group.

References

Breslow, N.E. & Day, N.E. (1980) *Statistical Methods in Cancer Research*, Vol. 1, *The Analysis of Case–Control Studies* (IARC Scientific Publications No. 32), Lyon, IARC*Press*

Breslow, N.E. & Day, N.E. (1987) *Statistical Methods in Cancer Research*, Vol. 2, *The Design and Analysis of Cohort Studies* (IARC Scientific Publications No. 82), Lyon, IARC*Press*

Cohen, S.M. & Ellwein, L.B. (1990) Cell proliferation in carcinogenesis. *Science*, **249**, 1007–1011

Gart, J.J., Krewski, D., Lee, P.N., Tarone, R.E. & Wahrendorf, J. (1986) *Statistical Methods in Cancer Research*, Vol. 3, *The Design and Analysis of Long-term Animal Experiments* (IARC Scientific Publications No. 79), Lyon, IARC*Press*

Hoel, D.G., Kaplan, N.L. & Anderson, M.W. (1983) Implication of nonlinear kinetics on risk estimation in carcinogenesis. *Science*, **219**, 1032–1037

Huff, J.E., Eustis, S.L. & Haseman, J.K. (1989) Occurrence and relevance of chemically induced benign neoplasms in long-term carcinogenicity studies. *Cancer Metastasis Rev.*, **8**, 1–21

IARC (1973–1996) *Information Bulletin on the Survey of Chemicals Being Tested for Carcinogenicity/Directory of Agents Being Tested for Carcinogenicity*, Numbers 1–17, Lyon, IARC*Press*

IARC (1976–1996), Lyon, IARC*Press*

    Directory of On-going Research in Cancer Epidemiology 1976. Edited by C.S. Muir & G. Wagner

    Directory of On-going Research in Cancer Epidemiology 1977 (IARC Scientific Publications No. 17). Edited by C.S. Muir & G. Wagner

    Directory of On-going Research in Cancer Epidemiology 1978 (IARC Scientific Publications No. 26). Edited by C.S. Muir & G. Wagner

    Directory of On-going Research in Cancer Epidemiology 1979 (IARC Scientific Publications No. 28). Edited by C.S. Muir & G. Wagner

    Directory of On-going Research in Cancer Epidemiology 1980 (IARC Scientific Publications No. 35). Edited by C.S. Muir & G. Wagner

    Directory of On-going Research in Cancer Epidemiology 1981 (IARC Scientific Publications No. 38). Edited by C.S. Muir & G. Wagner

    Directory of On-going Research in Cancer Epidemiology 1982 (IARC Scientific Publications No. 46). Edited by C.S. Muir & G. Wagner

    Directory of On-going Research in Cancer Epidemiology 1983 (IARC Scientific Publications No. 50). Edited by C.S. Muir & G. Wagner

    Directory of On-going Research in Cancer Epidemiology 1984 (IARC Scientific Publications No. 62). Edited by C.S. Muir & G. Wagner

    Directory of On-going Research in Cancer Epidemiology 1985 (IARC Scientific Publications No. 69). Edited by C.S. Muir & G. Wagner

    Directory of On-going Research in Cancer Epidemiology 1986 (IARC Scientific Publications No. 80). Edited by C.S. Muir & G. Wagner

    Directory of On-going Research in Cancer Epidemiology 1987 (IARC Scientific Publications No. 86). Edited by D.M. Parkin & J. Wahrendorf

    Directory of On-going Research in Cancer Epidemiology 1988 (IARC Scientific Publications No. 93). Edited by M. Coleman & J. Wahrendorf

    Directory of On-going Research in Cancer Epidemiology 1989/90 (IARC Scientific Publications No. 101). Edited by M. Coleman & J. Wahrendorf

    Directory of On-going Research in Cancer Epidemiology 1991 (IARC Scientific Publications No.110). Edited by M. Coleman & J. Wahrendorf

    Directory of On-going Research in Cancer Epidemiology 1992 (IARC Scientific Publications No. 117). Edited by M. Coleman, J. Wahrendorf & E. Démaret

Directory of On-going Research in Cancer Epidemiology 1994 (IARC Scientific Publications No. 130). Edited by R. Sankaranarayanan, J. Wahrendorf & E. Démaret

Directory of On-going Research in Cancer Epidemiology 1996 (IARC Scientific Publications No. 137). Edited by R. Sankaranarayanan, J. Wahrendorf & E. Démaret

IARC (1977) *IARC Monographs Programme on the Evaluation of the Carcinogenic Risk of Chemicals to Humans*. Preamble (IARC intern. tech. Rep. No. 77/002)

IARC (1978) *Chemicals with* Sufficient Evidence *of Carcinogenicity in Experimental Animals—* IARC Monographs *Volumes 1–17* (IARC intern. tech. Rep. No. 78/003)

IARC (1978–1993) *Environmental Carcinogens. Methods of Analysis and Exposure Measurement*, Lyon, IARC*Press*

   Vol. 1. Analysis of Volatile Nitrosamines in Food (IARC Scientific Publications No. 18). Edited by R. Preussmann, M. Castegnaro, E.A. Walker & A.E. Wasserman (1978)

   Vol. 2. Methods for the Measurement of Vinyl Chloride in Poly(vinyl chloride), Air, Water and Foodstuffs (IARC Scientific Publications No. 22). Edited by D.C.M. Squirrell & W. Thain (1978)

   Vol. 3. Analysis of Polycyclic Aromatic Hydrocarbons in Environmental Samples (IARC Scientific Publications No. 29). Edited by M. Castegnaro, P. Bogovski, H. Kunte & E.A. Walker (1979)

   Vol. 4. Some Aromatic Amines and Azo Dyes in the General and Industrial Environment (IARC Scientific Publications No. 40). Edited by L. Fishbein, M. Castegnaro, I.K. O'Neill & H. Bartsch (1981)

   Vol. 5. Some Mycotoxins (IARC Scientific Publications No. 44). Edited by L. Stoloff, M. Castegnaro, P. Scott, I.K. O'Neill & H. Bartsch (1983)

   Vol. 6. N-Nitroso Compounds (IARC Scientific Publications No. 45). Edited by R. Preussmann, I.K. O'Neill, G. Eisenbrand, B. Spiegelhalder & H. Bartsch (1983)

   Vol. 7. Some Volatile Halogenated Hydrocarbons (IARC Scientific Publications No. 68). Edited by L. Fishbein & I.K. O'Neill (1985)

   Vol. 8. Some Metals: As, Be, Cd, Cr, Ni, Pb, Se, Zn (IARC Scientific Publications No. 71). Edited by I.K. O'Neill, P. Schuller & L. Fishbein (1986)

   Vol. 9. Passive Smoking (IARC Scientific Publications No. 81). Edited by I.K. O'Neill, K.D. Brunnemann, B. Dodet & D. Hoffmann (1987)

   *Vol. 10. Benzene and Alkylated Benzenes (*IARC Scientific Publications No. 85). Edited by L. Fishbein & I.K. O'Neill (1988)

   Vol. 11. Polychlorinated Dioxins and Dibenzofurans (IARC Scientific Publications No. 108). Edited by C. Rappe, H.R. Buser, B. Dodet & I.K. O'Neill (1991)

   Vol. 12. Indoor Air (IARC Scientific Publications No. 109). Edited by B. Seifert, H. van de Wiel, B. Dodet & I.K. O'Neill (1993)

IARC (1979) *Criteria to Select Chemicals for* IARC Monographs (IARC intern. tech. Rep. No. 79/003)

IARC (1982) *IARC Monographs on the Evaluation of the Carcinogenic Risk of Chemicals to Humans*, Supplement 4, *Chemicals, Industrial Processes and Industries Associated with Cancer in Humans* (IARC Monographs, Volumes 1 to 29), Lyon, IARC*Press*

IARC (1983) *Approaches to Classifying Chemical Carcinogens According to Mechanism of Action* (IARC intern. tech. Rep. No. 83/001)

IARC (1984) *Chemicals and Exposures to Complex Mixtures Recommended for Evaluation in IARC Monographs and Chemicals and Complex Mixtures Recommended for Long-term Carcinogenicity Testing* (IARC intern. tech. Rep. No. 84/002)

IARC (1987a) *IARC Monographs on the Evaluation of Carcinogenic Risks to Humans*, Supplement 6, *Genetic and Related Effects: An Updating of Selected* IARC Monographs *from Volumes 1 to 42*, Lyon, IARCPress

IARC (1987b) *IARC Monographs on the Evaluation of Carcinogenic Risks to Humans*, Supplement 7, *Overall Evaluations of Carcinogenicity: An Updating of* IARC Monographs *Volumes 1 to 42*, Lyon, IARCPress

IARC (1988) *Report of an IARC Working Group to Review the Approaches and Processes Used to Evaluate the Carcinogenicity of Mixtures and Groups of Chemicals* (IARC intern. tech. Rep. No. 88/002)

IARC (1989) *Chemicals, Groups of Chemicals, Mixtures and Exposure Circumstances to be Evaluated in Future IARC Monographs, Report of an ad hoc Working Group* (IARC intern. tech. Rep. No. 89/004)

IARC (1991a) *A Consensus Report of an IARC Monographs Working Group on the Use of Mechanisms of Carcinogenesis in Risk Identification* (IARC intern. tech. Rep. No. 91/002)

IARC (1991b) *Report of an ad-hoc* IARC Monographs *Advisory Group on Viruses and Other Biological Agents Such as Parasites* (IARC intern. tech. Rep. No. 91/001)

IARC (1993) *Chemicals, Groups of Chemicals, Complex Mixtures, Physical and Biological Agents and Exposure Circumstances to be Evaluated in Future* IARC Monographs, *Report of an ad-hoc Working Group* (IARC intern. Rep. No. 93/005)

IARC (1998a) *Report of an ad-hoc* IARC Monographs *Advisory Group on Physical Agents* (IARC Internal Report No. 98/002)

IARC (1998b) *Report of an ad-hoc* IARC Monographs *Advisory Group on Priorities for Future Evaluations* (IARC Internal Report No. 98/004)

McGregor, D.B., Rice, J.M. & Venitt, S., eds (1999) *The Use of Short and Medium-term Tests for Carcinogens and Data on Genetic Effects in Carcinogenic Hazard Evaluation* (IARC Scientific Publications No. 146), Lyon, IARCPress

Montesano, R., Bartsch, H., Vainio, H., Wilbourn, J. & Yamasaki, H., eds (1986) *Long-term and Short-term Assays for Carcinogenesis—A Critical Appraisal* (IARC Scientific Publications No. 83), Lyon, IARCPress

Peto, R., Pike, M.C., Day, N.E., Gray, R.G., Lee, P.N., Parish, S., Peto, J., Richards, S. & Wahrendorf, J. (1980) Guidelines for simple, sensitive significance tests for carcinogenic effects in long-term animal experiments. In: *IARC Monographs on the Evaluation of the Carcinogenic Risk of Chemicals to Humans*, Supplement 2, *Long-term and Short-term Screening Assays for Carcinogens: A Critical Appraisal*, Lyon, IARCPress, pp. 311–426

Tomatis, L., Aitio, A., Wilbourn, J. & Shuker, L. (1989) Human carcinogens so far identified. *Jpn. J. Cancer Res.*, **80**, 795–807

Vainio, H., Magee, P.N., McGregor, D.B. & McMichael, A.J., eds (1992) *Mechanisms of Carcinogenesis in Risk Identification* (IARC Scientific Publications No. 116), Lyon, IARC*Press*

Vainio, H., Wilbourn, J.D., Sasco, A.J., Partensky, C., Gaudin, N., Heseltine, E. & Eragne, I. (1995) Identification of human carcinogenic risk in IARC Monographs. *Bull. Cancer,* **82**, 339–348 (in French)

Waters, M.D., Stack, H.F., Brady, A.L., Lohman, P.H.M., Haroun, L. & Vainio, H. (1987) Appendix 1. Activity profiles for genetic and related tests. In: *IARC Monographs on the Evaluation of Carcinogenic Risks to Humans*, Suppl. 6, *Genetic and Related Effects: An Updating of Selected IARC Monographs from Volumes 1 to 42*, Lyon, IARC*Press*, pp. 687–696

Wilbourn, J., Haroun, L., Heseltine, E., Kaldor, J., Partensky, C. & Vainio, H. (1986) Response of experimental animals to human carcinogens: an analysis based upon the IARC Monographs Programme. *Carcinogenesis*, **7**, 1853–1863

GENERAL INTRODUCTION

GENERAL INTRODUCTION

1. Introduction

Previous *Monographs* have considered distinct types of electromagnetic radiation: solar and ultraviolet radiation (IARC, 1992) and X- and γ-radiation (IARC, 2000). This volume is concerned with that region of the spectrum described as static and 'extremely low frequency' (ELF). Such electromagnetic energy occurs naturally or in association with the generation and transmission of electrical power and with the use of power in some appliances. It is different from electromagnetic fields and radiation consequent upon the transmission of radio and television signals and the operation of mobile phones (Figure 1).

Static electric and magnetic fields, as well as low-frequency fields, are produced by both natural and man-made sources. The natural fields are static or very slowly varying. The electric field in the air above the earth's surface is typically 100 V/m but during strong electric storms may increase 10-fold or more. The geomagnetic field is typically 50 μT (König *et al.*, 1981). Most man-made sources are at extremely low frequencies. The generation, transmission, distribution and use of electricity at 50 or 60 Hz result in widespread exposure of humans to ELF fields of the order of 10–100 V/m and 0.1–1 μT, and occasionally to much stronger fields (National Research Council, 1997; Portier & Wolfe, 1998; National Radiological Protection Board, 2001). Electrical power is an integral part of modern civilization. Because ELF fields can interact with biological systems, interest and concern about potential hazards are understandable.

The study of the effects of electric and magnetic fields on humans has a long history. Data pertaining to human health risks were first gathered in the 1960s from studies of workers with occupational exposure to ELF fields. The first data to address potential carcinogenic risks were obtained in a series of studies of adults and children with residential exposures from electrical facilities, appliances and external and internal wiring and grounding systems.

Experimental studies have expended considerable effort in the search for mechanism(s) that would predict the biological effects of the low-intensity fields that are found in residential and occupational environments. More recent investigations have focused on obtaining data to assess potential carcinogenic risks. Of particular importance are chronic bioassays that look for microscopic evidence of lesions and

Figure 1. Electromagnetic spectrum

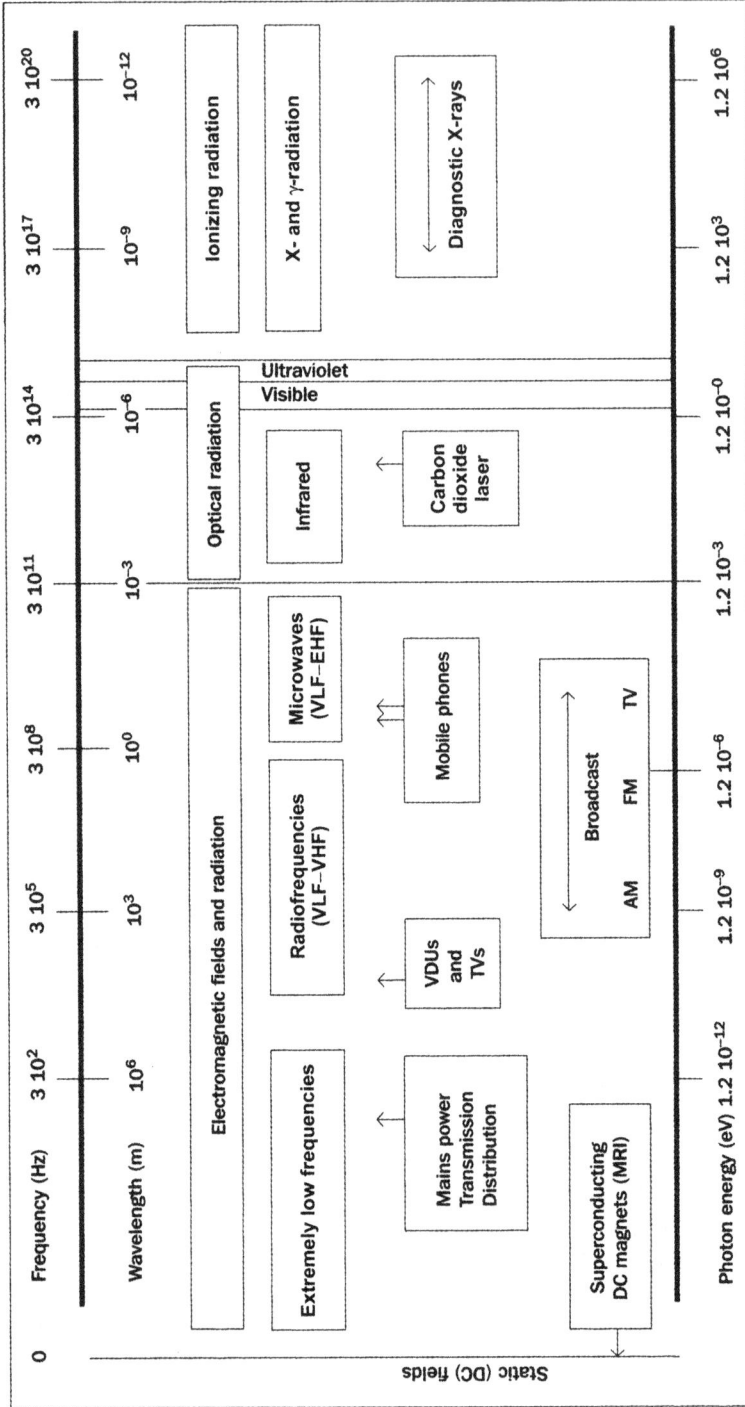

From National Radiological Protection Board (2001)

AM, amplitude modulation; DC, direct current; EHF, extremely high frequency; eV, electron volt; FM, frequency modulation; MRI, magnetic resonance imaging; TV, television; UHF, ultra high frequency; VDU, visual display unit; VHF, very high frequency; VLF, very low frequency tumours in rodents exposed over most of their lifespan. The default assumption is that the photon energies of ELF fields, as for other forms of non-ionizing radiation, are insufficient to ionize molecules or break chemical bonds in biological systems.

The genotoxicity of electric and magnetic fields has been tested both *in vivo* and *in vitro*. The possibility that these fields may be carcinogenic has been investigated in standard tumour promotion/co-promotion systems. These involve the exposure of rodents to ELF electric and magnetic fields following, or coincident with, the initiation of skin, liver and mammary tumours by chemical initiators.

In addition to a general assessment of the data and mechanisms of toxicity, the characterization of associations between exposure to ELF magnetic fields and cancer, particularly leukaemias and brain tumours, requires focused evaluations of experimental data regarding field effects on haematopoietic cells (leukaemia), the nervous system (brain tumours) and its function and neuroendocrine and immunological factors that might be suspected of influencing susceptibility to cancer.

2. Physical characteristics of electromagnetic fields

The main focus of this *Monographs* volume is on static and time-varying fields in the ELF range of 3–3000 Hz (IEEE, 1988). Ancillary ELF phenomena, such as transients, which have frequency components in excess of 3000 Hz have also been examined for their potential contribution to the possible hazard of ELF radiation.

At frequencies above those of interest here, electromagnetic fields propagate by means of tightly coupled electric and magnetic fields (radiation). In such cases, the magnitude of the electric field can be calculated exactly if the magnetic field is known and vice versa. However, in the ELF range, electric and magnetic fields are effectively uncoupled and can be evaluated separately as if they arose from independent sources. At the low frequencies where it is customary to use the quasi-static approximation, the wavelengths of electric and magnetic fields are very large (approximately 5000 km at 60 Hz[1]) in relation to the size and distances of objects of interest (National Radiological Protection Board, 2001). Under these 'near-field' conditions (less than one wavelength), electric and magnetic fields do not effectively 'radiate' away from the source nor do they occur together in an interrelated way. The field produced by a source is better described as a zone of influence in which the forces on electrical charges oscillate in time and space. More detailed information on physical characteristics may be found elsewhere (e.g. Polk & Postow, 1995).

---

[1] Wavelength (λ) is the distance in metres travelled by the wave in one period and is related to frequency (f) by $\lambda = c/f$, where c is the speed of light in a vacuum (3×10^8 m/s).

3. Definitions, quantities and units

3.1   Electric fields

An electric field **E** exists in a region of space if a charge experiences an electrical force **F**:

$$\mathbf{F} = q\mathbf{E}$$

where q is the unit positive charge (see Table 1). The direction of the field or force corresponds to the direction that a positive charge would move in the field. Vector quantities are characterized by magnitude and direction and are displayed in bold type. Electric fields are also characterized by the electric flux density or displacement vector **D**, where **D** = ε**E** and ε characterizes the material permittivity. The sources of electric fields are unbalanced electrical charges on conductors or other objects. For instance, on a dry winter's day, the action of pulling off a sweater can separate charges on the sweater and hair and the static electric field produced makes the wearer's hair stand on end. Electric utility facilities, including power lines, substations, appliances and building wiring are sources of time-varying electric fields. Electric fields from these sources arise from unbalanced electric charges on energized conductors. The source of the unbalanced charge is the voltage supplied by the power system.

Table 1. Quantities and units

Characteristic	Symbol[a]	SI unit	Symbol
Electric field intensity	**E**	Volt/metre	V/m
Magnetic field intensity	**B**	Tesla	T
Magnetic field	**H**	Ampere/metre	A/m
Current density	**J**	Ampere/metre squared	A/m^2
Frequency	f	Hertz	Hz
Charge density	ρ	Coulomb/metre cubed	C/m^3
Conductivity	σ	Siemens/metre	S/m
Current	I	Ampere	A
Charge	q	Coulomb	C
Force	**F**	Newton	N
Permittivity	ε	Farad per metre	F/m
Permeability	μ	Henry per metre	H/m
Permittivity of free space	ε_0	$\varepsilon_0 = 8.854 \times 10^{-12}$ F/m	
Permeability of free space	μ_0	$\mu_0 = 12.57 \times 10^{-7}$ H/m	

Note: 1 gauss (G) = 10^{-4} tesla (T); 1 oersted (Oe) = 1 gauss (G) in vacuum or air
[a] Vector quantities are displayed in bold type.

3.2   Current density

Electric fields exert forces on charged particles. In electrically conductive materials, including biological tissues, these forces cause an electric current to flow. The density of this current $\mathbf{J}$ across a cross-section of tissue is related to the electric field by σ, the electrical conductivity of the medium, as

$$\mathbf{J} = \sigma\mathbf{E}$$

where a homogeneous medium has been assumed.

3.3   Magnetic fields

A magnetic field $\mathbf{H}$ and associated magnetic flux density $\mathbf{B}$ exist only if electric charges are in motion, i.e. there is flow of electric current. The relationship between the two descriptions of the field is $\mathbf{B} = \mu\mathbf{H}$, where for most biological materials, $\mu = \mu_o$, where μ_o is the permeability of free space (vacuum, air). The term 'magnetic field' is used in this Monograph as being equivalent to 'magnetic flux density'. Magnetic fields in turn only exert forces on a moving charge, q, as given by the Lorenz force

$$\mathbf{F} = q\mathbf{v} \times \mathbf{B}$$

where the direction of the force is perpendicular to both the velocity $\mathbf{v}$ and the magnetic flux density $\mathbf{B}$.

Static magnetic fields are formed by unidirectional, direct currents (DCs), also called steady currents, and magnetic materials (permanent magnets). Time-varying magnetic fields are produced by the same types of alternating current (AC) sources as electric fields. Time-varying magnetic fields are also sources of electric fields. Similarly, electric fields produce magnetic fields.

3.4   Magnitude

Electric and magnetic fields are vector quantities characterized by a magnitude and direction. Electric fields are most commonly described in terms of the potential difference across a unit distance. A 240-volt source connected to parallel metal plates separated by 1 m produces a field of 240 volts per metre (V/m) between the plates. Large fields are expressed in units of kilovolts per metre (kV/m, 1000 volts per metre).

The magnitude of a magnetic field is described most often by its magnetic flux density $\mathbf{B}$ in terms of magnetic lines of force per unit area. The units for magnetic flux density are webers per square metre (Wb/m^2) or Système International (SI) units of tesla (T). The equivalent old unit, often seen in the earlier literature, is gauss (G). Units of μT (100 μT = 1 G) are often used to describe ambient strengths of the magnetic field. Less commonly in the biological literature, the magnetic field strength $\mathbf{H}$ is given in amperes per metre (A/m). Occasionally, the unit of oersted (Oe) is still used.

The magnitudes of electric and magnetic fields are customarily expressed as root-mean-square (rms) values. The rms values of single-frequency sinusoidally varying fields are obtained by dividing the peak amplitude by the square root of two.

3.5   Frequency

Electric and magnetic fields are further determined by the frequency characteristics of their sources. The voltages and currents of the electric power system oscillate at 50 times per second (Hz) (60 Hz in North America) and produce a sinusoidal rise and fall in the magnitude of the associated fields at the same frequency. Electrified rail transport is sometimes powered at frequencies of 25 Hz in the USA (National Research Council, 1997) and $16\,^2/_3$ Hz in many European countries (National Radiological Protection Board, 2001). The operation of some electrical devices in the power system also produces fields at other frequencies. For example, fields can occur at multiples (harmonics) of the fundamental frequency: at 100, 150, 200 Hz, etc., for a 50-Hz system and at 120, 180, 240 Hz, etc., for a 60-Hz system.

3.6   Polarization

Fields add vectorially: both the magnitude and direction of the field must be considered in combining fields from different sources. A single conductor produces a field vector that changes its direction along a straight line. This is a linearly polarized field and has been used most often in biological studies. Multiple sources that are in phase (synchronized voltage or current waves) also produce linearly polarized fields. However, fields from multiple conductors that are not in phase, such as a three-phase distribution or transmission lines, are not necessarily linearly polarized. In these cases, the field vector is not fixed in space but rotates during a cycle, tracing out an ellipse. The field is then polarized elliptically and the ratio of the minor to major field axis defines the ellipticity or degree of polarization of the field. When the two axes of the ellipse are of equal magnitude, the ellipse forms a circle and the field is described as a circularly polarized field; when one axis is zero the field is linearly polarized.

4.   Physical interactions with biological materials

To understand the effects of electric and magnetic fields on animals and humans, their electrical properties, as well as their size and shape, have to be considered with respect to the wavelength of the external field. At ELF, the size of all mammalian and other biological bodies is a very small fraction of the wavelength.

The electrical properties of the body, namely its permittivity and permeability, relate to its interaction with the electric and magnetic fields, respectively. Human and animal bodies consist of numerous tissues, whose electrical properties differ considerably.

The permittivity $\hat{\varepsilon}$ is often written as $\hat{\varepsilon}_r\varepsilon_0$, where $\hat{\varepsilon}_r$ is the relative permittivity and ε_0 is permittivity of the vacuum, $8.854.10^{-12}$ farad/m (WHO, 1984). The permittivity determines the interactions with the electric field and the dielectric constant defines the ability to store the field energy.

Similarly, the permeability, $\hat{\mu}$, can be written as $\hat{\mu} = \hat{\mu}_r\mu_0$.

Conductive materials, i.e. those that have free electric charges (e.g. electrons and ions) are also characterized by conductivity, σ. Free charges, if in motion, can interact with both electric and magnetic fields.

Most biological tissues have a permeability equal to that of free space (air, vacuum) (Foster & Schwan, 1989, 1996). Many animal species, including humans, are known to have minuscule amounts of biogenic magnetite (Fe_3O_4) in their brains and other tissues (with permeability $\mu_r \geq 1$) (Kirschvink et al., 1992).

The permittivity of biological tissues is to a large extent determined by water and electrolyte contents. Thus, tissues such as blood, muscle, liver and kidneys, which have a higher water content than tissues such as fat and lungs, have higher dielectric constants and conductivities. Both the permittivity and conductivity vary with frequency, and exhibit relaxation phenomena. The physical phenomenon responsible for the dispersion at low frequencies is counterion polarization (Foster & Schwan, 1989, 1996).

At ELF, biological bodies (e.g. humans or animals) can be considered as conductive dielectrics. Induced fields in tissues can be determined solely on the basis of their conductivity. To provide an idea of the range of conductivity values for biological tissues, Table 2 lists the most recently published conductivity measurements (Gandhi et al., 2001).

Table 2. Conductivities of various tissues assumed for power-frequency electric and magnetic fields

Tissue	σ (S/m)	Tissue	σ (S/m)
Bladder	0.2	Heart	0.5
Blood	0.7	Kidney	0.09
Bone (cancellous)	0.08	Liver	0.04
Bone (compact)	0.02	Lungs	0.07
Brain (white)	0.06	Muscle	0.24
Cerebrospinal fluid	2.0	Skin	0.04
Eye sclera	0.5	Spinal cord	0.07
Fat	0.02	Testes	0.42

From Gandhi et al. (2001)

4.1   Static fields

A static electric field does not penetrate human and animal bodies. The field is always perpendicular to the body surface and induces surface charge density. A sufficiently large charge density may be perceived through its interaction with body hair. Indirect effects associated with induction charges on objects are well known. These range from perception, to pain, to burn resulting from a direct contact or spark discharge. There are well-established thresholds for these effects for human populations (Bernhardt, 1988).

Static magnetic fields can interact with tissues by three mechanisms (Tenforde, 1990, 1992). Firstly, electrodynamic interactions occur with ionic currents, such as blood flow or nerve impulse conduction. This interaction leads to the induction of electric field and electrical potential, e.g. across a blood vessel. This type of interaction is significant only at high flux density (≥ 1 T).

The second interaction mechanism is a magneto-mechanical effect which involves the orientation of certain biological structures in strong magnetic fields (Tenforde, 1992). Sensitivity to low intensity fields is seen in several biological species, such as certain bacteria, fish and birds. Furthermore, magnetite domains have been found in some animals, e.g. bees, tuna, salmon, turtles, pigeons, dolphins and humans. The ability to use these fields for navigation has been demonstrated for some species, e.g. bees. Studies of humans have not yielded any evidence of direction-finding based on the geomagnetic field (Tenforde, 1992).

The third mechanism relates to the Zeeman effect, whereby a magnetic field changes the energy levels of certain molecules. One consequence of the Zeeman effect is to change the probability of recombination of pairs of radicals formed in certain biochemical processes. This may result in changes in the concentration of free radicals, which can be highly reactive. This 'radical-pair mechanism' is well established in magnetochemistry (Hamilton et al., 1988; McLauchlan, 1989; Cozens & Scaiano, 1993; Scaiano et al., 1994; Grissom, 1995; Mohtat et al., 1998), and the relevance to biological effects at low field strengths (e.g. below 500 μT) is currently under investigation (Brocklehurst & McLauchlan, 1996).

Strong static magnetic fields have several indirect effects, such as electromagnetic interference with implanted medical devices (e.g. cardiac pacemakers and defibrillators), and through forces exerted on external and implanted metallic objects. For instance, magnetic field gradients in magnetic resonance imaging facilities are known to turn metallic objects into potentially dangerous projectiles.

4.2   Extremely low-frequency (ELF) fields

The physical interactions between fields and tissues are governed by Maxwell's equations, but not all tissue components are equally interactive.

At ELFs, the photon energy is exceedingly small, thus a direct interaction causing breakage of chemical bonds and the resultant damage to DNA is not possible. At power frequencies (50 or 60 Hz), the photon energy is about 10^{-12} of the energy required to break the weakest chemical bond (Valberg *et al.*, 1997). It is generally agreed that whatever the interaction mechanism, it must be consistent with noise constraints. In principle, meaningful physiological changes can result only if the 'signal' produced by the field exceeds the 'noise' level present in the relevant biological system. For example, in the case of induced currents, the noise level is set by thermal processes (determined in part by kT, where k is Boltzmann's constant and T is the absolute temperature). However, a number of hypotheses listed below have sought to overcome this limitation (e.g. processes involving extremely narrow bandwidths).

The basic interaction mechanism of exposure to magnetic fields is the induction of current density in tissue: currents will always be induced in conductors exposed to time-varying magnetic fields, and current density increases with frequency and body size.

The spatial patterns of the currents induced by electric and magnetic fields are quite different from each other. In an upright human body exposed to a vertical electric field, the induced field and current flow are also vertical. Conversely, in the case of a magnetic field, the current flow forms closed loops, perpendicular to the direction of the magnetic field. General patterns of the current flow induced by exposure to magnetic fields are illustrated in Figure 2.

Figure 2. Induction of eddy currents in the human body perpendicular to (a) a vertical magnetic field and (b) a horizontal magnetic field

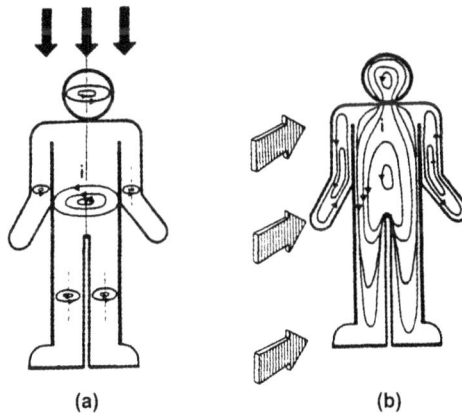

(a)   (b)

From Silny (1986)
In principle, the current density approaches zero at the centre of the loops.

Biological bodies perturb external electric fields. Because the tissue conductivity is low at low frequencies (see Table 2), the induced fields are approximately $10^5–10^8$ times lower than the external fields. The perturbation of the external electric field, like the static electric field, induces surface charge on the body surface. The time-varying surface charge may cause hair oscillation, particularly in some laboratory animals (Tenforde & Kaune, 1987; Tenforde, 1991). Humans can detect 60-Hz electric fields through hair stimulation at about 20 kV/m, while the threshold is lower for some furry rodents (Tenforde, 1991). In contrast, because tissue permeability to magnetic fields is the same as that of free space, these fields penetrate the body with virtually no distortion. The magnitude of the induced electric fields depends mostly on the body size and shape, and the field orientation. The conductivities of various tissues have a lesser influence on the induced electric fields. Extensive data are available on induced electric field and current density values for exposure to ELF electric and magnetic fields, as outlined in detail in section 1.3.2.

A well-established physical mechanism of interaction at the cellular level is the stimulation of excitable cells, such as those in nerves, muscles and the heart which occurs when the electric field in the tissue exceeds a threshold value of V_m (the potential across a cell membrane). Once this threshold is exceeded, the nerve or muscle cell propagates an action potential. The threshold V_m depends on cell type, dimension and shape as well as the signal frequency, duration and waveform (e.g. monopolar pulse, bipolar pulse, sinusoid, single pulse or repeated pulses). Cell excitation and action potential propagation are complex non-linear processes (Plonsey & Barr, 1988; Reilly, 1992; Malmivuo & Plonsey, 1995). A typical value for V_m is 20 mV for the optimal pulse shape, duration and an appropriate polarity causing depolarization (Reilly, 1992, 1998).

To induce neural or cardiac stimulation by 50- or 60-Hz fields, very strong external electric or magnetic fields are required: the reported thresholds are above 1 A/m² (Bailey et al., 1997).

Experimental evidence and thresholds have been determined in human volunteers for magnetic stimulation of the visual system causing phosphenes, which are weak visual sensations (Lövsund et al., 1979, 1980). The lowest threshold magnetic field strength is 8 mT (in darkness) at 20 Hz. The threshold increases for higher and lower frequencies of the magnetic field as well as when the background is illuminated. Phosphenes have also been produced by direct electrostimulation. Again, the threshold was observed for 20 Hz and increases at higher and lower frequencies. It is believed that the effect is a result of the interaction of the induced current with electrically excitable cells in the retina.

The above mechanisms have a well-understood physical basis.

Several other physical interactions have been examined theoretically. Forces exerted by the field on ions and charged molecules have been compared with forces generated by biological structures. For example, an electric field of 5 mV/m in tissue produces a force on a charged molecule of 2×10^{-5} piconewton (pN). In comparison, biological activity reported in various studies is associated with forces above 1 pN, and typically above 10 pN (Valberg et al., 1997).

At ELF, the radical-pair mechanism described above for static magnetic fields still applies. This is because the period of ELF fields, ~ 20 ms, is long compared to the lifetime of radical pairs (nano- to microseconds). Therefore, the radical pairs experience the instantaneous combined static and ELF magnetic fields.

In addition, a number of other mechanisms have been suggested, for which either the physical basis is not yet clear or the experimental evidence for relevance to the biological effects of ELF fields is still being sought.

From present knowledge, it is clear that there are a number of mechanisms theoretically capable of explaining the occurrence of biological effects at high field strengths. Electric fields induced in tissue are known to produce effects at levels corresponding to an external field of above 0.4 mT and 5–10 kV/m (Bernhardt, 1988).

5. Studies of ELF electric and magnetic fields relevant to carcinogenicity

Studies on the possible carcinogenic effects of ELF electric and magnetic fields are hampered by complications, or lack of information, at almost every level. Issues of EMF exposure which affect the interpretation of referent epidemiological studies are discussed in section 2.1.1. High-quality studies of cancer induced in experimental animals by ELF electric and magnetic fields have been initiated fairly recently. Attempts have been made to examine carcinogenesis associated with exposure to electric and magnetic fields in the context of a multistep process of cancer causation. However, the manner in which these fields might interact with other stimuli, including carcinogens such as X-rays, ultraviolet radiation and chemicals may include as yet undefined processes. Studies of carcinogenesis in experimental animals, and their implications are discussed in section 3.

Studies of the effects of ELF electric and magnetic fields on cells in culture, or using simpler biological systems have attempted to characterize particular molecular processes. Sufficient studies are available to identify other limitations on any simple correlation between 'conventional' carcinogens and the putative hazard under investigation. Firstly, it is apparent that many systems fail to provide any evidence of a treatment-related effect. Secondly, when such effects are observed, other laboratories have often been unable to reproduce the observation. Such failure to reproduce extends across a range of phenomena, and often involves numerous investigators. It is not, therefore, a simple difference between two centres and is unlikely to be explicable simply as the result of a lack of care by one or more of the laboratories concerned, but may be a consequence of an inability to reproduce exactly the electric and magnetic fields to which cells or whole animals have been exposed in a particular study.

The many parameters known to characterize electric and magnetic fields have been outlined above. Most often, an experimental field is described by reference to the intensity alone (usually specified in V/m for electric fields or Tesla for magnetic fields). Other as yet unspecified characteristics of the field may contribute, if not be critical to, particular experimental results.

6.   References

Bailey, W.H., Su, S.H., Bracken, T.D. & Kavet, R. (1997) Summary and evaluation of guidelines for occupational exposure to power frequency electric and magnetic fields. *Health Phys.*, **73**, 433–453

Bernhardt, J.H. (1988) The establishment of frequency dependent limits for electric and magnetic fields and evaluation of indirect effects. *Radiat. environ. Biophys.*, **27**, 1–27

Brocklehurst, B. & McLauchlan, K.A. (1996) Free radical mechanism for effects of environmental electromagnetic fields in biological systems. *Int. J. Radiat. Biol.*, **69**, 3–24

Cozens, F.L. & Scaiano, J.C. (1993) A comparative study of magnetic field effects on the dynamics of geminate and random radical pair processes in micelles. *J. Am. chem. Soc.*, **115**, 5204–5211

Foster, K.R. & Schwan, H.P. (1989) Dielectric properties of tissues and biological materials: a critical review. *Crit. Rev. biomed. Eng.*, **17**, 25–104

Foster, K.R. & Schwan, H.P. (1996) Dielectric properties of tissue. In: Polk, C. & Postow, E., eds, *Handbook of Biological Effects of Electromagnetic Fields*, 2nd Ed., New York, CRC Press

Gandhi, O.P., Kang, G., Wu, D. & Lazzi, G. (2001) Currents induced in anatomic models of the human for uniform and nonuniform power frequency magnetic fields. *Bioelectromagnetics*, **22**, 112–121

Grissom, C.B. (1995) Magnetic field effects in biology: a survey of possible mechanisms with emphasis on radical-pair recombination. *Chem. Rev.*, **95**, 3–24

Hamilton, C.A., Hewitt, J.P., McLauchlan, K.A. & Steiner, U.E. (1988) High resolution studies of the effects of magnetic fields on chemical reactions. *Mol. Phys.*, **65**, 423–438

IARC (1992) *IARC Monographs on the Evaluation of Carcinogenic Risks to Humans*, Vol. 55, *Solar and Ultraviolet Radiation*, Lyon, IARCPress

IARC (2000) *IARC Monographs on the Evaluation of Carcinogenic Risks to Humans*, Vol. 75, *Ionizing Radiation, Part 1: X- and Gamma (γ)-Radiation, and Neutrons*, Lyon, IARCPress

IEEE (Institute of the Electrical and Electronics Engineers) (1988) *IEEE Standard Dictionary of Electrical and Electronics Terms*, 4th Ed, New York

Kirschvink, J.L., Kobayashi-Kirschvink, A. & Woodford, B.J. (1992) Magnetite biomineralization in the human brain. *Proc. natl Acad. Sci.*, **89**, 7683–7687

König, H.L., Krueger, A.P., Lang, S. & Sönning, W. (1981) *Biologic Effects of Environmental Electromagnetism*, New York, Springer-Verlag

Lövsund, P., Öberg, P.Å. & Nilsson, S.E.G. (1979) Influence on vision of extremely low frequency electromagnetic fields. Industrial measurements, magnetophosphene studies in volunteers and intraretinal studies in animals. *Acta Ophthalmol.*, **57**, 812–821

Lövsund, P., Öberg, A. & Nilsson, S.E.G. (1980) Magneto- and electophosphenes: a comparative study. *Med. biol. Eng. Comput.*, **18**, 758–764

Malmivuo, J. & Plonsey, R. (1995) *Bioelectromagnetism*, New York, Oxford Press

McLauchlan, K.A. (1989) Magnetokinetics, mechanistics and synthesis. *Chemistry in Britain*, September, 895–898

Mohtat, N., Cozens, F.L., Hancock-Chen, T., Scaiano, J.C., McLean, J. & Kim, J. (1998) Magnetic field effects on the behavior of radicals in protein and DNA environments. *Photochem. Photobiol.*, **67**, 111–118

National Radiological Protection Board (NRPB) (2001) *ELF Electromagnetic Fields and the Risk of Cancer, Report of an Advisory Group on Non-ionizing Radiation* (Doc NRPB 12), Chilton, UK

National Research Council (1997) *Possible Health Effects of Exposure to Residential Electric and Magnetic Fields*, Committee on the Possible Effects of Electromagnetic Fields on Biologic Systems, Board on Radiation Effects Research, Commission on Life Sciences, Washington DC, National Academy Press

Plonsey, R. & Barr, R.C. (1988) *Bioelectricity, A Quantitative Approach*, Plenum Press, New York

Polk, C. & Postow, E., eds (1995) *Handbook of Biological Effects of Electromagnetic Fields*, 2nd Ed., New York, CRC Press

Portier, C.J. & Wolfe, M.D., eds (1998) *Assessment of Health Effects from Exposure to Power-line Frequency Electric and Magnetic Fields*, NIEHS Working Group Report (NIH Publication No. 98-3981), Research Triangle Park, National Institute of Environmental Health Sciences

Reilly, J.P. (1992) *Electrical Stimulation and Electropathology*, Cambridge University Press, New York

Reilly, J.P. (1998) *Applied Bioelectricity. From Electrical Stimulation to Electropathology*, Springer-Verlag, New York

Scaiano, J.C., Mohtat, N., Cozens, F.L., McLean, J. & Thansandote, A. (1994) Application of the radical pair mechanism to free radicals in organized systems: Can the effects of 60 Hz be predicted from studies under static fields? *Bioelectromagnetics*, **15**, 549–554

Silny, J. (1986) The influence thresholds of the time-varying magnetic fields in the human organism. *bga-Schriften*, **3**, 105–112

Tenforde, T.S. (1990) Biological effects of static magnetic fields. *Int. J. appl. Electromagn.*, **1**, 157–165

Tenforde, T.S. (1991) Biological interactions of extremely-low-frequency electric and magnetic fields. *Bioelectrochemistry Bioenergetics*, **25**, 1–17

Tenforde, T.S. (1992) Interaction mechanisms and biological effects of static magnetic fields. *Automedica*, **14**, 271–293

Tenforde, T.S. & Kaune, W.T. (1987) Interaction of extremely low frequency electric and magnetic fields with humans. *Health Phys.*, **53**, 585–606

Valberg, P.A., Kavet, R. & Rafferty, C.N. (1997) Can low-level 50/60 Hz electric and magnetic fields cause biological effects? *Radiat. Res.*, **148**, 2–21

WHO (1984) *Extremely Low Frequency (ELF) Fields* (Environmental Health Criteria Report 35), Geneva, World Health Organization

STATIC AND EXTREMELY LOW-FREQUENCY (ELF) ELECTRIC AND MAGNETIC FIELDS

1. Sources, Exposure and Exposure Assessment

1.1   Sources

1.1.1   *Natural magnetic and electric fields*

Humans are exposed daily to electric and magnetic fields from both natural and man-made sources. The strengths of fields from man-made sources can exceed those from natural sources by several orders of magnitude.

The existence of the geomagnetic field has been known since ancient times. The geomagnetic field is primarily dipolar in nature. The total field intensity diminishes from its maxima of about 60 µT at the magnetic poles, to a minimum of about 30 µT near the equator (König *et al.*, 1981). In temperate latitudes, the geomagnetic field, at sea-level, is approximately 45–50 µT whereas in regions of southern Brazil, flux densities as low as 24 µT have been reported (Hansson Mild, 2000).

The geomagnetic field is not constant but fluctuates continuously and is subject to diurnal, lunar and seasonal variations (Strahler, 1963; König *et al.*, 1981). More information on this subject is available (Dubrov, 1978) and in databases on the Web (e.g. National Geophysical Data Center).

There are also short-term variations associated with ionospheric processes. When the solar wind carries protons and electrons towards the earth, phenomena such as the Northern Lights, and rapid fluctuations in the intensity of the geomagnetic field occur. Figure 1 shows a 9-hour recording made at the Kiruna observatory in Sweden in January 2002. The variation may be large and can sometimes range from 0.1 µT to 1 µT within a few minutes. Such rapid variations are rare and correlated with the solar cycle. More commonly, variations of similar magnitude occur over a longer period of time. Despite these variations, the geomagnetic field should always be considered as a static field.

The atmosphere also has an electric field that is directed radially because the earth is negatively charged. The field strength depends to some extent on geographical latitude; it is lowest towards the poles and the equator and highest in the temperate latitudes. The average strength is around 100 V/m in fair weather, although it may range from 50–500 V/m depending on weather, altitude, time of day and season. During precipitation and bad weather, the values can change considerably, varying over a range of ± 40 000 V/m (König *et al.*, 1981). The average atmospheric electric field is not very different from that produced in most dwellings by typical 50- or 60-Hz electric field

Figure 1. Magnetogram recording from a geomagnetic research station in Kiruna, Sweden

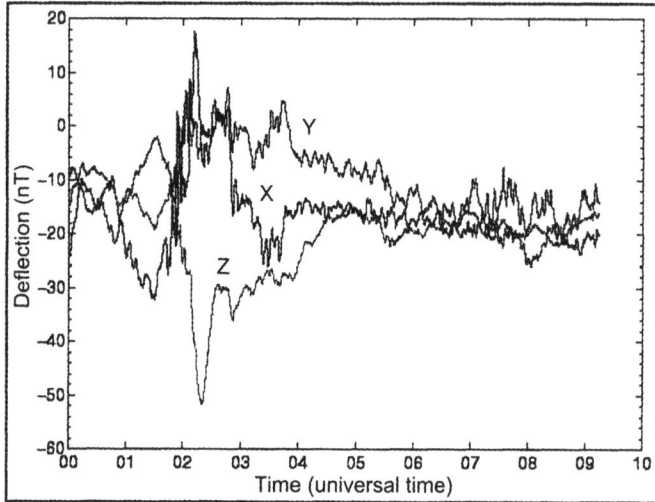

Kiruna magnetogram 2002-01-28, 09:13:35
Real-time geomagnetogram recordings can be seen at (http://www.irf.se/mag). The recordings are made in three axes: X, north, Y, east, and Z, down. The trace shown is the deflection from the mean value of the magnetic field at this location.

power sources (National Radiological Protection Board, 2001), except when measurements are made very close to electric appliances.

The electromagnetic processes associated with lightning discharges are termed *atmospherics* or '*sferics*' for short. They occur in the ELF range and at higher frequencies (König *et al.*, 1981). Each second, about 100 lightning discharges occur worldwide and can be detected thousands of kilometres away (Hansson Mild, 2000).

1.1.2   *Man-made fields and exposure*

People are exposed to electric and magnetic fields arising from a wide variety of sources which use electrical energy at various frequencies. Man-made sources are the dominant sources of exposure to time-varying fields. At power frequencies (a term that encompasses 50 and 60 Hz and their harmonics), man-made fields are many thousands of times greater than natural fields arising from either the sun or the earth.

When the source is spatially fixed and the source current and/or electrical potential difference is constant in time, the resulting field is also constant, and is referred to as static, hence the terms *magnetostatic* and *electrostatic*. Electrostatic fields are produced by fixed potential differences. Magnetostatic fields are established by permanent magnets and by steady currents. When the source current or voltage varies in time, for example, in a sinusoidal, pulsed or transient manner, the field varies proportionally.

In practice, the waveform may be a simple sinusoid or may be more complex, indicating the presence of harmonics. Complex waveforms are also observed when transients occur. Transients and interruptions, either in the electric power source or in the load, result in a wide spectrum of frequencies that may extend above several kHz (Portier & Wolfe, 1998).

Power-frequency electric and magnetic fields are ubiquitous and it is important to consider the possibilities of exposure both at work and at home. Epidemiological studies may focus on particular populations because of their proximity to specific sources of exposure, such as local power lines and substations, or because of their use of electrical appliances. These sources of exposure are not necessarily the dominant contributors to a person's time-weighted average exposure if this is indeed the parameter of interest for such studies. Various other metrics have been proposed that reflect aspects of the intermittent and transient characteristics of fields. Man-made sources and their associated fields are discussed more fully elsewhere (see National Radiological Protection Board, 2001).

(a) Residential exposure

There are three major sources of ELF electric and magnetic fields in homes: multiple grounded current-carrying plumbing and/or electric circuits, appliances and nearby power lines, including lines supplying electricity to individual homes (known as service lines, service drops or drop lines).

(i) Background exposure

Extremely low-frequency magnetic fields in homes arise mostly from currents flowing in the distribution circuits, conducting pipes and the electric ground, and from the use of appliances. The magnetic fields are partially cancelled if the load current matches the current returning via the neutral conductor. The cancellation is more effective if the conductors are close together or twisted. In practice, return currents do not flow exclusively through the associated neutral cable, but are able to follow alternative routes because of interconnected neutral cables and multiple earthing of neutral conductors. This diversion of current from the neutral cable associated with a particular phase cable results in unbalanced currents producing a net current that gives rise to a residual magnetic field. These fields produce the general background level inside and outside homes (National Radiological Protection Board, 2001). The magnetic fields in the home that arise from conductive plumbing paths were noted by Wertheimer *et al.* (1995) to "provide opportunity for frequent, prolonged encounters with 'hot spots' of unusually high intensity field — often much higher than the intensity cut-points around [0.2 or 0.3 µT] previously explored".

The background fields in homes have been measured in many studies. Swanson and Kaune (1999) reviewed 27 papers available up to 1997; other significant studies have been reported by Dockerty *et al.* (1998), Zaffanella and Kalton (1998), McBride *et al.* (1999), UK Childhood Cancer Study Investigators (UKCCSI) (1999) and Schüz *et al.* (2000). The distribution of background field intensities in a population is usually best characterized by a log-normal distribution. The mean field varies from country to country, as a consequence of differences in supply voltages, per-capita electricity consumption and wiring practices, particularly those relating to earthing of the neutral. Swanson and Kaune (1999) found that the distribution of background fields, measured over 24 h or longer, in the USA has a geometric mean of 0.06–0.07 µT, corresponding to an arithmetic mean of around 0.11 µT, and that fields in the United Kingdom are lower (geometric mean, 0.036–0.039 µT; arithmetic mean, approximately 0.05 µT), but found insufficient studies to draw firm conclusions on average fields in other European countries. Wiring practices in some countries such as Norway lead to particularly low field strengths in dwellings (Hansson Mild *et al.*, 1996).

In addition to average background fields, there is interest in the percentages of homes with fields above various cut-points. Table 1 gives the magnetic field strengths measured over 24 or 48 h in the homes of control subjects from four recent large epidemiological studies of children.

Few homes are exposed to significant fields from high-voltage power lines (see below). Even in homes with fields greater than 0.2 or 0.4 µT, high-voltage power lines are not the commonest source of the field.

The electric field strength measured in the centre of a room is generally in the range 1–20 V/m. Close to domestic appliances and cables, the field strength may increase to a few hundred volts per metre (National Radiological Protection Board, 2001).

Table 1. Measured exposure to magnetic fields in residential epidemiological studies

Study	Country	No. of control children having long-term measurements	Percentage of controls exposed to field strengths greater than	
			0.2 µT	0.4 µT
Linet *et al.* (1997)[a]	USA	530	9.2	0.9
McBride *et al.* (1999)[a]	Canada	304	11.8	3.3
UKCCSI (1999)[a]	United Kingdom	2224	1.5	0.4
Schüz *et al.* (2001a)[b]	Germany	1301	1.4	0.2

UKCCSI, UK Childhood Cancer Study Investigators
[a] Percentages calculated from data on geometric means from Ahlbom *et al.* (2000). (The results presented by Dockerty *et al.* (1999) have not been included as the numbers are too small to be meaningful at these field strengths.)
[b] Percentages calculated from medians from original data. The medians are expected to be very similar to the geometric means.

(ii)   *Fields from appliances*

The highest magnetic flux densities to which most people are exposed in the home arise close to domestic appliances that incorporate motors, transformers and heaters (for most people, the highest fields experienced from domestic appliances are also higher than fields experienced at work and outside the home). The flux density decreases rapidly with distance from appliances, varying between the inverse square and inverse cube of distance, and at a distance of 1 m the flux density will usually be similar to background levels. At a distance of 3 cm, magnetic flux densities may be several hundred microtesla or may even approach 2 mT from devices such as hair dryers and can openers, although there can be wide variations in fields at the same distance from similar appliances (National Radiological Protection Board, 2001).

Exposure to magnetic fields from home appliances must be considered separately from exposure to fields due to power lines. Power lines produce relatively low-intensity, small-gradient fields that are always present throughout the home, whereas fields produced by appliances are invariably more intense, have much steeper gradients, and are, for the most part, experienced only sporadically. The appropriate way of combining the two field types into a single measure of exposure depends critically on the exposure metric considered.

Various features of appliances determine their potential to make a significant contribution to the fields to which people are exposed, and epidemiological studies of appliances have focused on particular appliances chosen for the following reasons:

- Use particularly close to or touching the body. Examples include hair dryers, electric shavers, electric drills and saws, and electric can openers or food mixers.
- Use at moderately close distances for extended periods of time. Examples include televisions and video games, sewing machines, bedside clocks and clock radios and night storage heaters, if, for example, they are located close to the bed.
- Use while in bed, combining close proximity with extended periods of use. Examples include electric blankets and water beds (which may or may not be left on overnight).
- Use over a large part of the home. Examples include underfloor electric heating.

Table 2 gives values of broadband magnetic fields at various distances from domestic appliances in use in the United Kingdom (Preece *et al.*, 1997). The magnetic fields were calculated from a mathematical model fitted to actual measurements made on the numbers of appliances shown in the Table. Gauger (1985) and Zaffanella & Kalton (1998) reported narrow band and broadband data, respectively, for the USA. Florig and Hoburg (1990) characterized fields from electric blankets, using a three-dimensional computer model and Wilson *et al.* (1996) used spot measurements made in the home and in the laboratory. They reported that the average magnetic fields to which

Table 2. Resultant broadband magnetic field calculated at 5, 50 and 100 cm from appliances for which valid data could be derived on the basis of measured fields at 5, 30, 60 and 100 cm

| Appliance type | No. | Magnetic field (μT) at discrete distances from the surface of appliances computed from direct measurements | | | | | |
		5 cm	± SD	50 cm	± SD	100 cm	± SD
Television	73	2.69	1.08	0.26	0.11	0.07	0.04
Kettle, electric	49	2.82	1.51	0.05	0.06	0.01	0.02
Video-cassette recorder	42	0.57	0.52	0.06	0.05	0.02	0.02
Vacuum cleaner	42	39.53	74.58	0.78	0.74	0.16	0.12
Hair dryer	39	17.44	15.56	0.12	0.10	0.02	0.02
Microwave oven	34	27.25	16.74	1.66	0.63	0.37	0.14
Washing machine	34	7.73	7.03	0.96	0.56	0.27	0.14
Iron	33	1.84	1.21	0.03	0.02	0.01	0.00
Clock radio	32	2.34	1.96	0.05	0.05	0.01	0.01
Hi-fi system	30	1.56	4.29	0.08	0.14	0.02	0.03
Toaster	29	5.06	2.71	0.09	0.08	0.02	0.02
Central heating boiler	26	7.37	10.10	0.27	0.26	0.06	0.05
Central heating timer	24	5.27	7.05	0.14	0.17	0.03	0.04
Fridge/freezer	23	0.21	0.14	0.05	0.03	0.02	0.01
Radio	23	3.00	3.26	0.06	0.04	0.01	0.01
Central heating pump	21	61.09	59.58	0.51	0.47	0.10	0.10
Cooker	18	2.27	1.33	0.21	0.15	0.06	0.04
Dishwasher	13	5.93	4.99	0.80	0.46	0.23	0.13
Freezer	13	0.42	0.87	0.04	0.02	0.01	0.01
Oven	13	1.79	0.89	0.39	0.23	0.13	0.09
Shower, electric	12	30.82	35.04	0.44	0.75	0.11	0.25
Burglar alarm	10	6.20	5.21	0.18	0.11	0.03	0.02
Food processor	10	12.84	12.84	0.23	0.23	0.04	0.04
Extractor fan	9	45.18	107.96	0.50	0.93	0.08	0.14
Cooker hood	9	4.77	2.53	0.26	0.10	0.06	0.02
Speaker	8	0.48	0.67	0.07	0.13	0.02	0.04
Hand blender	8	76.75	87.09	0.97	1.05	0.15	0.16
Tumble dryer	7	3.93	5.45	0.34	0.42	0.10	0.10
Food mixer	6	69.91	69.91	0.69	0.69	0.11	0.11
Fish-tank pump	6	75.58	64.74	0.32	0.09	0.05	0.01
Computer	6	1.82	1.96	0.14	0.07	0.04	0.02
Electric clock	6	5.00	4.15	0.04	0.00	0.01	0.00
Electric knife	5	27.03	13.88	0.12	0.05	0.02	0.01
Hob	5	2.25	2.57	0.08	0.05	0.01	0.01
Deep-fat fryer	4	4.44	1.99	0.07	0.01	0.01	0.00
Tin/can opener	3	145.70	106.23	1.33	1.33	0.20	0.21
Fluorescent light	3	5.87	8.52	0.15	0.20	0.03	0.03
Fan heater	3	3.64	1.41	0.22	0.18	0.06	0.06
Liquidizer	2	3.28	1.19	0.29	0.35	0.09	0.12

Table 2 (contd)

| Appliance type | No. | Magnetic field (μT) at discrete distances from the surface of appliances computed from direct measurements | | | | | |
		5 cm	± SD	50 cm	± SD	100 cm	± SD
Bottle sterilizer	2	0.41	0.17	0.01	0.00	0.00	0.00
Coffee maker	2	0.57	0.03	0.06	0.07	0.02	0.02
Shaver socket	2	16.60	1.24	0.27	0.01	0.04	0.00
Coffee mill	1	2.47		0.28		0.07	
Shaver, electric	1	164.75		0.84		0.12	
Tape player	1	2.00		0.24		0.06	

From Preece *et al.* (1997)

the whole body is exposed are between 1 and 3 μT. From eight-hour measurements, Lee *et al.* (2000) estimated that the time-weighted average magnetic field exposures from overnight use of electric blankets ranged between 0.1 and 2 μT.

Measurements of personal exposure are expected to be higher than measurements of background fields because they include exposures from sources such as appliances. Swanson and Kaune (1999) found that in seven studies which measured personal exposure and background fields for the same subjects, the ratio varied from 1.0 to 2.3 with an average of 1.4.

(iii)   *Power lines*

Power lines operate at voltages ranging from the domestic supply voltage (120 V in North America, 220–240 V in Europe) up to 765 kV in high-voltage power lines (WHO, 1984). At higher voltages, the main source of magnetic field is the load current carried by the line. Higher voltage lines are usually also capable of carrying higher currents. As the voltage of the line and, hence, in general, the current carried, and the separation of the conductors decrease, the load current becomes a progressively less important source of field and the net current, as discussed in (i) above, becomes the dominant source. It is therefore convenient to treat high-voltage power lines (usually taken to mean 100 kV or 132 kV, also referred to as transmission lines) as a separate source of field (Merchant *et al.*, 1994; Swanson, 1999).

High-voltage power lines in different countries follow similar principles, but with differences in detail so that the fields produced are not identical (power-line design as it affects the fields produced was reviewed by Maddock, 1992). For example, high-voltage power lines in the United Kingdom can have lower ground clearances and can carry higher currents than those in some other countries, leading to higher fields under the lines. When power lines carry two or more circuits, there is a choice as to the physical distribution of the various wires on the towers. An arrangement called 'transposed phasing', in which the wires or bundles of wire — phases — in the circuit on one side of the tower have the opposite order to those on the other side, results in fields that decrease more rapidly with distance from the lines than the alternatives (Maddock, 1992). Transposed phasing is more common in the United Kingdom than, for example, in the USA.

In normal operation, high-voltage power lines have higher ground clearances than the minimum permitted, and carry lower currents than the maximum theoretically possible. Therefore, the fields present in normal operation are substantially lower than the maxima theoretically possible.

Electric fields

High-voltage power lines give rise to the highest electric field strengths that are likely to be encountered by people. The maximum unperturbed electric field strength immediately under 400-kV transmission lines is about 11 kV/m at the minimum clearance of 7.6 m, although people are generally exposed to fields well below this level. Figure 2 gives examples of the variation of electric field strength with distance from the centreline of high-voltage power lines with transposed phasing in the United Kingdom. At 25 m to either side of the line, the field strength is about 1 kV/m (National Radiological Protection Board, 2001).

Objects such as trees and other electrically grounded objects have a screening effect and generally reduce the strength of the electric fields in their vicinity. Buildings attenuate electric fields considerably, and the electric field strength may be one to three

Figure 2. Electric fields from high-voltage overhead power lines

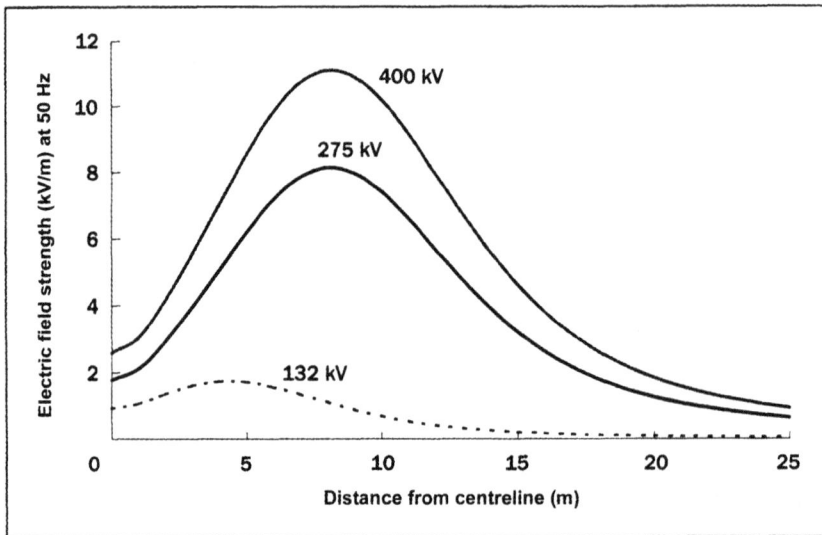

From National Radiological Protection Board (2001)

orders of magnitude less inside a building than outside it. Electric fields to which people are exposed inside buildings are generally produced by internal wiring and appliances, and not by external sources (National Radiological Protection Board, 2001).

Magnetic fields

The average magnetic flux density measured directly beneath overhead power lines can reach 30 μT for 765-kV lines and 10 μT for the more common 380-kV lines (Repacholi & Greenebaum, 1999). Theoretical calculations of magnetic flux density beneath the highest voltage power line give ranges of up to 100 μT (National Radiological Protection Board, 2001). Figure 3 gives examples of the variation of magnetic flux density with distance from power lines in the United Kingdom. Currents (and hence the fields produced) vary greatly from line to line because power consumption varies with time and according to the area in which it is measured.

Magnetic fields generally fall to background strengths at distances of 50–300 m from high-voltage power lines depending on the line design, current and the strength of background fields in the country concerned (Hansson Mild, 2000). Few people live so close to high-voltage power lines (see Table 3); meaning that these power lines are a major source of exposure for less than 1% of the population according to most studies (see Table 4).

In contrast to electric fields for which the highest exposure is likely to be experienced close to high-voltage power lines, the highest magnetic flux densities are

Figure 3. Magnetic fields from high-voltage overhead power lines

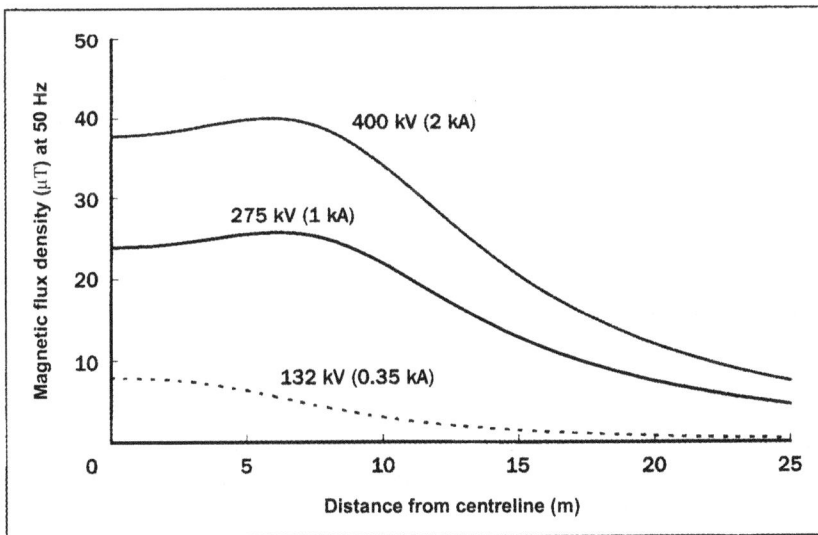

From National Radiological Protection Board (2001)

Table 3. Percentages of people in certain countries within various distances of high-voltage power lines

Country (reference)	No. of subjects	Voltages of power lines included (kV)	Distance (m)	Subjects within this distance	
				No.	%
Canada (McBride et al., 1999)	399[a]	≥ 50	50	4	1.00
			100	7	1.75
Denmark (Olsen et al., 1993)	6495[b]	132–150	75	28	0.43
		50–60	35	22	0.34
					0.46
		50–440	150	52	0.80
United Kingdom (Swanson, 1999)	22 million[c]	≥ 275	50		0.07
			100		0.21
United Kingdom (UKCCSI, 2000a)	3390[a]	≥ 66	50	9	0.27
			120	35	1.03
		≥ 275	50	3	0.09
			120	9	0.27
USA (Kleinerman et al., 2000)	405[a]	≥ 50[d]			
		power line	40	98	24.2
		transmission line	40	20	4.94

UKCCSI, UK Childhood Cancer Study Investigators
[a] Controls from epidemiological study of children
[b] Cases and controls from epidemiological study of children
[c] All homes in England and Wales (Source: Department of Transport, Local Government and the Regions; National Assembly for Wales, 1998, http://www.statistics.gov.uk/statbase/Expodata/Spreadsheets/D4524.xls)
[d] Not stated in Kleinerman et al. (2000), assumed to be the same as Wertheimer & Leeper (1979)

more likely to be encountered in the vicinity of appliances or types of equipment that carry large currents (National Radiological Protection Board, 2001).

Direct current lines

Some high-voltage power lines have been designed to carry direct current (DC), therefore producing both electrostatic and magnetostatic fields. Under a 500-kV DC transmission line, the static electric field can reach 30 kV/m or higher, while the magnetostatic field from the line can average 22 μT which adds vectorially to the earth's field (Repacholi & Greenebaum, 1999).

Table 4. Percentages of people in various countries living in homes in which high-voltage power lines produce magnetic fields in excess of specified values

Country (reference)	No. of subjects	Voltages of power lines included (kV)	Measured field (μT)	Subjects whose homes exceed the measured field	
				No.	%
Denmark (Olsen et al., 1993)	4788[a]	≥ 50	0.25	11	0.23
			0.4	3	0.06
Germany (Schüz et al., 2000)	1835[b]	≥ 123	0.2	8	0.44
United Kingdom (UKCCSI, 2000a)	3390[a]	≥ 66[c]	0.2	11	0.32
			0.4	8	0.24

UKCCSI, UK Childhood Cancer Study Investigators
[a] Controls from epidemiological study of children
[b] Cases and controls from epidemiological study of children
[c] Probably over 95% were ≥ 132 kV

(iv) *Substations*

Outdoor substations normally do not increase residential exposure to electric and magnetic fields. However, substations inside buildings may result in exposure to magnetic fields at distances less than 5–10 m from the stations (National Radiological Protection Board, 2001). On the floor above a station, flux densities of the order of 10–30 μT may occur depending on the design of the substation (Hansson Mild *et al.*, 1991). Normally, the main sources of field are the electrical connections (known as busbars) between the transformer and the other parts of the substation. The transformer itself can also be a contributory source.

(v) *Exposure to ELF electric and magnetic fields in schools*

Exposure to ELF electric and magnetic fields while at school may represent a significant fraction of a child's total exposure. A study involving 79 schools in Canada took a total of 43 009 measurements of 60-Hz magnetic fields (141–1543 per school). Only 7.8% of all the fields measured were above 0.2 μT. For individual schools, the average magnetic field was 0.08 μT (SD, 0.06 μT). In the analysis by use of room, only typing rooms had magnetic fields that were above 0.2 μT. Hallways and corridors were above 0.1 μT and all other room types were below 0.1 μT. The percentage of classrooms above 0.2 μT was not reported. Magnetic fields above 0.2 μT were mostly associated with wires in the floor or ceiling, proximity to a room containing electrical appliances or movable sources of magnetic fields such as electric typewriters, computers and overhead projectors. Eight of the 79 schools were situated near high-voltage power lines. The survey showed no clear difference in overall magnetic field strength between the schools and domestic environments (Sun *et al.*, 1995).

Kaune *et al.* (1994) measured power-frequency magnetic fields in homes and in the schools and daycare centres of 29 children. Ten public shools, six private schools and one daycare centre were included in the study. In general, the magnetic field strengths measured in schools and daycare centres were smaller and less variable than those measured in residential settings.

The UK Childhood Cancer Study Investigators (UKCCSI) (1999) carried out an epidemiological study of children in which measurements were made in schools as well as homes. Only three of 4452 children aged 0–14 years who spent 15 or more hours per week at school during the winter, had an average exposure during the year above 0.2 μT as a result of exposure at school.

In a preliminary report reviewed elsewhere (Portier & Wolfe, 1998), Neutra *et al.* (1996) reported a median exposure level of 0.08 μT for 163 classrooms at six California schools, with approximately 4% of the classrooms having an average magnetic field in excess of 0.2 μT. These fields were mainly due to ground currents on water pipes, with nearby distribution lines making a smaller contribution. [The Working Group noted that no primary publication was available.] The study was subsequently extended and an executive summary was published in an electronic form, which is available at www.dhs.ca.gov/ehib/emf/school_exp_ass_exec.pdf

(b)   Occupational exposure

Exposure to magnetic fields varies greatly across occupations. The use of personal dosimeters has enabled exposure to be measured for particular types of job. Table 5 (Portier & Wolfe, 1998) lists the time-weighted average exposure to magnetic fields for selected job classifications. In some cases the standard deviations are large. This indicates that there are instances in which workers in these categories are exposed to far stronger fields than the means listed here.

Floderus *et al.* (1993) investigated sets of measurements made at 1015 different workplaces using EMDEX (electric and magnetic field digital exposure system)-100 and EMDEX-C personal dosimeters. This study covered 169 different job categories and participants wore the dosimeters for a mean duration of 6.8 h. The distribution of all 1-s sampling period results for 1015 measurements is shown in Figure 4. The most common measurement was 0.05 μT and measurements above 1 μT were rare. It should be noted that the response of the EMDEX-C is non-linear over a wide frequency range. For example, the railway frequency in Sweden is 16 $^2/_3$ Hz, which means that the measurements obtained with the EMDEX are underestimates of the exposure.

It can be seen from Table 5 that workers in certain occupations are exposed to elevated magnetic fields. Some of the more significant occupations are considered below.

Table 5. Time-weighted average exposure to magnetic fields by job title

Occupational title	Average exposure (µT)	Standard deviation
Train (railroad) driver	4.0	NR
Lineman	3.6	11
Sewing machine user	3.0	0.3
Logging worker	2.5	7.7
Welder	2.0	4.0
Electrician	1.6	1.6
Power station operator	1.4	2.2
Sheet metal worker	1.3	4.2
Cinema projectionist	0.8	0.7

Modified from Portier & Wolfe (1998)
NR, not reported

Figure 4. Distribution of all occupational magnetic field samples

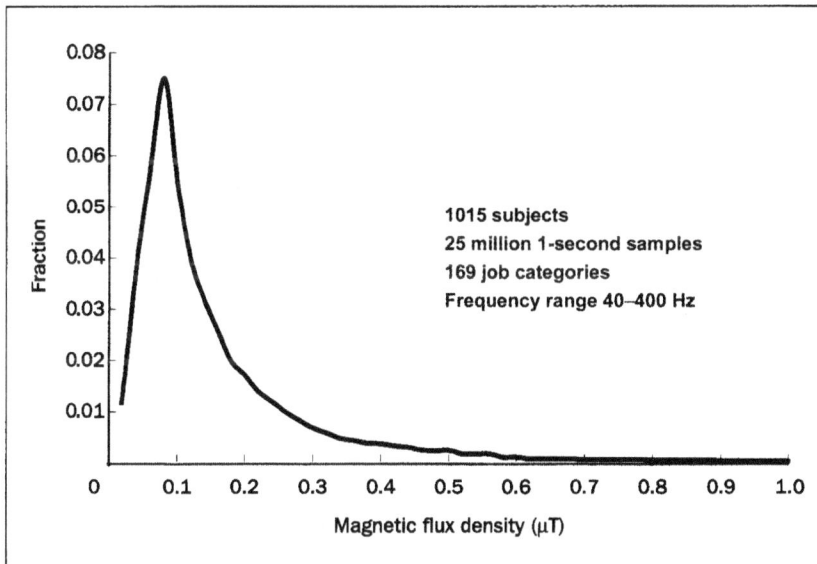

1015 subjects
25 million 1-second samples
169 job categories
Frequency range 40–400 Hz

Modified from National Radiological Protection Board (2001) (original figure from Floderus *et al.*, 1993)
The distribution should not be interpreted as a distribution of results for individuals.

(i)   *The electric power industry*

Strong magnetic fields are encountered mainly in close proximity to high currents (Maddock, 1992). In the electric power industry, high currents are found in overhead lines and underground cables, and in busbars in power stations and substations. The busbars close to generators in power stations can carry currents up to 20 times higher than those typically carried by the 400-kV transmission system (Merchant *et al.*, 1994).

Exposure to the strong fields produced by these currents can occur either as a direct result of the job, e.g. a lineman or cable splicer, or as a result of work location, e.g. when office workers are located on a power station or substation site. It should be noted that job categories may include workers with very different exposures, e.g. linemen working on live or dead circuits. Therefore, although reporting magnetic-field exposure by job category is useful, a complete understanding of exposure requires a knowledge of the activities or tasks and the location as well as measurements made by personal exposure meters.

The average magnetic fields to which workers are exposed for various jobs in the electric power industry have been reported as follows: 0.18–1.72 μT for workers in power stations, 0.8–1.4 μT for workers in substations, 0.03–4.57 μT for workers on lines and cables and 0.2–18.48 μT for electricians (Portier & Wolfe, 1998; National Radiological Protection Board, 2001).

(ii)   *Arc and spot welding*

In arc welding, metal parts are fused together by the energy of a plasma arc struck between two electrodes or between one electrode and the metal to be welded. A power-frequency current usually produces the arc but higher frequencies may be used in addition to strike or to maintain the arc. A feature of arc welding is that the welding cable, which can carry currents of hundreds of amperes, can touch the body of the operator. Stuchly and Lecuyer (1989) surveyed the exposure of arc welders to magnetic fields and determined separately the exposure at 10 cm from the head, chest, waist, gonads, hands and legs. Whilst it is possible for the hand to be exposed to fields in excess of 1 mT, the trunk is typically exposed to several hundred microtesla. Once the arc has been struck, these welders work with comparatively low voltages and this is reflected in the electric field strengths measured; i.e. up to a few tens of volts per metre (National Radiological Protection Board, 2001).

Bowman *et al.* (1988) measured exposure for a tungsten–inert gas welder of up to 90 μT. Similar measurements reported by the National Radiological Protection Board indicate magnetic flux densities of up to 100 μT close to the power supply, 1 mT at the surface of the welding cable and at the surface of the power supply and 100–200 μT at the operator position (National Radiological Protection Board, 2001). London *et al.* (1994) reported the average workday exposure of 22 welders and flame cutters to be much lower (1.95 μT).

(iii) *Induction furnaces*

Measurements on induction furnaces and heaters operating in the frequency range from 50 Hz to 10 kHz have been reported (Lövsund *et al.*, 1982) and are summarized in Table 6. The field strengths decrease rapidly with distance from the coils and do not reflect whole-body exposure. However, in some cases, whole-body exposure occurs. Induction heater operators experience short periods of exposure to relatively strong fields as the induction coils are approached (National Radiological Protection Board, 2001).

Table 6. Frequency and magnetic flux densities from induction furnaces

Type of machine	No.	Frequency band	Magnetic flux density (mT): measured ranges
Ladle furnace in conjunction with 1.6-Hz magnetic stirrer, measurements made at 0.5–1 m from furnace	1	1.6 Hz, 50 Hz	0.2–10
Induction furnace,			
at 0.6–0.9 m	2	50 Hz	0.1–0.9
at 0.8–2.0 m	5	600 Hz	0.1–0.9
Channel furnace, at 0.6–3.0 m	3	50 Hz	0.1–0.4
Induction heater, at 0.1–1.0 m	5	50 Hz–10 kHz	1–60

Modified from Lövsund *et al.* (1982)

(iv) *Electrified transport*

Electricity is utilized in various ways in public transport. The power is supplied as DC or at alternating frequencies up to those used for power distribution. Many European countries such as Austria, Germany, Norway, Sweden and Switzerland have systems that operate at $16\,^2/_3$ Hz. Most of these systems use a DC traction motor, and rectification is carried out either on-board or prior to supply. On-board rectification usually requires a smoothing inductor, a major source of static and 100-Hz alternating magnetic fields. For systems that are supplied with nominal DC there is little smoothing at the rectification stage, resulting in a significant alternating component in the 'static' magnetic fields (National Radiological Protection Board, 2001).

On Swedish trains, Nordenson *et al.* (2001) found values ranging from 25 to 120 µT for power-frequency fields in the driver's cab, depending on the type (age and model) of locomotive. Typical daily average exposures were in the range of 2–15 µT.

Other forms of transport, such as aeroplanes and electrified road vehicles are also expected to increase exposure, but have not been investigated extensively.

(v)   *Use of video display terminals*

Occupational exposure to ELF electric and magnetic fields from video display terminals has recently received attention. Video display terminals produce both power-frequency fields and higher frequency fields ranging from about 50 Hz up to 50 kHz (Portier & Wolfe, 1998). Sandström *et al.* (1993) reported median magnetic fields at ELF as 0.21 µT and 0.03 µT for frequencies between 15 kHz and 35 kHz. The median electric fields measured in the same frequency ranges were 20 V/m and 1.5 V/m, respectively.

(vi)   *Use of sewing machines*

Hansen *et al.* (2000) reported higher-than-background magnetic fields near industrial sewing machines, because of proximity to motors, with field strengths ranging from 0.32–11.1 µT at a position corresponding approximately to the sternum of the operator. The average exposure for six workers working a full work-shift in the garment industry ranged from 0.21–3.20 µT.

(c)   *Transients*

Transients occur in electrical systems mainly as a result of switching loads or circuits on and off. They can be produced deliberately, as in circuit testing, or occur accidentally, caused by sudden changes in current load following a short-circuit or lightning strike. Such disturbances invariably have a much higher frequency content than that of the signal that is interrupted (Kaune *et al.*, 2000).

A number of devices have been designed to record electric power transients (Deadman *et al.*, 1988; Héroux, 1991; Kaune *et al.*, 2000). These devices differ primarily in the range of frequencies used to define a transient and in their storage capacities. Kaune *et al.* (2000) examined magnetic transients within the range of 2–200 kHz that had threshold peak intensity levels, measured using a dual channel recorder, of either 3.3 or 33 nT. Recordings were made for a minimum of 24 h in each of 156 homes distributed at six different locations in the USA. Although the recordings of the less intense 3.3-nT transients might have been contaminated somewhat by nearby television sets, this was not the case for the recordings of the 33-nT transients. It was found that transient activity in homes has a distinct diurnal pattern, generally following variations in power use. Evidence was also presented indicating that the occurrence of the larger, 33-nT magnetic transients is increased ($p = 0.01$) in homes with well-grounded metal plumbing that is also electrically connected to an external water system. In contrast, the increased transient activity in the homes tested was not related to wire code.

1.2 Instrumentation and computational methods of assessing electric and magnetic fields

1.2.1 *Instruments*

Measurements of electric and magnetic fields are used to characterize emissions from sources and exposure of persons or experimental subjects. The mechanisms that define internal doses of ELF electric and magnetic fields and relate them to biological effects are not precisely known (Portier & Wolfe, 1998) with the exception of the well-studied neurostimulatory effects of electric and magnetic fields (Bailey *et al.*, 1997; Reilly, 1998). Therefore, it is important that investigators recognize the possible absence of a link between selected measured fields and a biological indicator of dose. The instrument best suited to the purpose of the investigation should be selected carefully. Investigators should evaluate the instrument and its proposed use before starting a study and calibrate it at appropriate intervals thereafter.

Early epidemiological and laboratory studies used simple survey instruments that displayed the maximum field measured along a single axis. More recent studies of magnetic fields have used meters that record the field along three orthogonal axes and report the resultant root-mean-square (rms) field as:

$$\text{Resultant} = \sqrt{(X^2 + Y^2 + Z^2)}$$

Survey meters are easy to use, portable and convenient for measuring field magnitudes over wide areas or in selected locations. Three-axis survey meters are capable of simple signal processing, such as computing the resultant field, storing multiple measurements in their memory or averaging measurements. It is important to note that the resultant field can be equal to, or up to 40% greater (for a circularly polarized field) than, the maximum field measured by a single-axis meter (IEEE, 1995a). Computer-based waveform capture measurement systems are designed to perform sophisticated signal processing and to record signals over periods ranging from a fraction of a second to several days. The instruments discussed here are those most commonly used for measuring fields in the environment or laboratory (Table 7). The measurement capabilities of selected instruments are summarized in Table 8. Less frequently used instruments designed for special purposes are described elsewhere (e.g. WHO, 1984, 1987). The operation of the electric and magnetic field meters recommended for use is described in IEEE (1995a) and IEC (1998).

Table 7. General characteristics of instruments

Meter type	Primary uses	Field parameters measured	Data-collection features	Cost	Ease of use	Data recording	Portability
Computer-based waveform measurement systems	Spot measurements Mapping Long-term measurements Waveform capture Transient capture	AC/DC field magnitude (x,y,z, resultant) AC field magnitude at each frequency of interest (x,y,z axes, resultant) AC field polarization AC–DC orientation Peak-to-peak	Full waveform capture Highest quantification content in data collection	Very high	High-level technical understanding required The vast quantities of data collected are difficult to manage (approximately 50 kbytes for an average spot measurement vs. 10 bytes with a three-axis AC-field recording meter)	Digitized recording features	Less portable than typical meters 5-kg 'portable' system commercially available
Three-axis AC field recording root-mean-square meter	Personal exposure Spot measurements Mapping Long-term measurements Exploratory measurements	AC field magnitude (x,y,z axes, resultant) in a bandwidth dependent upon model Some models can provide harmonic content	Many have software for mapping capabilities if used with mapping wheel	Medium–high	Almost no instruction required for accurate resultant measurements More difficult to use for exploratory measurements ('sniffing') than single-axis meters because of delay between readouts	Recording features	Small, portable

Table 7 (contd)

Meter type	Primary uses	Field parameters measured	Data-collection features	Cost	Ease of use	Data recording	Portability
Three-axis cumulative exposure meter with display	Personal exposure Spot measurements Exploratory measurements Long-term measurements for cumulative information	AC field magnitude (x,y,z axes, resultant) in a bandwidth dependent upon model	Most frequently used for personal exposure measurements	Medium	Almost no instruction required for accurate resultant measurements	Records accumulated data, rather than individual samples	Small, portable
Three-axis AC-field survey meter	Spot measurements	AC-field magnitude (x,y,z axes, resultant) in a bandwidth dependent upon model	Similar to three-axis recording meters, with recording capabilities	Medium	Almost no instruction required for accurate resultant measurement	No recording feature	Small, portable
	Exploratory measurements	Some models can provide total harmonic content			More difficult to use for exploratory measurements ('sniffing') than single-axis meters because of delay between readouts		
Single-axis AC-field survey meter	Exploratory measurements	AC field magnitude in one direction, in a bandwidth dependent upon model	Can be used to determine polarization	Low	Continuous readout provides easy source investigation	No recording feature	Small, portable
	Spot measurements	Some models can be switched from flat to linear response to provide rough data on presence of harmonics	Easy determination of direction of field Can be used with an audio attachment. For exploratory measurements		Maximum field must be 'found' by properly rotating the meter, or measuring in three orthogonal directions to calculate the resultant field		

AC, alternating current; DC, direct current
For further details and handling information, see IEC (1998).

Table 8. Characteristics of field meters

Model	Fields	Sensor	No. of axes	Frequency response (Hz)[a]	Maximum full-scale range (µT)	Output		Function	Comment
AMEX	B	C	1	–	12.5	TWA	AVG	P	
AMEX-3D	B	C	3	25 Hz–1.2 kHz	15	TWA	AVG	P	
EMDEX C	B, E	C, P	3,1	40–400 Hz	2550	D, DL	AVG	P	Built-in E field
EMDEX II	B	C	3	40–800 Hz	300	D, DL	RMS	P	Has harmonic capability
Positron	B, E, HF	C, P, F	3,1	50–60 Hz	50	D, DL	PEAK	P	Built-in E field
Monitor Ind.	B	C	1	40 Hz–1 kHz	250	A	RMS	S	
Multiwave	B	C, FG	3	0–10 kHz	500	D, DL	RMS	S	Waveform capture
Power frequency Meter MOD120	B, E	C, P	1	35–600 Hz	3000	A	AVG	S	
STAR[b]	B	C	3	60 Hz	51	D, DL	RMS	S	
MAG 01	B	FG	1	0–10 Hz	200	D	–	S	
IREQ	B	C	3	40 Hz–1 kHz	100	D, DL	RMS	S	
Field meter	B, E	D	1,1	25 Hz–10 MHz	–	–	–	S	Used by Hietanen & Jokela (1990)
BMM - 3000	B	C	3	5 Hz–2 kHz	2000	A	RMS	S	Frequency filters MPR/TC092 Band I testing
Sydkraft	B	C	1	50–60 Hz	20	D, DL	AVG	S	

Modified from Portier & Wolfe (1998)

E, electric; B, magnetic; HF, high frequency; C, coil, P (sensor), plate; F, conductive foam; FG, flux gate; D (sensor), active dipole; D (output), digital spot; A, analogue spot; DL, data logging; TWA, single readout of TWA; AVG, average; RMS, root-mean-square; P (function), personal monitor; S, survey

[a] Frequency response and maximum range refer only to the magnetic field measurement channel
[b] The specifications are for the original STAR meter that was produced only in limited quantities for non-commercial use. The commercial version of the instrument (Field StAR from Dexsil) has a range of 100 µT on each of three orthogonal axes.

(*a*)   *Electric fields*

(i)   *Survey meters*

The meters commonly used in occupational and environmental surveys of electric fields are small both for convenience and to minimize their effect on the electric field being measured. To measure the unperturbed field, the meter is suspended at the end of a long non-conductive rod or tripod to minimize interference with the measurement by the investigator. In an oscillating electric field, the current measured between two isolated conducting parts of the sensor is proportional to the field strength. The accuracy of the measurements obtained with these instruments is generally high, except under the following conditions:

- extremes of temperature and humidity;
- insufficient distance of the probe from the investigator;
- instability in meter position;
- loss of non-conductive properties of the supporting rod.

Electric fields can also be measured at fixed locations, e.g. under transmission lines or in laboratory exposure chambers by measuring the current collected by a flat conducting plate placed at ground level. For sinusoidal fields, the electric flux density can be calculated from the area of the plate (A), the permittivity of a vacuum (ε_0), the frequency (*f*) and the measured current induced in the plate (I_{rms}) in the expression below:

$$E = \frac{I_{rms}}{2\pi f \varepsilon_0 A}$$

Electric field meters can be calibrated by placing the probe in a uniform field produced between two large conducting plates for which the field strength can be easily calculated (IEEE, 1995a, b).

(ii)   *Personal exposure meters for measuring electric fields*

Personal exposure meters are instruments for measuring the exposure of a person to electric fields in various environments, e.g. work, home and travel (see below for personal exposure meters for measuring magnetic fields). However, wearing a meter on the body perturbs the electric field being measured in unpredictable ways. Typically, where exposure to electric fields of large groups of subjects is being measured, a meter is placed in an armband, shirt pocket or belt pouch (Male *et al.*, 1987; Bracken, 1993). Perturbation of the ambient field by the body precludes obtaining an absolute value of the field and, at best, the average value of such measurements reflects the relative level of exposure.

(b)   *Magnetic fields*

(i)   *Survey meters*

Magnetic fields can be measured with a survey meter, fixed location monitor or a wearable field meter. The simplest meter measures the voltage induced in a coil of wire. For a sinusoidally varying magnetic field, **B,** of frequency f, the voltage, V, induced in the coil is given by:

$$V = -2\pi f \mathbf{B}_0 A \cos\omega t$$

where f is the frequency of the field and $\omega = 2\pi f$, A is the area of the loop, and $\mathbf{B}_0$ is the component of **B** perpendicular to the loop.

The voltage induced by a given field increases with the addition of turns of wire or of a ferromagnetic core. To prevent interference from electric fields, the magnetic field probe must be shielded. If the meter is used for surveys or personal exposure measurements, frequencies lower than approximately 30 Hz must be filtered out to remove voltages induced in the probe by the motion of the meter in the earth's magnetic field.

The presence of higher frequencies, such as harmonics, can affect magnetic field measurements depending on the frequency response of the magnetic field meter. The frequency response of three different meters is illustrated in Figure 5 (modified from Johnson, 1998). These meters are calibrated so that a 60-Hz, 0.1-μT field reads as 0.1 μT on all three instruments. The narrow-band meter focuses on the 60-Hz magnetic field and greatly attenuates the sensitivity of the meter to higher and lower frequencies. The broadband meter provides an accurate measurement of the magnetic field across a wider frequency range because it has a flat frequency response between 40 Hz and 1000 Hz. The broadband meter with a linear response provides very different measurements in this range as the magnetic field reading is weighted by its frequency (Johnson, 1998).

Figure 5. Frequency response of linear broadband, flat broadband and narrow-band magnetic-field meters to a reference field of 0.1 μT

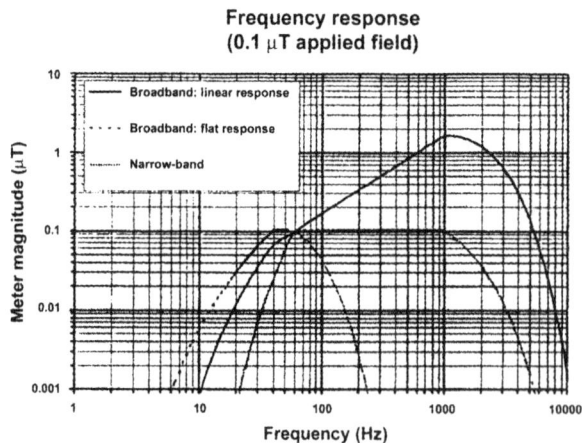

Modified from Johnson (1998)

Fluxgate magnetometers have adequate sensitivity for measuring magnetostatic fields in the range 0.1 µT–0.01 T. Above 100 µT, both AC and DC magnetic fields can be measured using a Hall effect sensor (IEEE, 1995b). The sensor is designed to measure the voltage across a thin strip of semiconducting material carrying a control current. The voltage change is directly related to the magnetic flux density of AC and DC magnetic fields (Agnew, 1992).

Early survey meters made average field readings and then extrapolated them to root-mean-square values by applying a calibration factor. These meters give erroneous readings when in the presence of harmonics and complex waveforms.

(ii) *Personal exposure meters for measuring magnetic fields*

Wearable meters for measuring magnetic fields have facilitated assessments of the personal exposure of individuals as they go about daily activities at home, school and work. A few instruments can also record electric-field measurements. The available personal exposure meters can integrate field readings in single or multiple data registers over the course of a measurement period. For a single-channel device, the result is a single value representing the integrated exposure over time in µT·h or (kV/m) h. Some meters classify and accumulate exposures into defined intensity 'bins'. Other personal exposure meters collect samples at fixed intervals and store the measurements in computer memory for subsequent downloading and analysis (see Table 9).

One of the most popular instruments used in occupational surveys and epidemio-logical studies is the electric and magnetic field digital exposure system (EMDEX). The EMDEX II data logger records the analogue output from three orthogonal coils or the computed resultant magnetic field. It can also record the electric field detected by a separate sensor. Different versions of the meter are used for environmental field ranges (0.01 µT–0.3 mT) and near high intensity sources (0.4 µT–12 mT) (data from the manu-facturer, 2001).

Smaller, lighter versions of the EMDEX are available to collect time series records over longer time periods (EMDEX Lite) or to provide statistical descriptors — mean, standard deviation, minimum, maximum and accumulated time above specified thresholds — of accumulated measurements (EMDEX Mate). The AMEX (average magnetic exposure)-3D measures only the average magnetic field over time of use. IEC (1998) has provided detailed recommendations for the use of instruments in measuring personal exposure to magnetic fields.

(iii) *Frequency response*

The bandwidth of magnetic field meters is generally between 40 Hz and 1000 Hz. Further differentiation of field frequency within this range is not possible unless filtered to a narrow frequency band of 50 or 60 Hz. However, a data logger, the SPECLITE®, was employed in one study to record the magnetic field in 30 frequency bins within this range at 1-min intervals (Philips *et al.*, 1995).

Table 9. Commercially available instruments for measuring ELF magnetic fields[a]

Company, location	Meter, field type	Frequency range
AlphaLab Inc. Salt Lake City, Utah, USA	TriField Meter (3-axis **E**, **M** & **RF**)	50 Hz–3 GHz
Bartington Instruments Ltd Oxford, England	MAG-01 (1-axis **M**) MAG-03 (3-axis **M**)	DC–a few kHz 0 Hz–3000 Hz
Combinova AB Bromma, Sweden	MFM 10 (3-axis **M** recording) MFM 1020 (3-axis **E**, **M** recording) FD 1 (**E**, 3-axis **M** survey) FD 3 (3-axis **M** recording)	20 Hz–2000 Hz 5 Hz–400 kHz 20 Hz–2000 Hz 20 Hz–2000 Hz
Dexsil Corp. Hamden, Connecticut, USA	Field Star 1000 (3-axis **M** recording) Field Star 4000 (3-axis **M** recording) Magnum 310 (3-axis **M** survey)	not specified not specified 40 Hz–310 Hz
Electric Research Pittsburgh, Pennsylvania, USA	MultiWave® System II (**E**, **M** 3-axis, waveform)	0–3000 Hz
Enertech Consultants Campbell, California, USA	EMDEX SNAP (3-axis **M** survey) EMDEX PAL (3-axis **M** limited recording) EMDEX MATE (3-axis **M** limited recording) EMDEX LITE (3-axis **M** recording) EMDEX II (3-axis **E** & **M** recording) EMDEX WaveCorder (3-axis **M** waveform) EMDEX Transient Counter (3-axis **M**)	40 Hz–1000 Hz 40 Hz–1000 Hz 40 Hz–1000 Hz 10 Hz–1000 Hz 40 Hz–800 Hz 10 Hz–3000 Hz 2000 Hz–220 000 Hz
EnviroMentor AB Mölndal, Sweden	Field Finder Lite (1-axis **M** & **E**) Field Finder (3-axis **M** & 1-axis **E**) ML-1 (3-axis **M**, 3-dimensional presentation) BMM-3000 (3-axis **M**, analysis program)	15 Hz–1500 Hz 30 Hz–2000 Hz 30 Hz–2000 Hz 5 Hz–2000 Hz
Holaday Industries, Inc. Eden Prairie, Minnesota, USA	HI-3624 (**M**) HI-3624A (**M**) HI-3604 (**E**, **M**) HI-3627 (3-axis **M**, recorder output)	30 Hz–2000 Hz 5 Hz–2000 Hz 30 Hz–2000 Hz 5 Hz–2000 Hz
Magnetic Sciences International Acton, Massachusetts, USA	MSI-95 (1-axis **M**) MSI-90 (1-axis **M**) MSI-25 (1-axis **M**)	25 Hz–3000 Hz 18 Hz–3300 Hz 40 Hz–280 Hz
Physical Systems International Holmes Beach, Florida, USA	FieldMeter (1-axis **E**, **M**) FieldAnalyzer (1-axis **E**, 3-axis **M**, waveform)	16 Hz–5000 Hz 1 Hz–500 Hz
Sypris Test and Measurement Orlando, Florida, USA	4070 (1-axis **M**) 4080 (3-axis **M**) 4090 (3-axis **M**) 7030 (3-axis **M**)	40 Hz–200 Hz 40 Hz–600 Hz 50 Hz–300 Hz 10 Hz–50 000 Hz
Tech International Corp. Hallandale, Florida, USA	CellSensor (1-axis **M** & **RF**)	~50 Hz–~835 MHz

Table 9 (contd)

Company, location	Meter, field type	Frequency range
Technology Alternatives Corp. Miami, Florida, USA	ELF Digital Meter (**M**) ELF/VLF Combination Meter (**M**)	20 Hz–400 Hz 20 Hz–2000 Hz ELF; 10.000 Hz–200 000 Hz VLF
Walker LDJ Scientific, Inc. Worcester, Massachusetts, USA	ELF 45D (1-axis **M**) ELF 60D (1-axis **M**) ELF 90D (3-axis **M**) BBM-3D (3-axis **M**, ELF & VLF)	30 Hz–300 Hz 40 Hz–400 Hz 40 Hz–400 Hz 12 Hz–50 000 Hz

Source: Microwave News (2002) and industry sources
E, electric; **M**, magnetic (50 or 60 Hz); **RF**, radiofrequency; **ELF**, extremely low frequency; **VLF**, very low frequency
[a] Some instruments are suitable for measuring both magnetic and electric fields.

Specialized wave-capture instruments, such as the portable MultiWave system, can measure static and time-varying magnetic fields at frequencies of up to 3 kHz (Bowman & Methner, 2000). The EMDEX WaveCorder can also measure and record the waveform of magnetic fields for display and downloading.

In addition to measuring power-frequency fields, the Positron meter was designed to detect pulsed electric and magnetic fields or high-frequency transients associated with switching operations in the utility industry (Héroux, 1991). Only after its use in two epidemiology studies was it discovered that the readings of the commercial sensors were erratic and susceptible to interference from radiofrequency fields outside the bandwidth specification of the sensor. The interference by radio signals from hand-held walkie-talkies and other communication devices was recorded (Maruvada et al., 2000).

The EMDEX Transient Counter, which has recently been developed, continuously measures changes in magnetic fields at frequencies between 2000 Hz and 220 000 Hz and reports the number of times that the change in amplitude exceeds thresholds of 5 nT and 50 nT (data from the manufacturer, 2001).

A list of some currently available instruments for measuring magnetic fields is given in Table 9.

1.2.2 Computation methods

For many sources, measurements are the most convenient way to characterize exposure to ELF electric and magnetic fields. However, unperturbed fields from sources such as power lines can also be easily characterized by calculations. Calculated electric field intensity and direction may differ from those that are measured because of the presence of conductive objects close to the source and/or near the location of interest.

The fields from power lines can be calculated accurately if the geometry of the conductors, the voltages and currents (amplitude and phase angle) in the conductors and return paths are known. The currents flowing in the conductors of power lines are typically logged at substations and historical line-loading data may be available. However, in some cases, currents do not all return to utility facilities and may flow into the earth or into any conductor which is at earth potential, such as a neutral wire, telephone wire, shield wire or buried piping. Because the magnitude and location of the currents on these paths are not known, it is difficult or impossible to include them in computations.

The simplest calculations assume that the conductors are straight, parallel and located above, and parallel to, an infinite flat ground plane. Balanced currents are also typically assumed. Calculations of magnetic fields that do not include the contribution of small induced currents in the earth are accurate near power lines, but may not be so at distances of some hundreds of metres (Maddock, 1992). Very accurate calculations of the maximum, resultant and vector components of electric and magnetic fields are possible if the actual operating conditions at the time of interest are known, including the current flow and the height of conductors, which vary with ambient temperature and line loading.

A number of computer programs have been designed for the calculation of fields from power lines and substations. These incorporate useful features such as the calculation of fields from non-parallel conductors. While the computation of simple fields by such programs may be quite adequate for their intended purpose, it may be difficult for other investigators to verify the methods used to calculate exposures. Epidemiological studies that estimated the historical exposures of subjects to magnetic fields from power lines by calculations did not usually report using documented computer programs or publish the details of the computation algorithms, e.g. Olsen *et al.* (1993), Verkasalo *et al.* (1993, 1996), Feychting and Ahlbom (1994), Tynes and Haldorsen (1997) and UK Childhood Cancer Study Investigators (2000a). However, for exposure assessment in these studies, it is likely that the uncertainty in the historical loading on the power lines would contribute much more to the overall uncertainty in the calculated field than all of the other parameters combined (Jaffa *et al.*, 2000).

Calculations are also useful for the calibration of electric and magnetic field meters (IEEE, 1995b) and in the design of animal and in-vitro exposure systems, e.g. Bassen *et al.* (1992), Kirschvink (1992), Mullins *et al.* (1993).

1.3   Exposure assessment

1.3.1   *External dosimetry*

(a)   *Definition and metrics*

External dosimetry deals with characterization of static and ELF electric and magnetic fields that define exposure in epidemiological and experimental studies. For static fields, the field strength (or flux density) and direction unambiguously describe exposure conditions. As with other agents, the timing and duration of exposure are important parameters, but the situation is more complex in the case of ELF fields. The difficulty arises, not from the lack of ability to specify complete and unique characteristics for any given field, but rather from the large number of parameters requiring evaluation, and, more importantly, the inability to identify the critical parameters for biological interactions.

Several exposure characteristics, also called metrics, that may be of biological significance have been identified (Morgan & Nair, 1992; Valberg, 1995). These include:

- intensity (strength) or the corresponding flux density, root mean square, average or peak value of the exposure field; or a function of the field strength such as field-squared;
- duration of exposure at a given intensity;
- time (e.g. daytime versus night-time);
- single versus repeated exposure;
- frequency spectrum of the field; single frequency, harmonic content, intermittency, transients;
- spatial field characteristics: orientation, polarization, spatial homogeneity (gradients);
- single field exposure, e.g. ELF magnetic versus combined electric and magnetic field components, and possibly their mutual orientation;
- simultaneous exposure to a static (including geomagnetic field) and ELF field, with a consideration of their mutual orientation;
- exposure to ELF fields in conjunction with other agents, e.g. chemicals.

The overall exposure of a biological system to ELF fields can be a function of the parameters described above (Valberg, 1995).

(b)　Laboratory exposure systems

Laboratory exposure systems have the advantage that they can be designed to expose the subjects to fields of specific interest and the fields created are measurable and controllable. Laboratory exposure systems for studying the biological effects of electric and magnetic fields are readily classified as *in vivo* or *in vitro*. Most studies of exposure *in vivo* have been in animals; few have involved humans. In-vitro studies of exposure are conducted on isolated tissues or cultured cells of human or animal origin.

One reason for studying the effects of very strong fields is the expectation that internal dose is capable of being biologically scaled. For this reason, many laboratory experiments have been performed at field strengths much higher than those normally measured in residential and occupational settings. This approach is usually used on the assumption that the amplitude of biological effects increases with field strength up to the maxima set in exposure guidelines, and the physical limitations of the exposure system.

(i)　In-vivo exposure systems

Many in-vivo studies have used magnetostatic fields (Tenforde, 1992; see also section 4). Both iron-core electromagnets and permanent magnets are routinely used in such studies. Although it is theoretically possible to obtain even larger DC magnetic fields from iron-core devices (up to approximately 2 T), there is a limitation on the size of the active volume between the pole faces where the field is sufficiently uniform. Experimental studies of fields greater than 1.5 T are difficult because limited space is available for exposing biological systems to reasonably uniform magnetic fields.

The most commonly used apparatus for studying exposure to electric fields consists of parallel plates between which an alternating voltage (50 or 60 Hz, or other frequencies) is applied. Typically, the bottom plate is grounded. When appropriate dimensions of the plates are selected (i.e. a large area in comparison to the distance between the plates), a uniform field of reasonably large volume can be produced between the plates. The distribution of the electric-field strength within this volume can be calculated. The field becomes less uniform close to the plate edges.

A uniform field in an animal-exposure system can be significantly perturbed by two factors. An unavoidable but controllable perturbation is due to the presence of test animals and their cages. Much information is available on correct spacing of animals to ensure similar exposure for all test animals and to limit the mutual shielding of the animals (Kaune, 1981a; Creim et al., 1984). Animal cages, drinking bottles, food and bedding cause additional perturbations of the electric field (Kaune, 1981a). One of the most important causes of artefactual results in some studies is induction of currents in the nozzle of the drinking-water container. If the induced currents are sufficiently large, animals experience electric microshocks while drinking. Corrective measures have been developed to alleviate this problem (Free et al., 1981). Perturbation of the exposure field by nearby metallic objects is easy to prevent. The faulty design, construction and use of the electric-field-exposure facility can result in unreliable exposure over and above the limitations that normally apply to animal bioassays.

A magnetic field in an animal-exposure experiment is produced by current flowing through an arrangement of coils. The apparatus can vary from a simple set of two Helmholtz coils (preferably square or rectangular to fit with the geometry of cages), to an arrangement of four coils (Merritt et al., 1983), to more complicated coil systems (Stuchly et al., 1991; Kirschvink, 1992; Wilson et al., 1994; Caputa & Stuchly, 1996). The main objectives in designing apparatus for exposure to magnetic fields are (1) to ensure the maximal uniformity of the field within as much as possible of the volume encompassed by the coils, and (2) to minimize the stray fields outside the coils, so that sham-exposure apparatus can be placed in the same room. Square coils with four windings arranged according to the formulae of Merritt et al. (1983) best satisfy the field-uniformity requirement. Limiting the stray fields is a challenge, as shielding magnetic fields is much more complex than shielding electric fields. Non-magnetic metal shields only slightly reduce the field strength. Only properly designed multilayer-shielding enclosures made of high-permeability materials are effective. An alternative solution relies on partial field cancellation. Two systems of coils placed side by side or one above the other form a quadrupole system that effectively decreases the magnetic field outside the exposure system (Wilson et al., 1994). An even greater reduction is obtained with a doubly compensating arrangement of coils. Four coils (each consisting of four windings) are arranged side by side and up and down; coils placed diagonally are in the same direction as the field, and the neighbouring coils are in the opposite direction (Caputa & Stuchly, 1996).

Likely artefacts associated with magnetic-field-exposure systems include heating, vibrations and audible or high-frequency (non-audible to humans) noise. These factors can be minimized (although not entirely eliminated) with careful design and construction, which can be costly. The most economical and reliable way of over-coming these problems is through essentially identical design and construction of the field- and sham-exposure systems except for the current direction in bifilarly wound coils (Kirschvink, 1992; Caputa & Stuchly, 1996). This solution provides for the same heating of both the control and exposed systems. Vibration and noise are usually not exactly the same but are similar. To limit the vibration and noise, the coil windings should be restricted mechanically in their motion.

Another important feature of a properly designed magnetic-field system is shielding against the electric field produced by the coils. Depending on the coil shape, the number of turns in the coil and the diameter of the wire, a large voltage drop can occur between the ends of the coils. Shielding of the coils can eliminate interference from the electric field.

(ii)   In-vitro exposure systems

Cell and tissue cultures can be exposed to the electric field produced between parallel plates in the same way that animals are exposed. In practice, this procedure is hardly ever used, because the electric fields in the in-vitro preparation produced this way are very weak, even for strong applied fields. For instance, an externally applied field of 10 kV/m at 60 Hz results in only a fraction of a volt per metre in the culture (Tobey et al., 1981; Lymangrover et al., 1983). Furthermore, the field strength is usually not uniform throughout the culture, unless the culture is thin and is placed perpendicular or parallel to the field. A practical solution involves the placement of appropriate electrodes in the cultures. Agar or other media bridges can be used to eliminate the problem of electrode contamination (McLeod et al., 1987). A comprehensive review of in-vitro exposure systems has recently been published (Misakian et al., 1993).

The shape and size of the electrodes determine the uniformity of the electric field and associated spatial variations of the current density. Either accurate modelling or measurements, or preferably both, should be performed to confirm that the desired exposure conditions are achieved. Additional potential problems associated with this type of exposure system are the heating of the medium and accompanying induced magnetic fields. Both of these factors can be evaluated (Misakian et al., 1993).

Coils similar to those used for animal studies can be used for in-vitro experiments (Misakian et al., 1993). The greatest uniformity is achieved along the axis within the volume enclosed in the solenoid. One great advantage of solenoids over Helmholtz coils is that the uniform region within the solenoid extends from the axis across the whole of the cross-sectional diameter.

In in-vitro studies, special attention should be paid to ambient levels of 50 or 60 Hz and to other magnetic fields. Magnetic flux densities from incubators unmodified for bioeffect studies may have background gradients of magnetic fields ranging from a few tenths of a microtesla to approximately 100 μT. Similarly, some other laboratory equipment with an electric motor might expose biological cells to high, but unaccounted for, magnetic flux densities. Specially designed in-vitro systems can avoid these problems. Exposure to magnetic fields that is unaccounted for or is at an incorrect level, as well as the critical influence of temperature and carbon dioxide concentration on some cell preparations, can lead to unreliable findings in laboratory experiments (Misakian et al., 1993).

In some in-vitro studies, simultaneous exposure to alternating and static magnetic fields is used in a procedure intended to test the hypothesis of possible 'resonant' effects. Almost all the requirements for controlled exposure to the alternating field apply to the static field. Some precautions are not required in static field systems. For example, static systems have no vibrations (with the possible exception of on and off switching) so prevention of vibrations is unnecessary. In experiments involving static magnetic fields, the earth's magnetic field should be measured and controlled locally.

1.3.2   Internal dosimetry modelling

(a)   Definition for internal dosimetry

At ELF, electric fields and magnetic fields can be considered to be uncoupled (Olsen, 1994). Therefore, internal dosimetry is also evaluated separately. For simultaneous exposure to both fields, internal measures can be obtained by superposition. Exposure to either electric or magnetic fields results in the induction of electric fields and associated current densities in tissue. The magnitudes and spatial patterns of these fields depend on the type of field (electric or magnetic), its characteristics (frequency, magnitude, orientation), and the size, shape and electrical properties of the exposed body (human, animal). Exposure to electric fields also results in an electric charge on the body surface.

The primary dosimetric measure is the induced electric field in tissue. The most frequently reported dosimetric measures are the average, root-mean-square and maximum induced electric field and current density values (Stuchly & Dawson, 2000). Additional measures include the 50th, 95th and 99th percentiles which indicate values not exceeded in a given volume of tissue, e.g. the 99th percentile indicates the dosimetric measure exceeded in 1% of a given tissue volume (Kavet et al., 2001). The electric field in tissue is typically expressed in μV/m or mV/m and the current density in $μA/m^2$ or mA/m^2. Some safety guidelines (International Commission on Non-Ionizing Radiation Protection (ICNIRP), 1998) specify exposure limits measured as the current density averaged over 1 cm^2 of tissue perpendicular to the direction of the current.

The internal (induced) electric field **E** and conduction current density **J** are related through Ohm's law:

$$\mathbf{J} = \sigma\mathbf{E}$$

where the bold symbols denote vectors and σ is the bulk tissue conductivity which may depend on the orientation of the field in anisotropic tissues (e.g. muscle).

(b)   Electric-field dosimetry

Early dosimetry models represented the human (or animal) body in a simplified way, as reviewed elsewhere (Stuchly & Dawson, 2000). During the past 10 years, several laboratories have developed sophisticated heterogeneous models of the human body (Gandhi & Chen, 1992; Zubal et al., 1994; Gandhi, 1995; Dawson et al., 1997; Dimbylow, 1997). These models partition the body into volumes of different conductivity. Typically, over 30 distinct organs and tissues are identified and represented by cubic cells (voxels) with 1–10-mm sides. Voxels are assigned a conductivity value based on the measured values reported by Gabriel et al. (1996). A model of the human body constructed from several geometrical bodies of revolution has also been used (Baraton et al., 1993; Hutzler et al., 1994).

Various methods have been used to compute induced electric fields in these high-resolution models. Because of the low frequency involved, exposures to electric and magnetic fields are considered separately and the induced vector fields are added, if needed. Exposure to electric fields is generally more difficult to compute than exposure to magnetic fields, since the human body significantly perturbs the electric field. Suitable numerical methods are limited by the highly heterogeneous electrical properties of the human body and the complex external and organ shapes. The methods that have been successfully used so far for high-resolution dosimetry are: the finite difference method in frequency domain and time domain, and the finite element method. The advantages and limitations of each method have been reviewed by Stuchly and Dawson (2000). Some of the methods and computer codes have been extensively verified by comparison with analytical solutions (Dawson & Stuchly, 1997).

Several numerical evaluations of the electric field and the current density induced in various organs and tissues have been performed (Dawson et al., 1998; Furse & Gandhi, 1998; Dimbylow, 2000; Hirata et al., 2001). Average organ (tissue) and maximum voxel values of the electric field and current density are typically reported. In the recent studies (Dimbylow, 2000; Hirata et al., 2001), the maximum current density was averaged over 1 cm^2 for excitable tissues. The latter computation is clearly aimed at testing compliance with the International Commission on Non-Ionizing Radiation Protection (ICNIRP) guideline (1998) and the commentary published thereafter (Matthes, 1998).

The induced electric fields computed in various laboratories are in reasonable agreement (Stuchly & Gandhi, 2000). As expected, smaller differences are observed between calculated electric fields than between calculated values for current density.

The observed differences can be explained by differences between the body models and the conductivity values allocated to different tissues.

The differences observed in the results of high-resolution models depend in part upon the conductivity values assumed (Dawson et al., 1998). In general, the lower the induced electric fields (the higher the current density) the higher the conductivity of tissue. The exceptions are those parts of the body associated with concave curvature, e.g. the tissue surrounding the armpits, where the electric field is enhanced. For the whole body, the computed average values do not differ by more than 2% (Stuchly & Dawson, 2000).

The resolution of the model influences the accuracy with which the induced fields are evaluated in various organs. Organs that are small in any dimension are poorly represented by large voxels. The maximum induced electric field is higher for the finer resolution. The differences are typically of the order of 50–190% for voxels of 3.6-mm sides compared to 7.2-mm voxels (Stuchly & Dawson, 2000).

The highest induced fields are found in a body that is in contact with perfect ground through both feet. The average values for the organs or tissues of a 'grounded' body are about two or three times those for a body in free space (Dawson et al., 1998), and intermediate values are obtained for various degrees of separation from the ground. This dependence on the contact with or separation from a perfect ground is in agreement with earlier experimental data (Deno & Zaffanella, 1982).

The main features of dosimetry for exposure of a person standing in an ELF electric field can be summarized as follows:

- The magnitudes of the electric fields in tissue are typically 10^{-5}–10^{-8} lower than the magnitude of the external field.
- For exposure to external fields from power lines, the predominant direction of the induced fields is also vertical.
- The largest fields in a human body are induced by a vertical electric field when the feet are in contact with a perfectly conducting ground plane.
- The weakest fields are induced in a body when it is in free space, i.e. infinitely far from the ground plane.
- The short-circuit current for a body in contact with perfect ground is determined by the size and shape of the body (including posture), rather than its tissue conductivity.

Table 10 summarizes the computed internal electric fields for a model of an adult body in a vertical field of 1 kV/m at 60 Hz (Kavet et al., 2001), where the body (1.77 m in height and 76 kg in weight) makes contact with a perfectly conducting ground with both feet. Table 11 summarizes computed internal electric fields for a model of a five-year-old child (1.10 m in height and 18.7 kg in weight) (Hirata et al., 2001). Selected conductivity values are given in Table 2 of the General Introduction. Dawson et al. (2001) demonstrated that the voxel maximum values are significantly overestimated, and the 99th percentiles are therefore more representative.

Table 10. Calculated electric fields (mV/m) in a vertical uniform electric field (60 Hz, 1 kV/m) induced in a model of a grounded adult human body[a]

Tissue/organ	E_{avg}	$E_{99\ percentile}$	E_{max}
Blood	1.4	8.9	24
Bone marrow	3.6	34	41
Brain	0.86	2.0	3.7
Cerebrospinal fluid	0.35	1.0	1.6
Heart	1.4	2.8	3.6
Kidneys	1.4	3.1	4.5
Lungs	1.4	2.4	3.6
Muscle	1.6	10	32
Prostate	1.7	2.8	3.1
Spleen	1.8	2.6	3.2
Testes	0.48	1.2	1.6

Modified from Kavet *et al.* (2001)
[a] Corresponding current densities can be computed from tissue conductivity values (see Table 2, General Introduction)

Table 11. Calculated electric fields (mV/m) in a vertical uniform electric field (60 Hz, 1 kV/m), induced in a model of the grounded body of a child

Tissue/organ	E_{avg}	$E_{99\ percentile}$	E_{max}
Blood	1.5	9.2	18
Bone marrow	3.7	35	42
Brain	0.7	1.6	3.1
Cerebrospinal fluid	0.28	0.87	1.4
Heart	1.6	3.1	3.7
Lungs	1.6	2.6	3.7
Muscle	1.7	10	31

Modified from Hirata *et al.* (2001)

Exposure in occupational situations, e.g. in a substation, where a person is close to a conductor at high potential, induces greater electric fields in certain organs (e.g. brain) than would be predicted from the measured exposure field 1.5 m above ground (Potter *et al.*, 2000). This is to be expected, since the external field increases above the ground.

(c)   *Magnetic-field dosimetry*

Early dosimetry models represented the body as a circular loop corresponding to a given body contour to determine the induced electric field or current density based on Faraday's law:

$$| \mathbf{J} | = \pi f \sigma r \, | \mathbf{B} |$$

where f is the frequency, r is the loop radius and $\mathbf{B}$ is the magnetic flux density vector perpendicular to the current loop. Similarly, ellipsoidal loops have been used to fit the body shape better.

More realistic models of the human body have been analysed by the numerical impedance method (Gandhi & De Ford, 1988; Gandhi & Chen, 1992; Gandhi *et al.*, 2001) and the scalar potential finite difference technique (Dawson & Stuchly, 1998; Dimbylow, 1998). The dosimetry data available for magnetic fields are more extensive than those for electric fields. The effects of tissue properties in general (and specifically muscle anisotropy), field orientation with respect to the body and, to a certain extent, body anatomy have been investigated (Dawson *et al.*, 1997; Dawson & Stuchly, 1998; Dimbylow, 1998). In the past, the largest loop of current fitted within a body part, e.g. head or heart, was often used to calculate the maximum current density in that body part. This is now known to underestimate the maximum induced field and the current densities (Stuchly & Dawson, 2000)

The main features of dosimetry for exposure to uniform ELF magnetic fields can be summarized as follows:

- The induced electric fields depend on the orientation of the magnetic field with respect to the body.
- For most organs and tissues, the magnetic field orientation perpendicular to the torso (front-to-back) gives maximum induced quantities.
- For the brain, cerebrospinal fluid, blood, heart, bladder, eyes and spinal cord, the strongest induced electric fields are produced by a magnetic field directed towards the side of the body.
- Magnetic fields oriented along the vertical body axis induce the weakest electric fields.
- Stronger electric fields are induced in bodies of larger size.

Table 12 lists the electric fields induced in certain organs and tissues by a 60-Hz, 1-µT magnetic field oriented front-to-back (Dawson *et al.*, 1997; Dawson & Stuchly, 1998; Kavet *et al.*, 2001).

The exposure of humans to relatively high magnetic flux densities occurs most often in occupational settings. Numerical modelling has been applied mostly to workers exposed to high-voltage power lines (Baraton & Hutzler, 1995, Stuchly & Zhao, 1996; Dawson *et al.*, 1999a,b,c). In these cases, current-carrying conductors can be represented as infinite straight-line sources. However, some of the exposure occurs in more complex scenarios, two of which have been analysed, and a more realistic representation of the source conductors based on finite line segments has been used

(Stuchly & Dawson, 2000). Table 13 lists calculated electric fields for the two repre-sentative exposure scenarios illustrated in Figure 6 (Stuchly & Dawson, 2000).

Table 12. Calculated electric fields (μV/m) in a uni-form magnetic field (60 Hz, 1 μT) oriented front-to-back induced in a model of an adult human

Tissue/organ	E_{avg}	$E_{99\ percentile}$	E_{max}
Blood	6.9	23	83
Bone marrow	16	93	154
Brain	11	31	74
Cerebrospinal fluid	5.2	17	25
Heart	14	38	49
Kidneys	25	53	71
Lungs	21	49	86
Muscle	15	51	147
Prostate	17	36	52
Spleen	41	72	92
Testes	15	41	73

Modified from Kavet *et al.* (2001)

Table 13. Calculated electric fields (mV/m) induced in a model of an adult human for the occupational exposure scenarios shown in Figure 6 (total current in conductors, 1000 A)

Tissue/organ	Scenario A		Scenario B	
	E_{max}	E_{rms}	E_{max}	E_{rms}
Blood	20	3.7	15	2.4
Bone	90	11	58	7.2
Brain	22	4.6	28	5.9
Cerebrospinal fluid	9.2	2.3	14	3.7
Heart	27	11	9.0	3.2
Kidneys	22	7.9	2.8	0.9
Lungs	31	10	9.9	2.9
Muscle	59	6.9	33	5.5
Prostate	5.5	1.9	2.6	1.2
Testes	18	5.5	2.7	1.2

Modified from Stuchly & Dawson (2000)

Figure 6. Body positions in two occupational exposure scenarios

Scenario A   Scenario B

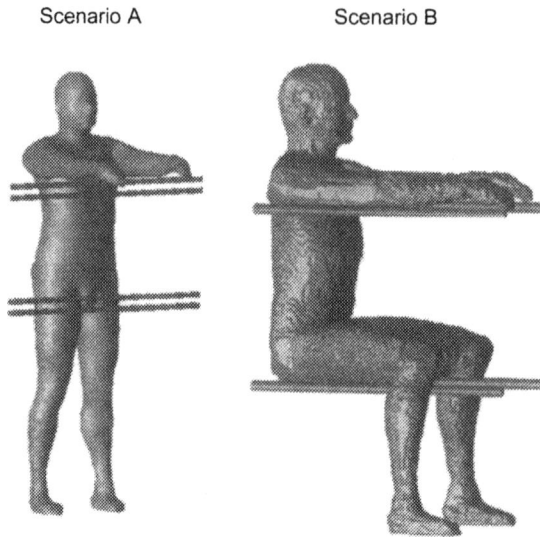

From Stuchly & Dawson (2000)
The current in each conductor is 250 A for a total of 1000 A in four conductors.

(*d*)   Contact-current dosimetry

Contact currents produce electric fields in tissue similar to those induced by external electric and magnetic fields. Contact currents are encountered in a dwelling or workplace when a person touches conductive surfaces at different potentials and completes a path for current flow through the body. The current pathway is usually from hand-to-hand and/or from a hand to one or both feet. Sources of contact current may include an appliance chassis or household fixture that, because of typical residential wiring practices, carries a small potential above a ground. Other sources of contact current are conductive objects situated in an electric field, such as a vehicle parked under a power line. The importance of contact currents has been suggested by Kavet *et al.* (2000). Recently, electric fields have been computed in a model of a child with electrodes on hands and feet simulating contact current (Dawson *et al.*, 2001). The most common source of exposure to contact current is touching an ungrounded object while both feet are grounded. The electric fields calculated to be induced in the bone marrow of the hand and arm of a child for a 1-μA contact current are shown in Table 14. Electric fields above 1 mV/m can be produced in the bone marrow of a child from a low contact current of 1 μA. In residential settings such a current could result from an open-circuit voltage of only 100 mV, which is not uncommon. A total resistance of 5–10 kΩ is representative (Kavet *et al.*, 2000). Provided that there is good contact to the ground, only 5–10 mV is needed to produce a current of 1–2 μA. Contact current with vehicles in an electric field

Table 14. Calculated electric field (mV/m) induced by a contact current of 60 Hz, 1 μA, in voxels of bone marrow of a child

Body part	E_{avg}	$E_{99\ percentile}$
Lower arm	5.1	14.9
Upper arm	0.9	1.4
Whole body	0.4	3.3

Modified from Dawson *et al.* (2001)

(e.g. under high-voltage power lines) typically ranges from 0.1 mA per 1 kV/m for a car to 0.6 mA per 1 kV/m for a large truck (Deno & Zaffanella, 1982).

(*e*)   *Biophysical relevance of induced fields*

The lowest electric field in tissue to be associated with well-documented biological effects (not necessarily harmful) has been estimated as 1 mV/m (Portier & Wolfe, 1998). It is interesting to compare the different exposure conditions that produce an internal field of this magnitude. Table 15 shows the average exposure to electric and magnetic fields required to induce a field of 1 mV/m in selected tissues (Stuchly & Dawson, 2000). Although the mechanisms for biological effects of fields at 1 mV/m are unclear, it is, nevertheless, interesting to compare the electric fields induced in humans by exposure to residential magnetic fields, electric fields and contact currents. Table 16 shows that the electric field induced in a model of a child's bone marrow (i.e. the tissue involved in leukaemia) is 10 times greater for the exposure to a contact current than for exposure to either the maximum electric or magnetic field encountered in a dwelling (Kavet *et al.*, 2000).

(*f*)   *Microscopic dosimetry*

Macroscopic dosimetry that describes induced electric fields in various organs and tissues can be extended to more spatially refined models of subcellular structures to predict and understand biophysical interactions. The simplest cellular model for considering linear systems requires evaluation of induced fields in various parts of a cell. Such models, for instance, have been developed to understand neural stimulation (Plonsey & Barr, 1988; Basser & Roth, 1991; Reilly, 1992; Malmivuo & Plonsey, 1995). Computations are available as a function of the applied electric field and its frequency. Because cell membranes have high resistivity and capacitance (nearly constant for all mammalian cells and equal to 0.5–1 μF/cm^2) (Reilly, 1992), at sufficiently low frequencies, high fields are produced at the two poles of the membrane. The field is nearly zero inside the cell, as long as the frequency of the applied field is below the membrane relaxation frequency (~ 1 MHz) (Foster & Schwan, 1995). The total

Table 15. Calculated electric (grounded model) or magnetic field (front to back) source levels needed to induce average (E_{avg}) and maximum (E_{max}) electric fields of 1 mV/m

Organ	Electric field exposure		Magnetic field exposure	
	E_{avg} = 1 mV/m	E_{max} = 1 mV/m	E_{avg} = 1 mV/m	E_{max} = 1 mV/m
Blood	0.72 kV/m	61 V/m	115 µT	15 µT
Bone	0.31 kV/m	22 V/m	46 µT	6.0 µT
Brain	1.2 kV/m	355 V/m	87 µT	26 µT
Cerebrospinal fluid	3.3 kV/m	901 V/m	233 µT	52 µT
Heart	0.93 kV/m	457 V/m	56 µT	20 µT
Kidneys	0.97 kV/m	412 V/m	43 µT	18 µT
Liver	0.79 kV/m	372 V/m	38 µT	11 µT
Lungs	0.99 kV/m	435 V/m	46 µT	14 µT
Muscle	0.76 kV/m	43 V/m	57 µT	6.9 µT
Prostate	0.68 kV/m	442 V/m	58 µT	28 µT
Testes	1.8 kV/m	769 V/m	53 µT	20 µT
Whole body	0.59 kV/m	21 V/m	49 µT	1.3 µT

Modified from Stuchly & Dawson (2000)

Table 16. Calculated average electric field (mV/m) induced by an electric field, magnetic field and contact current in child's bone marrow (model)

Exposure	Scenario	Intensity	Electric field (mV/m)
Magnetic field	Uniform, horizontal and frontal exposure	10 µT	0.2
Electric field	Uniform, vertical grounded	100 V/m	0.3
Contact current	Current injection into shoulders	18 µA	3.5

Modified from Kavet et al. (2000)

membrane resistance and capacitance define this frequency; thus, it depends on the cell size (total membrane surface). The larger the cell, the higher the induced membrane potential for the same applied field, but the larger the cell, the lower the membrane relaxation frequency.

Gap junctions connect many cells. A gap junction is an aqueous pore or channel through which neighbouring cell membranes are connected. Thus, cells can exchange ions, for example, providing local intercellular communication (Holder et al., 1993). In gap-junction-connected cells there is electrical coupling between the cytoplasm of adjoining cells and such systems have previously been modelled as leaky cables (Cooper, 1984). Simplified models have also been used, in which a group of gap-junction-connected cells is represented by a large cell of the same size (Polk, 1992).

Using such models, relatively large membrane potentials have been estimated, even for applied fields of only moderate intensity. A numerical analysis has been performed to compute membrane potentials in more realistic multiple-cell models (Fear & Stuchly, 1998). Simulations have indicated that simplified models such as a single cell or leaky-cable can be used only in some specific situations. Even when these models are appropriate, equivalent cells must be constructed, in which the cytoplasm properties are modified to account for the properties of gap-junctions. These models are reasonably accurate for very small assemblies of cells of certain shapes exposed at very low frequencies. As the size of the cell-assembly increases, the membrane potential, even at static fields, does not increase linearly with dimensions as it does for very short elongated assemblies. There is also a limit to the membrane potential for assemblies of other shapes.

From this linear model of gap-connected cells, it can be concluded that, at 50 or 60 Hz, an induced membrane potential of 0.1 mV is not attained in any organ or tissue of the human body exposed to a uniform magnetic flux density of up to 1 mT or to an electric field of 10 kV/m or less (Fear & Stuchly, 1998). These external field levels are much higher than those that elicit 1 mV/m in the bone marrow.

1.4   Biophysical mechanisms

Beyond the well-established mechanisms of interaction described above, such as the induction of currents from time-varying magnetic fields, a number of hypotheses have been advanced to explain ELF and static field interactions. These include radical-pair mechanisms; charge-to-mass signature; biogenic magnetite; etc.

1.4.1   *Induced currents*

The role of induced currents has been discussed by Adair (1991) who argued that because currents induced by ambient-level magnetic fields are comparable to, or smaller than, those resulting from thermal fluctuations, they must have little physiological significance. This argument is based on calculations of the thermal (or 'kT') noise developed in the cell membrane. The four major sources of electrical noise in biological membranes are:

— Johnson–Nyquist thermally-generated electrical noise;
— 1/f noise produced by ion current through membrane channels;
— 'shot' noise resulting from the discrete nature of ionic charges; and
— endogenous fields produced by electrically active organs such as the heart, muscles and the nervous system (Tenforde, 1995).

However, it must be remembered that the electrical characteristics of the membrane are different from those of the other regions of the cell. Taking this into consideration, conclusions have been reached concerning the potential effects of weak ELF magnetic fields. For example, Adair (1991) calculated that the theoretical threshold sensitivity for biological effectiveness due to Faraday induction by ELF magnetic fields was much larger. This threshold is much higher than those reported from a variety of laboratory experiments (Fitzsimmons *et al.*, 1995; Jenrow *et al.*, 1996; Harland & Liburdy, 1997; Zhadin *et al.*, 1999; Blackman *et al.*, 2001). If some of these experimental results are correct, the discrepancy between theoretical and experimental results indicates that the thermal-noise arguments have to be reconsidered. Indeed, low thresholds of 4 mV/m and 10 mV/m were calculated by Polk (1993) and Tenforde (1993), respectively, based on a redistribution of charges in the counterion layer rather than on changes in trans-membrane potential. Amplification due to the electric coupling of large arrays of cells must also be taken into account in the estimation of threshold values.

1.4.2   *Radical-pair mechanism*

Increasing attention is being paid to the possibility that static and ELF magnetic fields may affect enzymatic processes that involve radical pairs (radical-pair mechanism). The radical-pair mechanism is a well established physical mechanism for describing how applied magnetic flux densities as low as 0.1–1 mT can affect chemical or biochemical reactions nonthermally (Walleczek, 1995). The simplified radical-pair mechanism can be summarized as follows: according to Pauli's exclusion principle, two valence electrons of the same orbital differ in their quantum spin number and a pair can be represented with one electron having the spin up ($\uparrow$) and the other a spin down ($\downarrow$). When a molecular bond is broken, a pair of free radicals is produced in the so-called singlet state ($\uparrow\downarrow$) which can either recombine to the original molecule or separate into two free radicals. However, if the relative orientation of the spins is altered (inter-conversion from singlet to triplet), the kinetics of recombination are modified. Three types of process can change the orientation of the spins:

— hyperfine coupling (linked to the magnetic environment of the pair);
— differences in Larmor precession rates ('Δg' mechanism); and
— crossing from one energy level to another.

The first process, which corresponds to a decrease of the rate of the interconversion with increasing field strength, is the most likely to occur at low field-strength. Since the lifetime of the radical pair (nano- to microseconds) is much shorter than the period of the ELF signal (~ 20 ms), the ELF magnetic field can be considered as static when considering processes consisting of a single elementary chemical reaction. However, in biochemical systems involving enzymes, in which sequences of elementary reactions can lead to oscillations of concentrations of intermediate species occurring at ELFs, the external field could, in principle, couple to the system and have an effect even at low field-strength (Walleczek, 1995; Eichwald & Walleczek, 1997), possibly in the µT range, though arguments have been advanced that this could not occur at 5 µT (Adair, 1999). Experimental evidence for the radical-pair mechanism in biological processes at field strengths below 500 µT is still lacking (Brocklehurst & McLauchlan, 1996).

1.4.3   *Effects related to the charge-to-mass ratio of ions*

The results of several experimental studies suggest that consideration of some ELF magnetic field interactions requires that the static magnetic field be taken into account as well. The ion cyclotron resonance (ICR) model (Liboff, 1985) proposes that ion transfer through cell membranes is affected by cyclotron resonance when an alternating electric or magnetic field is superimposed on a static magnetic field, e.g. the geomagnetic field. It is based on the fact that the cyclotron resonance frequency of several physiologically important ions like Na^+, K^+, Mg^{2+} and Ca^{2+} falls into the ELF range. For example, for Mg^{2+} the resonance frequency would be 61.5 kHz in a static magnetic field of 50 μT, as can be calculated from the formula below (Liboff, 1985; Polk, 1995):

$$\omega_c = 2\,\pi f_c = \frac{q\,B_{DC}}{m}$$

where ω_c is the angular frequency of the alternating magnetic field, B_{DC} is the intensity of the static field, and q/m is the ionic charge-to-mass ratio. Despite the many reports (Thomas *et al.*, 1986; Rozek *et al.*, 1987; Smith *et al.*, 1987; Ross, 1990; Lerchl *et al.*, 1991; Liboff *et al.*, 1993; Smith *et al.*, 1993; Deibert *et al.*, 1994; Jenrow *et al.*, 1995; Zhadin *et al.*, 1999) that have indicated that such combinations of fields are effective in altering biological responses, there is no definitive experimental evidence and other authors have failed to replicate these effects (e.g. Parkinson & Hanks, 1989; Liboff & Parkinson, 1991; Parkinson & Sulik, 1992; Coulton & Barker, 1993).

Most importantly, there is no accepted explanation at either the microscopic or molecular level of how such field combinations could be effective. Therefore, this unique signature must, at present, be regarded as tentative and purely empirical in nature. There is some experimental evidence (Smith *et al.*, 1987) to indicate that higher frequency harmonics are also effective, following the allowed harmonic relation $f_n = (2n + 1)\,f_o$, $n = 0, 1, 2, 3, \ldots$. The same authors also observed that, if all other parameters remain the same, small changes in B_{DC} (intensity of the static field) could shift the charge-to-mass ratio given above from one ionic species to another, with a totally different resultant change in the expected biological response. The implication is that, for one specific value of B_{AC} (intensity of the alternating field), there may be markedly contrasting biological outcomes if exposure to ELF fields occurs in different static fields. The geomagnetic field varies substantially over the earth's surface, and from place to place within the same building due to local perturbations. If interaction hypotheses based upon the ion charge-to-mass ratios were valid and furthermore were a cause of cancer, then it might be difficult for epidemiological studies to capture associations with exposure to ELF magnetic fields (Smith *et al.*, 1987).

From a theoretical model, Lednev (1991) suggested that the cyclotron resonance frequency appears in the transition probability of an excited state of a charged oscillator (e.g. Ca^{2+}) located in one of the binding sites of a protein. This parametric resonance mechanism makes use of Zeeman splitting of the energy levels in a magnetic field. In addition to the ionic charge-to-mass ratio and the static field intensity, which are both well-defined experimental parameters, the transition probability p(B) is also dependent on B_{AC}, a feature that was not considered in the original hypothesis (Liboff, 1985).

The field-dependent part of the parametric resonance mechanism transition probability is to a first approximation:

$$p(B) = (-1)^n K J_n (nB_{AC}/B_{DC})$$

where K is a constant and J_n is the nth order Bessel function with argument (nB_{AC}/B_{DC}).

An alternative theoretical formulation, called the ion parametric resonance model (Blanchard & Blackman, 1994) is very similar to the parametric resonance mechanism model, except that it is not related to calcium-binding, but rather to enzyme activation. In the ion parametric resonance model, the transition probability becomes:

$$p(B) = (-1)^n K J_n (2nB_{AC}/B_{DC})$$

Exposure 'windows' are predicted in both models; the intensities at which these windows occur are entirely dependent on the respective arguments of the two Bessel functions. See section 4.3 for a description of experimental data in support of this formulation. By contrast, Adair (1992, 1998) gave reasons as to why these proposed mechanisms would not be expected to produce biological effects.

Other theoretical attempts to explain the experimental results have been made by Binhi (2000), using quantum mechanics to estimate the dissociation probability of an ion from a protein, and by Zhadin (1998), who hypothesized magnetically induced changes in the thermal energy distribution.

The hypothesis discussed above may explain the frequency 'windows' previously reported (Bawin & Adey, 1976; Blackman *et al.*, 1985). If so, the exposure conditions related to cyclotron resonance may have to be considered in a discussion of exposure to electric and magnetic fields taking into account the role of the local geomagnetic field.

1.4.4   *Biogenic magnetite*

Following the original discovery by Blakemore (1975) that certain bacteria use iron-rich intracytoplasmic inclusions for orientational purposes, such domain-sized magnetite (Fe_3O_4) particles have been found in other biological systems, notably the human brain (Kirschvink *et al.*, 1992). Kirschvink suggested that weak ELF magnetic fields coupling to biogenic magnetite might be capable of producing coherent biological signals. However, the number of magnetite crystals is exceeded by that of neurons by a factor of about 10 (Malmivuo & Plonsey, 1995) and, moreover, no experimental evidence exists to support this hypothesis. Based on a mathematical model, Adair (1993) has estimated that a 60-Hz magnetic field weaker than 5 µT could not generate a sufficiently large signal to be detectable in a biological system by interaction with magnetite. According to Polk (1994), reported experimental results indicate effects in mammals of 50-Hz fields at the 1-μT level. Rather strong static magnetic fields are required to affect the orientation behaviour of honey bees, which depends, in part, upon the influence of the geomagnetic field on magnetite in the bee's abdomen (Kirschvink *et al.*, 1997).

1.4.5   *Other mechanisms*

Electric fields can increase the deposition of charged airborne particles on surfaces. It has been suggested that this well-known phenomenon could lead to increased exposure of the skin or respiratory tract to ambient pollutants close to high-voltage AC power lines (Henshaw *et al.*, 1996; Fews *et al.*, 1999a). It is also known that the high-voltage power-lines emit corona ions, which can affect the ambient distribution of electrical charges in the air (Fews & Henshaw, 2000). Fews *et al.* (1999b) have suggested that this could enhance the deposition of airborne particles in the lung. The relevance of these suggestions to health has not been established (Jeffers, 1996; Stather *et al.*, 1996; Jeffers, 1999; Swanson & Jeffers, 1999; Fews & Henshaw, 2000; Swanson & Jeffers, 2000).

2. Studies of Cancer in Humans

2.1 Exposure assessment in epidemiological studies

2.1.1 *Considerations in assessment of exposure to electric and magnetic fields relevant to epidemiology*

Electric and magnetic fields are complex and many different parameters are necessary to characterize them completely. These parameters are discussed more fully in section 1. In general, they include transients, harmonic content, resonance conditions, peak values and time above thresholds, as well as average levels. It is not known which of these parameters or what combination of parameters, if any, are relevant for carcinogenesis. If there were a known biophysical mechanism of interaction for carcinogenesis, it would be possible to identify the critical parameters of exposure, including relevant timing of exposure. However, in the absence of a generally accepted mechanism for carcinogenesis, most exposure assessments in epidemiological studies are based on a time-weighted average of the field, a measure that is also related to many other characteristics of the fields (Zaffanella & Kalton, 1998).

Exposure to electric and magnetic fields and approaches for exposure assessment have been described in detail in section 1. Some of the characteristics of exposure to electric and magnetic fields which make exposure assessment for the purposes of epidemiological studies particularly difficult are listed below:

- *Prevalence of exposure.* Everyone in the population is exposed to some degree to ELF electric and magnetic fields and therefore exposure assessment has to separate the more from the less exposed individuals, as opposed to the easier task of separating individuals who are exposed from those who are not.
- *Inability of subjects to identify exposure.* Exposure to electric and magnetic fields, whilst ubiquitous, is neither detectable by the exposed person nor memorable, and hence epidemiological studies cannot rely solely on questionnaire data to characterize past exposures adequately.
- *Lack of clear contrast between 'high' and 'low' exposure.* The difference between the electric and magnetic fields to which 'highly exposed' and 'less highly exposed' individuals in a population are subjected is not great. The typical average magnetic fields in homes appear to be about 0.05–0.1 μT. Pooled analyses of childhood leukaemia and magnetic fields, such as that by Ahlbom

et al. (2000), have used ≥ 0.4 µT as a high-exposure category. Therefore, an exposure assessment method has to separate reliably exposures which may differ by factors of only 2 or 4. Even in most of the occupational settings considered to entail 'high exposures' the average fields measured are only one order of magnitude higher than those measured in residential settings (Kheifets *et al.*, 1995).

- *Variability of exposure over time: short-term*. Fields (particularly magnetic fields) vary over time-scales of seconds or longer. Assessing a person's exposure over any period involves using a single summary figure for a highly variable quantity.

- *Variability of exposure over time: long-term*. Fields are also likely to vary over time-scales of seasons and years. With the exception of historical data on loads carried by high-voltage power lines, data on such variation are rare. Therefore, when a person's exposure at some period in the past is assessed from data collected later, an assumption has to be made. The usual assumption is that the exposure has not changed. Some authors (e.g. Jackson, 1992; Petridou *et al.*, 1993; Swanson, 1996) have estimated the variations of exposure over time from available data, for example, on electricity consumption. These apply to population averages and are unlikely to be accurate for individuals.

- *Variability of exposure over space*. Magnetic fields vary over the volume of, for example, a building so that, as people move around, they may experience fields of varying intensity. Personal exposure monitoring captures this, but other assessment methods generally do not.

People accumulate exposure to fields in different settings, such as at home, at school, at work, while travelling and outdoors, and there can be great variability of fields between these environments. Current understanding of the contributions to exposure from different sources and in different settings is limited. Most studies make exposure assessments within a single environment, typically at home for residential studies and at work for occupational studies. Some recent studies have included measures of exposure from more than one setting (e.g. Feychting *et al.*, 1997; UK Childhood Cancer Study Investigators, 1999; Forssén, 2000).

In epidemiological studies, the distribution of exposures in a population has consequences for the statistical power of the study. Most populations are characterized by an approximately log-normal distribution with a heavy preponderance of low-level exposure and much less high-level exposure. Pilot studies of exposure distribution are important for developing effective study designs.

2.1.2   *Assessing residential exposure to magnetic fields*

(*a*)   *Methods not involving measurement*

(i)   *Distance*

The simplest possible way of assessing exposure is to record proximity to a facility (such as a power line or a substation) which is likely to be a source of field. This does provide a very crude measure of exposure to both electric and magnetic fields from that source, but takes no account of other sources or of how the fields vary with distance from the source (which is different for different sources). Distances reported by study subjects rather than measured by the investigators tend to be unreliable.

(ii)   *Wire code*

Wire coding is a non-intrusive method of classifying dwellings on the basis of their distance from visible electrical installations and the characteristics of these installations. This method does not take account of exposure from sources within the home.

Wertheimer and Leeper (1979) devised a simple set of rules to classify residences with respect to their potential for having a higher than usual exposure to magnetic fields. Their assumptions were simple:

— the field strength decreases with distance from the source;
— current flowing in power lines decreases at every pole from which 'service drop' wires deliver power to houses;
— if both thick and thin conductors are used for lines carrying power at a given voltage, and more than one conductor is present, it is reasonable to assume that more and thicker conductors are required to carry greater currents; and
— when lines are buried in a conduit or a trench, their contribution to exposure can be neglected. This is because buried cables are placed close together and the fields produced by currents flowing from and back to the source cancel each other much more effectively than when they are spaced apart on a cross beam on a pole.

Wertheimer and Leeper (1979) used these four criteria to define two and later four (Wertheimer & Leeper, 1982) then five (Savitz *et al.*, 1988) classes of home: VHCC (very high current configuration), OHCC (ordinary high current configuration), OLCC (ordinary low current configuration), VLCC (very low current configuration) and UG (underground, i.e. buried). The houses with the higher classifications were assumed to have stronger background fields than those with lower classifications.

Wire coding, in the original form developed by Wertheimer and Leeper, has been used in a number of studies. Although some relationship between measured magnetic fields and the wire-coding classification is seen in all studies (see for example Table 17 for studies of childhood leukaemia), wiring codes generally misclassify many homes although they do differentiate between high-field homes and others (Kheifets *et al.*, 1997a).

Table 17. Typical mean values of time-weighted average magnetic fields (μT) — and percentage of houses > 0.2 μT — associated with wire-code exposure classes from childhood leukaemia studies

Reference, country	Classification	Underground (UG)	Very low (VLCC)	Low (LCC)	High (HCC)	Very high (VHCC)
Savitz et al. (1988)[a] USA	No. of observations	133	27	174	88	12
	mean (μT)	0.05	0.05	0.07	0.12	0.21
	% > 0.2 μT	3	0	6	21	60
Tarone et al. (1998)[b] USA	No. of homes	150	221	262	170	55
	mean (μT)	0.06	0.08	0.12	0.14	0.2
	% > 0.2 μT	3	6	15	20	40
McBride et al. (1999)[c] Canada	No. of residences	127	137	131	164	43
	mean (μT)	0.09	0.08	0.11	0.17	0.26
Green et al. (1999a)[d] Canada	No. of measurements	66	9	25	19	6
	mean (μT)	0.07	0.04	0.14	0.18	0.38
	SD	0.06	0.02	0.1	0.2	0.3
London et al. (1991)[e] USA	No. of measurements	19	20	94	108	50
	mean (μT)	0.05	0.05	0.07	0.07	0.12
	% > 0.25 μT	0.3	3.7	11.6	6.4	16.6

[a] Childhood cancer. Magnetic fields measured under low power use conditions
[b] 24-h magnetic field measurements in the bedroom
[c] 24-h magnetic field measurements (child's bedroom); % > 0.2 μT not reported
[d] Personal monitoring of controls; SD, standard deviation
[e] 24-h measurements

The concept of wire coding, that is, assessing residential exposure on the basis of the observable characteristics of nearby electrical installations, has been shown to be a usable surrogate when tailored to local wiring practices. However, the so-called Wertheimer and Leeper wire code may not be an adequate surrogate in every environment (see Table 17). In general, wire codes have been used only in North American studies, as their applicability is limited in other countries where power drops to homes are mostly underground.

(iii)   *Calculated historical fields*

Feychting and Ahlbom (1993) carried out a case–control study nested in a cohort of residents living in homes within 300 m of power lines in Sweden. The geometry of the conductors on the power line, the distance of the houses from the power lines and historical records of currents, were all available. This special situation allowed the investigators to calculate the fields to which the subjects' homes were exposed at various times (e.g. prior to diagnosis) (Kheifets *et al.*, 1997a).

The common elements between wire coding and the calculation model used by Feychting and Ahlbom (1993) are: the reliance on the basic physical principles that the field increases with the current and decreases with the distance from the power line, and the fact that both neglect magnetic-field sources other than visible power lines. There is, however, one important difference: in the Wertheimer and Leeper code, the line type and thickness are a measure of the *potential* current carrying capacity of the line. In the Feychting and Ahlbom (1993) study, the approximate yearly average current was obtained from utility records; thus the question of temporal stability of the estimated fields did not even arise: assessment carried out for different times, using different load figures, yielded different estimates.

The approach of Feychting and Ahlbom (1993) has been used in various Nordic countries and elsewhere, although the likely accuracy of the calculations has varied depending in part on the completeness and precision of the available information on historical load. The necessary assumption that other sources of field are negligible is reasonable only for subjects relatively close to high-voltage power lines. The validity of the assumption also depends on details such as the definition of the population chosen for the study and the size of average fields from other sources to which the relevant population is exposed.

There is some evidence from Feychting and Ahlbom (1993) that their approach may work better for single-family homes than for apartments. When Feychting and Ahlbom (1993) validated their method by comparing calculations of present-day fields with present-day measurements, they found that virtually all homes with a measured field < 0.2 μT, whether single-family or apartments, were correctly classified by their calculations. However, for homes with a measured field > 0.2 μT, the calculations were able to classify correctly [85%] of single-family homes, but nearly half of the apartments were misclassified.

The difference between historical calculations and contemporary measurements was also evaluated by Feychting and Ahlbom (1993) who found that calculations using contemporary current loads resulted in a [45%] increase in the fraction of single-family homes estimated to have a field > 0.2 μT, compared with calculations based on historical data. If these calculations of historical fields do accurately reflect exposure, this implies that present-day spot measurements overestimate the number of exposed homes in the past.

(b)   *Methods involving measurement*

Following the publication of the Wertheimer and Leeper (1979, 1982) studies, doubt was cast on the reported association between cancer and electrical wiring configurations on the grounds that exposure had not been 'measured'. Consequently, many of the later studies included measurements of various types.

All measurements have the advantage that they capture exposure from whatever sources are present, and do not depend on prior identification of sources, as wire codes and calculated fields do. Furthermore, because measurements can classify fields on a continuous scale rather than in a limited number of categories, they provide greater scope for investigating different thresholds and exposure–response relationships.

(i)   *Spot measurements in the home*

The simplest form of measurement is a reading made at a point in time at one place in a home. To capture spatial variations of field, some studies have made multiple spot measurements at different places in or around the home. In an attempt to differentiate between fields arising from sources inside and outside the home, some studies have made spot measurements under 'low-power' (all appliances turned off) and 'high-power' (all appliances turned on) conditions. Neither of these alternatives truly represents the usual exposure conditions in a home, although the low-power conditions are closer to the typical conditions.

The major drawback of spot measurements is their inability to capture temporal variations. As with all measurements, spot measurements can assess only contemporary exposure, and can yield no information about historical exposure, which is an intrinsic requirement for retrospective studies of cancer risk. An additional problem of spot measurements is that they give only an approximation even for the contemporary field, because of short-term temporal variation of fields, and unless repeated throughout the year do not reflect seasonal variations.

A number of authors have compared the time-stability of spot measurements over periods of up to five years (reviewed in Kheifets *et al.*, 1997a; UK Childhood Cancer Study Investigators, 2000a). The correlation coefficients reported were from 0.7–0.9, but even correlation coefficients this high may result in significant misclassification (Neutra & DelPizzo, 1996).

(ii)   *Longer-term measurements in homes*

Because spot measurements capture short-term temporal variability poorly, many studies have measured fields at one or more locations for longer periods, usually 24–48 h, most commonly in a child's bedroom, which is an improvement on spot measurements. Comparisons of measurements have found only a poor-to-fair agreement between long-term and short-term measurements. This was mainly because short-term increases in fields caused by appliances or indoor wiring do not affect the average field measured over many hours (Schüz *et al.*, 2000).

Measurements over 24–48 h cannot account for longer-term temporal variations. One study (UK Childhood Cancer Study Investigators, 1999) attempted to adjust for longer-term variation by making 48-h measurements, and then, for subjects close to high-voltage power lines, modifying the measurements by calculating the fields using historical load data. In a study in Germany, Schüz *et al.* (2001a) identified the source of elevated fields by multiple measurements, and attempted to classify these sources as to the likelihood of their being stable over time. Before beginning the largest study in the USA (Linet *et al.*, 1997), a pilot study was conducted (Friedman *et al.*, 1996) to establish the proportion of their time children of various ages spent in different parts of the home. These estimates were used to weight the individual room measurements in the main study (Linet *et al.*, 1997) for the time-weighted average measure. In addition, the pilot study documented that magnetic fields in dwellings rather than schools accounted for most of the variability in children's exposure to magnetic fields.

(iii) *Personal exposure monitoring*

Monitoring the personal exposure of a subject by a meter worn on the body is attractive because it captures exposure to fields from all sources. Because all sources are included, the average fields measured tend to be higher than those derived from spot or long-term measurements. However, the use of personal exposure monitoring in case–control studies could be problematic, due to age- or disease-related changes in behaviour. The latter could introduce differential misclassification in exposure estimates. However, personal exposure monitoring can be used to validate other types of measurements or estimates.

(c) *Assessment of exposure to ELF electric and magnetic fields from appliances*

The contribution to overall exposure by appliances depends, among other things, on the type of appliance, its age, its distance from the person using it, and the pattern and duration of use. Epidemiological studies have generally relied on questionnaires, sometimes answered by proxies such as other household members (Mills *et al.*, 2000). These questionnaires ascertain some (but not usually all) of these facts, and are subject to recall bias. It is not known how well data from even the best questionnaire approximate to the actual exposure. Mezei *et al.* (2001) reported that questionnaire-based information on appliance use, even when focused on use within the last year, has limited value in estimating personal exposure to magnetic fields. Some limited attempts have been made (e.g. UK Childhood Cancer Study Investigators, 1999) to include some measurements as well as questionnaire data.

Because exposure to magnetic fields from appliances tends to be short-term and intermittent, the appropriate method for combining assessments of exposure from different appliances and chronic exposure from other sources would be particularly dependent on assumptions made about exposure metrics. Such methods have yet to be developed.

2.1.3 *Assessing occupational exposure to magnetic fields*

Following Wertheimer and Leeper's report of an association between residential magnetic fields and childhood leukaemia, Milham (1982, 1985a,b) noted an association between cancer and some occupations (often subsequently called the 'electrical occu-pations') intuitively expected to involve proximity to sources of electric and magnetic fields. However, classification based on job title is a very coarse surrogate. Critics (Loomis & Savitz, 1990; Guénel *et al.*, 1993a; Thériault *et al.*, 1994) have pointed out that, for example, many electrical engineers are basically office workers and that many electricians work on disconnected wiring.

Intuitive classification of occupations by investigators can be improved upon by taking account of judgements made by appropriate experts (e.g. Loomis *et al.*, 1994a), and by making measurements in occupational groups (e.g. Bowman *et al.*, 1988).

A further improvement is a systematic measurement programme to characterize exposure in a range of jobs corresponding as closely as possible to those of the subjects in a study, thus creating a 'job–exposure matrix', which links measurement data to job titles.

Despite the improvements in exposure assessment, the ability to explain exposure variability in complex occupational environments remains poor. Job titles alone explain only a small proportion of exposure variability. A consideration of the work environment and of the tasks undertaken by workers in a specific occupation leads to a more precise estimate (Kelsh *et al.*, 2000). Harrington *et al.* (2001) have taken this approach one stage further by combining job information with historical information not only on the environment in general but on specific power stations and substations. The within-worker and between-worker variability which account for most of the variation are not captured using these assessments.

It should be noted that even the limited information that is available on occupational exposure is confined almost entirely to the so-called electrical occupations and the power utility workforce. There is evidence (Zaffanella & Kalton, 1998) that workers in some non-electrical occupations are among those most heavily exposed to magnetic fields.

In addition to the need for correct classification of jobs, the quality of occupational exposure assessment depends on the details of work history available to the investigators. The crudest assessments are based on a single job (e.g. as mentioned on a death certificate). This assessment can be improved by identifying the job held for the longest period, or even better, by obtaining a complete job history which would allow for the calculation of the subject's cumulative exposure often expressed in μT–years.

2.1.4   *Assessing exposure to electric fields*

Assessment of exposure to electric fields is generally even more difficult and less well developed than the assessment of exposure to magnetic fields. All of the difficulties encountered in assessment of exposure to magnetic fields discussed above also apply to electric fields. In addition, electric fields are easily perturbed by any conducting object, including the human body. Therefore, the very presence of subjects in an environment means that they are not being exposed to an 'unperturbed field' although most studies that have assessed electric fields have attempted to assess the unperturbed field.

2.2 Cancer in children

2.2.1 *Residential exposure*

(*a*) *Descriptive studies*

In an ecological study in Taiwan, Lin and Lee (1994) observed a higher than expected incidence of childhood leukaemia in five districts in the Taipei Metropolitan Area where a high-voltage power line passed over at least one elementary school campus (standardized incidence ratio [SIR], 1.5; 95% CI, 1.2–1.9; based on 67 cases) for the period 1979–88. In a re-analysis, Li *et al.* (1998) focused on the three districts densely scattered with high-voltage power lines during the period 1987–92 and found an SIR of 2.7 (95% CI, 1.1–5.6) on the basis of seven observed cases versus 2.6 expected cases in all children in Taiwan, living within a distance of 100 m from an overhead power line.

Milham and Ossiander (2001) hypothesized that the emergence of the peak in incidence of acute lymphoblastic leukaemia in children aged 3–4 years may be due to exposure to ELF electric and magnetic fields. The authors examined state mortality rates in the USA during the years 1928–32 and 1949–51 and related this to the percentage of residences within each state with an electricity supply. The peak incidence of acute lymphoblastic leukaemia in children appeared to have developed earlier in those states in which more homes were connected earlier to the electricity supply.

(*b*) *Cohort study*

The only cohort study of childhood cancer and magnetic fields (see Table 18) was conducted by Verkasalo *et al.* (1993) in Finland. The study examined the risk of cancer in children living at any time from 1970–89 within 500 m of overhead high-voltage power lines (110–400 kV), where average magnetic fields were calculated to be ≥ 0.01 µT. The cohort comprised 68 300 boys and 66 500 girls under the age of 20 (contributing 978 100 person–years). During the observation period of 17 years, a total of 140 patients with childhood cancer (35 children with leukaemia, 39 with a tumour of the central nervous system, 15 with a lymphoma and 51 with other malignant tumours) were identified by the Finnish Cancer Registry. Historical magnetic fields were estimated for each year from 1970–89 by the Finnish power company. The dwellings of each child were ascertained from the central population registry and the shortest distance to nearby power lines was calculated by using exact coordinates of homes and power lines. Additional variables used in the calculation of the magnetic field strength were the current flow and the location of phase conductors of each power line. Point estimates of average annual currents for 1984–89 were generated by a power system simulator; information on existing line load was available for 1977–83; and data on power consumption from 1977, corrected for year of construction of power lines, were used to estimate current flow for the years 1970–76. Cumulative exposure was defined as the average exposure per year multiplied by the number of years of exposure (µT–years).

Table 18. Cohort study of childhood cancer and exposure to ELF magnetic fields

Study size, number of cases	Exposure	SIR (95% CI) by cancer site									
		Leukaemia	No. of cases	CNS	No. of cases	Lymphoma	No. of cases	Other sites	No. of cases	All cancers	No. of cases
68 300 boys, 66 500 girls, aged 0–19 years; 140 incident cancer cases diagnosed 1970–89	**Calculated historical magnetic fields**										
	< 0.01 μT (baseline)	1.0		1.0		1.0		1.0		1.0	
	0.01–0.19 μT	0.89 (0.61–1.3)	32	0.85 (0.59–1.2)	34	0.91 (0.51–1.5)	15	1.1 (0.79–1.4)	48	0.94 (0.79–1.1)	129
	≥ 0.2 μT	1.6 (0.32–4.5)	3	2.3 (0.75–5.4)	5	0 (0.0–4.2)	0	1.2 (0.26–3.6)	3	1.5 (0.74–2.7)	11
	Calculated cumulative magnetic fields (μT–years)										
	< 0.01 (baseline)	1.0		1.0		1.0		1.0		1.0	
	0.01–0.39	0.90 (0.62–1.3)	32	0.82 (0.56–1.2)	32	0.88 (0.48–1.5)	14	1.1 (0.80–1.4)	47	0.93 (0.78–1.1)	125
	≥ 0.4	1.2 (0.26–3.6)	3	2.3 (0.94–4.8)	7	0.64 (0.02–3.6)	1	1.0 (0.27–2.6)	4	1.4 (0.77–2.3)	15

From Verkasalo *et al.* (1993), Finland
SIR, standardized incidence ratio; CI, confidence interval; CNS, central nervous system
Expected numbers calculated in sex-specific five-year age groups; no further adjustments. SIRs for highest exposure categories for CNS tumours are questionable, since one boy with three primary tumours was counted three times.

The cut-points chosen to indicate high exposure were ≥ 0.2 µT for average exposure and ≥ 0.4 µT–years for cumulative exposure. The expected number of cases was calculated in five-year age groups by multiplying the stratum-specific number of person–years by the corresponding cancer incidence in Finland. No effect modifiers were considered. Standardized incidence ratios for children exposed to magnetic fields of ≥ 0.2 µT were 1.6 (95% CI, 0.32–4.5) for leukaemia, 2.3 (95% CI, 0.75–5.4) for tumours of the central nervous system (all in boys) and 1.5 (95% CI, 0.74–2.7) for all cancers combined. No child exposed to magnetic fields was diagnosed with lymphoma versus 0.88 expected. The corresponding SIRs with cumulative exposure of ≥ 0.4 µT–years were 1.2 (95% CI, 0.26–3.6) for leukaemia, 2.3 (95% CI, 0.94–4.8) for tumours of the central nervous system, 0.64 (95% CI, 0.02–3.6) for lymphoma and 1.4 (95% CI, 0.77–2.3) for all cancers, respectively. The SIRs in the intermediate category for each metric (0.01–< 0.2 µT, average exposure; 0.01–< 0.4 µT–years, cumulative exposure) were below unity. The SIRs for tumours of the central nervous system require careful interpretation, since one 18-year-old boy with three primary brain tumours and neuro-fibromatosis type 2 was counted as three cases. If this child were considered as one case, the number of cases of tumours of the central nervous system in exposed children would be reduced from five to three.

(c)   Case–control studies

A number of case–control studies of childhood leukaemia and ELF electric and magnetic fields have been published.

The results of these studies by tumour type (leukaemia and central nervous system) and by magnetic and electric fields are shown in Tables 19–21. The tables show only studies that contributed substantially to the overall summary and only the results of a-priori hypotheses are presented.

The first study of ELF electric and magnetic fields and childhood cancer was conducted in Denver, CO, USA (Wertheimer & Leeper, 1979). The population base consisted of children born in Colorado who resided in the greater Denver area between 1946 and 1973. The cases were all children aged less than 19 years who had died from cancer between 1950 and 1973 ($n = 344$), including 155 children with leukaemias and 66 with brain tumours, 44 with lymphomas and 63 with cancers of other sites. The controls ($n = 344$) were selected from two sources: Denver-area birth certificates and listings of all births in Colorado during the time period. Exposure was assessed by using diagrams to characterize electrical wiring configurations near the dwelling occupied by the child at birth and that occupied two years prior to death, or the corresponding dates for matched controls. The wiring was classified as having a high or low current configuration (HCC or LCC). Potential confounding was evaluated by examining the results within strata by age, birth order, social class, urban versus suburban, and heavy traffic areas versus lighter traffic areas. Point estimates were not reported, but p values calculated from chi-square tests were given. The percentage of children living in HCC homes two years before death was 41%, 41% and 46% for

Table 19. Case–control studies of childhood leukaemia and exposure to ELF magnetic fields[a]

Reference, area	Study size (for analyses)	Exposure	No. of cases	Risk estimates: odds ratio (95% CI)	Comments
Wertheimer & Leeper (1979), Denver, CO, USA	155 deceased cases, 155 controls, aged 0–19 years	**Wire code** LCC	92 (126 controls)		No risk estimates presented; lack of blinding for the exposure assessment; hypothesis-generating study
		HCC	63 (29 controls)		
London et al. (1991), Los Angeles County, CA, USA	*Wire code:* 211 cases, 205 controls; *24-h measurements:* 164 cases, 144 controls, aged 0–10 years	**Wire code** UG/VLCC (baseline)	31	1.0	Matched analysis, no further adjustments; low response rates for measurements; no wire coding of subjects who refused to participate
		OLCC	58	0.95 (0.53–1.7)	
		OHCC	80	1.4 (0.81–2.6)	
		VHCC	42	2.2 (1.1–4.3)	
		Mean magnetic fields (24-h bedroom measurement)			
		< 0.067 µT (baseline)	85	1.0	
		0.068–0.118 µT	35	0.68 (0.39–1.2)	
		0.119–0.267 µT	24	0.89 (0.46–1.7)	
		≥ 0.268 µT	20	1.5 (0.66–3.3)	
Feychting & Ahlbom (1993), Sweden (corridors along power lines)	39 cases, 558 controls, aged 0–15 years	**Calculated historical magnetic fields** < 0.1 µT (baseline)	27	1.0	Adjusted for sex, age, year of diagnosis, type of house, Stockholm county (yes/no); in subsequent analysis also for socioeconomic status and air pollution from traffic; no contact with subjects required
		0.1–0.19 µT	4	2.1 (0.6–6.1)	
		≥ 0.2 µT	7	2.7 (1.0–6.3)	

Table 19 (contd)

Reference, area	Study size (for analyses)	Exposure	No. of cases	Risk estimates: odds ratio (95% CI)	Comments
Olsen et al. (1993), Denmark	833 cases, 1666 controls, aged 0–14 years	**Calculated historical magnetic fields** < 0.1 μT (baseline) 0.1–0.24 μT ≥ 0.25 μT	829 1 3	1.0 0.5 (0.1–4.3) 1.5 (0.3–6.7)	Adjusted for sex and age at diagnosis; socioeconomic status, distribution similar between cases and controls; no contact with subjects required
Tynes & Haldorsen (1997), Norway (census wards crossed by power lines)	148 cases, 579 controls, aged 0–14 years	**Calculated historical magnetic fields** < 0.05 μT (baseline) 0.05–< 0.14 μT ≥ 0.14 μT	139 8 1	1.0 1.8 (0.7–4.2) 0.3 (0.0–2.1)	Adjusted for sex, age and municipality, also for socioeconomic status, type of house, and number of dwellings; no contact with subjects required
Michaelis et al. (1998), Lower Saxony and Berlin (Germany)	176 cases, 414 controls, aged 0–14 years	**Median magnetic fields (bedroom 24-h measurement)** < 0.2 μT (baseline) ≥ 0.2 μT	167 9	1.0 2.3 (0.8–6.7)	Adjusted for sex, age and part of Germany (East, West), socioeconomic status and degree of urbanization; information on a variety of potential confounders was available; low response rates

Table 19 (contd)

Reference, area	Study size (for analyses)	Exposure	No. of cases	Risk estimates: odds ratio (95% CI)	Comments
McBride *et al.* (1999), five Canadian provinces, subjects living within 100 km of major cities, Canada	*Personal monitoring:* 293 cases, 339 controls, aged 0–14 years	**Personal monitoring (48-h)**			Adjusted for age, sex, province, maternal age at birth of child, maternal education, family income, ethnicity and number of residences since birth; information on a variety of potential confounding factors was available; relatively low response rates for the personal monitoring portion; children with Down syndrome excluded from this study
		< 0.08 µT (baseline)	149	1.0	
		0.08–< 0.15 µT	67	0.57 (0.37–0.87)	
		0.15–< 0.27 µT	45	1.1 (0.61–1.8)	
		≥ 0.27 µT	32	0.68 (0.37–1.3)	
	Wire code: 351 cases, 362 controls	**Wire code**			
		UG (baseline)	79	1.0	
		VLCC	73	0.70 (0.41–1.2)	
		OLCC	77	0.76 (0.45–1.3)	
		OHCC	83	0.64 (0.38–1.1)	
		VHCC	39	1.2 (0.58–2.3)	
UKCCSI (1999), England, Wales and Scotland	1073 cases, 1073 controls, aged 0–14 years	**Time-weighted average magnetic fields (1.5–48-h measurement)**			Adjusted for sex, date of birth and region, also for socioeconomic status; information on a variety of potential confounders was available; low reponse rates
		< 0.1 µT (baseline)	995	1.0	
		0.1–< 0.2 µT	57	0.78 (0.55–1.1)	
		≥ 0.2 µT	21	0.90 (0.49–1.6)	
		0.2–< 0.4 µT	16	0.78 (0.40–1.5)	
		≥ 0.4 µT	5	1.7 (0.40–7.1)	

Table 19 (contd)

Reference, area	Study size (for analyses)	Exposure	No. of cases	Risk estimates: odds ratio (95% CI)	Comments
Schüz et al. (2001a), West Germany	514 cases, 1301 controls, aged 0–14 years	**Median magnetic fields (24-h bedroom measurement)**			Adjusted for sex, age, year of birth, socioeconomic status and degree of urbanization; information on a variety of potential confounders was available; low response rates; relatively long time lag between date of diagnosis and date of the measurement
		< 0.1 μT (baseline)	472	1.0	
		0.1–< 0.2 μT	33	1.2 (0.73–1.8)	
		0.2–< 0.4 μT	6	1.2 (0.43–3.1)	
		≥ 0.4 μT	3	5.8 (0.78–43)	
		Night-time magnetic fields			
		< 0.1 μT (baseline)	468	1.0	
		0.1–< 0.2 μT	34	1.4 (0.90–2.2)	
		0.2–< 0.4 μT	7	2.5 (0.86–7.5)	
		≥ 0.4 μT	5	5.5 (1.2–27)	

Table 19 (contd)

Reference, area	Study size (for analyses)	Exposure	No. of cases	Risk estimates: odds ratio (95% CI) Unmatched	No. of cases	Risk estimates: odds ratio (95% CI) Matched	Comments
Linet et al. (1997), nine mid-western and mid-Atlantic states, USA	*Wire code:* 408 cases, 408 controls, aged 0–14 years; *24-h measurements:* 638 cases, 620 controls	**Time-weighted average (24-h bedroom measurement plus spot measurements in two rooms)**					Unmatched analysis additionally adjusted for age, sex, mother's education and family income; information on a variety of potential confounding factors was available; wire coding of subjects who refused to participate; relatively low response rates for the measurements in controls; only acute lymphoblastic leukaemia; children with Down syndrome excluded from this study (Schüz et al., 2001a)
		< 0.065 µT (baseline)	267	1.0	206	1.0	
		0.065–0.099 µT	123	1.1 (0.81–1.5)	92	0.96 (0.65–1.4)	
		0.100–0.199 µT	151	1.1 (0.83–1.5)	107	1.2 (0.79–1.7)	
		≥ 0.200 µT	83	1.2 (0.86–1.8)	58	1.5 (0.91–2.6)	
		Wire code					
		UG/VLCC (baseline)			175	1.0	
		OLCC			116	1.1 (0.74–1.5)	
		OHCC			87	0.99 (0.67–1.5)	
		VHCC			24	0.88 (0.48–1.6)	

UG, underground wires; VLCC, very low current configuration; OLCC, ordinary low current configuration; OHCC, ordinary high current configuration; VHCC, very high current configuration; LCC, low current configuration; HCC, high current configuration; UKCCSI, UK Childhood Cancer Study Investigators

[a] In these tables, only studies that contributed substantially to the overall summary were considered; only results that were part of the analysis strategy defined above are presented; exposure metrics and cut-points vary across studies, for a better comparison, please refer to Table 23.

Table 20. Case–control studies of childhood tumours of the central nervous system and exposure to ELF magnetic fields

Reference, area	Study size	Exposure	No. of cases	Risk estimates: odds ratio (95% CI)	Comments
Wertheimer & Leeper (1979), Denver, CO, USA	66 deceased cases, 66 controls, aged 0–19 years	**Wire code** LCC HCC	36 (49 controls) 30 (17 controls)		No risk estimates presented; lack of blinding for the exposure assessment; hypothesis-generating study
Feychting & Ahlbom (1993), Sweden (corridors along power lines)	33 cases, 558 controls, aged 0–15 years	**Calculated historical magnetic fields** < 0.1 µT (baseline) 0.1–0.19 µT ≥ 0.2 µT	29 2 2	1.0 1.0 (0.2–3.8) 0.7 (0.1–2.7)	Adjusted for sex, age, year of diagnosis, type of house, Stockholm county (yes/no); in subsequent analysis also for socioeconomic status and air pollution from traffic; no contact with subjects required
Olsen et al. (1993), Denmark	624 cases, 1872 controls, aged 0–14 years	**Calculated historical magnetic fields** < 0.1 µT (baseline) 0.1–0.24 µT ≥ 0.25 µT	621 1 2	1.0 1.0 (0.1–9.6) 1.0 (0.2–5.0)	Adjusted for sex and age at diagnosis; socioeconomic distribution similar among cases and controls; no contact with subjects required
Preston-Martin et al. (1996a), Los Angeles County, CA, USA	298 cases, 298 controls, aged 0–19 years	**Mean magnetic fields (24-h bedroom)** 0.010–0.058 µT (baseline) 0.059–0.106 µT 0.107–0.248 µT 0.249–0.960 µT Wire code UG VLCC/OLCC (baseline) OHCC VHCC	48 29 16 13 39 114 97 31	1.0 1.5 (0.7–3.0) 1.2 (0.5–2.8) 1.6 (0.6–4.5) 2.3 (1.2–4.3) 1.0 0.8 (0.6–1.2) 1.2 (0.6–2.2)	Adjusted for age, sex, birth year, socioeconomic status, maternal waterbed use; low response rates for measurements

Table 20 (contd)

Reference, area	Study size	Exposure	No. of cases	Risk estimates: odds ratio (95% CI)	Comments
Gurney et al. (1996), Seattle, WA, USA	133 cases, 270 controls, aged 0–19 years	**Wire code** UG (baseline) VLCC OLCC OHCC VHCC	47 39 11 19 4	1.0 1.3 (0.7–2.1) 0.7 (0.3–1.6) 1.1 (0.6–2.1) 0.5 (0.2–1.6)	Unadjusted, but evaluated for confounding by age, sex, race, county, reference year, mother's education, family history of brain tumours, passive smoking, farm residence, history of head injury, X-rays, epilepsy
Tynes & Haldorsen (1997), Norway (census wards crossed by power lines)	156 cases, 639 controls, aged 0–14 years	**Historical calculated magnetic fields** < 0.05 μT (baseline) 0.05–< 0.14 μT ≥ 0.14 μT	144 8 4	1.0 1.9 (0.8–4.6) 0.7 (0.2–2.1)	Adjusted for sex, age and municipality, also for socioeconomic status, type of house, and number of dwellings; no contact with subjects required
UKCCSI (1999), England, Wales and Scotland	387 cases, 387 controls, aged 0–14 years	**Time-weighted average magnetic fields (1.5–48-h measurement)** < 0.1 μT (baseline) 0.1–< 0.2 μT ≥ 0.2 μT 0.2–< 0.4 μT	359 25 3 3	1.0 2.4 (1.2–5.1) 0.46 (0.11–1.9) 0.70 (0.16–3.2)	Adjusted for sex, date of birth, and region, also for socioeconomic status; information on a variety of potential confounders was available; low response rates; no exposure to magnetic fields ≥ 0.4 μT
Schüz et al. (2001b), Lower Saxony and Berlin (Germany)	64 cases, 414 controls, aged 0–14 years	**Median magnetic fields (24-h bedroom measurement)** < 0.2 μT (baseline) ≥ 0.2 μT	62 2	1.0 1.7 (0.32–8.8)	Adjusted for sex, age, part of Germany (East, West), socioeconomic status and degree of urbanization; information on a variety of potential confounders was available; low response rates; same control group as for leukaemia cases (Michaelis et al., 1998)

UG, underground wires; VLCC, very low current configuration; OLCC, ordinary low current configuration; OHCC, ordinary high current configuration; VHCC, very high current configuration; UKCCSI, UK Childhood Cancer Study Investigators

Table 21. Case–control studies of childhood leukaemia and exposure to ELF electric fields

Reference, area	Study size (for analyses)	Exposure	No. of cases	Risk estimates: odds ratio (95% CI)	Comments
London et al. (1991), Los Angeles County, CA, USA	**Spot measurements (child's bedroom)** 136 cases, 108 controls, aged 0–10 years	< 50th percentile (baseline) 50–74th percentile (baseline) 75–89th percentile (baseline) ≥ 90th percentile (baseline)	NR	1.0 0.66 (0.36–1.2) 1.1 (0.58–2.6) 0.44 (0.19–1.0)	Matched analysis, no further adjustments; low response rates for measurements; no wire coding of subjects who refused to participate
McBride et al. (1999), five Canadian provinces, subjects living within 100 km of major cities, Canada	**Personal monitoring** 274 cases, 331 controls, aged 0–14 years	< 12.2 V/m (baseline) 12.2–< 17.2 V/m 17.2–< 24.6 V/m 24.6–64.7 V/m	143 64 39 28	1.0 0.79 (0.51–1.2) 0.76 (0.45–1.2) 0.82 (0.45–1.5)	Adjusted for age, sex, province, maternal age at birth of child, maternal education, family income, ethnicity and number of dwellings since birth; children with Down syndrome excluded from this study

NR, not reported

children with leukaemia, lymphoma and brain tumours, respectively, compared with 19%, 25% and 26% in the controls. Forty-four per cent of 109 cases and 20.3% of 128 controls with stable dwellings had HCC wiring configurations ($p < 0.001$). The results were similar when birth addresses were used. When exposure was further subdivided into the categories of substation (highest exposure), other HCC, LCC except end poles and end poles, the percentage of children with cancer declined with lower wiring configuration. The results also appeared to be fairly consistent within broad categories of potential confounding variables. [The Working Group noted that the wire-coding technicians in this study were not blinded as to the status of cases or controls leading to potential bias in exposure assessment.]

Fulton *et al.* (1980) conducted a case–control study in Rhode Island, USA. Patients with leukaemia aged between 0 and 20 years were identified from the records of the Rhode Island Hospital from 1964–78. Out of 155 cases, a total of 119 were selected and 36 cases who had resided out of the state for part of the eight years preceding diagnosis were excluded. The analysis was based on dwellings, not individuals, and 209 case dwellings were included. Two hundred and forty control addresses were selected from Rhode Island birth certificates. Two controls were matched to each case on year of birth. The authors obtained complete address histories for cases, but not for controls. Exposure assessment consisted of mapping the power lines situated within 50 yards (45.72 m) of each residence and categorizing the expected current according to a method of wire coding. A total of 95% of case and 94% of control addresses were successfully mapped. The association between exposure category and childhood leukaemia was tested by means of chi-square tests. The analysis showed no relationship between childhood leukaemia and exposure category. [The Working Group noted that the shortcomings of this study include lack of comparability of cases and controls, analysis by dwelling as opposed to by individual and the lack of control for confounding.]

The first European study on childhood leukaemia and exposure to magnetic fields was carried out in Sweden (Tomenius, 1986). The study included children aged between 0 and 18 years with benign and malignant tumours. The children had been born and diagnosed in the county of Stockholm and were registered with the Swedish Cancer Registry during the years 1958–73. A total of 716 children of whom 660 had a malignant tumour were included. For each case, a control matched for sex, age and church district of birth was selected from birth registration records held in the same parish office. The controls had been born just before or after the child with a tumour and still lived in Stockholm county at the date of diagnosis of the corresponding case. The analysis was based on dwellings rather than individuals. A total of 1172 dwellings were included for cases and 1015 dwellings for controls. Almost all (96%) dwellings were visited to determine their proximity to different types of electrical installation (200-kV power lines, 6–< 200-kV power lines, substations, transformers, electric railways, underground railways). Spot measurements of peak magnetic field were conducted outside the entrance door. From magnetic field measurements, the odds ratios in those children exposed to magnetic fields of ≥ 0.3 μT were 0.3 for leukaemia, 1.8 for lymphoma, 3.7 for cancer of the central nervous system ($p < 0.05$) and 2.1 for all tumours combined ($p < 0.05$). An excess tumour risk was also reported for children living less than 150 m from a 200-kV power line, but this was because the observed number of case dwellings located 100–150 m from a power line was higher than expected, while the number of case dwellings located within 100 m from a power line was as expected. [The Working Group noted that outdoor spot measurements, sometimes made more than 30 years after the etiologically relevant time period, are a poor proxy for an individual's exposure to magnetic fields. Another limitation is that the analyses are based on dwellings rather than individuals, and that the numbers of dwellings were different for cases and controls.]

Savitz *et al*. (1988) carried out a second study in the Denver, CO, area during the 1980s. The population base consisted of all children < 15 years of age residing in the 1970 Denver Standard Metropolitan Statistical Area. A total of 356 cases of childhood cancers diagnosed from 1976–1983 were identified from the Colorado Central Cancer Registry and from records of area hospitals. A total of 278 potential controls were identified through random digit dialling and were matched to the cases by age (± 3 years), sex and telephone exchange. The cases had been diagnosed up to nine years prior to selection, so controls had to be restricted to those who still lived in the same residence as they had done at the time the case was diagnosed. Exposure to electric fields was assessed by means of spot measurements under 'high-power' conditions (when selected appliances and lights were turned on) and exposure to magnetic fields by means of spot measurements under both 'high-power' and 'low-power' conditions (when most appliances and lights were turned off). The measurements were made in the current dwelling of the case if it was also the dwelling occupied prior to diagnosis, and homes were classified by Wertheimer-Leeper wire codes. A total of 252 (70.8%) of the cases were interviewed; spot measurements of fields were made for 128 cases (36%) and wire coding was completed for 319 (89.6%). Two hundred and twenty-two controls were interviewed, giving a final response rate from the random-digit dialling phase of 63%; a total of 207 (74.5%) of the controls had spot measurements made in their homes and the homes of 259 (93.2%) were wire coded. Potential confounding variables included year of diagnosis and residential stability; electric load at the time of measurement; parental age, race, education and income; traffic density, and various in-utero exposures. For low-power conditions, the odds ratios for magnetic field measurements of ≥ 0.2 μT versus < 0.2 μT were 1.4 (95% CI, 0.63–2.9) for all cancers combined, 1.9 (95% CI, 0.67–5.6) for leukaemia, 2.2 (95% CI, 0.46–10) for lymphoma and 1.0 (95% CI, 0.22–4.8) for brain cancer. The odds ratios for high-power conditions were near unity for most cancer sites, and those for high electric fields were mostly below unity. For assessment of the influence of wire codes, underground wires were considered to have the lowest magnetic fields. The odds ratios for 'high' (HCC and VHCC) versus 'low' (underground, VLCC and LCC) wire codes were 1.5 (95% CI, 1.0–2.3) for all cancers combined, 1.5 (95% CI,

0.90–2.6) for leukaemia, 0.8 (95% CI, 0.29–2.2) for lymphoma and 2.0 (95% CI, 1.1–3.8) for brain cancer. For VHCC versus buried wires, the odds ratios were 2.2 (95% CI, 0.98–5.2) for all cancers, 2.8 (95% CI, 0.94–8.0) for leukaemia, 3.3 (95% CI, 0.80–14) for lymphoma and 1.9 (95% CI, 0.47–8.0) for brain cancer. [The Working Group noted the differential residential requirements for controls compared with cases, leading to a possible selection bias. Spot measurements were taken only in dwellings that were still occupied by the cases, although many years might have elapsed since the date of diagnosis (measurements were made for only 36% of the eligible cases) and in many instances, measurements were made years after the etiological time period of interest.]

Coleman *et al.* (1989) conducted a registry (Thames Cancer Registry)-based case–control study in the United Kingdom. The study included leukaemia patients of all ages diagnosed between 1965 and 1980 and resident in one of four adjacent London boroughs. Two tumour controls were selected randomly from the same registry and matched to each case for sex, age and year of diagnosis. The childhood study population comprised 84 leukaemia cases and 141 cancer controls under the age of 18. Only one case and one control lived within 100 m of an overhead power line; thus, no risk estimates were presented for proximity to power lines. No clear pattern was seen for children living within 100 m of a substation.

The second case–control study in the United Kingdom was carried out by Myers *et al.* (1990) on children born within the boundaries of the Yorkshire Health Region and registered in the period 1970–79. A total of 419 cases and 656 controls were identified, but some could not be located, and 374 cases (89%) and 588 (90%) controls were finally analysed. Exposure was assessed by calculations of historical magnetic fields due to the load currents of overhead power lines at the birth addresses of the children, on the basis of line-network maps and load records. Risk estimates were presented for all cancers, and separately for non-solid tumours (mostly leukaemia and lymphoma) and solid tumours (all brain tumours, neuroblastomas and tumours of other sites). For all cancers combined, for children in the group calculated to have the highest exposure to magnetic fields, i.e. ≥ 0.1 µT, the resulting odds ratio was 0.4 (95% CI, 0.04–4.3). For the two diagnostic subgroups, non-solid tumours and solid tumours, a cut-point of ≥ 0.03 µT was chosen and the respective odds ratios were 1.4 (95% CI, 0.41–5.0) and 3.1 (95% CI, 0.31–32). The distance analysis with a cut-point of < 25 m gave an odds ratio of 1.1 (95% CI, 0.47–2.6) for all cancers combined.

The first North American study to include long-term measurements of ELF magnetic fields was carried out by London *et al.* (1991, 1993) in Los Angeles, CA. The study population consisted of children from birth to the age of 10 years who had resided in Los Angeles County. A total of 331 cases of childhood leukaemia were identified by the Los Angeles County Cancer Surveillance Program from 1980–87. A total of 257 controls were identified, using a combination of friends of the patients and random digit dialling. The cases and controls were individually matched on age, sex and ethnicity. Exposure assessment consisted of spot measurements of electric and magnetic fields in three or four locations inside the home and three locations outside, a 24-h magnetic field measurement made in the child's bedroom and wire coding. Lifetime residential histories were obtained, and measurements were sought for at least one dwelling per subject. Spot measurements were made in multiple residences when possible. Latency was considered in the design phase by defining an 'etiological time-period' that extended from birth up to a reference date that depended upon the child's age at diagnosis. The same reference date was used for each matched control. The response rates for cases for the various parts of the study were approximately 51% for the 24-h measurement and about 66% for wire coding. [The Working Group noted that it was not possible to calculate accurate response rates for the controls.] Twenty-four-hour measurements for both cases and controls were analysed according to percentile cut-points (< 50th (< 0.07 μT), 50–74th, 75–89th and ≥ 90th (≥ 0.27 μT)). When compared with the referent group of < 50th percentile, the odds ratios for each category were 0.68 (95% CI, 0.39–1.2), 0.89 (95% CI, 0.46–1.7) and 1.5 (95% CI, 0.66–3.3) in relation to the arithmetic mean of 24-h measurements of magnetic field in the child's bedroom. When compared with a referent group with VLCC and underground wire codes, the odds ratios for OLCC, OHCC and VHCC were 0.95 (95% CI, 0.53–1.7), 1.4 (95% CI, 0.81–2.6) and 2.2 (95% CI, 1.1–4.3). Adjustment for confounding variables reduced the estimate for VHCC from 2.2 to 1.7 (95% CI, 0.82–3.7), but the trend was still statistically significant. There was no significant association of childhood leukaemia with spot measurements of magnetic or electric fields. [The Working Group considered that the limitations of this study include somewhat low response rates for the measurement component of the study.]

Ebi *et al.* (1999) re-analysed the Savitz *et al.* (1988) and London *et al.* (1991) studies using the 'case-specular method'. This method compared the wire codes of subjects' homes with those of 'specular residences': imaginary homes constructed as a mirror image of the true home, symmetrical with respect to the centre of the street. This method is intended to discriminate between the 'neighbourhood variables', which are normally the same for the true home and its mirror image, and the effects of power lines that are normally not placed symmetrically in the centre of the street. The study confirmed the association reported in the original studies. [The Working Group noted that this study did not correct for limitations noted for the original studies and did not address selection bias.]

Feychting and Ahlbom (1993) conducted a population-based nested case–control study in Sweden. The study base consisted of all children aged less than 16 years who had lived on a property at least partially located within 300 m of any 220- or 400-kV power lines from 1960–85. They were followed from the time they moved into the corridor until the end of the study period. The Swedish population registry was used to identify individuals who had lived on the respective properties, and record linkage to the Swedish Cancer Registry was performed to identify patients with childhood cancer among this group. A total of 142 children with cancer were identified within the power-line corridors, 39 of whom were diagnosed as having leukaemia, 33 as having a tumour of the central nervous system, 19 as having a lymphoma and 51 as having some other type of cancer. Approximately four controls per case, matched according to age, sex, parish of residence during the year of diagnosis or during the last year before the case moved to a new home, and proximity to the same power line, were selected randomly from the study base, providing a total of 558 controls. Exposure to magnetic fields was assessed by calculated historical fields, calculated contemporary fields and spot measurements under low-power conditions within the dwelling. To calculate historical fields, information on the average power load on each power line was obtained for each year. Spot measurements were made with a meter, constructed specifically for the purpose of this study, in the home within the power-line corridor where the child lived at the time closest to diagnosis. Spot measurements were conducted 5–31 years after diagnosis [the participation rate was 63% among cases and 62% among controls]. In the analysis, the authors placed most emphasis on calculated historical fields using a three-level exposure scale with categories of < 0.1 µT, $0.1–< 0.2$ µT and ≥ 0.2 µT. The relative risks were calculated by using a logistic regression model stratified according to age, sex, year of diagnosis, type of house (single-family house or apartment), and whether or not the subject lived within the county of Stockholm. Other potential effect modifiers considered in the analysis were socioeconomic status taken from the population censuses made closest to the year of birth and closest to the year of diagnosis of cancer, and air pollution from traffic estimated by the Swedish Environmental Protection Board. Cancer risk in relation to calculated magnetic fields closest in time to diagnosis at ≥ 0.2 µT compared with < 0.1 µT was elevated for childhood leukaemia (odds ratio, 2.7; 95% CI, 1.0–6.3; 7 cases) but not for tumours of the central nervous system (odds ratio, 0.7; 95% CI, 0.1–2.7; 2 cases), lymphoma (odds ratio, 1.3; 95% CI, 0.2–5.1; 2 cases) or all cancers combined (odds ratio, 1.1; 95% CI, 0.5–2.1; 12 cases). At ≥ 0.3 µT compared with < 0.1 µT, the increased risk for leukaemia was more pronounced with an odds ratio of 3.8 (95% CI, 1.4–9.3; 7 cases), while the risks for the other types of cancer were only slightly altered. Subgroup analysis revealed the highest odds ratios for children aged 5–9 years at date of diagnosis and for children living in single-family homes. Spot measurements showed a good agreement with calculated contemporary fields, demonstrating that calculated fields could predict residential magnetic fields, but agreement with calculated historical fields was poor. Based on a distance of ≤ 50 m compared with > 100 m to nearby power lines, the odds ratio was 2.9 (95% CI, 1.0–7.3; 6 exposed cases, 34 controls) for leukaemia, 1.0 (95% CI, 0.5–2.2; 9 cases, 34 controls) for all cancers and 0.5 (95% CI, 0.0–2.8; 1 case, 34 controls) for tumours of the central nervous system.

The results of a similar population-based case–control study were published in the same year by Olsen *et al.* (1993). The study population included all Danish children under the age of 15 years who had been diagnosed as having leukaemia, a tumour of the central nervous system or malignant lymphoma during the period 1968–86. A total of 1707 patients was identified from the Danish Cancer Registry, of whom 833 had a leukaemia, 624 had a tumour of the central nervous system and 250 had a malignant lymphoma. Two controls for each patient with leukaemia, three controls for each patient with a tumour of the central nervous system and five controls for each patient with a lymphoma were drawn at random from the files of the Danish central population registry. The matching criteria chosen were sex and date of birth within one year. The total number of controls was 4788. The residential histories of each family were ascertained restrospectively from the date of diagnosis to nine months before the child's birth. Each address was checked against maps of existing or former 50–400-kV power lines, and areas of potential exposure to magnetic fields ≥ 0.1 µT were defined. For all dwellings outside the potential exposure areas, the magnetic fields were assumed to be zero. For other dwellings, historical fields were calculated taking into account the annual average current flow of the line, the type of pylon, the category of the line, the ordering of the phases and any reconstructions. The basic measure of exposure was the average magnetic field generated from a power line to which the child was ever exposed. Exposure was categorized into the groups < 0.1 µT, $0.1–< 0.25$ µT and ≥ 0.25 µT. The cumulative dose of magnetic fields was obtained by multiplying the number of months of exposure by the average magnetic field at the dwelling (µT–months). Odds ratios were derived from logistic regression models adjusted for sex and age at diagnosis. The distribution of cases and controls according to socio-economic group did not differ. Odds ratios at calculated field levels of ≥ 0.25 µT compared with < 0.1 µT were 1.5 (95% CI, 0.3–6.7) for leukaemia based on three exposed cases and four exposed controls, 1.0 (95% CI, 0.2–5.0) for tumours of the central nervous system (2 cases, 6 controls) and 5.0 (0.3–82) for malignant lymphoma (1 case, 1 control). In a post-hoc analysis comparing calculated field levels of ≥ 0.4 µT with the same baseline, the respective odds ratios were 6.0 (95% CI, 0.8–44; 3 cases) for leukaemia, 6.0 (95% CI, 0.7–44; 2 cases) for tumours of the central nervous system, 5.0 (95% CI, 0.3–82; 1 case) for malignant lymphoma and 5.6 (95% CI, 1.6–19) for the combined groups based on only six exposed cases and three exposed controls. The distribution functions of cumulative doses of magnetic fields for cases and controls showed that doses were generally higher among cases; but there was never any significant association with an increase in risk.

Coghill *et al.* (1996) conducted a small case–control study in the United Kingdom involving 56 patients with leukaemia and 56 controls. Patients with leukaemia diagnosed between 1985 and 1995 were recruited by media advertising, personal introduction and with the support of the Wessex Health Authority and various self-help groups. Controls of the same age and sex were suggested by the parents of the patients (friend controls). Measurements of the magnetic field (mean, 0.070 µT for the cases, 0.057 µT for the controls) were conducted in the child's bedroom between 20:00 and 08:00. The main result of a conditional logistic regression analysis was an association between leukaemia and electric fields of ≥ 20 V/m with an odds ratio of 4.7 (95% CI, 1.2–28) and somewhat weaker associations in groups with intermediate exposures of 10–19 V/m (odds ratio, 2.4; 95% CI, 0.79–8.1) and of 5–9 V/m (odds ratio, 1.5; 95% CI, 0.47–5.1), suggesting a dose–response relationship. No association was seen between leukaemia and magnetic fields. [The Working Group noted the potential lack of comparability of cases and controls.]

A study of 298 children with brain tumours (ICD-9 191, 192) and 298 control children was carried out in Los Angeles, CA, by Preston-Martin et al. (1996a). The study subjects were aged 19 years or younger, resident in Los Angeles County and diagnosed between 1984 and 1991. Controls were identified by random digit dialling and matched on age and sex. The response rates were about 70% for both cases (298/437) and controls (298/433). Cases and controls were accrued prospectively from 1989 onwards, but retrospectively from 1984 to 1988. The authors attempted to obtain exterior spot measurements and wire codes for all the homes occupied by subjects from conception until diagnosis of brain tumour. The 596 (298 cases, 298 controls) study subjects reported living in a total of 2000 homes; of these, some measurements were made or wire codes were assigned for 1131. No measurements or wire coding were attempted if the former home was outside Los Angeles County. Interior 24-h measurements were made in the child's bedroom and one other room if at the time of interview the child still occupied a home lived in prior to diagnosis. Interior measurements were available for 110 cases (37% of those interviewed) and 101 controls (34% of those interviewed). Wire codes were available for at least one residence for 292 cases (98%) and 269 controls (90%) and exterior spot measure-ments, made at the front door, were available for 255 cases (86%) and 208 controls (70%). There was no association between living in a home with a wire code of VHCC at diagnosis and childhood brain tumours (odds ratio for VHCC versus VLCC and OLCC, 1.2; 95% CI , 0.6–2.2; 31 cases), adjusted for age, sex, year of birth, socio-economic status and maternal use of a water bed during pregnancy. The data showed an increased risk for subjects living in homes with underground wiring (odds ratio, 2.3; 95% CI, 1.2–4.3; 39 cases); but this increased risk was apparent only in cases diagnosed before 1989 (odds ratio, 4.3; 95% CI, 1.6–11; 28 cases). There was no increased risk associated with the measurements of magnetic field taken outside the home occupied at the time of diagnosis. The odds ratio was 0.7 (95% CI, 0.3–1.5) for front door measurements > 0.2 μT (13 cases) and 0.9 (95% CI, 0.3–3.2) for fields over 0.3 μT (7 cases). For 24-h means over 0.2 μT in the child's bedroom, the odds ratio was 1.2 (95% CI, 0.5–2.8; 16 cases) and, for fields over 0.3 μT, the odds ratio was 1.7 (95% CI, 0.6–5.0; 12 cases). [The Working Group considered that the limitations of this study included relatively low response rates for both cases and controls, in particular for the interior measurement portion of the study. There is also some indication of bias in the control selection process, as manifested by the different results for underground wiring between cases diagnosed before and after 1989.]

Gurney et al. (1996) studied childhood brain tumours in relation to ELF magnetic fields in the Seattle, WA, area of the USA. Patients under 20 years of age were identified through a population-based registry. One hundred and thirty-three of a total of 179 identified cases (74%), and 270 of 343 controls (79%), identified through random digit dialling, participated in the study. Magnetic field exposure was assessed by wire coding of homes occupied by cases and controls at the diagnosis or reference date. There was no evidence of an association between risk for brain tumour and wire codes. The odds ratio for VHCC homes was 0.5 (95% CI, 0.2–1.6; 4 cases) and, when wire codes were classified into high and low exposure categories, the odds ratio was 0.9 (95% CI, 0.5–1.5; 23 cases). When wire codes for the homes of eligible non-participants were included in the calculations, the odds ratios were similar. [The Working Group noted that cases had lived in their homes for an average of 12 months longer than controls, suggesting the possibility of selection bias.]

In a hospital-based case–control study on childhood leukaemia in Greece, the possible relationship between childhood leukaemia and residential proximity to power lines was investigated (Petridou et al., 1997). The study population comprised 117 out of 153 (76%) incident leukaemia cases in children under the age of 15 years diagnosed in 1993–94 and identified by a nationwide network of paediatric oncologists, and two controls per case (202/306; 66%). For every study participant, the Public Power Cooperation of Greece specified the distance between the centre of the dwelling and the two closest power lines between 0.4 and 400 kV (blindly as to status as case or control). From this information, the voltage of each of the two closest power lines was divided by the distance (V/m) and the maximum of the two values was taken as the subject's exposure (modified distance measure). The same procedure was performed by dividing the voltage level of each power line by the square of the distance (V/m^2) and the cube of the distance (V/m^3). Compared with the first quintile, odds ratios for V/m^2 were 0.58 (95% CI, 0.24–1.4), 0.65 (95% CI, 0.26–1.6), 1.5 (95% CI, 0.58–3.8) and 1.7 (95% CI, 0.67–4.1) for the 2nd, 3rd, 4th and 5th quintile, respectively (p value for trend = 0.08). The Wertheimer-Leeper wire code was adapted to conditions in Greece. The odds ratios for the four levels in ascending order of magnetic field strength were 0.99 (95% CI, 0.54–1.8), 1.8 (95% CI, 0.26–13), 4.3 (95% CI, 0.94–19) and 1.6 (95% CI, 0.26–9.4); p value for trend = 0.17. [The Working Group noted that the main limitation of the study was the crude exposure assessment.]

Tynes and Haldorsen (1997) reported the results from a nested case–control study in Norway. The cohort comprised children under the age of 15 years who had lived in a census ward crossed by a power line (45 kV or greater in urban areas, 100 kV or greater in rural areas) during at least one of the years 1960, 1970, 1980, 1985, 1987 or 1989. Cancer cases occurring in the study area between 1965 and 1989 were identified by record linkage with the Cancer Registry of Norway. Out of 532 children with cancer, 500 were included in the study, of whom 148 had a leukaemia, 156 had a brain tumour, 30 had a lymphoma and 166 had cancer at another site. For each case, one to five controls (depending on eligibility) matched for sex, year of birth and municipality were selected at random from the cohort, resulting in a total of 2004 controls. Exposure was assessed by calculating historical magnetic fields based on the historical current load on the line, the height of the towers and ordering and distance between phases for every power line of 11 kV or greater. The exposure metric was validated by comparing calculated contemporary fields with magnetic fields measured for 65 schoolchildren who wore personal dosimeters for 24 h. The validation study showed a good agreement between the measured and the calculated magnetic fields. Time-weighted average (TWA) exposure to calculated magnetic fields as well as calculated magnetic fields closest in time to diagnosis were categorized into the groups < 0.05 µT, 0.05–< 0.14 µT and ≥ 0.14 µT. The odds ratio was computed by conditional logistic regression models for matched sets. Effect modifiers considered in additional analyses included socioeconomic status based on the occupation of the father (from the National Central Bureau of Statistics), type of building and number of dwellings. The risk for all cancers combined at TWA exposure ≥ 0.14 µT was estimated to be 0.9 (95% CI, 0.5–1.8) based on 12 exposed cases and 51 exposed controls. The odds ratios for the different types of cancer were 0.3 (95% CI, 0.0–2.1; 1 case) for leukaemia, 0.7 (95% CI, 0.2–2.1; 4 cases) for brain tumour, 2.5 (95% CI, 0.4–16; 2 cases) for lymphoma and 1.9 (95% CI, 0.6–6.0; 5 cases) for cancers at other sites. At a TWA exposure of ≥ 0.2 µT, the odds ratio for children with leukaemia was 0.5 (95% CI, 0.1–2.2; 2 cases). On the basis of the magnetic field exposure closest in time to diagnosis, the odds ratios at ≥ 0.14 µT were generally close to unity (brain tumour, 1.1 (95% CI, 0.5–2.5; 9 cases), lymphoma, 1.2 (95% CI, 0.2–6.4; 2 cases), leukaemia, 0.8 (95% CI, 0.3–2.4; 4 cases), all cancers combined, 1.3 (95% CI, 0.8–2.2; 24 cases), with the exception of cancers at other sites where the odds ratio was 2.5 (95% CI, 1.1–5.9; 9 cases). The cancer risk in relation to calculated magnetic fields of ≥ 0.14 µT during the first year of a child's life was 0.8 (95% CI, 0.1–7.1; 1 case) for leukaemia, 2.3 (0.8–6.6; 7 cases) for tumours of the central nervous system and 2.0 (0.9–4.2; 12 cases) for all cancers combined. A distance from nearby power lines of ≤ 50 m was associated with a significantly enhanced odds ratio of 2.8 (95% CI, 1.5–5.0; 23 cases) for tumours at other sites, but the odds ratio was not significantly different for leukaemia (0.6; 95% CI, 0.3–1.3; 9 cases), brain tumour (0.8; 95% CI, 0.4–1.6; 14 cases) or lymphoma (1.9; 95% CI, 0.6–6.4; 5 cases). The exposure to electric fields was also calculated, but since shielding between houses and power lines was not accounted for, the figures were not used in the risk analysis.

Linet et al. (1997) conducted a large study of acute lymphoblastic leukaemia in children in nine mid-western and mid-Atlantic states in the USA between 1989 and 1994 (the NCI (National Cancer Institute)/CCG (Children's Cancer Group) study). The eligible patients were less than 15 years of age and resided in one of the nine states at the time of diagnosis. Controls were selected by random-digit dialling and matched to the cases on age, ethnicity and telephone exchange. Exposure assessment consisted of spot measurements of magnetic fields in three rooms under normal and low-power conditions and outside the front door, and a 24-h measurement made under the child's bed. Wire codes were also noted. For children under the age of five years, measurements were taken in homes that they had occupied for at least six months, if the residences available for measurements collectively accounted for at least 70% of the child's lifetime from conception to the date of diagnosis. For children over the age of five years, measurements were made for a maximum of two homes occupied during the five years prior to diagnosis, and these homes had to account for at least 70% of the five-year 'etiological time period'. For the wire-coding portion of the study, one dwelling was selected that accounted for at least 70% of the child's lifetime (children < 5 years) or at least 70% of the five years prior to diagnosis for children ≥ 5 years. Thus, this study focused on residentially stable children, particularly for the wire-code part of the study. Measurements were made in the homes of 638 cases and 620 controls (78% and 63% response rates, respectively, according to the eligibility criteria described by Kleinerman et al., 1997). The homes of 408 matched pairs were wire coded. Subjects who refused to participate further in the study after the telephone interview that collected data on residential history were included in the wire-coding portion of the study if they had an eligible current or former dwelling. The main exposure metric consisted of a TWA summary measure based on the 24-h measurement and the indoor spot measurements taken in multiple residences, if applicable. The measurements were weighted by an estimate of the time spent in each room, made in a separate personal dosimetry study (Friedman et al., 1996). The metric was divided into four a-priori cut-points based on the distribution of measurements in the control group. When compared with children who were exposed to magnetic fields < 0.065 μT, the odds ratios for exposure to 0.065–0.099 μT, 0.10–0.199 μT and ≥ 0.2 μT were 1.1 (95% CI, 0.81–1.5; 123 cases), 1.1 (95% CI, 0.83–1.5; 151 cases) and 1.2 (95% CI, 0.86–1.8; 83 cases), respectively, using unmatched analyses. Matched analyses resulted in a slightly higher estimate for the highest exposure category (odds ratio, 1.5; 95% CI, 0.91–2.6; 58 cases). The risk was elevated when the category of magnetic fields of 0.3 μT and above was considered (odds ratio, 1.7; 95% CI, 1.0–2.9; 45 cases), but the trend was not statistically significant, and the odds ratio for magnetic fields of ≥ 0.5 μT was near unity in the matched analysis. There were no significantly elevated risks when exposure during pregnancy was considered. Measurements in the homes that were occupied during pregnancy were made for 257 cases and 239 controls. There was no positive association between wire codes and childhood leukaemia (odds ratio for VHCC versus underground and VLCC, 0.88; 95% CI, 0.48–1.6; 24 cases). [The Working Group noted that the low response rate of the controls was a limitation of this study.]

Hatch et al. (2000) conducted a re-analysis of the National Cancer Institute/ Children's Cancer Group study to evaluate internal evidence for selection bias. Certain characteristics of the subjects who did not allow in-home measurements or interviews (partial participants) were compared with those of subjects who did allow a data collector inside their homes (complete participants). The partial participants were found to be more likely to have annual incomes of < $ 20 000 (23% versus 12%), mothers who were unmarried (25% versus 10%), a lower education level (46% versus 38%) and were less likely to live in a single-family home (58% versus 83%) than complete participants. When partial participants were excluded from the analysis of measured fields, the odds ratios for magnetic fields of ≥ 0.3 μT increased from 1.6

(95% CI, 0.98–2.6) to 1.9 (95% CI, 1.1–3.3). When partial participants were excluded from the wire-code analysis, the odds ratios for VHCC (versus UG/VLCC) increased from 1.0 (95% CI, 0.62–1.6) to 1.2 (95% CI, 0.74–2.0). If the non-participants had similar characteristics to partial participants, the National Cancer Institute/Children's Cancer Group study may have overestimated risk estimates due to selection bias. [The Working Group noted that this publication included both complete and partial participants. The risk estimates differed slightly due to small differences in the study populations included and to differences in the variables adjusted for.]

Auvinen *et al.* (2000) carried out an exploratory analysis of the National Cancer Institute/Children's Cancer Group study data using alternative magnetic field exposure metrics. The analysis was restricted to 515 cases of acute lymphoblastic leukaemia and 516 controls who had lived in one home for at least 70% of the time-period of interest. Subjects with Down syndrome were excluded. Measures of the central tendency, peak values, the percentage of time above various thresholds and the short-term variability of the 24-h bedroom measurements were also assessed. A weak positive association was found between acute lymphoblastic leukaemia and measures of the central tendency, particularly when night-time exposure was assessed. For example, when the 30th percentile values of the 24-h measurements were examined, the odds ratios for the highest versus the lowest category (90th% versus < 50th%) were 1.4 (95% CI, 0.87–2.2) for the 24-h measurements and 1.7 (95% CI, 1.1–2.7) for the night-time measurements. Little evidence for any association with peak exposure, thresholds or variability was found.

Kleinerman *et al.* (2000) examined data from the National Cancer Institute/Children's Cancer Group study in relation to distance from power lines and an exposure index which took into account both distance and relative load for high-voltage and three-phase primary power lines. Most of the subjects (601/816; 74%) had lived more than 40 m from a high-voltage or three-phase primary power line. The odds ratio for living within 14 m of a potentially high-exposure line was 0.79 (95% CI, 0.46–1.3) and that for the highest category of the exposure index (mean magnetic field in homes, 0.213 µT), described above, was 0.98 (95% CI, 0.59–1.6).

Measurements of magnetic fields were included in a population-based case–control study of leukaemia in children under the age of 15 years in Germany. The study area was at first restricted to north-western Germany (i.e. Lower Saxony) (Michaelis *et al.*, 1997), but was extended to include the metropolitan area of Berlin (Michaelis *et al.*, 1998), before the first part was completed. Patients in whom leukaemia was diagnosed between 1988 and 1993 (for Lower Saxony) or 1991 and 1994 (for Berlin) were identified by the nationwide German Childhood Cancer Registry. In the Lower Saxony part of the study, two controls per case were selected randomly from the files for registration of residents. One control was matched for sex, date of birth and community; a second control was matched only for sex and date of birth, but drawn at random from any community in Lower Saxony, taking the population size of each community into account. In the Berlin part of the study, one control per case matched according to sex, date of birth and district within the city was randomly selected from the Berlin population registry. Measurements of the magnetic field were also made for patients with tumours of the central nervous system (Schüz et al., 2001b), but no controls were selected specifically for this diagnostic group. A total of 176 children with leukaemia, 64 with tumours of the central nervous system and 414 controls participated in the study (Michaelis et al., 1998; Schüz et al., 2001b). The response rates were 62% (176/283) for cases and 45% (414/919) for controls. In both parts of the study, measurements of the magnetic field over 24 h were performed in the child's bedroom and in the living room of the dwelling where the child had lived for longest before the date of diagnosis. Additional spot measurements were made in all dwellings where the child had lived for more than one year. All measurements were made between 1992 and 1996. The main analysis was based on the median magnetic field in the child's bedroom, with 0.2 μT as a cut-point. Post-hoc exposure metrics included the mean of the spot measurements, and the magnetic field during the night (22:00 to 06:00, extracted from the 24-h measurement). The odds ratios were derived from a logistic regression analysis stratified for age, sex and part of Germany (East-Berlin versus West-Berlin and Lower Saxony) and were adjusted for socioeconomic status and degree of urbanization. For the analysis of tumours of the central nervous system, the sample of controls selected for the leukaemia cases was used in unconditional logistic regression models adjusted for age, sex, socioeconomic status and degree of urbanization. Information on a variety of potential confounders was available. The odds ratio for median magnetic fields ≥ 0.2 μT compared with fields of < 0.2 μT was 2.3 (95% CI, 0.8–6.7; 9 exposed cases and 8 exposed controls) for leukaemia and 1.7 (95% CI, 0.3–8.8); 2 exposed cases) for tumours of the central nervous system. The association with leukaemia was more pronounced for children aged four years or younger (odds ratio, 7.1 (95% CI, 1.4–37; 7 cases, 2 controls) and for all children exposed to median magnetic fields ≥ 0.2 μT during the night (odds ratio, 3.8; 95% CI, 1.2–12; 9 cases, 5 controls). No association was seen with spot measurements; spot measurements and 24-h measurements showed a poor agreement. It is also of interest that more of the stronger magnetic fields were caused by low-voltage field sources than by overhead power lines. [The Working Group noted that selection bias is a cause for concern due to the high proportion of non-participants.]

To assess the risk of childhood cancer from exposure to ELF electric and magnetic fields, Dockerty et al. (1998) conducted a population-based case–control study in New Zealand. The study base consisted of children under the age of 15 years diagnosed from 1990–1993 with leukaemia or a solid tumour. Children with cancer were identified from the New Zealand Cancer Registry, the New Zealand Children's Cancer Registry or the computerized records of admissions and discharges from public hospitals. The controls were selected at random from national birth records and one control was matched to each case on age (same quarter of the birth year) and sex. Only cases and controls resident in New Zealand and not adopted were included. Altogether, 344 children with cancer were eligible for inclusion in this study, 131 of whom had leukaemia. Household measurements were made for 115 leukaemia patients and 117 controls, resulting in 113 matched pairs (86%). The response rate among first-choice controls was 69%. Measurements of the magnetic field and the electrical field were conducted over 24 h in two rooms of the dwelling; one was the room in which the child slept at night and one was the room in which the child spent most of his or her day. The two measurements were taken on subsequent days so that the parents had to move the measurement instrument from one room to another. A log sheet was used to record the times and dates on which the instrument was started and moved. Analyses using conditional logistic regression models were performed for thirds of the empirical distributions of the exposure metrics for electric fields and, for magnetic field measurements, for categories 0.1–< 0.2 µT and ≥ 0.2 µT compared with < 0.1 µT. The confounders considered in the analyses included mother's education, maternal smoking during pregnancy, residence of the child on a farm, home ownership status, number of people in the household, residential mobility, mother's marital status and season of the measurement. Risk estimates were presented for a subset of 40 matched pairs for which, two years before the date of diagnosis, both the leukaemia case and the matched control lived in the same house in which the measurements were subsequently made. For leukaemia, the adjusted odds ratios for magnetic fields ≥ 0.2 µT compared with < 0.1 µT were 16 (95% CI, 1.1–224; based on 5 exposed cases and 1 exposed control) for the bedroom measurement and 5.2 (95% CI, 0.9–31; based on 7 exposed cases and 3 exposed controls) for the daytime room measurement. The respective odds ratios for the highest third electric field (≥ 10.75 V/m) compared with the lowest third (< 3.64 V/m) were 2.3 (95% CI, 0.4–13) and 2.5 (95% CI, 0.3–18). Dockerty et al. (1999) re-analysed the above data by combining daytime and night-time magnetic fields to produce TWA magnetic fields. The odds ratio for magnetic fields ≥ 0.2 µT decreased to 3.3 (95% CI, 0.5–24) based on the same 40 matched pairs. The analyses of all 113 matched pairs showed no association with exposure to magnetic fields ≥ 0.2 µT (odds ratio, 1.4; 95% CI, 0.3–6.3). [The Working Group noted that risk estimates for leukaemia were presented for only 35% of the matched pairs included in the study.]

McBride et al. (1999) conducted a prospective case–control study of childhood leukaemia in five Canadian provinces (Alberta, British Columbia, Manitoba, Quebec and Saskatchewan) from 1990–95. Cases were identified through paediatric oncology treatment centres in each province and provincial cancer registries for all provinces except Quebec. Children under 15 years of age in whom leukaemia had been diagnosed and who resided in census tracts within 100 km of major cities were eligible for the study. A total of 445 potentially eligible cases were identified, and 399 of them were interviewed (90%). In-home measurements were made for 67% of the total eligible cases. The controls were identified from health insurance rolls (and family allowance rolls for the first two years of the study period in Quebec) and were matched by age, sex and area to cases. Of the 526 eligible controls, 399 were interviewed (76%) and in-home measurements were made for 65% of the total. Exposure was assessed by personal monitoring for 48 h as well as by 24-h stationary measurements in the child's bedroom if the home lived in before diagnosis was still occupied by the child at the date of interview. Wire codes were assigned and outdoor measurements at current and former dwellings (except for apartments more than four storeys high) were made. The personal exposure of subjects in their former dwellings was assessed from outdoor (perimeter) measurements in conjunction with wire codes. The model was based on analyses from currently occupied residences. For children under three years of age, homes occupied for at least three months, and for children over three years of age, homes lived in for six months or more were eligible for exposure assessment. Cases were ascertained retrospectively for one year, and prospectively thereafter; thus, most of the measurements were taken relatively close in time to the diagnosis or reference date. The potential confounding variables that were assessed included outdoor temperature at the time of measurement, family history of cancer, occupational and recreational exposure of parents, exposure to ionizing radiation and socioeconomic factors. The results of personal monitoring gave no indication of a positive association between risk for leukaemia and increasing exposure to magnetic fields, whether based on contemporaneous measures, a measure of estimated exposure two years before the reference date, or estimated lifetime exposure. For the highest exposure category of the contemporary measures ($\geq$ 90th percentile or ≥ 0.27 μT versus < 50th percentile or < 0.08 μT), the unadjusted odds ratio was 0.78 (95% CI, 0.46–1.3), based on 32 exposed cases and 37 exposed controls. Similar results were found when estimated exposure from former residences was included in the exposure assessment. For the 24-h bedroom measurements, the odds ratio for $\geq$ 90th percentile was 1.3 (95% CI, 0.69–2.3) compared to < 50th percentile. For wire codes, when VHCC was compared with UG and VLCC, the adjusted odds ratio was 1.2 (95% CI, 0.58–2.3), using the residence at the reference date. There was also no association found between childhood leukaemia and measured electric fields. [The Working Group considered that the limitations of this study include relatively low response rates for controls and a higher proportion of controls than of cases who had not moved home since diagnosis.]

Green *et al.* (1999b) carried out a case–control study of childhood leukaemia in the greater Toronto area of Ontario, Canada. Eligible children were under 15 years of age at diagnosis, treated at the Hospital for Sick Children (the only children's hospital in the greater Toronto area) between 1985 and 1993 and still resident in the study catchment area when the study was conducted (1992–95). Patients were identified through a paediatric oncology registry in Ontario. All subjects had to have lived in the study area at the diagnosis or reference date, to ensure comparability in terms of residential stability. A total of 298 children were identified of whom 256 were approached and 203 [68%] interviewed. Controls were selected from a random sample of 10 000 published telephone numbers. A total of 4180 numbers were called and 1133 households were found to have eligible children and be willing to participate. [The number of eligible persons who refused at this stage was not stated.] Of the 1133 potential controls, 645 (two controls per case) were randomly selected and matched by age and sex. A total of 419 (65%) of the 645 controls approached were interviewed. The assessment of exposure to magnetic fields included spot measurements made in the child's bedroom under normal-power conditions and in two other rooms frequently used by the child, outside measurements around the perimeter of the house, personal monitoring and wire codes. Personal monitoring and in-home measurements were used only if the current residence was occupied before diagnosis or the comparable reference date for the controls. Bedroom measurements were taken for 152 cases [51% of those originally identified] and 300 controls (47% of those approached). The results were analysed using conditional logistic regression and measurements were divided into quartiles according to the distribution among the controls. For bedroom measurements, the adjusted odds ratio for all leukaemias for the highest quartile (≥ 0.13 µT versus < 0.03 µT) was 1.1 (95% CI, 0.31–4.1). Similarly, for the average of interior measurements, the odds ratio was 1.5 (95% CI, 0.44–4.9) for the highest versus lowest quartile. For the exterior measurements, which were taken for a greater number of residences (183 cases and 375 controls), the odds ratios for the second quartile (4.1; 95% CI, 1.3–13 for 0.03–0.07 µT) and for the fourth quartile (3.5; 95% CI, 1.1–11 for ≥ 0.15 µT) were elevated compared with the lowest quartile. There was no association between wire-code and leukaemia incidence. The results for the personal exposure monitoring, based on only 88 cases (34%) and 113 controls (18%) were published separately (Green *et al.*, 1999a). There was a significantly increased risk for all childhood leukaemias for the third (0.07 µT–0.14 µT) and fourth (≥ 0.14 µT) quartiles of magnetic field exposure, compared with the lowest quartile (< 0.03 µT). The odds ratios were 4.0 (95% CI, 1.1–14) and 4.5 (95% CI, 1.3–16) for the third and fourth quartile, respectively, after adjustment for average power consumption, family income, residential mobility, exposure of the child to chemicals and birth order. The odds ratios for electric fields measured by personal dosimetry were mostly below unity. [The Working Group noted that the limitations of the study include the low response rates, especially for the personal monitoring part of the study, and that measurements were taken many years after the time period of interest. The use of published telephone listings raises concern about the comparability of cases and controls.]

The United Kingdom Childhood Cancer Study (UKCCS) was a population-based case–control study covering the whole of England, Wales and Scotland (UK Childhood Cancer Study Investigators, 1999). The study population was defined as children under the age of 15, registered with one of the Family Health Service Authorities (England and Wales) or with one of the Health Boards (Scotland). The prospective collection of cases with a pathologically confirmed malignant disease began in 1992 (except in Scotland where it began in 1991) and ended in 1994 (except in England and Wales, where cases with leukaemia were collected throughout 1996 and cases with non-Hodgkin lymphoma throughout 1995). For each case, two controls, matched for sex and date of birth, were selected randomly from the list of the same Family Health Service Authorities or Health Board as the case. For the study of electric and magnetic fields, only one control per case was chosen. At first, the family of the control with the lower identification number of the two controls was approached and, in case of non-participation or ineligibility, a second control family was chosen. Case and control families were ineligible for the electric and magnetic field part of the study if they had moved house during the year before diagnosis or lived in a mobile home. A total of 3838 cases (87% of all eligible cases in the UK Childhood Cancer Study) were included, and at least one of the parents was interviewed. A total of 7629 controls were included and the participation rate was 64%. Measurements were made for 2423 cases and 2416 controls; 2226 matched pairs (50% of all cases [37% of half of the controls]) were available for analysis. Of the 2226 cases, 1073 had leukaemia, 387 had cancer of the central nervous system and 766 had another malignant disease. The protocol for exposure assessment was specifically designed to estimate the average magnetic fields to which the subjects had been exposed in the year before diagnosis and measurements were made in the homes of all participants during the first phase of exposure assessment. These first-phase measurements comprised a 1.5-h stationary measurement in the centre of the main family room and three spot measurements at different places in the child's bedroom, which were repeated after the 1.5-h measurement. During the household visits, the parents were asked about potential sources of exposure, e.g. night storage heaters, and about the amounts of time the child spent in his or her room and at school. An exposure assessment was carried out in the child's school where relevant. In the second phase of exposure assessment, a measurement over a period of 48 h was conducted in homes where the first measurement had indicated magnetic fields $\geq 0.1\ \mu T$, where a potential source of exposure had been identified during the first visit or where an external source of exposure had been reported on a questionnaire that had been completed by the regional electricity companies. If the potential field source was assumed to have a seasonal variability, the dwelling was revisited during the winter months. The dwellings of the families of matched cases or controls where there were potential sources of exposure were also revisited in the second phase. An algorithm was developed to calculate the TWA exposure to magnetic fields on an individual basis, including the magnetic field strengths measured in the bedroom, in other rooms of the dwelling and at school, but with different weightings for each child according to the amount of time he or she had spent in each place. For participants for whom only first-phase measurements had been made, exposure in the bedroom was estimated from the spot measurements and exposure outside the bedroom was estimated from the 1.5-h measurement made in the main family room. For participants for whom long-term measurements had been made, exposure in the bedroom was estimated from the 48-h measurement and exposure outside the bedroom from the 1.5-h measurement conducted during the first visit. To allow for changes in line-loading and circuit configuration between the year of diagnosis and the time of measurement, the exposure measurements were adjusted to take into account calculated historical magnetic fields. Exposure was divided into four groups ($< 0.1\ \mu T$, $0.1-< 0.2\ \mu T$, $0.2-< 0.4\ \mu T$ and $\geq 0.4\ \mu T$) and into three categories with cut-points at 0.1 μT and 0.2 μT, respectively. Additional adjustments were performed for a census-derived deprivation index based on unemployment, over-crowding and car ownership in the appropriate district. The odds ratios were presented separately for leukaemia, cancers of the central nervous system, other malignant diseases and all cancers combined. At magnetic fields ≥ 0.2 μT, all the adjusted odds ratios were below unity (leukaemia, odds ratio, 0.90 (95% CI, 0.49–1.6), cancer of the central nervous system, odds ratio, 0.46 (95% CI, 0.11–1.9), other types of cancer, odds ratio, 0.97 (95% CI, 0.46–2.1), all cancers combined, odds ratio, 0.87 (95% CI, 0.56–1.4)). At ≥ 0.4 μT, the odds ratio for leukaemia was slightly elevated (1.7; 95% CI, 0.40–7.1), based on five exposed cases and three exposed controls. No association with magnetic fields of ≥ 0.4 μT was seen for cancer of the central nervous system (no exposed cases) or other malignant diseases (odds ratio, 0.71; 95% CI, 0.16–3.2; three exposed cases). For the intermediate category 0.2–< 0.4 μT, all odds ratios were below unity or, in the case of other malignant diseases, very close to unity (leukaemia, odds ratio, 0.78 (95% CI, 0.40–1.5), cancer of the central nervous system, odds ratio, 0.70 (95% CI, 0.16–3.2), other malignant disease, odds ratio, 1.1 (95% CI, 0.45–2.6)). Adjustment for deprivation index had only a small effect on the risks and risk did not vary according to age. [The Working Group considered that the main limitation of this study was the low proportion of subjects for whom fields were measured.]

In a second approach, the study examined distance from external sources of electric and magnetic fields (UK Childhood Cancer Study Investigators, 2000a). These data were available for nearly 90% of the children eligible for the study of electric and magnetic fields. Separate odds ratios were calculated for different types of overhead power line (11- and 20-kV, 33-kV, 66-kV, 132-kV, 275-kV and 400-kV) and for different types of underground high-voltage cable (33-kV, 132-kV and 275-kV), substations and low-voltage circuits. The only association seen was between leukaemia and 66-kV overhead power lines (odds ratio, 3.2; 95% CI, 1.0–9.7; 5 cases), although associations with other sources of field, including stronger ones, were close to unity. The magnetic fields associated with power lines were also calculated for all dwellings on the basis of line-load data for the period of interest. For magnetic fields of ≥ 0.4 μT, the odds ratio was decreased for leukaemia (0.27; 95% CI, 0.03–2.2), but only one case and eight controls were classified as being exposed. No excess risk for any type of malignancy was seen with exposure to magnetic fields ≥ 0.2 μT.

Bianchi *et al.* (2000) conducted a small case–control study in Italy. The study areas were the municipalities within the Province of Varese that were crossed by high-voltage power lines. A total of 103 children under the age of 15 years diagnosed with leukaemia between 1976 and 1992 were identified by the Lombardy Cancer Registry and four healthy controls per case were selected randomly from the 1996 lists of Health Service Archives. A total of 101 cases and 412 controls were available for analysis. The average magnetic fields for subjects living within 150 m of a power line were calculated from data on the power load for the year 1998. In addition, spot measurements at the entrance of the dwelling were conducted in a validation study.

Odds ratios obtained from logistic regression analysis stratified for age and sex, revealed considerable increases in leukaemia risk from exposure to magnetic fields in the ranges 0.001–0.1 μT (odds ratio, 3.3; 95% CI, 1.1–9.7, 6 cases) and > 0.1 μT (odds ratio, 4.5; 95% CI, 0.88–23, 3 cases), compared to exposure to fields < 0.001 μT (92 cases). [The Working Group noted that current data on power load were used to esti-mate historical magnetic fields up to 22 years in the past. The highest exposure group was defined at a very low cut-point (> 0.1 μT) and even then comprised only a few subjects. Furthermore, cases and controls were enrolled from two different recruitment periods.]

Schüz et al. (2001a) reported the results of a large-scale population-based case–control study covering the whole of the former West Germany. A total of 514 patients with acute leukaemia aged less than 15 years were identified from the German Child-hood Cancer Registry from 1990–94, and 1301 controls from population registration files were included. Measurements of magnetic fields were made in 1997–99. [Overall participation rates were 51% among cases and 41% among controls]. Of those families who were asked for permission to conduct measurements, 66% (520/783) responded. The exposure assessment was similar to that of the first German study (Michaelis et al., 1998; see above), except that spot measurements were conducted only to identify the source of strong magnetic fields and were not part of the risk analysis. The main exposure metrics were the median magnetic field over 24 h and the median magnetic field during the night. Odds ratios were calculated using logistic regression models adjusted for sex, age, year of birth, social class and degree of urbanization. The odds ratio for 24-h median magnetic fields ≥ 0.2 μT was 1.6 (95% CI, 0.65–3.7; 9 exposed cases and 18 exposed controls). An elevated risk for leukaemia was observed for night-time exposure with an odds ratio of 3.2 (95% CI, 1.3–7.8; 12 exposed cases and 12 exposed controls). At a cut-point of ≥ 0.4 μT, the odds ratio for median magnetic fields increased to 5.8 (95% CI, 0.78–43), but was based on only three exposed cases and three exposed controls. Odds ratios were altered only slightly when the analyses were restricted to residentially stable children. The association was strongest for children aged four years or younger. Two exposed children with Down syndrome had median and night-time exposure > 0.2 μT. The exclusion of children with Down syndrome from the analyses led to a decrease in the odds ratio at ≥ 0.2 μT to 1.3 (95% CI, 0.49–3.2) (7 exposed cases) for median magnetic fields and to 2.8 (95% CI, 1.1–7.0) (10 exposed cases) for night-time expo-sure, the increase in the latter was still statistically significant. [The Working Group noted that the study had two limitations, the low participation rate and the very long time lag between date of diagnosis and date of measurement.]

A pooled analysis of the two German studies (Michaelis et al., 1998; Schüz et al., 2001a) resulted in an increase in the odds ratios for leukaemia in children exposed to 24-h median magnetic fields ≥ 0.4 μT to 3.5 (95% CI, 1.0–12; 7 cases). No associations were seen for the intermediate exposure categories of 0.1–< 0.2 μT (odds ratio, 1.1; 95% CI, 0.73–1.6; 43 cases) and 0.2–< 0.4 μT (odds ratio, 1.2; 95% CI, 0.55–2.6;

11 cases), compared with the baseline < 0.1 μT (629 cases). A dose–response relation-ship was observed for median magnetic fields during the night, with respective odds ratios of 1.3 (95% CI, 0.90–2.0; 44 cases), 2.4 (95% CI, 1.1–5.4; 14 cases) and 4.3 (95% CI, 1.3–15; 7 cases) for the exposure categories 0.1–< 0.2 μT, 0.2–< 0.4 μT and ≥ 0.4 μT, respectively (p value for trend < 0.01) (Schüz et al., 2001a).

The study also examined exposure to $16 \, ^2/_3$-Hz magnetic fields, which is the frequency used by the German railway system (Schüz et al., 2001c). Magnetic fields ≥ 0.2 μT at this frequency were measured in less than 1% of all dwellings. Considering this additional exposure in the main analysis changed the results only marginally, thus, neglecting magnetic fields at this frequency is not likely to affect studies of residential electric and magnetic fields.

(d)   Pooled analyses

(i)   Ahlbom et al. (2000) reported a pooled analysis of studies that examined the relation between childhood leukaemia and residential magnetic fields (Table 22). They included all studies except one (London et al., 1991) in which long-term indoor measurements had been reported and that were completed before 2000 (Linet et al., 1997; Michaelis et al., 1998; Dockerty et al., 1998; 1999; McBride et al., 1999; UK Childhood Cancer Study Investigators, 1999) and all studies that reported calculations of historical exposure to ELF magnetic fields (Feychting & Ahlbom, 1993; Olsen et al., 1993; Verkasalo et al., 1993; Tynes & Haldorsen, 1997). The analysis strategy was defined a priori. The greatest emphasis was placed on the geometric mean of the child's exposure measured in the bedroom in the most recent home inhabited before or at diagnosis. Exposure was categorized into the groups < 0.1 μT, 0.1–< 0.2 μT, 0.2–< 0.4 μT and ≥ 0.4 μT. The potential effect modifiers that were considered in an additional analysis included type of house, residential mobility, social group (or mother's education or family income), degree of urbanization and exposure to car exhaust. The study population comprised 3247 children with leukaemia, of whom 2704 had acute lymphoblastic leukaemia, and 10 400 controls, all under the age of 15 years. Due to the study protocol described above, the results for the single studies within this pooled analysis sometimes differed from the results originally reported for the same study. These differences were greatest for the US study (Linet et al., 1997), the Canadian study (McBride et al., 1999) and the United Kingdom study (UK Childhood Cancer Study Investigators, 1999). The pooled analysis modified the data of Linet et al. (1997) as follows: homes in which 24-h measurements had not been made were excluded; expo-sure measured in the year prior to diagnosis, rather than five years immediately prior to diagnosis were used, and arithmetic means were replaced by geometric means. The changes to the original Canadian results (McBride et al., 1999) made for the pooled analysis meant that exposure assessments from fixed-location in-home measurements were used instead of measures of exposure recorded with personal dosimeters. The original United Kingdom results (UK Childhood Cancer Study Investigators, 1999) modified for the pooled analysis used the geometric mean from the 1.5/48-h

Table 22. Pooled analysis of total leukaemia in children

Type of study	0.1–<0.2 µT	0.2–<0.4 µT	≥0.4 µT	O	E	Continuous analysis
Measurement studies						
Canada (McBride et al., 1999)	1.3 (0.84–2.0)	1.4 (0.78–2.5)	1.6 (0.65–3.7)	13	10.3	1.2 (0.96–1.5)
Germany (Michaelis et al., 1998)	1.2 (0.58–2.6)	1.7 (0.48–5.8)	2.0 (0.26–15)	2	0.9	1.3 (0.76–2.3)
New Zealand (Dockerty et al., 1998, 1999)	0.67 (0.20–2.2)	4 cases/0 controls	0 cases/0 controls	0	0	1.4 (0.40–4.6)
United Kingdom (UKCCSI, 1999)	0.84 (0.57–1.2)	0.98 (0.50–1.9)	1.0 (0.30–3.4)	4	4.4	0.93 (0.69–1.3)
USA (Linet et al., 1997)	1.1 (0.81–1.5)	1.0 (0.65–1.6)	3.4 (1.2–9.5)	17	4.7	1.3 (1.0–1.7)
Calculated field studies						
Denmark (Olsen et al., 1993)	2.7 (0.24–31)	0 cases/8 controls	2 cases/0 controls	2	0	1.5 (0.85–2.7)
Finland (Verkasalo et al., 1993)	0 cases/19 controls	4.1 (0.48–35)	6.2 (0.68–57)	1	0.2	1.2 (0.79–1.7)
Norway (Tynes & Haldorsen, 1997)	1.8 (0.65–4.7)	1.1 (0.21–5.2)	0 cases/10 controls	0	2.7	0.78 (0.50–1.2)
Sweden (Feychting & Ahlbom, 1993)	1.8 (0.48–6.4)	0.57 (0.07–4.7)	3.7 (1.2–11.4)	5	1.5	1.3 (0.98–1.7)
Summary						
Measurement studies	1.1 (0.86–1.3)	1.2 (0.85–1.5)	1.9 (1.1–3.2)	36	20.1	1.2 (1.0–1.3)
Calculated field studies	1.6 (0.77–3.3)	0.79 (0.27–2.3)	2.1 (0.93–4.9)	8	4.4	1.1 (0.94–1.3)
All studies	1.1 (0.89–1.3)	1.1 (0.84–1.5)	2.0 (1.3–3.1)	44	24.2	1.2 (1.0–1.3)

From Ahlbom et al. (2000)

The results of the pooled analysis show relative risks (95% CI) by exposure level and with exposure as continuous variable (relative risk per 0.2 µT) with adjustment for age, sex and socioeconomic status (measurement studies) and residence (in East or West Germany). The reference level is < 0.1 µT. Observed (O) and expected (E) case numbers at ≥ 0.4 µT are shown, with expected numbers given by modelling the probability of membership of each exposure category based on distribution of controls including covariates.
UKCCSI, UK Childhood Cancer Study Investigators

measurements rather than the TWA of the measurement protocol. The investigators of the Finnish cohort study (Verkasalo et al., 1993) provided a sample of 1027 controls drawn from the cohort.

To estimate a summary relative risk across centres in this pooled analysis, a logistic regression model was applied to the raw data, with study centres represented as effect modifiers. This was performed separately for measurement studies and studies of calculated fields, but also across all studies. Across the measurement studies, the summary relative risk was estimated at 1.9 (95% CI, 1.1–3.2) in the highest exposure category (≥ 0.4 μT). The two intermediate categories had relative risks close to unity (0.1–< 0.2 μT: relative risk, 1.1; 95% CI, 0.86–1.3; 0.2–< 0.4 μT: relative risk, 1.2; 95% CI, 0.85–1.5). The corresponding summary relative risks for the studies of calculated fields were 1.6 (95% CI, 0.77–3.3) in the category 0.1–< 0.2 μT, 0.79 (95% CI, 0.27–2.3) in the category 0.2–< 0.4 μT, and 2.1 (95% CI, 0.93–4.9) in the category ≥ 0.4 μT. The summary relative risks across all studies were also close to unity (0.1–< 0.2 μT: relative risk, 1.1; 95% CI, 0.89–1.3; 0.2–< 0.4 μT: relative risk, 1.1; 95% CI, 0.84–1.5), but in the highest category (≥ 0.4 μT), the summary relative risk was 2.0 (95% CI, 1.3–3.1) with a respective p value < 0.01. A similar analysis was conducted on continuous expo-sure, and the resulting relative risk per 0.2 μT interval was 1.2 (95% CI, 1.0–1.3). A homogeneity test based on the continuous analysis across all nine centres revealed that the variation in point estimates between the studies was not larger than would be expected from random variability. Subsequent sensitivity analysis confirmed that the observed association between leukaemia and stronger magnetic fields was not due to the choice of exposure metric (geometric mean) or the definition of cut-points, and was not strongly influenced by any of the studies. Consideration of potential confounders did not materially affect the risk estimates. The summary relative risks for acute lymphoblastic leukaemia only were similar to those obtained for total leukaemia. While the relative risks for the intermediate exposure categories were 1.1 (95% CI, 0.88–1.3) for the category 0.1–< 0.2 μT and 1.1 (95% CI, 0.84–1.5) for the 0.2–< 0.4 μT category, the relative risk for the highest exposure category (≥ 0.4 μT) showed a twofold increase (2.1; 95% CI, 1.3–3.3).

A comparison was made in the pooled analysis between the number of observed cases and the number of expected cases under the null hypothesis at ≥ 0.4 μT. In three studies, no excess leukaemia cases were observed; these were the United Kingdom study (4 observed, 4.4 expected cases), the Norwegian study (0 observed, 2.7 expected) and the New Zealand study (0 observed, 0 expected). The summary numbers across all studies were 44 observed cases compared with 24.2 expected cases.

Another finding of this pooled analysis related to the so-called wire code paradox. In earlier reviews, it had been observed that there was a stronger association between surrogates for exposure to ELF electric and magnetic fields and leukaemia risk than between direct measurements and leukaemia risk. The new studies did not support this. The summary relative risk of the US (Linet et al., 1997) and Canadian studies

(McBride *et al.*, 1999) combined for the highest wire-code category was 1.2 (95% CI, 0.82–1.9) which was lower than that in the measurement or calculated field studies.

(ii) Greenland *et al.* (2000) reported a pooled analysis of 16 studies of childhood leukaemia and residential magnetic fields, based on either magnetic field measurements or wire codes. In contrast to the pooled analysis by Ahlbom *et al.* (2000), this analysis also included studies that relied only on wire codes for exposure assessment as well as some of the earlier studies which were smaller and less methodologically sound than more recent studies. The additional studies not included by Ahlbom *et al.* (2000) were those by Wertheimer and Leeper (1979), Fulton *et al.* (1980), Tomenius (1986), Savitz *et al.* (1988), London *et al.* (1991), Coghill *et al.* (1996), Fajardo-Gutiérrez *et al.* (1997) and Green *et al.* (1999a,b). The study carried out in the United Kingdom (UK Childhood Cancer Study Investigators, 1999) was not included in this pooled analysis, and from the study by Green *et al.* (1999a,b), only wire-code data were included. Eight of the studies (Coghill *et al.*, 1996; Linet *et al.*, 1997; London *et al.*, 1991; Michaelis *et al.*, 1998; Savitz *et al.*, 1988; Tomenius, 1986; Dockerty *et al.*, 1998; McBride *et al.*, 1999) provided some direct measurements of magnetic fields; four studies from the Nordic countries (Feychting & Ahlbom, 1993, Sweden; Olsen *et al.*, 1993, Denmark; Verkasalo *et al.*, 1993, Finland; Tynes & Haldorsen, 1997, Norway) were based upon calculated historical fields. Most studies provided multiple measurements. The a-priori measurement chosen for this pooled analysis was the best approximation of TWA exposure up to three months before diagnosis. Magnetic field strengths were categorized into groups ≤ 0.1 µT, $> 0.1–\leq 0.2$ µT, $> 0.2–\leq 0.3$ µT and > 0.3 µT. Data were analysed using maximum likelihood logistic regression and tabular methods. For the wire code analyses, the referent group consisted of low wire codes (underground [UG], VLCC and ordinary low current [OLCC] combined). For the measurement analysis, the combined results of the 12 studies gave relative risks of 1.01 (95% CI, 0.84–1.2), 1.06 (95% CI, 0.78–1.4) and 1.7 (95% CI, 1.2–2.3) for $> 0.1–\leq 0.2$, $> 0.2–\leq 0.3$ and > 0.3 µT compared with < 0.1 µT, respectively, using Mantel-Haenszel summary estimates and adjusting for study, age and sex. Restricting the studies to those with complete covariate data resulted in very similar estimates. The relative risks were 1.01 (95% CI, 0.82–1.3), 0.94 (95% CI, 0.65–1.4) and 2.1 (95% CI, 1.4–3.0) for $> 0.1–\leq 0.2$, $> 0.2–\leq 0.3$ and > 0.3 µT, respectively, using Mantel-Haenszel summary estimates, and adjusting for age, sex and socioeconomic variables. For the analysis of wire codes, summary estimates were not given for all of the studies, because of a great deal of heterogeneity within the study results, ranging in relative risks for VHCC of < 1 in three studies to > 2 in three studies (homogeneity $p = 0.005$). Eliminating the two earliest studies, which had extreme results, the summary relative risks were 1.02 (95% CI, 0.87–1.2) and 1.5 (95% CI, 1.2–1.9) for OHCC and VHCC, respectively, based on six studies with wire code data. Covariate adjustment had little effect on these results. As with the pooled analysis of Ahlbom *et al.* (2000), the 'wire-code paradox' was not evident, since measured fields showed stronger associations with childhood leukaemia than did wire codes. The two pooled analyses reached similar conclusions.

2.2.2   *Exposure to ELF electric and magnetic fields from electrical appliances* (Table 23)

Seven studies have examined the relationship between use of household electrical appliances and all childhood cancers, childhood leukaemia or tumours of the brain and nervous system. The first study, as described above (Savitz *et al.*, 1988), was conducted in Denver, CO, USA (Savitz *et al.*, 1990). A total of 252 children with cancer, identified through a tumour registry and area hospitals, and 222 controls, identified by random-digit dialling, were interviewed. The response rates were 70.8% for cases (252/356 eligible cases) and 79.9% for eligible controls (222/278). Maternal use of appliances during pregnancy and the use of appliances by the children in the study were assessed. Results for four appliances were presented: electric blankets, heated water beds, bedside electric clocks and bed-heating pads. For ever-use of electric blankets during pregnancy, the adjusted odds ratio was 1.7 (95% CI, 0.8–3.6; 13 exposed cases) for leukaemia and 2.5 (95% CI, 1.1–5.5; 11 exposed cases) for brain cancer in children. Slightly stronger effects were noted when use during the first trimester of pregnancy was considered (leukaemia, odds ratio, 2.3 (95% CI, 1.0–5.8), 9 cases; brain cancer, odds ratio, 4.0 (95% CI, 1.6–9.9), 9 cases) and for more hours of use, i.e. > 8 h versus < 8 h (leukaemia, odds ratio, 11 (95% CI, 1.8–67), 4 cases; brain cancer, odds ratio, 4.6 (95% CI, 0.5–39), 1 case). No significant associations were found for childhood use of electric appliances. Electric blankets had been used in childhood by only 13 cancer cases and eight control children; the odds ratio was 1.5 (95% CI, 0.5–5.1) for leukaemia and 1.2 (95% CI, 0.3–5.7) for brain cancer. Odds ratios for use of electrically heated water beds and hair dryers were mostly below one and those for bedside electric clocks were slightly elevated, but not significant (odds ratio for total cancer, 1.3; 95% CI, 0.8–2.2). [The Working Group noted that a potential problem of the study, in addition to the possible selection bias described previously, is that parents of cases and controls were interviewed many years after the time period of interest.]

The second study to include use of electric appliances as part of assessment of exposure to magnetic fields was conducted in Los Angeles, CA, USA (London *et al.*, 1991, 1993). Two hundred and thirty-two children with leukaemia and 232 matched controls were interviewed. There was no indication of any important associations between maternal use of electrical appliances during pregnancy and risk for childhood leukaemia, but there were several significantly elevated odds ratios for use of appliances during childhood. Exposure during childhood was defined as use at least once per week in comparison to no use of the appliance. For black-and-white televisions, the odds ratio was 1.5 (95% CI, 1.0–2.2); that for use of hair dryers was 2.8 (95% CI, 1.4–6.3). Elevations in risk were also seen for use of electric dial clocks (odds ratio, 1.9; 95% CI, 0.97–3.8), curling irons (odds ratio, 6.0; 95% CI, 0.72–105), electric blankets (odds ratio, 7.0; 95% CI, 0.86–122) and video games (odds ratio, 1.6; 95% CI, 0.8–3.3). [The Working Group noted that because of the small numbers of appliance users, no attempt had been made to define high- or low-exposure groups.]

Table 23. Case–control studies of childhood cancer in relation to use of electrical appliances

Reference, area	Study size and cancer site	Exposure	No. of cases	Risk estimates: odds ratios (95% CI)	Comment
Savitz et al. (1990), Denver, CO, USA	252 cases, 222 controls, aged 0–14 years, diagnosed 1976–83	**Electric blankets**			Unadjusted odds ratios. Evidence of effect modification by income; no consistent evidence for increased risks with water beds, bedside electric clocks or heating pads; study vulnerable to selection bias due to differential residential restrictions placed on cases versus controls
	All cancers (233 cases) (244 cases)	Prenatal use	38	1.1 (0.7–1.8)	
		Postnatal use	13	1.5 (0.6–3.4)	
	Leukaemia (70 cases) (73 cases)	Prenatal use	13	1.3 (0.7–2.6)	
		Postnatal use	4	1.5 (0.5–5.1)	
	Brain cancer (45 cases) (47 cases)	Prenatal use	11	1.8 (0.9–4.0)	
		Postnatal use	2	1.2 (0.3–5.7)	
London et al. (1991), Los Angeles, CA, USA	232 cases of leukaemia, 232 controls, aged 0–10 years, diagnosed 1980–87	**Electric blankets**			No evaluation by frequency and/or duration of use of appliances; assessment of use made many years after etiological time-period
		Prenatal use	23	1.2 (0.66–2.3)	
		Postnatal use	7	7.0 (0.86–122)	
		Water beds			
		Prenatal use	14	0.67 (0.34–1.3)	
		Postnatal use	12	1.0 (0.45–2.3)	
		Television (black and white)			
		Postnatal use	64	1.5 (1.0–2.2)	
		Hair dryer			
		Postnatal use	31	2.8 (1.4–6.3)	
McCredie et al. (1994), Australia	82 cases of brain tumour, 164 controls, aged 0–14 years, diagnosed 1985–89	**Postnatal**			No assessment of dose–response and only two appliances considered
		Electric blankets	6	0.4 (0.2–1.2)	
		Water beds	1	0.2 (0–1.5)	

Table 23 (contd)

Reference, area	Study size and cancer site	Exposure	No. of cases	Risk estimates: odds ratios (95% CI)	Comment
Preston-Martin et al. (1996a), Los Angeles, CA	298 cases of brain tumour, 298 controls, aged 0–19 years, diagnosed 1984–91	**Electric blankets**			Slightly, non-significantly elevated risks for electric heat: prenatal (odds ratio, 1.6; 95% CI, 0.8–3.0), postnatal (odds ratio, 1.3; 95% CI, 0.7–2.4); some indication of higher risks among cases diagnosed in earlier time-period, suggesting possible control selection bias
		Prenatal use	20	1.2 (0.6–2.2)	
		Postnatal use	11	1.2 (0.5–3.0)	
		Water beds			
		Prenatal use	23	2.1 (1.0–4.2)	
		Postnatal use	8	2.0 (0.6–6.8)	
		Television (black and white)			
		Postnatal use	20	0.7 (0.4–1.4)	
		Hair dryer			
		Postnatal use	55	1.2 (0.7–2.1)	
Gurney et al. (1996), Seattle, WA	133 cases of brain tumour, 270 controls; aged 0–19 years, diagnosed 1984–90	Electric blankets			Some elevated odds ratios for childhood use of digital clocks, black-and-white television, incubators, and baby monitors; no association for electric heat
		Prenatal use	20	0.9 (0.5–1.6)	
		Postnatal use	6	0.5 (0.2–1.4)	
		Water beds			
		Prenatal use	20	0.7 (0.4–1.3)	
		Postnatal use	8	0.8 (0.3–1.9)	

Table 23 (contd)

Reference, area	Study size and cancer site	Exposure	No. of cases	Risk estimates: odds ratios (95% CI)	Comment
Hatch et al. (1998), 9 mid-western and mid-Atlantic states	651 cases of acute lymphoblastic leukaemia, 651 matched controls, aged 0–14 years, diagnosed 1989–93	**Electric blankets**			Dose–response trends by frequency and duration of use of appliances were not apparent; results may have been affected by recall bias
		Prenatal use	91	1.6 (1.1–2.3)	
		Postnatal use	45	2.8 (1.5–5.0)	
		Sewing machines			
		Prenatal use	198	0.76 (0.59–0.98)	
		Television (<4 ft vs ≥6 ft [1.2 vs ≥1.8 m] from TV)			
		Prenatal use	17	1.9 (0.79–4.5)	
		Postnatal use	166	1.6 (1.1–2.4)	
		≥6 h vs <2 h/day			
		Postnatal use	178	2.4 (1.5–3.8)	
		Hair dryer			
		Postnatal use	266	1.6 (1.2–2.1)	
Dockerty et al. (1998), New Zealand	303 cancer cases, 303 controls, aged 0–14 years, diagnosed 1990–93	**Electric blankets**			Adjusted odds ratios. No assessment of dose–response trends by amount of use
	Leukaemia (121 cases)	Prenatal use	30	0.8 (0.4–1.6)	
		Postnatal use	17	2.2 (0.7–6.4)	
	Central nervous system cancer (58 cases)	Prenatal use	18	1.6 (0.6–4.3)	
		Postnatal use	8	1.6 (0.4–7.1)	
	Other solid tumours (124 cases)	Prenatal use	35	1.8 (0.9–3.5)	
		Postnatal use	26	2.4 (1.0–6.1)	

TV, television

McCredie *et al.* (1994) included an assessment of use of electric blankets and water beds in a study of childhood brain tumours (ICD-9, 191, 192) in New South Wales, Australia. A total of 97 eligible children aged 0–14 years and diagnosed with a brain tumour between 1985 and 1989, were identified from a population-based cancer registry for the areas of Sydney, Wollongong and Newcastle. Eighty-two (85%) of the mothers of children with cancer were interviewed. Potential control mothers were identified from electoral rolls in a two-phase selection process. Sixty per cent (400/672) of the mothers of eligible control children agreed to be interviewed and 164 of them were interviewed. Childhood use of electric blankets and water beds were the only potential sources of exposure to magnetic field assessed in this study. The odds ratio was 0.4 (95% CI, 0.2–1.2) for regular use of an electric blanket and 0.2 (95% CI, 0–1.5) for regular use of a water bed.

Preston-Martin *et al.* (1996b) studied the use of electrical appliances in relation to risk for childhood brain tumours (ICD-9 191, 192). Children aged 0–19 years, with brain tumours diagnosed between 1984 and 1991 were identified from three population-based cancer registries on the West Coast of the USA (Los Angeles County, five counties of the San Francisco area and 13 counties in Washington State including Seattle) for the years 1984–91. Controls were identified by random-digit dialling and were frequency-matched by age and sex to the case group. Mothers of a total of 540/739 cases (73%) and 801/1079 controls (74%) were interviewed about their use of electric blankets and electrically heated water beds during pregnancy and about use by the child after birth. No association of brain tumours with in-utero exposure to electric blankets (odds ratio, 0.9; 95% CI, 0.6–1.2) or use by the child (odds ratio, 1.0; 95% CI, 0.6–1.7) was found. There was also no effect of in-utero exposure resulting from use of water beds by the mother (odds ratio, 0.9; 95% CI, 0.6–1.3) or of use of water beds by children (odds ratio, 1.2; 95% CI, 0.7–2.0). When the analysis was restricted to the Los Angeles county (Preston-Martin *et al.* (1996a), the odds ratios for electric blankets were 1.2 (95% CI, 0.6–2.2) and 1.2 (95% CI, 0.5–3.0) for in-utero and postnatal exposure, respectively, and for water beds, the odds ratios were 2.1 (95% CI, 1.0–4.2) and 2.0 (95% CI, 0.6–6.8), respectively.

In a subset of the study population from the Seattle area (98 cases and 208 controls), Gurney *et al.* (1996) reported small, but non-significant elevations in risk for brain tumours associated with childhood use of portable black-and-white televisions (odds ratio, 1.6; 95% CI, 0.6–3.9), bedside digital clocks (odds ratio, 1.8; 95% CI, 0.9–3.3) and incubators (odds ratio, 1.5; 95% CI, 0.8–3.1), but no elevations in risk were associated with maternal use of appliances during pregnancy. In another subset of 133 cases and 270 controls, no association was seen for prenatal or postnatal exposure to electric blankets or water beds.

Hatch *et al.* (1998) examined both prenatal and postnatal use of appliances in the National Cancer Institute/Children's Cancer Group Study in the USA as described above. Interview data on the use of electrical appliances was available for 788 children, aged 0–14 years, with acute lymphoblastic leukaemia diagnosed between 1989 and 1993

[88% response] and 699 controls [64% response], providing 651 matched pairs. The use of several appliances during the prenatal period was significantly associated with the occurrence of acute lymphoblastic leukaemia, but there was no evidence of a dose–response effect. For ever- versus never-use by the mother during pregnancy, the odds ratios for the offspring were 1.6 (95% CI, 1.1–2.3) for electric blankets, 1.5 (95% CI, 1.0–2.1) for bed-heating pads, 1.4 (95% CI, 1.0–2.0) for humidifiers and 0.76 (95% CI, 0.59–0.98) for sewing machines. Some significant associations with childhood leukaemia were also found with use of electrical appliances during childhood, based on the mother's report. Ever-use of an electric blanket prior to the reference date was associated with an odds ratio of 2.8 (95% CI, 1.5–5.0), but the highest risk was found for the shortest duration of use in years (odds ratio for < 1 year of use, 5.5; 95% CI, 1.1–26). Similarly, the odds ratio for ever-use of a hair dryer was 1.6 (95% CI, 1.2–2.1), but the highest risk was for children who had used one hair dryer for less than one year (odds ratio, 2.5; 95% CI, 1.3–4.9). There was some suggestion of effects for video arcade games (odds ratio, 1.7; 95% CI, 1.2–2.3) and video games connected to televisions (odds ratio, 1.9; 95% CI, 1.4–2.7), but no indication of increased risks associated with use of a personal computer (odds ratio, 1.2; 95% CI, 0.83–1.7). The risk increased with increasing amount of time spent watching television (odds ratio for ≥ 6 h per day versus < 2 h per day, 2.4; 95% CI, 1.5–3.8), but these effects were seen regardless of the reported distance that the child sat from the television.

Dockerty *et al.* (1998) included assessment of exposure to electrical appliances in a nationwide study of childhood cancer in New Zealand (described above) (303 cases, 303 controls). There was little evidence for any relationship between maternal use of electrical appliances in pregnancy and childhood cancer. The odds ratios for use of electric blankets were 0.8 (95% CI, 0.4–1.6) for leukaemia, 1.6 (95% CI, 0.6–4.3) for cancers of the central nervous system and 1.8 (95% CI, 0.9–3.5) for other solid tumours. For childhood use of appliances, there was some suggestion of an increased risk associated with the use of an electric blanket. The odds ratios were 2.2 (95% CI, 0.7–6.4) for leukaemia, 1.6 (95% CI, 0.4–7.1) for tumours of the central nervous system and 2.4 (95% CI, 1.0–6.1) for other solid tumours. There was also the suggestion of an effect for electric heating, but only in the room occupied during the day (odds ratio, 1.8; 95% CI, 0.9–3.5), not in the child's bedroom (odds ratio, 1.0; 95% CI, 0.5–2.3).

2.2.3   *Parental exposure to ELF electric and magnetic fields*

(a)   *Cohort study*

Feychting *et al.* (2000) conducted a cohort study on occupational exposure of parents to magnetic fields and cancer in offspring. Children born in Sweden in 1976, 1977, 1981 and 1982 were followed until 1993, and those who developed cancer before the age of 15 years were identified. A total of 522 children with cancer including 161 with leukaemias and 162 with cancer of the central nervous system were identified. The occupations of their mothers and fathers were taken from data recorded in the 1975 and

1980 censuses. The percentages of parents without a recorded job were 27.1% for mothers and 5.4% for fathers. The likelihood of occupational exposure to electric and magnetic fields was quantified through use of a job–exposure matrix. For children whose mothers had been exposed to magnetic fields ≥ 0.19 μT (third quartile) or ≥ 0.26 μT (90th percentile), the relative risks for all types of tumour were close to unity (relative risk, 1.1 (95% CI, 0.7–1.4) and relative risk, 1.1 (95% CI, 0.7–1.7), respectively). For children whose fathers had been exposed to magnetic fields ≥ 0.3 μT, the risk for leukaemia was elevated (relative risk, 2.0; 95% CI, 1.1–3.5) and the risk for cancers of the central nervous system was less than unity (relative risk, 0.5; 95% CI, 0.3–1.0).

    (b)   Case–control studies

    In a case–control study of 157 children, less than 15 years of age, who had died of neuroblastoma during 1964–78 in Texas, USA, Spitz and Johnson (1985) reported an elevated risk for neuroblastoma (odds ratio, 2.1; 95% CI, 1.1–4.4) among the children of electrical workers. Data on parental occupation at birth of the child were abstracted from the birth certificate, and exposure was inferred from occupational title.

    A subsequent hospital-based study on the incidence of neuroblastoma in Ohio, USA of 101 incident cases of neuroblastoma in children < 15 years old born during 1942–67 and 404 controls (Wilkins & Hundley, 1990) made use of information on paternal occupation from birth certificates to infer exposure, but found no association between employment of the father in an electrical occupation and risk of neuroblastoma in the offspring.

    Bunin *et al.* (1990) conducted a small case–control study of neuroblastoma in 104 children diagnosed from 1970–79 at two hospital-based tumour registries in North-east USA. One hundred and four controls were selected by random-digit dialling. Data on parental occupation were obtained by telephone interview and exposure to electric and magnetic fields was classified using the same scheme as that used by Spitz and Johnson (1985). No association was seen between neuroblastoma in offspring, and exposure of fathers employed as electricians, insulation workers or power utility workers during the preconception period (odds ratio, 0.3; 95% CI, 0.1–1.2) or mother's exposure during pregnancy (same occupational groups as for fathers) (0.3; 95% CI, 0.1–1.3).

    Nasca *et al.* (1988) conducted a case–control study of children with cancer and parental occupation. Three hundred and thirty-eight children (aged 0–14 years) with a primary tumour of the central nervous system diagnosed between 1968 and 1977 in 53 New York counties were included. Six hundred and seventy-six controls matched by age and geographical location were also selected. Parents were interviewed by telephone to obtain job information. Exposure was classified according to occupational title. No association was seen between cancer of the central nervous system in offspring and parental exposure to electric and magnetic fields before the birth of the child (odds ratio, 1.6; 95% CI, 0.83–3.1).

Wilkins and Koutras (1988) in Ohio, USA, conducted a case–control study of morta-lity from brain cancer during 1959–78. The study population included 491 offspring (< 20 years of age) of men whose job title suggested occupational exposure to electric and magnetic fields. An elevated risk of brain cancer was seen in the children of men involved in electrical assembly, installation and repairing occupations (odds ratio, 2.7; 95% CI, 1.2–6.1).

Johnson and Spitz (1989) conducted a mortality case–control study of all children under the age of 15 years who had died in Texas, USA from 1964–80 of intracranial and spinal cord tumours (499 cases, 998 controls). Data on parental occupation collected at birth of the children were used to infer exposure. For all occupational categories thought to involve potential exposure of parents to ELF electric and magnetic fields, the risk was marginally elevated (odds ratio, 1.6; 95% CI, 0.96–2.8) in the offspring.

Parental occupation as a risk factor for astrocytoma in children aged 0–14 was exa-mined by Kuijten et al. (1992). The patients were identified through tumour registries in eight hospitals in Pennsylvania, New Jersey and Delaware (USA) and included all cases diagnosed from 1980–86. Controls were selected by random-digit dialling and were pair-matched to cases by age, race and telephone exchange. The mothers and fathers of the 158 case–control pairs were interviewed by telephone, and exposure to electric and magnetic fields was inferred from job title. In general no associations with childhood astrocytoma were seen; however, in a sub-analysis, men reported as being 'electrical repairing workers' during the preconception period had a significantly elevated risk of fathering a child who later developed astrocytoma (odds ratio, 8.0; 95% CI, 1.1–356).

Wilkins and Wellage (1996) identified 94 patients aged 20 years or less with tumours of the central nervous system who were diagnosed during the years 1975–82 through the Columbus Children's Hospital Tumor Registry (USA). Random-digit dialling was used to select 166 controls from the 48-county referral area of the registry. For fathers who had occupations presumed to have resulted in exposure to electric and magnetic fields during the period before conception, no elevated risk of cancer of the central nervous system was noted in their offspring. However, exposure of the father working in welding-related jobs during preconception was associated with an elevated risk (odds ratio, 3.8; 95% CI, 0.95–16).

2.3   Cancer in adults

2.3.1   *Residential exposure to ELF electric and magnetic fields*

In addition to the many methodological considerations discussed in other sections, including the lack of studies that have included a comprehensive assessment of expo-sure, residential studies of adults present unique difficulties. These problems are:

— the contribution of occupational exposure — not considered in most studies;
— the lack of assessment of other sources of exposure likely to be important for adults who spend only a fraction of their time at home;

—   the long latency period for most adult malignancies, often necessitating assessment (owing to residential mobility) in several residences;

—   the need to use proxy response for deceased cases; and

—   low participation rates.

The assessment of exposure in most of the following studies was based either on proximity to electrical installations or on simple questions regarding appliance use. Few studies included spot measurements in several locations. Even long-term residential measurements are unlikely to capture the strength or variability of daily exposure for working adults. In a 1000-person study, Zaffanella and Kalton (1998) found that occupational exposure was often significantly higher and more variable than other sources of exposure; the highest mean and median exposure occurs at work, followed by exposure at home and during travel. Since most people spend much of their time at home, ignoring exposure either at home or at work is likely to lead to a large misclassification. In a small study of the use of household appliances, Mezei *et al.* (2001) found that a large proportion of total exposure for most adults is accumulated at home. Similarly, the 1000-person study found exposure at home to be moderately predictive of 24-h average exposure or of time spent in magnetic fields above 0.4 μT, but completely uncorrelated with maximum fields or with field changes.

The long latency of cancers in adults and the unknown biological mechanism necessitate estimation of exposure over long time periods, an exceptionally difficult task owing to the mobility and behavioural changes likely to occur with time. The situation is even more difficult for rapidly fatal diseases such as brain cancer about which information is generally obtained from numerous proxies.

Following the publication of the study by Wertheimer and Leeper (1979) suggesting an association between residential exposure to ELF magnetic fields and cancer in children (see p. 105), many studies have investigated the possible carcinogenic effects of electric and magnetic fields. Most of the epidemiological studies have focused on cancer in children (see section 2.2). Studies of adults have looked primarily at occupational exposure, but some have investigated residential settings. As shown in Table 24, which lists studies of residential adult cancer by exposure category, several studies have investigated links between the use of electric blankets and breast cancer. Many studies have examined proximity to power lines and cancer, focusing particularly on leukaemia and brain cancer, but studies in which a sophisticated assessment of exposure has been made are few.

The first study on residential exposure to ELF magnetic fields and adult cancer was conducted by Wertheimer and Leeper (1982) in the USA. [The Working Group noted that this was a hypothesis-generating paper, but its usefulness for hypothesis testing was compromised because of unblinded exposure assessment, potential overmatching for the Denver cases and the unusual and complex method for selection of cases and controls.]

Table 24. Residential studies of adult cancer by exposure category

Outcome	Exposure					
	Electric blanket	Other appliances	Proximity	Calculated fields	Spot measurements	Combined occupational and residential
Leukaemia						
Wertheimer & Leeper (1987)	—	—	✓	—	—	—
McDowall (1986)	—	—	✓	—	—	—
Coleman et al. (1989)	—	—	✓	—	—	—
Youngson et al. (1991)	—	—	✓	—	—	—
Schreiber et al. (1993)	—	—	✓	—	—	—
Severson et al. (1988)	✓	—	✓	—	—	—
Feychting & Ahlbom (1994)	—	—	✓	✓	✓	—
Feychting et al. (1997)	—	—	—	✓	✓	✓
Verkasalo et al. (1996)	✓	—	✓	✓	—	—
Li et al. (1997)	—	—	✓	—	—	—
Preston-Martin et al. (1988)	✓	—	✓	—	—	—
Lovely et al. (1994)	—	✓	—	—	✓	—
Sussman & Kheifets (1996)	—	✓	—	—	—	—
Brain						
Wertheimer & Leeper (1982, 1987)	—	—	✓	—	—	—
Schreiber et al. (1993)	—	—	✓	✓	—	✓
Feychting & Ahlbom (1994)	—	—	✓	✓	—	—
Feychting et al. (1997)	—	—	—	✓	—	—
Verkasalo et al. (1996)	—	—	✓	—	—	—
Li et al. (1997)	—	—	✓	—	—	—
Wrensch et al. (1999)	✓	✓	—	—	✓	—
Ryan et al. (1992)	✓	✓	—	—	—	✓
Mutnick & Muscat (1997)	—	—	—	—	—	—

Table 24 (contd)

Outcome	Exposure					
	Electric blanket	Other appliances	Proximity	Calculated fields	Spot measurements	Combined occupational and residential
Breast						
Wertheimer & Leeper (1982; 1987)	—	—	✓	—	—	—
McDowall (1986)	—	—	✓	—	—	—
Schreiber et al (1993)	—	—	✓	—	—	—
Verkasalo et al (1996)	—	—	—	✓	—	—
Li et al. (1997)	—	—	✓	✓	—	—
Coogan & Aschengrau (1998)	✓	✓	✓	✓	—	✓
Feychting et al. (1998)	—	—	—	✓	—	✓
Forssén et al. (2000)	—	—	—	—	—	✓
Vena et al. (1991, 1994, 1995)	✓	—	—	—	—	—
Gammon et al. (1998)	✓	—	—	—	—	—
Laden et al. (2000)	✓	—	—	—	—	—
Zheng et al. (2000)	✓	✓	—	—	—	—
Other cancers						
Wertheimer & Leeper (1982, 1987)	—	—	✓	—	—	—
Verkasalo et al. (1996)	✓	—	—	✓	—	—
Zhu et al. (1999)	✓	—	—	—	—	—

(a)   *Leukaemia*

Early studies of leukaemia focused mostly on the potential association between proximity to power lines and cancer development. From 1971–83, McDowall (1986) followed a cohort of 7631 people in East Anglia, England, who lived within 50 m of a substation or other electrical installation, or within 30 m of overhead power lines at the time of the 1971 census. Coleman *et al.* (1989) conducted a case–control study of leukaemia and residential proximity to electric power facilities in four London boroughs. Seven hundred and seventy-one leukaemia cases diagnosed between 1965 and 1980 were identified from a population-based cancer registry. In a matched case–control study, Youngson *et al.* (1991) investigated adult haematological malignancies in relation to overhead power lines; the study included 3144 adults with leukaemia identified from regional cancer registries in north-west England and Yorkshire; controls were selected from hospital discharge listings. Schreiber *et al.* (1993) investigated mortality and residence near electric power facilities in a retrospective cohort study of 3549 people who lived for five consecutive years between 1956 and 1981 in an urban quarter of Maastricht, The Netherlands. Koifman *et al.* (1998) investigated cancer clusters near power lines in Brazil; small numbers and other methodological problems make the study uninformative for evaluation, and it is mentioned here only for completeness. [The Working Group noted that although some of these studies indicated a small, non-significant elevation of risk, they are based on small numbers, low potential exposures and very crude exposure assessment methods. The overall results are non-informative.]

Several studies of adult leukaemia deserve special mention, including Severson *et al.* (1988), Feychting and Ahlbom (1994), with a follow-up study by Feychting *et al.* (1997), and studies by Verkasalo (1996), Verkasalo *et al.* (1996) and Li *et al.* (1997) (see Table 25).

Severson *et al.* (1988) conducted a case–control study of 164 adults, both living and deceased, diagnosed with acute non-lymphocytic leukaemia in the USA. The patients studied were aged from 20–79 years, diagnosed between 1981 and 1984 and recorded in a population-based cancer registry in western Washington State. The response rate was 70%. For controls, the response rate was 65%. One hundred and fourteen patients (or the next-of-kin if the patient had died) and 133 controls completed detailed questionnaires on residential history and use of electrical appliances. Three different methods were used to assess exposure. (1) The wire-coding scheme of Wertheimer and Leeper (1979) was used to classify all homes in the study area in which a subject had lived in the previous 15 years. Residential magnetic fields were also estimated according to a method developed by Kaune *et al.* (1987) using wiring configuration maps of dwellings. (2) Single measurements of indoor and outdoor magnetic fields were made at the time of the interview in a subject's home if the subject had lived there continuously for one year or longer immediately preceding the reference date (controls) or the date of diagnosis (cases). Measurements were made in the kitchen, the subject's bedroom and the family room, under both low-power (all possible appliances

Table 25. Design and results of epidemiological studies of residential exposure to ELF magnetic fields and adult leukaemia

Reference, country	Study base and subject identification	Exposure metrics	Results				Comments
Severson *et al.* (1988) USA	**Case selection**: ANLL cases aged 20–79 years, resident in western Washington state, from cancer registry (1981–84). 114 cases included in analyses (91 AML) **Control selection**: controls from random-digit dialling, matched on geographical area and frequency matched on age and sex. 133 controls included in analyses	Wertheimer and Leeper wire-coding. Estimation of magnetic fields from maps and wire coding — method of Kaune *et al.* (1987). Single measurements of 60-Hz magnetic fields inside (kitchen, bedroom, family room in HPC and LPC) and outside house; 24-h measurements in sample of houses. Electric appliance use from questionnaire	Mean exposure, low-power configuration					Refusal rate for measurements much higher among controls than cases. Single measurements made in only 56% of houses as many subjects had moved recently
				Ref.: ≤ 0.05 µT		OR (95% CI)	
			Single measurements	0.051–0.199 µT ≥ 0.2 µT		1.2 (0.52–2.6) 1.5 (0.48–4.7)	
			Weighted mean	0.051–0.199 µT ≥ 0.2 µT		1.2 (0.54–2.5) 1.0 (0.33–3.2)	
Feychting & Ahlbom (1994) Sweden	**Case selection**: All incident cancer cases from cancer registry (1960–85), from cohort of Swedish population aged ≥ 16 years, living on a property located within 300 m of any 220- or 400-kV power lines. 325 cases analysed (72 AML, 57 CML, 14 ALL and 132 CLL) **Control selection**: Two controls per case from same cohort. Matched on age, sex, parish and residence near same power line. 1091 controls in analysis	Distance to power lines from residence. In-home magnetic-field spot measurements under low- and high-power use conditions. Calculations were made of magnetic fields generated by power lines at the time of spot measurements (calculated contemporary fields) and for the year closest in time to diagnosis (calculated historical fields).	Calculated fields closest to time of diagnosis					Matched and unmatched analyses, adjusted or not for age and socioeconomic status were carried out. No information on other sources of residential exposure to electric and magnetic fields
				Ref.: ≤ 0.09 µT	No.	OR (95% CI)	
			All leukaemia	0.10–0.19 µT ≥ 0.2 µT	20 26	0.9 (0.5–1.5) 1.0 (0.7–1.7)	
			AML	0.10–0.19 µT ≥ 0.2 µT	5 9	1.0 (0.4–2.5) 1.7 (0.8–3.5)	
			CML	0.10–0.19 µT ≥ 0.2 µT	2 7	1.4 (0.5–3.3) 1.7 (0.7–3.8)	
			CLL	0.10–0.19 µT ≥ 0.2 µT	8 7	0.8 (0.4–1.7) 0.7 (0.3–1.4)	
Feychting *et al.* (1997) Sweden	Same as Feychting and Ahlbom (1994)	Same as above for residential. Occupational exposure from job–exposure matrix [developed from workday measurements made for another study] and information on occupation held in the year preceding the reference date	Subjects with both residential and occupational exposure					Same as above. Job–exposure matrix. Relevance especially for females unclear
				Ref.: ≤ 0.1 µT res. and < 0.13 µT occ.	No.	OR (95% CI)	
			All leukaemia	≥ 0.2 µT for both	9	3.7 (1.5–9.4)	
			AML	≥ 0.2 µT for both	3	6.3 (1.5–26)	
			CML	≥ 0.2 µT for both	3	6.3 (1.5–27)	
			CLL	≥ 0.2 µT for both	2	2.1 (0.4–10)	

Table 25 (contd)

Reference, country	Study base and subject identification	Exposure metrics	Results				Comments

Verkasalo et al. (1996) Finland

Study base: **Cohort** consisting of 383 700 persons (189 300 men) aged 20 years or older who contributed 2.5 million person–years of follow-up between 1970 and 1989

Case selection: All primary leukaemia cases (1974–89) living within 500 m of overhead power lines. 203 cases identified

Exposure metrics: Cumulative exposure. Estimates based on residential history, distance to 110–400 kV power line in 500 m corridor and calculated average annual magnetic fields for each building presumed to be $\geq 0.01\,\mu T$. Took into account current, typical locations of phase conductors and distance.

Cumulative exposure

Ref.: general population

	No.	SIR (95% CI)
All leukaemia $< 0.20\,\mu T$	156	0.96 (0.82–1.1)
$0.20–0.39\,\mu T$	23	1.1 (0.68–1.6)
$0.40–0.99\,\mu T$	15	0.87 (0.49–1.4)
$1.00–1.99\,\mu T$	5	0.81 (0.26–1.9)
$\geq 2.0\,\mu T$	4	0.71 (0.19–1.8)

Comments: Cohort study, SIRs. No information on other sources of residential exposure to electric and magnetic fields. No direct information from study subjects

Verkasalo (1996) Finland

Study base: **Case selection:** Same as Verkasalo et al. (1996): 196 leukaemia cases included (60 AML, 12 ALL, 30 CML, 73 CLL and 21 other or unknown subtype)

Control selection: 10 controls per case from cohort. Matched on sex and age at diagnosis and alive in the year of diagnosis of the case

Exposure metrics: Cumulative exposure: total and within 0–4, 5–9 and ≥ 10 years of diagnosis. Annual average magnetic fields 1–20 years prior to diagnosis. Highest annual average magnetic field ever and in time windows before diagnosis. Age at first exposure to annual average magnetic field greater than a specific level. Duration and time since exposure to annual averages above that level

Cumulative exposure

Ref.: $< 0.2\,\mu T$–years

	No.	OR (95% CI)
All leukaemia $\geq 2.0\,\mu T$–years	4	0.77 (0.28–2.2)
ALL $\geq 2.0\,\mu T$–years	none	
AML $\geq 2.0\,\mu T$–years	none	
CML $\geq 2.0\,\mu T$–years	none	
CLL $\geq 2.0\,\mu T$–years	3	1.7 (0.48–5.8)

Li et al. (1997)

Study base: **Case selection:** Pathologically confirmed incident cases of leukaemia from northern Taiwan from cancer registry (1987–92). 870 cases included in analyses

Control selection: One control per case from cancer registry excluding cancers of the brain and breast, haematopoietic and reticulo-endothelial system, skin, ovary, fallopian tube and broad ligament. Matched on date of birth, sex and date of diagnosis. 889 controls included in analyses

Exposure metrics: Distance from lines. Average and maximum magnetic fields assessed using distance from the lines, distance between wires, height of wires above the ground, annual and maximum loads along the lines from 1987–92, current phase and geographical resistivity of earth

Calculated exposure in year of diagnosis

Ref.: $< 0.1\,\mu T$

	No.	OR (95% CI)
All leukaemia $0.1–0.2\,\mu T$	47	1.3 (0.8–1.9)
$> 0.2\,\mu T$	97	1.4 (1.0–1.9)
ALL $0.1–0.2\,\mu T$	8	1.5 (0.7–3.2)
$> 0.2\,\mu T$	17	1.7 (1.0–3.1)
AML $0.1–0.2\,\mu T$	28	1.5 (0.9–2.5)
$> 0.2\,\mu T$	41	1.1 (0.7–1.7)
CML $0.1–0.2\,\mu T$	2	0.3 (0.1–1.2)
$> 0.2\,\mu T$	22	1.5 (0.9–2.6)
CLL $0.1–0.2\,\mu T$	4	2.8 (0.9–9.3)
$> 0.2\,\mu T$	3	0.6 (0.1–2.6)

Comments: Limited information on confounders because of restrictions on interview

ANNL, acute non-lymphocytic leukaemia; AML, acute myeloid leukaemia; ALL, acute lymphoblastic leukaemia; CLL, chronic lymphocytic leukaemia; SIR, standardized incidence ratio; HPC, high-power configuration; LPC, low-power configuration; res., residential; occ., occupational; Ref.: reference group with exposure level indicated; OR, odds ratio; CI, confidence interval; CML, chronic myeloid leukaemia;

turned off that could be (without overly disrupting the household) and high-power conditions (all such appliances switched on). (3) In a limited sample of dwellings, 24-h measurements were made. [However, neither details of the 24-h measurements nor the relevant results were given.] Cases tended to be of lower socioeconomic status than controls and were more likely to smoke or to have smoked in the past; these factors were adjusted for in subsequent analyses. No association was found between acute non-lymphocytic leukaemia and wire codes, either in the dwelling occupied for the longest period in the 3–10 years before the reference date or in the dwelling occupied closest to the reference date. There was also no association with TWA exposure to residential magnetic fields. For single measurements, available for only 56% of homes since many subjects had moved house after the reference date, a non-significant increase in odds ratio was found for mean exposures of $\geq 0.2 \, \mu T$ in both low-power (odds ratio, 1.5; 95% CI, 0.48–4.7) and high-power conditions (odds ratio, 1.6; 95% CI, 0.49–5.0). When weighted mean exposure was considered, the increase was no longer apparent in low-power conditions and was reduced in high-power conditions (odds ratio, 1.3; 95% CI, 0.35–4.5). [The Working Group noted that the participation rates in this study were low.]

Feychting and Ahlbom (1992a,b; 1994) conducted a nested case–control study of leukaemia and cancer of the central nervous system in a Swedish population who had lived for at least one year within 300 m of overhead 220- and 400-kV power lines between 1960 and 1985. The adult study population included 400 000 people ≥ 16 years of age, identified from the Population Registry, who lived on properties designated using maps from the Central Board for Real Estate Data, as being located within the power-line corridor. From this cohort, leukaemia cases were identified by record linkage with the Swedish Cancer Registry. Two controls for each case were selected at random from members of the cohort who had lived in the power-line corridor at least one year before the reference date (year of diagnosis of the case) and lived near the same power line as the corresponding case. Cases and controls were matched on age (within five years), sex, parish of residence and year of diagnosis. A total of 325 cases of leukaemia and 1091 controls were included in the analysis. Seventy-two of the cases had acute myeloid leukaemia, 57 chronic myeloid leukaemia, 14 acute lymphoblastic leukaemia, 132 chronic lymphocytic leukaemia and 50 had other types of leukaemia. In addition to spot measurements and distance from power lines, exposure metrics included estimated magnetic fields within residences as a function of their proximity to the lines. These fields were calculated from an engineering model that took into account past exposure (dating back to 1947, over more than three decades), physical dimensions of lines and their distance from a dwelling. The model served as the primary exposure index. Magnetic field strengths were estimated from calculations for the year of diagnosis, or the year closest to diagnosis if the subject had moved, as well as for one, five and 10 years before diagnosis. Cumulative exposure was also calculated by summing yearly averages for exposure to magnetic fields assigned to each of the 15 years before diagnosis. The study included information on age, sex, year of diagnosis, whether or not the subject resided in the county of Stockholm, type of housing and socioeconomic status. Some types of leukaemia were positively associated with fields calculated from the historical model and with proximity to the power line, but not with spot measurements. There was no association between the risk for all leukaemias and calculated exposure to magnetic fields closest to the time of diagnosis. For acute and chronic myeloid leukaemias, however, odds ratios were non-significantly increased for fields ≥ 0.2 µT compared with fields ≤ 0.09 µT. For acute myeloid leukaemia, the odds ratio, based on nine exposed cases, was 1.7 (95% CI, 0.8–3.5); for chronic myeloid leukaemia, the odds ratio, based on seven exposed cases, was 1.7 (95% CI, 0.7–3.8). For analyses based on calculated cumulative exposure during the 15 years preceding diagnosis, the odds ratios for all leukaemias were 1.0 (95% CI, 0.6–1.8) for cumulative exposures of 1.0–1.9 µT–years (16 cases), 1.5 (95% CI, 1.0–2.4) for ≥ 2.0 µT–years (29 cases) and 1.5 (95% CI, 0.9–2.6) for ≥ 3.0 µT–years (19 cases), in comparison with ≤ 0.99 µT–years. Odds ratios were increased for exposure ≥ 2.0 µT–years for acute myeloid leukaemia (odds ratio, 2.3; 95% CI, 1.0–4.6) (nine cases) and for exposure > 3.0 µT–years for chronic myeloid leukaemia (odds ratio, 2.7; 95% CI, 1.0–6.4) (6 cases). Adjustment for age and socioeconomic status had little effect on the results. Also, the results of matched analyses were similar to those of the unmatched analyses. For analyses based on spot measurements, odds ratios were close to unity for all categories of exposure and for all leukaemia subtypes, except for chronic myeloid leukaemia in the ≥ 0.2-µT category (odds ratio, 1.5; 95% CI, 0.7–3.2) (10 cases). [The Working Group noted that exposure assessment for leukaemia was complicated by the long time-period covered by the study, which necessitated estimation of field strengths going back 25 years or more.]

Feychting *et al.* (1997) conducted a follow-up study using the same study base together with information on occupation taken from censuses performed by Statistics Sweden every five years. For the occupation held in the year before the reference date, they assessed exposure based on a job–exposure matrix from a previous study (Floderus *et al.*, 1993, 1996). In that study, workday measurements had been made for a large number of jobs held by a sample of the general male population; consequently, no information was available on the occupations of 43% of the women. Combined analysis of residential and occupational exposure showed that subjects who had only residential exposure in the highest category (compared with 'unexposed' subjects with residential exposure < 0.1 µT and occupational exposure < 0.13 µT) had the following odds ratios: for acute myeloid leukaemia, 1.3 (95% CI, 0.4–5.0) (3 cases), and for chronic myeloid leukaemia, 0.5 (95% CI, 0.1–3.9) (1 case). [The very small number of cases prevents any interpretation of these results.] The odds ratios for subjects who had both high occupational and high residential exposure were much higher: for acute myeloid leukaemia, the odds ratio was 6.3 (1.5–26) and for chronic myeloid leukaemia the odds ratio was 6.3 (95% CI, 1.5–27), based on three exposed cases of each subtype). [The Working Group noted that the limitations of the previous study also apply to this one. The information on occupational exposure was difficult to interpret because of the limited applicability of the job–exposure matrix to this population.]

In a nationwide cohort study of 383 700 adults in Finland, Verkasalo et al. (1996) investigated cancer risk and exposure to magnetic fields in homes near high-voltage power lines. The cohort included all adults who had lived within 500 m of overhead power lines in homes with calculated magnetic field strengths of ≥ 0.01 µT at any time between 1970 and 1989. Through record linkage between nationwide data files (from the Finnish Cancer Registry, the Central Population Register, the 1970 Population Census, and the five Finnish power companies), information was obtained on cancer cases, residential history and residential exposure to magnetic fields. Follow-up took place from January 1974 until December 1989. Verkasalo (1996) presented a detailed case–control analysis of leukaemia. Of a total of 196 patients with leukaemia included in the study, 60 had acute myeloid leukaemia, 12 acute lymphoblastic leukaemia, 30 chronic myeloid leukaemia, 73 chronic lymphocytic leukaemia and 21 other, or unknown, subtypes. For each case, 10 controls were selected from the cohort and matched on sex, age at diagnosis of the case (within one year) and whether they were alive in the year of diagnosis. Several exposure measures were used. These included cumulative exposure and exposure 0–4, 5–9 and ≥ 10 years before diagnosis; annual average magnetic fields 1–20 years before diagnosis; highest annual average magnetic field 0–4, 5–9 and ≥ 10 years before diagnosis; age at first exposure to an annual average magnetic field greater than a specified strength; and duration of exposure and time since exposure to annual averages above that strength. No association was seen between the risk for all leukaemias or for specific subtypes and cumulative exposure or highest annual average exposure. Adjustment for type of housing or for occupational exposure (none versus possible or probable, based on expert judgement) did not affect the results. On the basis of three exposed cases, the study showed a significant increase in risk for chronic lymphocytic leukaemia with dichotomized cumulative exposure of ≥ 0.2 µT–years and ≥ 0.4 µT–years for ≥ 10 years before diagnosis (odds ratios, 2.8 (95% CI, 1.1–7.4) (9 cases) and 4.6 (95% CI, 1.4–15) (6 cases), respectively) and for duration of exposure to fields of ≥ 0.1 µT for ≥ 12 years (odds ratio, 4.8; 95% CI, 1.5–15) (3 cases). No association was observed for other types of leukaemia. [The Working Group noted that no measurements were made to validate the calculated fields in this study, and that the lack of information on other sources of residential exposure to electric and magnetic fields might have resulted in substantial exposure misclassification.]

Li et al. (1997) conducted a case–control study of leukaemia and other cancers in adults living in northern Taiwan. Cases and controls were ≥ 15 years of age and diagnosed with leukaemia between 1987 and 1992 and were selected from the National Cancer Registry of Taiwan. Controls were adults with cancers other than those potentially related to exposure to magnetic fields. Each case was matched with one control based on age, sex and date of diagnosis. Maps showing the location of each dwelling were available for only 69% of the study area; the lack of such maps was the primary reason for exclusion from the study. Power-company maps showed that 121 high-voltage power lines (69–345 kV) were operating in the study area between 1987 and 1992. The distance between each dwelling occupied by a study subject at the time of diagnosis and the nearest power line was measured from the maps with a precision of 10 m. Residential exposure was calculated from data supplied by the Taiwan Power Company that included distance between wires, height of wires above the ground, annual average and maximum loads and current phase. Calculated magnetic fields were validated by indoor measurements made with an EMDEX meter under low-power conditions (household power turned off) for 30–40 min in 407 residences. Questionnaire data on age, weight, height, educational level, smoking habits and previous exposure to X-rays were available for approximately one-third of study subjects. Information was obtained on potential confounding factors including urbanization (which took into consideration local population density), age, mobility, economic activity and family income, educational level and sanitation facilities. Of 1135 initial cases 870 incident cases of leukaemia were included in the analysis. [Not enough detail was provided to estimate the participation rate for controls.] The numbers of controls for cases of leukaemia living within 100 m and 50 m from the power lines were 10.9% and 5.4%, respectively. Of the controls, 9.9% had a calculated exposure of ≥ 0.2 µT and 5.6% had a calculated exposure of ≥ 0.5 µT. When the results were grouped into three exposure categories (< 0.1 µT, 0.1–0.2 µT and > 0.2 µT), the agreement (κ) between arithmetic means for measured and calculated fields was 0.64 (95% CI, 0.50–0.78). Compared with subjects living ≥ 100 m from the power lines, subjects who lived within 50 m of the lines had an odds ratio for leukaemia of 2.0 (95% CI, 1.4–2.9). For subjects whose homes were 50–99 m from the lines, the odds ratio was 1.5 (95% CI, 1.1–2.3). For calculated magnetic fields, the odds ratios for leukaemia were moderately elevated in the middle and highest exposure categories in the year of diagnosis: odds ratio, 1.3 (95% CI, 0.8–1.9) for exposure to 0.1–0.2 µT and odds ratio, 1.4 (95% CI, 1.0–1.9) for > 0.2 µT, compared with < 0.1 µT. A test for trend with increased exposure to magnetic fields was statistically significant ($p = 0.04$). [The Working Group noted the use of other cancer cases as controls and the low participation rate. Information on the power distribution systems near the dwellings of the study subjects was apparently unavailable. The ± 10-m precision of distance could have had a significant impact on calculations for dwellings within 20 m of power lines, but would contribute less error for those further away. Because the study was based on the dwelling occupied at the time of diagnosis, cumulative estimates of exposure to magnetic fields could not be calculated. Although examination of potential confounders in a subset of control subjects indicated little confounding from education, smoking, exposure to X-rays and reproductive factors, the authors were unable to adequately adjust for these risk factors for leukaemia.]

— *Appliance use*

A case–control study of leukaemia and use of electric blankets in the USA conducted by Preston-Martin *et al.* (1988) included patients aged 20–69 years, identified through the population-based Los Angeles County cancer registry, who had been diagnosed with histologically confirmed acute or chronic myeloid leukaemia between July 1979 and June 1985. Of 858 eligible cases, 485 who were still living were chosen, and permission to contact 415 of them was obtained from their physicians. [The Working Group noted that inclusion of only living patients might lead to bias, if exposure influences survival.] Completed questionnaires were available for 295 of the 415 patients, resulting in a participation rate of 61%. Each case was matched with one neighbourhood control on sex, race and birth year (within five years). [The authors did not give the response rate for controls, but stated that controls could not be found in three neighbourhoods.] In all, 293 matched pairs, including 156 cases of acute myeloid leukaemia and 137 of chronic myeloid leukaemia, participated in the study. Because questions on use of electric blankets were added after the study had begun, information on their use was available for only 224 matched pairs. The results indicated that use of electric blankets was not related to risk of leukaemia. For acute myeloid leukaemia, the odds ratio was 0.9 (95% CI, 0.5–1.6) and that for chronic myeloid leukaemia was 0.8 (95% CI, 0.4–1.6). Cases and controls did not differ with regard to average duration of use, year of first regular use, or number of years since last use. Adjustment for other significant risk factors did not change the results. [The Working Group noted that the study did not indicate whether blankets had been used only for pre-warming the bed or continuously throughout the night.]

The study by Severson et al. (1988), described above, used questionnaires to obtain information on ownership and use of 32 [Lovely et al., 1994] electrical domestic appliances. The study showed no association between risk of leukaemia and use of electric blankets, water-bed heaters or heated mattress pads. [The Working Group noted that participation rates in this study were low and limited information was available on use of electric blankets.]

The data from the study by Severson et al. (1988) were reanalysed by Lovely et al. (1994) and Sussman & Kheifets (1996). The bias due to the use of proxy respondents was noted by Sussman & Kheifets (1996) in the positive findings for the use of an electric razor (> 7.5 minutes/day) (odds ratio, 2.4; 95% CI, 1.1–5.5) reported by Severson et al. (1988).

(b)   Brain cancer

Few studies, summarized in Table 26, have investigated the potential association between adult brain cancer and residential exposure to ELF magnetic fields. [Although several studies of adult cancers have examined cancer of the brain or nervous system as a subtype, results have been unremarkable (Wertheimer & Leeper, 1982; Schreiber et al., 1993)]. Studies by Feychting and Ahlbom (1992a,b, 1994), Feychting et al. (1997), Verkasalo et al. (1996) and Li et al. (1997), which are described in detail in section (a), also analysed brain cancer risk.

The population-based, nested case–control study of Feychting and Ahlbom (1992a,b, 1994) investigated exposure to magnetic fields from high-voltage power lines

Table 26. Design and results of epidemiological studies of residential exposures to ELF magnetic fields and adult brain cancer

Reference, country	Study base and subject identification	Exposure metrics	Results	Comments
Feychting & Ahlbom (1994) Sweden	**Case selection**: All incident cancer cases from cancer registry (1960–85), from cohort of Swedish population aged ≥ 16 years, living on a property located within 300 m of any 220- or 400-kV power lines. 223 cases in analysis (66 astrocytoma I–II, 157 astrocytoma III–IV) **Control selection**: Two controls per case from same cohort. Matched on age, sex, parish and residence near same power line; 1091 controls in analysis	Distance to power lines from dwelling. In-home spot measurements of magnetic fields under low- and high-power use conditions. Calculations of the magnetic fields generated by the power lines at the time spot measurements were made (calculated contemporary fields), and for the year closest in time to diagnosis (calculated historical fields).	(see table below)	Matched and unmatched analyses, adjusted or not for age and socioeconomic status were carried out. No information on other sources of residential exposure to electric and magnetic fields
Feychting et al. (1997) Sweden	Same as Feychting and Ahlbom (1994)	Same as above for residential. Occupational exposure from job–exposure matrix [developed from workday measurements made for another study] and information on occupation held in the year before the reference date	(see table below)	Same as above. Job–exposure matrix. Relevance especially for females unclear
Verkasalo et al. (1996) Finland	**Cohort:** 383 700 persons (189 300 men) aged ≥ 20 who contributed 2.5 million person–years of follow-up between 1970 and 1989 **Case selection**: All primary brain cancer cases (1974–89) living within 500 m of overhead power lines; 301 cases identified	Cumulative exposure. Estimates based on residential history, distance to 110–400-kV power line in 500-m corridor and calculated average annual magnetic fields for each building presumed to be ≥ 0.01 µT. Takes into account current, typical locations of phase conductors and distance.	(see table below)	Cohort study, SIRs. No information on other sources of residential exposure to electric and magnetic fields. No direct information from study subjects. ICD-7 code 193

Feychting & Ahlbom (1994) — Calculated fields closest to time of diagnosis

		Ref.: ≤ 0.09 µT	No. of cases	OR (95% CI)
All CNS	0.10–0.19 µT		18	1.1 (0.7–2.0)
	≥ 0.2 µT		12	0.7 (0.4–1.3)
Astrocytoma I–II	0.10–0.19 µT		3	0.6 (0.1–1.8)
	≥ 0.2 µT		2	0.4 (0.1–1.3)
Astrocytoma III–IV	0.10–0.19 µT		15	1.4 (0.8–2.5)
	≥ 0.2 µT		10	0.8 (0.4–1.7)

Feychting et al. (1997) — Subjects with both residential and occupational exposure

		Ref.: ≤ 0.1 µT res. and < 0.13 µT occ.	No. of cases	RR (95% CI)
All CNS	≥ 0.2 µT for both		3	1.3 (0.3–4.8)
Astrocytoma I–II	≥ 0.2 µT for both		0	
Astrocytoma III–IV	≥ 0.2 µT for both		3	2.2 (0.6–8.5)

Verkasalo et al. (1996) — Cumulative exposure

	Ref.: general population	No. of cases	SIR (95% CI)
Nervous system	< 0.20 µT	238	0.94 (0.82–1.1)
	0.20–0.39 µT	35	1.1 (0.77–1.5)
	0.40–0.99 µT	16	0.64 (0.37–1.0)
	1.00–1.99 µT	5	0.55 (0.18–1.3)
	≥ 2.0 µT	7	0.92 (0.37–1.9)

Table 26 (contd)

Reference, country	Study base and subject identification	Exposure metrics	Results	Comments
Li et al. (1997)	**Case selection:** Pathologically confirmed incident cases of brain cancer from northern Taiwan from cancer registry (1987–92). 577 cases included in analyses. **Control selection:** One control per case from cancer registry excluding cancers of brain and breast, of the haematopoietic and reticulo-endothelial system, skin, ovary, fallopian tube, and broad ligament. Matched on date of birth, sex and date of diagnosis. 552 controls included in analyses	Distance from lines. Average and maximum magnetic fields assessed using distance from the lines, distance between wires, height of wires above the ground, annual and maximum loads along the lines from 1987–92, current phase and geographical resistivity of the earth	Calculated exposure in year of diagnosis Ref.: < 0.1 µT |		No. of cases	OR (95% CI) | All brain tumours: 0.1–0.2 µT — 23 — 0.9 (0.5–1.7); > 0.2 µT — 71 — 1.1 (0.8–1.6) Astrocytoma: 0.1–0.2 µT — 4 — 0.6 (0.2–1.8); > 0.2 µT — 16 — 0.8 (0.5–1.5) Glioblastoma: 0.1–0.2 µT — 8 — 1.3 (0.5–2.9); > 0.2 µT — 19 — 1.1 (0.6–2.0) Oligodendro-glioma: 0.1–0.2 µT — 3 — 2.8 (0.8–10.4); > 0.2 µT — 2 — 0.6 (0.1–2.5)	Limited information on confounders because of restrictions on interview
Wrensch et al. (1999)	**Case selection** Study of adult glioma in the San Francisco Bay Area. 492 newly diagnosed cases between 1991 and 1994 identified through the Northern California Cancer Center. **Control selection** 462 controls identified through random-digit dialling. Controls were matched to cases on age, sex and ethnicity.	For current dwellings and for all other California dwellings occupied during the 7 years before the study, exposure was assessed through spot measurements, wire codes and characterization of electrical facilities located within 150 feet [46 m] of the dwelling	Calculated exposure in year of diagnosis Ref.: < 0.1 µT No. of cases — OR (95% CI) Glioma: 0.1–0.2 µT — 62 — 0.97 (0.7–1.4); 0.2–0.3 µT — 15 — 0.6 (0.3–1.1); > 0.3 µT — 20 — 1.7 (0.8–3.6)	Information was obtained from a proxy for 47% of the cases. 85% of the gliomas were glioblastomas multiforme or astrocytomas.

CNS, central nervous system; SIR, standardized incidence ratio; OR, odds ratio; ref, reference exposure; ICD, International Classification of Disease; res., residential; occ., occupational; Ref.:, reference group with exposure level indicated and risk for tumours of the central nervous system. The study examined 223 patients with brain tumours, including 66 with glioma (astrocytoma I and II) and 157 with glioblastoma (astrocytoma III and IV). There was no evidence of any association, whether exposure was assessed by spot measurements or by calculation of magnetic fields from power lines.

Feychting *et al.* (1997) combined residential and occupational exposure by incorporating estimates of occupational exposure to magnetic fields into their earlier residential study (Feychting & Ahlbom, 1994). They estimated residential exposure from calculated magnetic fields and occupational exposure from census information linked to a job–exposure matrix based on magnetic field measurements. Adults exposed to stronger magnetic fields both at home and at work showed no association between occupational or residential exposure and tumours of central nervous system. [The study also found no association for calculated residential exposure after exclusion of subjects who were not exposed at home but were exposed to field strengths ≥ 0.2 µT at work.] There was also no association when analyses were restricted to people who had only residential exposure (≥ 0.2 µT) (odds ratio, 0.7; 95% CI, 0.3–1.7; 7 exposed cases). [The Working Group comments on the limitations of this study are given in section (*a*). However, this study is important in that it attempted to incorporate both residential and occupational exposure.]

Verkasalo *et al.* (1996), in their study of a cohort of 383 700 persons, investigated 301 cases of tumour of the nervous system and found no difference in incidence between members of the cohort and the general Finnish population. They also observed no association with calculated cumulative exposure to magnetic fields. The SIRs with respect to the general population were 0.94 (95% CI, 0.8–1.1; 238 cases) for exposures < 0.2 µT, 1.1 (95% CI, 0.77–1.5; 35 cases) for 0.2–0.39 µT, 0.64 (95% CI, 0.37–1.0; 16 cases) for 0.4–0.99 µT, 0.55 (95% CI, 0.18–1.3; 5 cases) for 1.00–1.99 µT and 0.92 (95% CI, 0.37–1.9; 7 cases) for ≥ 2.0 µT. Although the authors analysed gliomas and meningiomas separately, they reported only that the results were consistent with those for tumours of the nervous system as a whole. [See comments on the limitations of this study in section *(a)*.]

The case–control study of Li *et al.* (1997) described in section (*a*) examined 705 histologically confirmed incident cases of brain tumour (ICD[1]-9 191) in 45 districts of northern Taiwan. After exclusion of subjects residing in 14 of the districts because maps were not available, 577 cases and 552 controls remained. The study found no association between brain tumours and calculated exposure to magnetic fields in the year of diagnosis. Compared with the < 0.1 µT exposure category, the odds ratio for exposure of 0.1–0.2 µT was 0.9 (95% CI, 0.5–1.7; 23 cases) and that for exposure > 0.2 µT was 1.1 (95% CI, 0. 8–1.6; 71 cases). In analyses by tumour subtype, the odds ratios ranged from 0.6–2.8. [See comments on the limitations of this study in section *(a)*.]

---

[1] International Classification of Diseases

A large study by Wrensch *et al.* (1999) investigated adult glioma and residential exposure to electric and magnetic fields in six San Francisco Bay Area counties. The eligible cases were all adults newly diagnosed with glioma between 1 August 1991 and 31 March 1994. The study included 492 cases (82% of 603 eligible) and 462 controls (63% of 732 eligible), identified through random-digit dialling. Controls were frequency-matched to cases on age, sex and ethnicity. The average age of subjects was 54 years; 83% were white and 57% were male. Interviews were conducted in person in the homes of consenting patients (or their proxies) and controls. The interviewers asked about all dwellings occupied by subjects for three months or more for 15 years before either diagnosis (for cases) or interview (for controls). They also enquired about the subject's family and personal medical history, occupation, diet, smoking habits and alcohol use. The original diagnosis for 85% of cases was glioblastoma multiforme or astrocytoma; the remainder had other types of glioma. Proxy interviews were conducted for 233 cases (47%): 50% of the proxies were spouses of the cases, 30% were their children, 9% were siblings, 4% were parents and 7% had other relationships to the cases. Questionnaires covered a 15-year exposure period, for which 954 subjects reported 2995 dwellings. Usable addresses for all dwellings occupied during the seven years prior to diagnosis or study entry were obtained for 81.7% of cases and 84.2% of controls giving 1723 dwellings in California. Exposure assessment for electric and magnetic fields was completed for 81% of case and 86% of control dwellings. Exposure was assessed using indoor and outdoor spot measurements; characterization of power lines, transformers and substations located within 150 feet [46 m] of the dwelling, and Wertheimer–Leeper and Kaune–Savitz wire codes. To determine wire-codes, trained field workers made standardized drawings of all power lines within 150 feet [46 m] of each dwelling. Within this distance, they categorized up to three lines as to highest current type. For houses, they determined the shortest distance from the lines to the house. For apartments, they measured the distance from the nearest power line to the nearest boundary wall of each unit. For index dwellings, defined as the current dwelling for controls or the dwelling at time of diagnosis for cases, spot measurements were made in the centre of the kitchen, family room and bedroom, at the front door, and at the four outdoor corners of the dwelling. In addition, each subject selected a room in which a meter ran during the in-home interview. Spot measurements were also made with EMDEX meters under up to three power lines within 150 feet [46 m] of both current and previous dwellings and at the front doors of previous dwellings. The odds ratio for the longest-occupied dwellings with high compared with low Kaune–Savitz wire codes was 0.9 (95% CI, 0.7–1.3). For spot measurements at the front door (longest-occupied dwelling), the odds ratios for exposures of 0.1–0.2 µT, 0.2–0.3 µT and > 0.3 µT compared with ≤ 0.1 µT were 1.0 (95% CI, 0.7–1.4), 0.6 (95% CI, 0.3–1.1) and 1.7 (95% CI, 0.8–3.6), respectively. Adjusting for age, sex, ethnicity and whether subjects owned their homes did not meaningfully change the results, nor did restricting analyses to the subjects' highest wire-coded or index dwellings, or to single-family homes. The authors pointed out that there was no difference between cases and controls in the cumulative distribution of average front door, average indoor or maximum EMDEX readings. [The Working Group noted that the use of random-digit dialling for control selection may have resulted in a control group that was not fully representative of the base population from which the cases arose. Information was obtained from proxies for 47% of the cases.]

— *Appliance use*

Two studies investigated whether the risk for adult brain tumours might be associated with the use of electric blankets and other domestic appliances. In an Australian brain tumour study, Ryan *et al.* (1992) used a questionnaire to obtain information on 110 incident cases of glioma and 60 of meningioma diagnosed in 1987–90, and 417 controls. The questionnaire was designed to examine the risk factors for brain tumour associated with the use of electric blankets and electrically heated water beds. Proxy or assisted interviews were necessary for 41% of cases and 7% of controls. [The data for direct and proxy interviews were not presented separately, but the authors stated that they found no important differences.] A non-significant excess risk (odds ratio, 1.5; 95% CI, 0.83–2.6) associated with the use of electric blankets was reported for glioma, but not for meningioma (odds ratio, 0.86; 95% CI, 0.39–1.9). The opposite was true for electrically-heated water beds (odds ratio, 0.67 (95% CI, 0.18–2.5) and 1.3 (95% CI, 0.25–6.4), for glioma and meningioma, respectively). [The power of the study for this exposure is not known, as the prevalence of use of electrically heated bedding was not given.] A second report (Mutnick & Muscat, 1997) presented a preliminary summary of the data collected so far in a hospital-based, case–control study of 328 patients with primary brain cancers (284 controls) in the USA (New York University Medical Center, Memorial Sloan-Kettering Cancer Center and Rhode Island Hospital). The authors reported no risk associated with regular use of a number of electrical appliances, including computers, electric blankets, hair dryers, razors, and bedside dial clocks. [The Working Group noted that aspects of the methodology (e.g. the low participation rate, the need for proxies, etc.) render this study uninformative.]

(c)   *Breast cancer* (see Table 27)

The studies by Wertheimer and Leeper (1982, 1987), McDowall (1986) (in women) and Schreiber *et al.* (1993) (in women) also reported on breast cancer. The Working Group found the results of these studies uninformative.

Verkasalo *et al.* (1996) assessed the risk of breast cancer in their nationwide cohort study of Finnish adults described in the section on Leukaemia on p. 152. Of 194 400 women in the cohort, 1229 had been diagnosed with breast cancer. The SIRs were 1.1 (95% CI, 0.98–1.1; 945 cases) for exposure to fields of < 0.20 µT, 1.1 (95% CI, 0.88–1.3; 130 cases) for 0.20–0.39 µT, 0.89 (95% CI, 0.71–1.1; 87 cases) for 0.40–0.99 µT, 1.2 (95% CI, 0.89–1.6; 44 cases) for 1.00–1.99 µT and 0.75 (95% CI, 0.48–1.1; 23 cases) for ≥ 2.0 µT. [For comments on the limitations of this study, see section *(a)*.]

Table 27. Design and results of epidemiological studies of residential exposures to magnetic fields and breast cancer

Reference, country	Study base and subject identification	Exposure metrics	Results			Comments
Verkasalo et al. (1996) Finland (women)	**Cohort:** 383 700 persons (194 400 women) aged 20 or older who contributed 2.5 million person–years of follow-up between the years 1970 and 1989	Cumulative exposure. Estimates based on residential history, distance to 110–400-kV power line in 500-m corridor and calculated average annual magnetic fields for each building presumed to be ≥ 0.01 μT. Took into account current, typical locations of phase conductors and distance.	Cumulative exposure			Cohort study, SIRs. No information on other sources of residential exposure to electric and magnetic fields. No direct information from study subjects.
			Ref.: general population		SIR (95% CI)	
				No. of cases		
	Case selection: All primary breast cancer cases (1974–89) living within 500 m of overhead power lines: 1229 cases identified		Breast cancer	> 0.20 μT	945	1.1 (0.98–1.1)	
				0.20–0.39 μT	130	1.1 (0.88–1.3)	
				0.40–0.99 μT	87	0.89 (0.7–1.1)	
				1.00–1.99 μT	44	1.2 (0.89–1.6)	
				≥ 2.0 μT	23	0.75 (0.48–1.1)	
Li et al. (1997) Taiwan (women)	**Case selection:** Pathologically confirmed incident cases of breast cancer from northern Taiwan from cancer registry (1990–92). 1980 cases included in analyses	Distance from lines. Average and maximum magnetic fields assessed using distance from the lines, distance between wires, height of wires above the ground, annual and maximum loads along the lines from 1987–92, current phase and geographical resistivity of earth	Calculated exposure in year of diagnosis			Limited information on confounders because of restrictions on interview
			Ref.: < 0.1 μT		OR (95% CI)	
				No. of cases		
	Control selection: One control per case from cancer registry excluding cancers of the brain and breast, of the haematopoietic and reticulo-endothelial system, skin, ovary, fallopian tube and broad ligament. Matched on date of birth, sex and date of diagnosis: 1880 controls included in analyses		All breast cancers	0.1–0.2 μT	107	1.1 (0.8–1.5)	
				> 0.2 μT	224	1.1 (0.9–1.3)	
			Group I	0.1–0.2 μT	89	1.0 (0.8–1.4)	
				> 0.2 μT	193	1.0 (0.8–1.2)	
			Group II	0.1–0.2 μT	0	–	
				> 0.2 μT	7	0.9 (0.6–1.3)	
			Group III	0.1–0.2 μT	3	1.3 (0.4–4.2)	
				> 0.2 μT	8	1.5 (0.7–3.2)	
Coogan & Aschengrau (1998) USA (women)	**Case selection:** Cases diagnosed between 1983 and 1986 in Cape Cod. Of 334 cases reported, 259 were included in the analysis.	Use of electrically heated bedding, occupational history since age 18 years and residential history from 1943. Residential exposure was determined from proximity (within 152 m) to power lines and substations for dwellings on Cape Cod. Occupations were assigned to one of three categories (high, medium and no exposure).	Proximity to power lines/substations			Adjusted OR. No measurement data are presented, no sources are cited. The grouping of occupations differs from that used by most other investigators.
				Years	OR (95% CI)	
					No. of cases		
	Control selection: Controls identified from three sources — random-digit dialling, lists of Medicare beneficiaries and death certificates. The 738 controls were matched on age, vital status (and if deceased, on year of death).		Breast cancer	1–5	7	1.3 (0.5–3.6)	
				> 5	4	1.7 (0.4–6.3)	

Table 27 (contd)

Reference, country	Study base and subject identification	Exposure metrics	Results				Comments
			Calculated fields closest to time at diagnosis				Highest OR, 7.4 (1.0–178) for ER+ and less than 50 years of age
				Ref.: ≤ 0.09 µT	No. of cases	OR (95% CI)	
Feychting *et al.* (1998a) Sweden (men and women)	**Case selection**: All incident cancer cases from cancer registry (1960–85), from cohort of Swedish population aged ≥ 16 years, living in single family homes, located within 300 m of any 220- or 400- kV power lines: 699 women, 9 men	Distance to power lines from dwelling. Calculations of the magnetic fields generated by the power lines	Women				
			All ages	0.10–0.19 µT	57	1.2 (0.8–1.8)	
				≥ 0.2 µT	54	1.0 (0.7–1.5)	
			< 50 years	0.10–0.19 µT	14	1.2 (0.6–2.8)	
				≥ 0.2 µT	15	1.8 (0.7–4.3)	
	Control selection: One control per case from same cohort. Matched on age, sex, parish and residence near same power line: 699 controls		≥ 50 years	0.10–0.19 µT	43	1.2 (0.7–1.9)	
				≥ 0.2 µT	39	0.9 (0.5–1.4)	
			Men				
				≥ 0.2 µT	2	2.1 (0.3–14)	
			Subjects with both residential and occupational exposure				Number of cases with ER– is zero.
			Ref.: < 0.1 µT res. and < 0.12 µT occ.		No.	OR (95% CI)	
Forssén *et al.* (2000) Sweden (women)	Same as Feychting *et al.* (1998a), but expanded to include apartments; 1767 cases and 1766 controls	Same as above for residential. Occupational exposure from job–exposure matrix [developed from workday measurements made for another study] and information on occupation held in the year before the reference date	All ages	≥ 0.1 µT res., ≥ 0.12 µT occ.	8	0.9 (0.3–2.7)	
			< 50 years	≥ 0.1 µT res., ≥ 0.12 µT occ.	4	7.3 (0.7–78.3)	
			≥ 50 years	≥ 0.1 µT res., ≥ 0.12 µT occ.	4	0.4 (0.1–1.4)	
			ER+	≥ 0.1 µT res., ≥ 0.12 µT occ.	6	1.6 (0.3–9.9)	

OR, odds ratio; CI, confidence interval; SIR, standardized incidence ratio; res., residential; occ., occupational; ER+, estrogen-receptor-positive; ER–, estrogen-receptor-negative; Ref., reference group with exposure level indicated

The case–control study of residential exposure to magnetic fields and adult cancer in Taiwan (Li *et al.*, 1997) included 2407 histologically confirmed, incident cases of breast cancer in women, of which 1980 were included in the analysis. No association was found between breast cancer and residence less than 50 m from power lines, compared to residence ≥ 100 m from the lines (odds ratio, 1.0; 95% CI, 0.8–1.3; 156 cases). For calculated exposure to magnetic fields, there was no increase in risk among the highest exposure group (> 0.2 μT) (odds ratio, 1.1; 95% CI, 0.9–1.3; 224 cases). [For comments on the limitations of this study see section *(a)*.]

Electric and magnetic fields were considered among a wide variety of environmental and behavioural factors evaluated in a large study seeking reasons for the higher-than-expected breast cancer rates in women resident in the Cape Cod, MA, area in the USA (Coogan & Aschengrau, 1998). The study found a small, non-significant association between breast cancer risk and exposure to magnetic fields. [The Working Group noted that the poor exposure assessment, together with other design flaws, render this study largely uninformative.]

In a population-based case–control study on the effects of exposure to magnetic fields from high-voltage power lines in Sweden, Feychting *et al.* (1998a) also investigated the risk for breast cancer. Men and women who lived within 300 m of a 220- or 400-kV power line for at least one year between 1960 and 1985 were eligible for the study. All male breast cancer cases were included, but only women living in single-family homes were included. Cases were identified from the Swedish National Cancer Registry. A total of 699 female patients matched 1:1 with controls and nine male patients matched 1:8 with controls were included in the analysis. Controls who lived near the same power line as the case with whom they were matched were selected at random from the study base from people who had lived in the power-line corridor for at least one year before the reference date. Controls were matched to cases on age (within five years), sex and parish of residence in the year of the diagnosis. Information from medical records on the estrogen-receptor status of tumours was available for only 102 of the 699 cases. The study showed no overall increase in risk for female breast cancer with increasing estimates of magnetic field exposure; adjusting for socio-economic status did not change this result. When exposure was defined as the average calculated exposure during the year closest in time to the diagnosis date, with categories of < 0.1 μT, 0.1–0.19 μT and > 0.2 μT, the odds ratio for the highest exposure group was 1.0 (95% CI, 0.7–1.5) for all women. For women aged 50 years or younger, the odds ratio was 1.8 (95% CI, 0.7–4.3); for older women the odds ratio was 0.9 (95% CI, 0.5–1.4). Analyses of cumulative exposure showed a non-significantly elevated risk among women with cumulative exposure ≥ 3.0 μT–years in the six years immediately preceding diagnosis (odds ratio, 1.6; 95% CI, 0.8–3.2; 25 cases). Among estrogen-receptor-positive women, the odds ratio for exposure to ≥ 0.1 μT was 1.6 (95% CI, 0.6–4.1; 17 cases). For estrogen-receptor-positive women aged under 50 years, the odds ratio was 7.4 (95% CI, 1.0–178), based on six exposed cases and one control. For men, a non-significant increase in risk (odds ratio, 2.1; 95% CI, 0.3–14) was observed for calculated exposure to magnetic fields of ≥ 0.2 µT during the year closest in time to the diagnosis, based on two exposed cases. [The Working Group reiterated the limitations of this study as described in section *(a)*. Additionally, because all the data were obtained from registry and hospital files, no information was available on important risk factors for breast cancer and information on estrogen-receptor status was available for only a few cases.]

The study by Feychting *et al.* (1998a) of breast cancer in Sweden was expanded to combine assessments of residential and occupational exposure (Forssén *et al.*, 2000). Unlike the previous breast cancer analysis, which had been limited to single-family homes, this study included all types of dwelling. Cases of breast cancer were identified from the national cancer registry, and one matched control per case was selected at random from the general population. The assessment of occupational exposure was based on census-derived information about occupation that was linked to a job–exposure matrix developed for another study (Floderus *et al.*, 1993). For residential exposure to magnetic fields ≥ 0.10 µT for the year closest in time to diagnosis, and occupational exposure < 0.12 µT, the estimated odds ratio was 0.5 (95% CI, 0.1–2.9; 5 cases) and women aged less than 50 years at diagnosis had an odds ratio of 2.4 (95% CI, 0.1–50; 1 case). The highest risk (odds ratio, 7.3; 95% CI, 0.7–78; 4 cases) was for younger women (< 50 years) with higher occupational (≥ 0.12 µT) and residential (≥ 0.1 µT) exposures. [The Working Group noted the very small number of subjects in some subgroups. Occupational exposure was estimated for only 43% of subjects. The study included no information on reproductive risk factors for breast cancer.]

— *Use of electric blankets*

Because of the potential for prolonged exposure to increased electric and magnetic fields, the use of electric blankets has been examined as a risk factor for breast cancer in several recent investigations (Table 28). Vena *et al.* (1991) reported a case–control study that examined the use of electric blankets among 382 women with breast cancer and 439 randomly selected community controls in western New York state in the USA from 1987–89. The study was limited to postmenopausal women and included newly diagnosed, histologically confirmed cases aged 41–85 years admitted to hospitals in the study area between 1987 and 1989. Controls living in the study area were randomly selected from New York drivers' licence records if they were aged under 65 years and from Health Care Financing Administration rosters if they were older. Cases and controls were matched on age. The participation rate was 56% among cases and 46% among controls. The histories of use of electric blankets were obtained through home interviews, using a questionnaire. Information sought included any use of electric blankets in the past 10 years, seasonal pattern of use and mode of use. The study found no significant association with any level of exposure and no dose–response effect. When the results were adjusted for age and education, the odds ratio for breast cancer with use of electric blankets was 0.89 (95% CI, 0.66–12). Further adjustment for risk factors for postmenopausal breast cancer (body mass index, age at first pregnancy,

Table 28. Use of electric blankets and risk for breast cancer in women

Study	Subjects	No. of cases/ controls	Ever use[a]			Daily use			Use through the night[b]			Long-term use[c]		
			OR	No. of cases	95% CI	OR	No. of cases	95% CI	OR	No. of cases	95% CI	OR	No. of cases	95% CI
Vena et al. (1991)	Postmenopausal	382/439	0.89	126	0.66–1.2	0.97	NR	0.70–1.4	1.5	68	0.96–2.2	1.4	32	0.77–2.4
Vena et al. (1994)	Premenopausal	290/289	1.2	115	0.83–1.7	1.3	84	0.86–1.9	1.4	75	0.94–2.2	1.1	24	0.59–2.1
Vena et al. (1995)	Pre- and post-menopausal	672/728	1.1	242	0.85–1.4	1.2	179	0.90–1.5	1.5	143	1.1–1.9	1.2	56	0.81–1.9
Coogan & Aschengrau (1998)[d]	Mostly post-menopausal	259/738	NR	NR	NR	1.0	112	0.7–1.4	NR	112	NR	1.2	23	0.7–2.2
Gammon et al. (1998)[e]	<45 years	1645/1498	1.01	780	0.86–1.2	NR	NR	NR	1.0	630	0.88–1.2	0.96	155	0.74–1.3
	Postmenopausal (45–54 years)	261/250	0.97	143	0.67–1.4	NR	NR	NR	NR	NR	NR	NR	NR	NR
Laden et al. (2000)[f]	Premenopausal	95 cases; 41 585 person–years	1.1	42	0.71–1.7	NR	NR	NR	NR	NR	NR	0.88	15	0.49–1.6
	Postmenopausal	797 cases; 233 130 person–years	1.1	354	0.92–1.2	NR	NR	NR	NR	NR	NR	1.1	82	0.85–1.4
Zheng et al. (2000)	Pre- and postmenopausal	608/609	0.90	241	0.7–1.1	NR	NR	NR	0.9	147	0.7–1.2	0.8	96	0.6–1.1

OR, odds ratio; CI, confidence interval; NR, not reported
[a] Defined as any use during the last 10 years by Vena; ever use by Gammon
[b] Defined as use through the night by Vena and Zheng; on most of the time by Gammon
[c] Defined as use through the night in-season for 10 years by Vena; longer than 8 years for women aged < 45 years by Gammon; ≥ 20 years by Coogan; longer than 3 years by Zheng; ≥ 10 years for premenopausal women, ≥ 20 years for postmenopausal women by Laden
[d] Sleep with 'electric heating device'
[e] Women aged < 45 years included women from New Jersey, Washington and Atlanta; women aged 45–54 years were from Atlanta only.
[f] Prospective follow-up

number of pregnancies, age at menarche, family history of breast cancer and history of benign breast disease) resulted in an odds ratio of 1.0. There was no trend with increasing number of years of use or with frequency of use. A slightly increased risk was observed for women who reported using electric blankets continuously throughout the night compared with those who never used them (odds ratio adjusted for all risk factors, 1.5; 95% CI, 0.96–2.2; $n = 68$). For the heaviest users who had used electric blankets continuously throughout the night every night during the cold season over the previous 10 years (only 8% of cases and 6% of controls), further analyses showed a slightly increased risk (odds ratio, 1.4; 95% CI, 0.77–2.4). [The Working Group noted that the very low response rates, particularly among controls and the lack of information on the type and age of electric blankets and on other sources of exposure to electric and magnetic fields hamper the interpretation of this study.]

In a second, similar study, Vena et al. (1994) again examined use of electric blankets in western New York state, this time among premenopausal women aged 40 years or more. The study included 290 premenopausal women with breast cancer diagnosed between 1986 and 1991 and 289 age-matched controls selected randomly from drivers' licence records in the same geographical area. The response rate was 66% for cases and 62% for controls. The participants were interviewed in their homes using a questionnaire that included questions about use of electric blankets; dietary, medical and reproductive histories, and lifestyle and environmental factors. Use of electric blankets during the previous 10 years was reported by 40% of cases (115 women) and 37% of controls (106 women). After adjustment for age, education, age at first pregnancy, number of pregnancies, family history of breast cancer and other risk factors, the odds ratio for use of an electric blanket at any time in the previous 10 years was 1.2 (95% CI, 0.83–1.7). There was no dose–response relationship between number of years of blanket use and risk of breast cancer. A slight increase in risk was observed among women who used electric blankets daily during the cold season compared with those who never used them (odds ratio, 1.3; 95% CI, 0.86–1.9) and among those who used them continuously throughout the night (odds ratio, 1.4; 95% CI, 0.94–2.2). For the women with the most hours of electric blanket use (continuously throughout the night every night during the cold season for the previous 10 years), the odds ratio was 1.1 (95% CI, 0.59–2.1). [The Working Group considered that this study was limited by the lack of any direct assessment of exposure to electric and magnetic fields, the potential for recall bias and misclassification and the low response rates. Information on the type and age of electric blankets and on other sources of exposure to electric and magnetic fields was also lacking.]

Following a suggestion by Stevens (1995), the two previous studies by Vena et al. (1991, 1994) were reanalysed using the combined data (Vena et al., 1995). The odds ratio was 1.1 (95% CI, 0.85–1.4) for use of an electric blanket at any time in the previous 10 years, 1.2 (95% CI, 0.90–1.5) for daily use and 1.5 (95% CI, 1.1–1.9) for continuous use throughout the night. Although the results reported for such use showed a significantly increased risk, there was no evidence of a dose–response effect. In the highest exposure group, which included women who had used electric blankets in the cold season and continuously throughout the night for 10 years, the odds ratio was less elevated, and the confidence interval included the null value (odds ratio, 1.2; 95% CI, 0.81–1.9). Analysis by duration of continuous use throughout the night showed no association with breast cancer except for women who had used the blankets continuously throughout the night for 3–5 years (odds ratio, 2.0; 95% CI, 1.1–3.8).

More recently, in a larger population-based case–control study, Gammon *et al.* (1998) examined use of electric blankets, mattress pads or heated water beds and breast cancer risk. The study included 1645 women under the age of 45 years with breast cancer newly diagnosed between 1990 and 1992 in one of three geographical regions of the United States with tumour registries (Atlanta, GA, five counties in New Jersey and the Puget Sound area in Washington State). The 1498 controls were frequency-matched to cases by five-year age group and geographical area. Also included in the study were 261 postmenopausal women aged 45–55 years and 250 matched controls. The data for postmenopausal women were based solely on Atlanta residents. Although exposure to electric and magnetic fields was not a primary focus of this study, all women were asked about their use of electric bed-heating equipment. Study results indicated that ever having used electric blankets, mattress pads or heated water beds did not increase the risk of breast cancer among premenopausal women (< 45 years old) (odds ratio, 1.0; 95% CI, 0.86–1.2) or postmenopausal women (45–54 years old) (odds ratio, 0.97; 95% CI, 0.67–1.4).

In their study described above, Coogan and Aschengrau (1998) examined the use of electric heating devices during sleep. There was no increase in breast cancer risk associated with regular use (odds ratio, 1.0; 95% CI, 0.7–1.4; 112 cases). [Although the authors did not stratify the results by menopausal status, most of the participants (more than 88%) were postmenopausal.]

A large cohort study, the Nurses' Health Study, in the USA also examined breast cancer and use of electric blankets (Laden *et al.*, 2000). The parent study began in 1976, when 121 700 female registered nurses completed a postal questionnaire. Diagnoses of breast cancer were reported on follow-up questionnaires and confirmed by medical records (for most cases). A question on use of electric blankets was added in 1992. The prospective (1992–96) analysis was restricted to 78 614 women not diagnosed with cancer before 1992 who had answered this question (954 breast cancer cases). The retrospective analyses (1976–92) included 85 474 women who had answered this question (2426 breast cancer cases) and were cancer free at the start of the study. The reported relative risks for ever having used electric blankets were 1.1 (95% CI, 0.95–1.2; 426 cases) and 1.0 (95% CI, 0.92–1.1; 1041 cases), based on prospective and retrospective follow-up, respectively. After adjusting for known risk factors for the disease, there was little indication of a trend in risk associated with number of years of electric blanket use. Similar results were obtained for pre- and postmenopausal women and for women with estrogen-receptor-positive breast cancer.

Zheng *et al.* (2000) analysed data from a case–control study of breast cancer in Connecticut, USA, between 1994 and 1997. The study included two separate sources of cases (31–85 years old) and controls. One group included incident cases identified from the surgical pathology department of Yale-New Haven Hospital (432/561; 77% participation) and hospital controls who had undergone breast surgery for benign breast disease or had histologically confirmed normal tissue (404/569; 71% participation). A second group comprised cases resident in Tolland County identified through the Connecticut Tumor Registry (176; 74% participation). The controls for this group were selected by random-digit dialling (152; 64% participation) or, for those over 65 years of age, from Health Care Administration records (53; 54% participation). Information on use of electric blankets and other electrical appliances was obtained by interviewing the participants. Around 40% of cases and controls reported regular use of electric blankets and the odds ratio was 0.9 (95% CI, 0.7–1.1). The risk did not vary with age at first use, duration of use, or menopausal or estrogen-receptor status and was the same for subjects who used electric blankets regularly throughout the night. Similarly unremarkable results were obtained for use of other common domestic appliances.

(*d*)   *Other cancers*

In the 1996 Finnish cohort study, described in section *(a)*, Verkasalo *et al.* (1996) examined the relationship between cancer risk and exposure to magnetic fields from high-voltage power lines. Overall, 8415 cancer cases were identified (4082 were men). No association was found between cumulative exposure and the risk for all cancers (SIR, 0.98; 95% CI, 0.96–1.0 per μT–year) or for any specific type of cancer studied. Only for skin melanoma was the risk slightly increased throughout the three highest cumulative exposure categories: SIRs were 0.87 (95% CI, 0.54–1.3; 21 cases), 1.5 (95% CI, 0.98–2.1; 28 cases), 1.5 (95% CI, 0.69–2.7; 10 cases) and 1.2 (95% CI, 0.48–2.5; 7 cases) for exposure to fields of 0.20–0.39 μT, 0.40–0.99 μT, 1.0–1.99 μT and ≥ 2.0 μT, respectively. The SIR for multiple myeloma showed a marginally significant increase in men (SIR, 1.2; 95% CI, 1.0–1.5 per μT–year) and a non-significant decrease in women (SIR, 0.87; 95% CI, 0.57–1.3 per μT–year). For colon cancer, the risk was marginally increased in women (SIR, 1.2; 95% CI, 1.0–1.3 per μT–year), but not in men. [For the comments of the Working Group on this study, see section *(a)*.]

Finally, in a population-based study of 175 men with prostate cancer aged 40–69 years and 258 controls, Zhu *et al.* (1999) reported a relative risk of 1.4 (95% CI, 0.9–2.2) for ever having used an electric blanket or heated water bed, but the risk did not appear to increase with increasing duration of use.

2.3.2   *Occupational exposure to ELF electric and magnetic fields*

(*a*)   *Proportionate mortality or incidence studies*

Proportionate mortality studies should be interpreted with caution because apparent mortality excesses, particularly those of moderate size, can be the result of a deficit of mortality from other causes (Monson, 1990).

The studies described below were designed mainly for generating hypotheses, especially by use of record linkage with routinely collected data.

In early studies on the relation between electric and magnetic fields and cancer, exposure to electric and magnetic fields was inferred from the job title only, on the assumption that 'electrical workers' were exposed to higher than background electric and magnetic fields. The first list of 'electrical' occupations which supposedly entailed high exposure to electric and magnetic fields was established by Milham (1982). This list, or a modified version thereof, was used as the basis for exposure assessment in subsequent studies. The job titles most generally considered to denote 'electrical' occupations were electronic technicians and engineers, radio and telegraph operators, electricians, power and telephone linemen, television and radio repairers, power-station operators, aluminium workers, welders (see IARC, 1990) and flame cutters, and motion-picture projectionists.

Milham (1982) conducted a study in 1950–79 based on death certificates in Washington State, USA, for white men $\geq$ 20 years of age. The study was later updated for the period 1950–82 (Milham, 1985b). Employment in one of nine electrical occupations as listed on the death certificate, was used as a surrogate for exposure to electric and magnetic fields. Significantly elevated proportionate mortality ratios (PMRs) were observed for all leukaemia (PMR, 1.4 [95% CI, 1.1–1.6]; 146 cases), acute leukaemia (PMR, 1.6 [95% CI, 1.3–2.1]), malignant brain tumours (PMR, 1.2 [95% CI, 1.0–1.5]) and other lymphomas (PMR, 1.6 [95% CI, 1.2–2.2]), as well as malignant tumours of the pancreas (PMR, 1.2 [95% CI, 1.0–1.4]) and lung (PMR, 1.1 [95% CI, 1.1–1.2]).

In a study in Los Angeles County, USA, Wright *et al.* (1982) looked at incident cases of leukaemia that occurred from 1972–79 among white men employed in one of 11 electrical occupations. The proportionate incidence ratios (PIRs) were 1.3 [95% CI, 0.9–1.8] for all leukaemia (35 cases), 1.7 [95% CI, 1.1–2.6] for acute leukaemia and 2.1 [95% CI, 1.3–3.1] for acute myeloid leukaemia.

To evaluate leukaemia mortality in men employed in one of 10 electrical occupations, McDowall (1983) re-analysed data routinely collected in England and Wales for a report on occupational mortality in 1970–72. On the basis of all deaths in men aged 15–74 years, the PMRs of these occupations taken together were not significantly different from those expected, either for all leukaemia (PMR, 0.98; 85 cases) or for any specific subtype of leukaemia.

To evaluate leukaemia incidence in electrical occupations further, Coleman *et al.* (1983) used the routinely collected records of the South Thames Cancer Registry in England to calculate the proportionate registration ratio (PRR) for leukaemia for the period 1961–79 for men aged 15–74 employed in one of 10 electrical occupations. Eight of the 10 occupations showed an excess of all leukaemia with a significantly increased PRR of 1.2 ($p < 0.05$) for all 10 electrical occupations taken together (113 cases). There was no overall excess of chronic myeloid leukaemia, but non-significant excesses occurred in acute lymphoblastic (PRR, 1.5), chronic lymphocytic (PRR, 1.3) and acute myeloid (PRR, 1.2) leukaemia.

The death certificates of men aged ≥ 20 years were used in Wisconsin, USA, during 1963–78 to analyse leukaemia deaths in relation to 10 electrical occupations (Calle & Savitz, 1985). The PMR for all leukaemia in electrical occupations was 1.0 ([95% CI, 0.82–1.3]; 81 cases) and that for acute leukaemia was 1.1 [95% CI, 0.81–1.5].

Data from five descriptive studies that examined either the PMR (Milham, 1982; McDowall, 1983; Calle & Savitz, 1985; Milham, 1985b) or the PIR (Wright et al., 1982; Coleman et al., 1983) for leukaemia in workers in electrical occupations, i.e. workers with suspected high exposure to ELF electric and magnetic fields were pooled by Stern (1987). This data set which included a total of 449 cases of all leukaemia, yielded a relative risk of 1.1 [95% CI, 1.0–1.3] for all occupations combined. For the subgroups of acute leukaemia and acute myeloid leukaemia, the relative risk estimates were 1.4 [95% CI, 1.2–1.6] and 1.4 [95% CI, 1.1–1.7], respectively.

Death certificates from 1950–84 for men aged ≥ 20 years were used in a study conducted in British Columbia, Canada (Gallagher et al., 1990). The PMR for all leukaemia in men working in one of nine electrical occupations was [1.1; 95% CI, 0.8–1.4] (65 cases). Using the same data, the PMR for brain cancer in workers employed in the same nine occupations was 1.3 (95% CI, 0.93–1.6; 55 cases) for men aged 20–65 years (Gallagher et al., 1991).

In a study on mortality data for white men > 15 years old collected from 14 states in the USA for one or more years from 1979–85, the PMR for all leukaemia in 11 electrical occupations was 1.2 (95% CI, 1.0–1.4; 183 observed) and 1.1 (95% CI, 0.85–1.5) for acute myeloid leukaemia (Robinson et al., 1991).

To evaluate PRRs of cancer among electrical workers, Fear et al. (1996) used routinely collected data reported to the national cancer registration scheme in England during 1981–87 on more than 1 million cancers in individuals aged 20–74 years. The analysis, however, was based on only 36% of registrations for which valid occupational information was provided. Twelve job groups out of a total of 194 were identified as electrical occupations. For these job groups combined, and for both sexes jointly, significantly raised risks were seen for all brain and meningeal cancers combined (PRR, 1.2; 95% CI, 1.0–1.3; 281 cases), malignant brain cancer alone (PRR, 1.2; 95% CI, 1.0–1.4; 204 cases), all leukaemias combined (PRR, 1.2; 95% CI, 1.1–1.4; 217 cases) and acute myeloid leukaemia alone (PRR, 1.3; 95% CI, 1.0–1.6; 80 cases). For several types of cancer, most notably malignant brain cancer and acute myeloid leukaemia, the increased PRRs were most evident in men < 65 years old.

A study in São Paulo, Brazil, based on death certificates obtained from a sample of electricity utility workers in 1975–85 was conducted by Mattos and Koifman (1996).

The PMR in the group of workers expected to have been exposed to strong electric and magnetic fields was increased for brain cancer (ICD-9 191, 192) (PMR, 3.8; 95% CI, 1.0–9.7), Hodgkin disease (PMR, 5.6; 95% CI, 1.1–16) and bladder cancer (PMR, 4.2; 95% CI, 1.4–9.7).

(*b*)   *Cohort studies*

Table 29 presents selected results of studies that have looked at occupational exposure to static and ELF magnetic fields and leukaemia, brain cancer and breast cancer, the malignancies on which most attention has focused. A few studies have reported excesses of other cancers such as malignant melanoma, non-Hodgkin lymphoma and lung cancer for which the majority of other studies could not reproduce the findings. These studies are not shown in the Table but are described in the text.

(i)   *Workers exposed to strong static magnetic fields*

There are certain industries (aluminium reduction (see IARC, 1984, 1987a) and production of chlorine by electrolysis in chloralkali plants) in which workers are exposed to static magnetic fields, usually created by strong rectified alternating current. The aluminium reduction process involves exposure to static magnetic fields from direct currents passing through the anodes during electrolysis (reduction). The magnetic field to which the workers are exposed has been estimated to be between 4 and 50 mT (Kowalczuk *et al.*, 1991). The process also involves potential exposure to a mixture of volatiles from coal-tar pitch (see IARC, 1985, 1987b) and petroleum coke.

Barregård *et al.* (1985) studied cancer mortality and cancer incidence in a group of 157 male workers at a Swedish chloralkali plant. The employees had all worked regularly or permanently for at least one year during the period 1951–83 in the cell room where the electrolysis process took place. These workers had been exposed to strong static magnetic fields (average, 14 mT). The investigators reported no excess incidence of, or mortality from, cancer.

In a cohort study of 27 829 aluminium workers employed for ≥ 5 years between 1946 and 1977 in 14 reduction plants in the USA, Rockette and Arena (1983) reported indications of higher than expected mortality from pancreatic, genitourinary and lymphohaematopoietic cancers. Deaths from lymphohaematopoietic cancer were not confined to one subcategory of disease, or to one industrial process. [The Working Group noted that these results cannot be interpreted in relation to exposure to magnetic fields.]

A cohort study was carried out in British Columbia, Canada, of 4213 workers with ≥ 5 years of work experience at an aluminium reduction plant between 1954 and 1985 (Spinelli *et al.*, 1991). The static magnetic fields usually generated in the plant were approximately 1 mT. The potential exposure to magnetic fields and to coal-tar pitch volatiles was determined for each job by industrial hygienists using a job–exposure matrix. The standardized mortality ratio (SMR) in the total cohort was 2.2 (90% CI, 1.2–3.7) for tumours of the brain and central nervous system (ICD-9 191, 192) and 1.8

Table 29. Cohort studies of leukaemia, brain and breast cancer in occupational groups with assumed or documented exposure to static or ELF magnetic fields

Country (reference)	Population, design (number), recruitment period; follow-up period	Assessment of exposure to ELF magnetic fields[a]	Cancer	No. of cases	Relative risk (95% CI)	Comments
Sweden (Einhorn et al., 1980; Wiklund et al., 1981)	Telephone operators SIR (14 480; 14 180 women and 300 men), 1960 census; 1961–73	Not estimated	Leukaemia	12	1.0	[0.6–1.8][b]	Unadjusted for potential occupational confounders
Sweden (Olin et al., 1985)	Electrical engineers SMR (1243 men), 1930–59; 1930–79	Not estimated	Leukaemia Brain	2 2	0.9 1.0	(0.1–3.2) (0.1–3.7)	Unadjusted for potential occupational confounders
Sweden (Vågerö et al., 1985)	Telecommunications equipment workers SIR (2914; 2047 men and 867 women), 1956–60; 1958–79	Not estimated	**Men** Nervous system **Women** Breast cancer	5 7	1.0 0.6	(0.3–2.3) (0.3–1.3)	Unadjusted for potential occupational confounders
Sweden (Törnqvist et al., 1986)	Power linesmen SIR (3358 men), 1960 census; 1961–79	Not estimated	Leukaemia Nervous system	10 13	1.3 1.5	(0.7–2.1) (0.9–2.4)	90% CI. Unadjusted for potential occupational confounders
Sweden (Törnqvist et al., 1986)	Power station operators SIR (6703 men), 1960 census; 1961–79	Not estimated	Leukaemia Nervous system	16 17	1.0 1.0	(0.6–1.5) (0.6–1.5)	90% CI. Unadjusted for potential occupational confounders
Sweden (Törnqvist et al., 1991)	Workers in electrical occupations SIR (133 687 men), 1960 census; 1961–79	Median during working hours: < 0.04–16.5 µT (50 measurements)	Leukaemia Brain tumours	334 250	SIR slightly raised SIR close to unity		Unadjusted for potential confounders

Table 29 (contd)

Country (reference)	Population, design (number), recruitment period; follow-up period	Assessment of exposure to ELF magnetic fields[a]	Cancer	No. of cases	Relative risk (95% CI)		Comments
USA (Garland et al., 1990)	US Navy personnel SIR (> 4 million person–years white men), 1974–84; 1974–84	Not estimated	**Leukaemia** All cohort Electrician's mate	102 7	0.9 2.4	(0.8–1.1) (1.0–5.0)	Unadjusted for potential occupational confounders
Finland (Juutilainen et al., 1990)	Male industrial workers SIR [number not given] 1970 census; 1971–80	No measurements	**Leukaemia** No exposure Possible exposure Probable exposure	117 94 10	1.0 1.4 1.9	(baseline) (1.1–1.8) (1.0–3.5)	Unadjusted for potential confounders
			Brain tumours No exposure Possible exposure Probable exposure	204 149 13	1.0 1.3 1.3	(baseline) (1.0–1.6) (0.7–2.3)	
Norway (Tynes & Andersen, 1990)	Workers in electrical occupations SIR (37 952 men), 1960 census; 1961–85	No measurements	Male breast	12	2.1	(1.1–3.6)	Unadjusted for potential confounders
Norway (Tynes et al., 1992)	Workers in electrical occupations SIR (37 945 men), 1960 census; 1961–85	No measurements	Leukaemia AML CLL CML Brain	107 38 27 19 119	1.1 1.3 0.97 1.5 1.1	(0.89–1.3) [0.88–1.7][b] (0.64–1.4) (0.90–2.3) (0.90–1.4)	Unadjusted for potential confounders

Table 29 (contd)

Country (reference)	Population, design (number), recruitment period; follow-up period	Assessment of exposure to ELF magnetic fields[a]	Cancer	No. of cases	Relative risk (95% CI)	Comments
USA (Matanoski et al., 1991)	Telephone workers SIR (4547 male cable splicers, 9561 central office technicians), 1976–80; 1976–80	Personal monitoring of a sample of workers	Male breast Cable splicer Central office technicians	0 2	– 6.5	– (0.79–24)	Electromechanical switches environment in central office
Canada, British Columbia (Spinelli et al., 1991)	Aluminium reduction plant workers SIR (4213 men), 1954–85; 1954–85	No personal measurements (magnetic fields around 1 mT generated during industrial process)	Leukaemia Brain	3 8	0.76 1.9	[0.15–2.2][b] [0.84–3.8][b]	No significant association with estimated cumulative exposure to strong static magnetic fields (values not given)
Denmark (Guénel et al., 1993b)	Workers in electrical occupations SIR (255 000; 172 000 men, 83 000 women), 1970 census; 1970–87	Continuously above 0.3 µT	**Men** Leukaemia – acute – other Brain and nervous system Breast	39 16 23 23 2	1.6 1.6 1.7 0.69 1.4	(1.2–2.2) (0.90–2.6) (1.1–2.5) (0.44–1.0) (0.16–4.9)	Unadjusted for potential confounders
			Women Leukaemia Brain and nervous system Breast	2 9 55	0.56 1.2 0.88	(0.07–2.0) (0.56–2.3) (0.68–1.2)	

Table 29 (contd)

Country (reference)	Population, design (number), recruitment period; follow-up period	Assessment of exposure to ELF magnetic fields[a]	Cancer	No. of cases	Relative risk (95% CI)	Comments
		Intermittently above 0.3 µT	**Men**			
			Leukaemia	282	0.94	(0.84–1.1)
			– acute	119	1.0	(0.84–1.2)
			– other	164	0.90	[0.77–1.1][b]
			Brain and nervous system	339	0.94	(0.85–1.1)
			Breast	23	1.2	(0.77–1.8)
			Women			
			Leukaemia	94	0.92	(0.75–1.1)
			– acute	47	0.93	(0.70–1.2)
			– other	47	0.91	(0.68–1.2)
			Brain and nervous system	198	1.1	(0.93–1.2)
			Breast	1526	0.96	(0.91–1.0)
Sweden (Floderus et al., 1994)	Male railroad workers SIR [not given], 1960 census; 1961–79	Spot measurements	**All leukaemia** Engine drivers			
			1961–69	6	1.6	(0.7–3.6)
			1970–79	8	1.0	(0.5–2.1)
			All railway workers			
			1961–69	17	1.2	(0.7–1.9)
			1970–79	26	0.9	(0.6–1.3)
			CLL Engine drivers			
			1961–69	4	2.7	(1.0–7.4)
			1970–79	4	1.1	(0.4–2.9)

Table 29 (contd)

Country (reference)	Population, design (number), recruitment period; follow-up period	Assessment of exposure to ELF magnetic fields[a]	Cancer	No. of cases	Relative risk (95% CI)	Comments
			Breast cancer			
			Engine drivers			
			1961–69	2	8.3 (2.0–34)	
			1970–79	0	0.0 (0.0–6.0)	
			Railway workers			
			1961–69	4	4.3 (1.6–12)	
			1970–79	0	0.0 (0.0–1.6)	
			Brain tumours			
			Engine drivers			
			1961–69	8	1.1 (0.6–2.2)	
			1970–79	10	0.9 (0.5–1.6)	
			Railway workers			
			1961–69	31	1.2 (0.8–1.6)	
			1970–79	39	0.9 (0.7–1.3)	
Norway (Tynes et al., 1994a)	Workers in 8 power companies SIR (5088 men), 1920–85; 1953–91	Spot measurements and assessment of cumulative exposure to ELF magnetic fields	Leukaemia	11	0.90 (0.45–1.6)	
			Duration of employment			
			< 10 years	1	0.56 NR	
			10–29 years	6	1.2 NR	
			≥ 30 years	4	0.73 NR	
			Cumulative exposure			
			< 5 μT–years	2	0.95 NR	
			5–35 μT–years	4	0.74 NR	
			> 35 μT–years	5	1.0 NR	

Table 29 (contd)

Country (reference)	Population, design (number), recruitment period; follow-up period	Assessment of exposure to ELF magnetic fields[a]	Cancer	No. of cases	Relative risk (95% CI)		Comments
			Brain tumours	13	0.88	(0.47–1.5)	
			Duration of employment				
			< 10 years	3	0.91	NR	
			10–29 years	7	1.0	NR	
			≥ 30 years	3	0.65	NR	
			Cumulative exposure				
			< 5 µT–years	6	1.8	NR	
			5–35 µT–years	5	0.71	NR	
			> 35 µT–years	2	0.44	NR	
USA (Savitz & Loomis, 1995)	Utility workers SMR (138 905 men), 1950–86; 1950–88	Job–exposure matrix based on measurements of magnetic fields	*Magnetic fields (µT–years)* **Leukaemia**				
			0.6–< 1.2	34	1.0	(0.66–1.6)	
			1.2–< 2.0	35	1.1	(0.70–1.8)	
			2.0–< 4.3	27	0.95	(0.56–1.6)	
			≥ 4.3	14	1.1	(0.57–2.1)	
			AML				
			0.6–< 1.2	12	1.3	(0.59–2.8)	
			1.2–< 2.0	7	0.94	(0.36–2.4)	
			2.0–< 4.3	5	0.72	(0.24–2.2)	
			≥ 4.3	5	1.6	(0.51–5.1)	
			CLL				
			0.6–< 1.2	8	1.3	(0.49–3.6)	
			1.2–< 2.0	13	2.0	(0.77–5.1)	
			≥ 2.0	5	0.55	(0.17–1.8)	

Table 29 (contd)

Country (reference)	Population, design (number), recruitment period; follow-up period	Assessment of exposure to ELF magnetic fields[a]	Cancer	No. of cases	Relative risk (95% CI)		Comments
			Brain cancer				
			0.6–< 1.2	34	1.6	(0.99–2.6)	
			1.2–< 2.0	26	1.5	(0.84–2.6)	
			2.0–< 4.3	27	1.7	(0.92–3.0)	
			≥ 4.3	16	2.3	(1.2–4.6)	
Sweden (Alfredsson et al., 1996)	Male railway engine drivers and conductors SIR (9738), 1976–90; 1976–90	Not estimated	Leukaemia	20	1.2	(0.7–1.9)	
			Lymphocytic leukaemia				
			All ages	14	1.6	(0.9–2.6)	
			20–64 years	10	2.3	(1.3–3.2)	
Denmark (Johansen & Olsen, 1998)	Utility workers SIR (32 006; 26 135 men, 5871 women), 1908–93; 1968–93	24-h measurements and job-exposure matrix for ELF magnetic fields	**Men**				
			All leukaemias	60	0.92	(0.7–1.2)	
			acute	20	0.87	(0.5–1.4)	
			chronic lymphoblastic	27	0.92	(0.6–1.3)	
			chronic myeloid	6	0.65	(0.2–1.4)	
			Brain	57	0.79	(0.6–1.0)	
			Breast	2	0.50	(0.1–1.8)	
			Women				
			All leukaemias	3	0.50	(0.1–1.5)	
			Brain	15	1.3	(0.7–2.2)	
			Breast	96	1.1	(0.9–1.3)	
China (Petralia et al., 1998)	Female population in urban Shanghai SIR (population size not given), 1980–84; 1980–84	Exposure estimated from occupation at diagnosis	*Breast*				
			Exposure probability				
			Low	683	1.0	(0.9–1.0)	
			Medium	72	1.1	(0.9–1.4)	
			High	72	0.9	(0.7–1.2)	

Table 29 (contd)

Country (reference)	Population, design (number), recruitment period; follow-up period	Assessment of exposure to ELF magnetic fields[a]	Cancer	No. of cases	Relative risk (95% CI)		Comments
			Exposure level				
			Low	602	1.0	(0.9–1.1)	
			Medium	0	–		
			High	130	1.0	(0.8–1.2)	
			Any exposure	827	1.0	(0.9–1.0)	
Sweden (Floderus et al., 1999)	Large proportion of national working population SIR (1 596 959 men and 806 278 women), 1970 census; 1971–1984	Measurements and job–exposure matrix; upper exposure tertile (men = 0.116 µT; women = 0.138 µT) versus all subjects	**Men**				
			All leukaemias	648	1.1	(1.0–1.2)	
			AML	199	1.1	(0.9–1.4)	
			CML	116	1.1	(0.8–1.4)	
			ALL	32	1.5	(0.9–2.7)	
			CLL	301	1.1	(0.9–1.2)	
			Brain	1100	1.1	(1.0–1.2)	
			Breast	37	1.2	(0.7–1.9)	
			Women				
			All leukaemias	263	1.1	(1.0–1.4)	
			AML	107	1.1	(0.8–1.5)	
			CML	57	0.8	(0.6–1.2)	
			ALL	12	1.1	(0.5–2.4)	
			CLL	87	1.7	(1.2–2.4)	
			Brain	598	0.9	(0.8–1.1)	
			Breast	4866	1.1	(1.0–1.1)	

Table 29 (contd)

Country (reference)	Population, design (number), recruitment period; follow-up period	Assessment of exposure to ELF magnetic fields[a]	Cancer	No. of cases	Relative risk (95% CI)	Comments
Norway (Kliukiene et al., 1999)	Female population of country SIR (1 177 129), 1960, 1970, 1980 censuses; 1961–92	Measurements and job–exposure matrix	Breast **Work hours** 1–899 900–999 1000–1999 ≥ 2000 **Exposure µT–years** 0.1–0.8 0.9–1.4 1.5–3.0 > 3.0	NR NR NR NR NR NR NR NR	1.00 1.07 1.08 1.14 1.00 (baseline) 1.07 (1.03–1.12) 1.12 (1.07–1.17) 1.08 (1.01–1.16)	Adjusted for age, time-period and socioeconomic status
Italy (Pira et al., 1999)	Geothermal power plant workers SMR (3946 men), 1950–90; 1950–90	No formal evaluation	Leukaemia Brain	8 11	0.79 (0.34–1.6) 1.2 (0.57–2.1)	
Norway (Rønneberg et al., 1999)	Aluminium smelter workers; production worker subcohort SIR (2888 men), 1953–93	Measurements and job–exposure matrix	Brain	7	0.71 (0.29–1.5)	Strong static magnetic fields
England and Wales (Harrington et al., 2001)	Electricity generation and transmission workers, SMR (72 954 men and 11 043 women); 1973–82; 1973–97	Job-exposure matrix based on measurements of magnetic fields	**Leukaemia** *SMR*: Total *Period from hire* 0–9 years 10–19 years 20–29 years ≥ 30 years	111 6 34 37 34	0.84 (0.69–1.0) 0.51 (0.19–1.1) 1.1 (0.73–1.5) 0.91 (0.64–1.3) 0.71 (0.49–1.0)	

Table 29 (contd)

Country (reference)	Population, design (number), recruitment period; follow-up period	Assessment of exposure to ELF magnetic fields[a]	Cancer	No. of cases	Relative risk (95% CI)		Comments
			RR: cumulative exposure (μT–years)				
			0–2.4	60	1.0		
			2.5–4.9	18	1.5	(0.87–2.5)	
			5.0–9.9	20	0.99	(0.59–1.7)	
			10.0–19.9	17	0.96	(0.55–1.7)	
			≥ 20.0	9	1.4	(0.68–2.8)	
Switzerland (Minder & Pfluger, 2001)	Railway workers, SMR (18 070 men), 1972–93; 1972–93	Measurements at the workplaces with estimates of historical exposure	**Leukaemia**				
			Station masters	6	1.0	(baseline)	
			Line engineers	19	2.4	(0.97–6.1)	
			Shunting yard engineers	3	2.0	(0.50–8.1)	
			Train attendants	9	1.1	(0.39–3.1)	
			μT–years				
			0–4.9	6	1.0	(baseline)	
			5–74.9	9	0.78	(0.72–2.2)	
			≥ 75	22	1.6	(0.64–4.2)	

Table 29 (contd)

Country (reference)	Population, design (number), recruitment period; follow-up period	Assessment of exposure to ELF magnetic fields[a]	Cancer	No. of cases	Relative risk (95% CI)		Comments
			Brain tumours				
			Station masters	3	1.0	(baseline)	
			Line engineers	4	1.0	(0.23–4.6)	
			Shunting yard engineers	5	5.1	(1.2–21)	
			Train attendants	11	2.7	(0.75–9.6)	
			μT–years				
			0–4.9	1	1.0	(baseline)	
			5–74.9	11	2.8	(0.35–23)	
			≥ 75	11	2.4	(0.29–19)	

AML, acute myeloid leukaemia; CML, chronic myeloid leukaemia; ALL, acute lymphoblastic leukaemia; CLL, chronic lymphocytic leukaemia; SMR, standardized mortality ratio; SIR, standardized incidence ratio; NR, not reported

[a] The studies by Spinelli et al. (1991) and Rønneberg et al. (1999) deal with exposure to static magnetic fields.

[b] Calculated by the Working Group

(90% CI, 0.8–3.3) for leukaemia (ICD-9, 204–208). For cancer incidence ascertained from 1970 onwards, the SIR was 1.9 (90% CI, 0.97–3.5) for brain cancer (ICD-9 191), and 0.76 (90% CI, 0.21–2.0) for leukaemia. However, no individual cause of cancer death or incident cancer was related to cumulative exposure to magnetic fields, as estimated from the job–exposure matrix.

Rønneberg *et al.* (1999) studied cancer incidence in a population composed of 2647 male short-term workers and two cohorts of men employed for at least four years (2888 production workers and 373 maintenance workers) at an aluminium smelter in Norway. Data on all men first hired at an hourly wage and with at least six months of continuous employment were obtained from company files dating back to 1946. Of the 5962 men who initially satisfied the inclusion criteria, six had died before the observation period started in 1953 and 48 were lost to follow-up. The remaining 5908 men were linked to the files of the Norwegian Cancer Registry and followed up from 1953 (or date of first employment) until date of death or emigration, or the end of 1993. Exposure to magnetic fields, ranging from 2–10 mT for static magnetic fields and from 0.3–10 μT for time-varying magnetic fields (mainly 50–300 Hz), was estimated from a survey of other Norwegian smelters (Thommesen & Bjølseth, 1992). Cumulative exposure for each worker was calculated as the product of the estimated exposure intensity and duration, summed for all jobs held at the smelter. Overall, the cancer incidence was not elevated in any of the three cohorts when compared with the expected incidence calculated on the basis of the age- and calendar year-specific cancer incidence of all men in Norway applied to the person–years at risk among cohort members. There was no association observed in the entire cohort of 2888 production workers between level of exposure to static magnetic fields or ELF magnetic fields and cancers of the brain or lymphatic and haematopoietic tissue, on the basis of seven and 32 observed cases, respectively. No excess of the latter cancers was observed among the highly exposed power-plant and rectifier workers of the production cohort, where two cases were observed as against 1.9 expected. Separate estimates of risk for leukaemia were not given.

(ii)   *Workers exposed to electric and magnetic fields (not strong static magnetic fields)*

Following the detection of four cases of leukaemia among telephone operators in the city of Gothenburg, Sweden during 1969–74, a retrospective record linkage study of the entire national population was undertaken (Einhorn *et al.*, 1980; Wiklund *et al.*, 1981). A total of 14 480 telephone operators (14 180 women and 300 men) were identified from the 1960 population census in Sweden. Data linkage with the files of the nationwide Swedish Cancer Registry for the period 1961–73 revealed a total of 12 cases of leukaemia when 11.7 cases were expected on the basis of national age- and sex-specific incidence rates of the disease. [The Working Group noted that no effort was made by the authors to estimate potential job-related exposure to ELF electric and magnetic fields.]

In a study in Sweden based on a large-scale linkage of occupational data from the 1960 census and data from the national cancer registry for the years 1961–73, Vågerö and Olin (1983) investigated 54 624 men and 18 478 women, aged 15–64 years, working in the electronics or electrical manufacturing industry. They found significantly increased risks, of 15% in men and 8% in women, for cancer at all sites combined, compared with those of the working population in general. These increases were due mainly to significant increases in the risk for cancers of the pharynx, larynx and lung, and for malignant melanoma among men, and cervical cancer among women.

In a mortality study by Olin et al. (1985) of 1254 male electrical engineers, all of whom had graduated from one university in Sweden during 1930–59, 11 were lost to follow-up. The remaining 1243 cohort members were followed until date of death or the end of 1979. When compared with the age- and calendar time-adjusted mortality rates of the general Swedish male population, the SMR for cancer at all sites combined was 0.5 (95% CI, 0.3–0.7) on the basis of 24 observed cases. Three deaths from malignant melanoma were observed as opposed to 0.9 expected (SMR, 3.2; 95% CI, 0.7–9.4), and two deaths from leukaemia occurred as opposed to 2.3 expected (SMR, 0.9; 95% CI, 0.1–3.2). [The Working Group noted that no effort had been made by the authors to estimate potential job-related exposure to ELF electric and magnetic fields.]

Vågerö et al. (1985) evaluated cancer incidence in 2918 subjects (2051 men, 867 women) employed for at least six months during the period 1956–60 by a Swedish company at one of three worksites producing telecommunications equipment. All but four subjects, for whom personal data could not be verified, were linked to the files of the Swedish Cancer Registry for the period 1958–79. The observed numbers of cancers among cohort members were compared with the expected numbers, calculated on the basis of age-, sex- and calendar year-specific incidence rates for the entire Swedish population. Overall, 102 cancers were observed among men and 37 among women, yielding SIRs of 1.03 (95% CI, 0.8–1.2) and 0.98 (95% CI, 0.7–1.4), respectively. An increased risk for malignant melanoma was seen in both men (SIR, 2.5; 95% CI, 1.1–4.9; 8 cases) and women (SIR, 2.8; 95% CI, 0.8–7.2; 4 cases); the highest risk estimates were seen for departments associated with soldering work, but this observation was based on a total of only four observed cases in men and two in women (SIR, 3.9; 95% CI, 1.4–8.5) (both sexes combined). Two cases of Brill-Symmer disease (a nodular lymphoma) were seen in men as opposed to 0.1 expected. [The Working Group noted that no effort was made by the authors to estimate the potential job-related exposure to ELF electric and magnetic fields.]

From the 1960 population census in Sweden, Törnqvist et al. (1986) identified a total of 3358 male power linemen and 6703 male power-station operators in the electric power industry, who were all aged 20–64 years at the time of the census. Cohort members were linked to the files of the national Swedish Cancer Registry and followed up for cancer incidence until the end of 1979. The observed numbers of cancers were compared with the expected numbers, calculated on the basis of age-, county- and calendar year-specific cancer incidence rates for all men classified as blue collar workers by the census. Overall, 236 cancers were observed among power linemen and 463 cancers among power station operators, yielding SIRs of 1.1 (90% CI, 1.0–1.2) and 1.0 (90% CI, 0.9–1.0), respectively. For none of the specific cancer sites included in the analysis did the lower confidence limit exceed unity. The SIR for leukaemia among power linemen was 1.3 (90% CI, 0.7–2.1) and that among power station operators was 1.0 (90% CI, 0.6–1.5) on the basis of 10 and 16 observed cases, respectively. None of the major subgroups of leukaemia showed an excess risk in either occupational group. The SIR estimate for tumours of the nervous system was 1.5 (90% CI, 0.9–2.4) among power linemen and 1.0 (90% CI, 0.6–1.5) among power-station operators on the basis of 13 and 17 observed cases, respectively. [The Working Group noted that no effort was made by the authors to estimate potential job-related exposure to ELF electric and magnetic fields.]

This cohort was later extended (Törnqvist *et al.*, 1991) to include all men in Sweden aged 20–64 who had worked in one of 11 electrically related occupations according to the job title recorded in the 1960 census, giving a total of 133 687 [or 7% of all Swedish working men]. In addition, the magnetic field exposure was estimated by five occupational hygienists, according to Swedish working conditions in the selected occupational categories. However, these estimates were based on only 50 measurements conducted over 4–8 h (except for two measurements made over nearly 18 h). A total of 334 cases of leukaemia was observed during 1961–79 which was only slightly in excess of the expected number [figure not stated] calculated on the basis of the incidence rates of cancer among all Swedish working men. Similarly, a total of 250 cases of brain tumour were identified which was approximately equal to the expected number [figure not stated]. Although significant, or marginally significant, positive associations were seen between one or more of the occupations under consideration (or industry subgroups thereof) and specific subtypes of leukaemia (acute myeloid, chronic myeloid, acute lymphoblastic, chronic lymphocytic, all subtypes combined) or brain tumour (glioma, glioblastoma, all subtypes combined), the authors concluded that no homogeneous pattern of increased risks associated with occupations for which there is presumed to be exposure to high ELF magnetic fields, was found. The authors noted that the occupation of the study subjects was known only on the census date in 1960, and that the estimates of ELF electric and magnetic fields were based on a small number of measurements.

De Guire *et al.* (1988) carried out a cohort study of 9590 workers employed between 1976 and 1983 in the Montreal plants of a telecommunications company. During the study period, 10 cases of malignant melanoma of the skin were diagnosed among men and none among women. Using reference rates for malignant melanomas in the Greater Montreal area during the same period, the SIR for men was 2.7 (95% CI, 1.3–5.0). [If the total cohort is considered, the expected number of cases is five (combined 95% CI, 1.1–3.7)] [The Working Group noted that the study was conducted in response to the observation of a cluster of melanomas among workers at one plant. No data were available on job titles or on specific types of exposure among these workers.]

In a cohort study of US Navy personnel (4 072 502 person–years), 102 cases of leukaemia were diagnosed among men on active duty in 1974–84 (Garland *et al.*, 1990). The overall incidence of leukaemia was close to that of the population of the United States. In the analysis by occupation, seven cases were reported among electrician's mates (111 944 person–years) with possible high exposure to 60-Hz electric and magnetic fields, yielding an increased risk for leukaemia in this group (SIR, 2.4; 95% CI, 1.0–5.0). No increased risk for leukaemia was observed in other occupational groups, in particular in naval workers with probable exposure to electric and magnetic fields at frequencies higher than 60 Hz. [The Working Group noted that there was no assessment of exposure to electric and magnetic fields by occupation.]

From the 1970 population census in Finland, Juutilainen *et al.* (1990) selected all Finnish men, aged 25–64 years, during the period 1971–80 who were classified as industrial workers according to their self-reported occupation. The occupations were grouped into three exposure categories according to the probability of exposure to ELF magnetic fields, i.e. no exposure, possible exposure and probable exposure. Cohort members were linked with the Finnish mortality files for determination of vital status through 1980 and to the files of the Finnish Cancer Registry for verification of incident cases of leukaemia and tumours of the central nervous system during the period 1971–80. Using all industrial workers with no exposure to ELF magnetic fields as the comparison group, the authors found relative risks for leukaemia of 1.4 (95% CI, 1.1–1.8; 221 cases) and 1.9 (95% CI, 1.0–3.5) for workers with possible and probable exposure, respectively, and, for brain tumour, the relative risks were 1.3 (95% CI, 1.0–1.6) and 1.3 (95% CI, 0.7–2.3), respectively.

From the 1960 population census in Norway, Tynes and Andersen (1990) identified a cohort of approximately 38 000 men aged $\geq$ 20 years who at that time had held jobs in which they might have been exposed to electric and magnetic fields. Cohort members were linked to the files of the Norwegian Cancer Registry and followed up for breast cancer incidence from 1961–85. A total of 12 cases of breast cancer was observed when 5.81 were expected on the basis of age- and calendar year-specific breast cancer incidence rates for all economically active men in Norway according to the census, yielding an SIR of 2.1 (95% CI, 1.1–3.6).

From the 1960 occupational cohort, a group was selected (37 945 men, aged 20–70 years) which was followed up to investigate other types of cancer in a second study (Tynes *et al.*, 1992). The jobs held by cohort members were categorized into one of 12 occupational groups, each of which was classified in turn according to the anticipated type of exposure to electric and magnetic fields, i.e. (i) radiofrequency, (ii) heavy magnetic, electric, (iii) intermediate magnetic, (iv) weak magnetic, electric, and (v) weak magnetic. No supporting field measurements were made. Cohort members were linked to the files of the national Norwegian Cancer Registry and followed up for cancer incidence from 1961 until the date of death or emigration, or to the end of 1985. Overall, 3806 incident cases of cancer were observed when 3583.7 were expected on the basis of age- and calendar year-specific cancer incidence rates for all economically active men in Norway according to the census, yielding an SIR of 1.06 (95% CI, 1.03–1.09). A total of 107 cases of leukaemia were reported, yielding a slightly increased SIR of 1.1 (95% CI, 0.9–1.3) with SIRs of 1.3 (95% CI, 0.88–1.2), 0.97 (95% CI, 0.64–1.4) and 1.5 (95% CI, 0.90–2.3) for acute myeloid, chronic lymphocytic and chronic myeloid subtypes of leukaemia, respectively. A total of 119 brain tumours were observed, which also resulted in a slightly increased SIR of 1.1 (95% CI, 0.90–1.4). In a separate analysis of the subgroup of cohort members still economically active at the time of the 1970 census, the SIR for leukaemia was 1.4 (95% CI, 1.1–1.8). On the basis of this subgroup, the authors found a tendency towards a dose–response relationship for leukaemia with SIRs of 1.8 (95% CI, 1.1–2.8), 1.4 (95% CI, 0.81–2.2) and 1.1 (95% CI, 0.70–1.6) for the exposure categories heavy magnetic, electric, intermediate magnetic and weak magnetic, respectively. No such tendency was apparent for brain tumours.

Matanoski *et al.* (1991) reported the results of a cohort study of telephone workers in the United States aged < 65 years. Central office technicians (9561), exposed to mean magnetic field strengths of 0.25 μT, had an SIR of 6.5 (95% CI, 0.79–24) for male breast cancer, based on two observed cases. These technicians were working in a central office with electromechanical switches, which produced a complex field environment. No men with breast cancer were observed among other telephone workers, in particular among cable splicers (4547) who had a mean exposure to magnetic fields of 0.43 μT. No results for leukaemia were reported in this study.

From the 1970 population census in Denmark, Guénel *et al.* (1993b) identified a total of 3932 combinations among men and 1885 combinations among women of a specific industry and a specific occupation. Only combinations in which ≥ 10 persons were involved were included. Each of these industry–occupation combinations was coded for potential occupational exposure (no exposure, probable exposure) to 50-Hz electric and magnetic fields using a threshold level of 0.3 μT, and the appropriate code was applied to the 2.8 million economically active Danes aged 20–64 years at the time of the census in 1970. Men and women judged to be occupationally exposed to intermittent magnetic fields (154 000 men, 79 000 women) and to continuous magnetic fields (18 000 men, 4000 women) were followed up in the Danish Cancer Registry until 1987. The numbers of first primary cancers observed in the exposed cohorts were compared with those expected, calculated on the basis of age-, sex- and calendar year-specific rates of primary cancer incidence among all persons who were economically active according to the census. The incidence of leukaemia was increased in men with probable continuous exposure to magnetic fields (SIR, 1.6; 95% CI, 1.2–2.2) on the basis of 39 observed cases. The excess risk was the same for acute leukaemia and for other leukaemias. The incidence of leukaemia was not increased in women with continuous exposure (SIR, 0.56; 95% CI, 0.07–2.0), but only two cases were observed. Both men and women with probable intermittent exposure to magnetic fields had leukaemia risks close to the average for all economically active persons. No significant result was found for breast cancer, brain tumours or malignant melanoma.

In a study from Sweden, Floderus *et al.* (1994) used the records from the 1960 Swedish population census to select all men, aged 20–64, who, in 1960, had been employed as workers by the Swedish railways. [The size of the cohort was not given; however, the study group appears identical to one of the 11 subcohorts included in a previous census-linkage study from Sweden (Törnqvist *et al.*, 1991).] Using spot measurements, exposure to electric and magnetic fields was estimated to be of the order of 4.03–0.58 μT (engine drivers), 0.61–0.36 μT (conductors), 0.30–0.25 μT (station masters and train dispatchers) and 0.59–0.37 μT (railroad assistants and linemen). [It should be noted that the exposure assessment was made with a device having a less-than-flat frequency response (Floderus *et al.*, 1993); the values quoted may therefore be underestimates.] Cohort members were linked to the national Swedish Cancer Registry and all cases of cancer notified to the cancer registry in 1961–79 were identified. All economically active men, aged 20–64 in 1960, acted as the reference population. No significantly increased relative risks were seen for leukaemia (all subtypes combined), for any of the subcohorts or for the combined cohort. In their consideration of the subtypes of leukaemia, the authors observed an increased risk for chronic lymphocytic leukaemia among engine drivers during the first decade of follow-up (1961–69) (SIR, 2.7; 95% CI, 1.0–7.4), but not in the second decade (1970–79) (SIR, 1.1; 95% CI, 0.4–2.9), but the relative risk estimates were based on only four cases of chronic lymphocytic leukaemia for each decade. No excess risks were seen for subtypes of brain tumour (astrocytoma I-II; astrocytoma III-IV) or for all subtypes combined (ICD-7 193 and astrocytoma only). But increased relative risks were observed for breast cancer, predominantly among engine drivers (SIR, 8.3; 95% CI, 2.0–34; 2 cases) and railway workers (SIR, 4.3; 95% CI, 1.6–12; 4 cases), and for tumours of the pituitary gland, predominantly among conductors (SIR, 3.3; 95% CI, 1.5–7.6; 6 cases) and railway workers (SIR, 2.9; 95% CI, 1.6–5.3; 11 cases). These results, however, were based on only a few observed cases and were seen only during the first decade of follow-up (1961–69). [The Working Group noted that calculation of person–years at risk was not corrected for elimination due to death, either in the study cohort or in the reference population. This implies that the relative risk estimates in the case of differential mortality in the study groups may have been distorted.]

In another study from Norway, Tynes *et al.* (1994a) studied the incidence of cancer in 5088 male workers in eight large hydroelectric power companies. From employment records available for each company, cohort members were selected on the following criteria: job title that indicated exposure to ELF electric and magnetic fields, employment for at least one year and first employment between 1920 and 1985. The average duration of employment among cohort members was 22 years. Spot measurements of magnetic fields were made at the two largest power companies and a job title–magnetic field exposure matrix was constructed. The matrix was applied to the work histories of the study subjects to provide calculated estimates of exposure to ELF electric and magnetic fields (μT–years) for each worker covering the period from first employment until date of retirement or the end of the study. Crude estimates of job-related exposure to solvents, herbicides, asbestos and cable oils were also made. Cohort members were linked to the files of the national Norwegian Cancer Registry and follow-up for cancer incidence was undertaken over the period 1953–91. Overall, 486 new cases of cancer were observed which matched the number of cases expected on the basis of person–years at risk among cohort members combined with the age- and calendar year-specific cancer incidence rates of Norwegian men (SIR, 1.0; 95% CI, 0.92–1.1). No significant deviation in risk from unity was seen for cancer at any site, including leukaemia (SIR, 0.90; 95% CI, 0.45–1.6) and brain tumours (ICD-9 193) (all tumours of the central nervous system and malignant tumours of the peripheral nervous system) (SIR, 0.88; 95% CI, 0.47–1.5) with 11 and 13 observed cases, respectively. In a sub-analysis, no trends in risks for leukaemia or brain tumours with increasing time since first employment or duration of employment were observed. Also, no association with cumulative exposure to magnetic fields was seen for leukaemia while brain tumour showed a tendency towards a negative correlation. An excess risk was seen for malignant melanoma at cumulative exposures above 35 μT–years (11 cases); however, the data showed no continuous exposure–response trend.

A mortality study was conducted in a cohort of workers at five electric utility companies in the USA (Savitz & Loomis, 1995). All men employed full-time continuously for at least six months between 1950 and 1986 were included. Vital status until 31 December 1988 was ascertained leading to the identification of 20 733 workers who had died out of a total of 138 905 workers. Exposure to magnetic fields was estimated from a job–exposure matrix, elaborated from exposure measurements made on workers randomly selected within occupational groups in each company. These measurements were taken using the AMEX meter which yields a TWA exposure. In total, 2842 usable measurements of a one-day work shift were collected. These were aggregated in five occupational groups to obtain maximum internal precision of the mean magnetic field within a group and maximum variability of mean magnetic field between groups. Occupational exposure to solvents and polychlorinated biphenyls was estimated for each occupational category through expert judgement. In the initial study, these analyses were restricted to total mortality (20 733 cases), total cancer (4833 cases), leukaemia (164 cases) and brain cancers (ICD-9 191, 192) (144 cases). A slight increased risk for brain cancer was apparent for workers employed in highly exposed occupations. The risk for leukaemia was increased for workers who had been employed for 20 years or more as electricians: SMR, 2.5 (95% CI, 1.1–5.8; 6 cases), but not in other exposed occupations. The risk for total cancer was slightly increased with indices of exposure to magnetic fields. Brain cancer, but not leukaemia, was associated with total exposure to magnetic fields, with a relative risk adjusted for potential occupational confounders of 2.3 (95% CI, 1.2–4.6) in the highest exposure category (≥ 4.3 μT–years, 90th percentile). The association with brain cancer was more apparent for recent exposure to magnetic fields, i.e. for exposure in the interval 2–10 years before death, suggesting a relatively short latency period: SMR, 1.2 (95% CI, 0.66–2.1), 1.4 (95% CI, 0.75–2.6), 1.5 (95% CI, 0.76–2.8) and 2.6 (95% CI, 1.4–4.9) for 0–< 0.2, 0.2–< 0.4;

0.4–< 0.7 and ≥ 0.7 μT–years, respectively. The relationship of brain cancer mortality to cumulative exposure to magnetic fields was not sensitive to the method used to treat historical exposure, the choice of exposure-time lags and windows, and the cut-points used to categorize the exposure variables (Loomis *et al.*, 1998). Using a case–cohort approach and a refined job–exposure matrix with more precise job definitions, Savitz *et al.* (2000) found that the rate ratios for brain cancer were essentially unchanged; a weak positive association with leukaemia was apparent. Mortality from non-Hodgkin lymphoma, Hodgkin disease and multiple myeloma in this cohort were investigated by Schroeder and Savitz (1997). Weak associations between total exposure to magnetic fields and non-Hodgkin lymphoma were observed at intermediate exposure levels, with a weaker association in the highest exposure category. No association was observed with Hodgkin disease or multiple myeloma.

Mortality from lung cancer in this cohort in relation to magnetic fields has also been reported (Savitz *et al.*, 1997). The rate ratio for lung cancer in the highest category of cumulative exposure to magnetic fields (4.28–15.45 μT–years, 90th percentile) was 1.1 (95% CI, 0.89–1.3). Modest associations were observed with exposure estimated in several time windows before death, or for duration of employment over 20 years in specific occupational groups exposed to strong 60-Hz magnetic fields, such as electricians or power plant operators. [The Working Group noted that adjustment for tobacco smoking was not feasible.]

In another study from Sweden, Alfredsson *et al.* (1996) investigated the incidence of cancer in 7466 railway engine drivers and 2272 conductors, who were employed by the Swedish State Railways on 1 January 1976 or who had started their employment there at any time during the period 1976–90. Information on date of hire, date of leaving and type of job was obtained from registers kept by the State Railways. No measurements of exposure to magnetic fields were made. Cohort members were followed up for cancer in the Swedish National Cancer Registry from date of first hire or 1 January 1976, whichever came first, until date of death or diagnosis, or until the end of 1990. The observed numbers of cohort members diagnosed with a first primary cancer were compared with the expected numbers calculated on the basis of person–years of follow-up among cohort members and cancer incidence rates of the general male population of Sweden. A total of 630 workers with cancers at all sites combined was observed (486 among railway engine drivers and 144 among conductors) yielding a relative risk of 0.9 (95% CI, 0.8–1.0). For railway engine drivers and conductors combined, the relative risk for acute lymphoblastic or chronic lymphocytic leukaemia was 1.6 (95% CI, 0.9–2.6). In a supplementary analysis where follow-up was restricted to workers in the age range 20–64 years, the authors found that the relative risk estimate was further increased: 2.3 (95% CI, 1.3–3.2). No clear association was seen for the other subtypes of leukaemia. For astrocytoma, the relative risk was close to one. [The Working Group noted that there must have been a substantial overlap between this study population and that studied by Floderus *et al.* (1994). It was not clear whether cancer reference rates including multiple cancers in individuals were used for the calculation of the expected numbers of cancers.]

Johansen and Olsen (1998) evaluated the incidence of cancer in a study population composed of 32 006 men and women with at least three months of employment at the 99 utility companies that supply Denmark with electricity. Personal data were obtained from manual files kept by the electricity companies, the Danish Supplementary Pension Fund and the public payroll administration; the date of first employment ranged from 1908–93. On the basis of a series of 24-h personal measurements and the judgements of four engineers, each of a total of 475 combinations of job title and work area for employees were assigned an average level of exposure to ELF electric and magnetic fields during a working day. These were in turn grouped into one of five categories according to exposure level: background (< 0.09 µT), low (0.1–0.29 µT), medium (0.3–0.99 µT), high (> 1.0 µT) and unknown exposure. A rough estimate was also made of exposure to asbestos. Cohort members were linked to the files of the national Danish Cancer Registry and follow-up for cancer was from 1968 until date of death, date of emigration or the end of 1993. Overall, 3008 cohort members with cancer were observed, as against 2825 expected on the basis of person–years at risk among cohort members combined with age-, sex- and calendar year-specific cancer incidence rates for the Danish population, yielding an SIR of 1.06 (95% CI, 1.03–1.10). No excess risk was seen for leukaemia [SIR, 0.88] or tumours of the brain [SIR, 0.86]; the overall reduction in the relative risk for brain tumours was due to a reduced risk of borderline significance (SIR, 0.79; 95% CI, 0.6–1.0) in men. Similarly, no excess was seen for any of the major subtypes of leukaemia, and no trends in risk could be distinguished for leukaemia or tumours of the brain in relation to time since first employment. Finally, there was no indication of a link between cumulative exposure to ELF electric and magnetic fields (duration of work combined with level of exposure) and the risk for any of these tumour types. Only two cases of breast cancer were seen in men, as against four expected, while the relative risk for breast cancer in female employees was slightly elevated (SIR, 1.1; 95% CI, 0.9–1.3), but breast cancer in women showed no correlation with cumulative exposure to ELF electric and magnetic fields. Increased risks for cancers of the lung and pleural cavity were seen mainly in workers whose jobs involved exposure to asbestos.

Petralia et al. (1998) carried out a study in urban Shanghai, the People's Republic of China, where all incident cases of breast cancer in women ≥ 30 years old in 1980–84 were identified. The incidence rates of breast cancer were calculated using the 1982 census data for the same population and SIRs were calculated with these rates as a reference. The extent of exposure to electric and magnetic fields was estimated through a job–exposure matrix using scores for exposure probability (high, medium, low or none) and exposure levels (high, medium, low or none). Using the occupation at the time of diagnosis for classifying women into exposure groups, electric and magnetic fields were not found to be related to breast cancer incidence; SIRs were close to 1.0 in all exposure categories for any exposure index.

In a linkage study from Sweden, Floderus *et al.* (1999) used data from the 1970 population census to evaluate overall and site-specific cancer incidence among 1 596 959 men and 806 278 women, aged 20–64 years in 1970, who had all been employed in a job the title of which had been included in a previously established job–exposure matrix for ELF electric and magnetic fields. This job–exposure matrix gave estimates of magnetic field exposure for the 100 most common jobs in Sweden according to the 1990 census, and for 10 specifically selected occupations that were less common, but more heavily exposed to ELF electric and magnetic fields (Floderus *et al.*, 1993, 1996). This job–exposure matrix formed the basis for allocation of levels of exposure to magnetic fields to the jobs included in the study. Cohort members were linked to the files of the national Swedish Cancer Registry and followed up for cancer incidence from 1971 through 1984; follow-up was discontinued when members reached 70 years of age. Cumulative incidence rates, adjusted for age, but unadjusted for mortality during follow up, were calculated for all men and women employed in jobs categorized as having medium exposure (men, 0.084–0.115 μT; women, 0.067–0.129 μT) and high exposure (men, ≥ 0.116 μT; women, ≥ 0.138 μT) [and presumably compared with those of all men and women included in the study]. The risk ratios for cancer at all sites combined in the investigators' medium and high exposure categories, respectively, were 1.1 (95% CI, 1.1–1.1) and 1.1 (95% CI, 1.1–1.1) for men, and 1.1 (95% CI, 1.0–1.1) and 1.1 (95% CI, 1.0–1.1) for women. Similar results were seen for brain tumours (nervous system) and leukaemia with risk ratios of 1.1 (95% CI, 1.0–1.2) and 1.1 (95% CI, 1.0–1.2), respectively, in the highest exposure tertile for men and 0.9 (95% CI, 0.8–1.1) and 1.1 (95% CI, 1.0–1.4) for women. Also, in the highest tertile, male breast cancer was non-significantly increased (risk ratio, 1.2; 95% CI, 0.7–1.9). The risk ratios for cancer at several other sites were slightly increased in the upper exposure tertile, including malignant melanoma among men (risk ratio, 1.4; 95% CI, 1.2–1.5) and women (risk ratio, 1.2; 95% CI, 1.1–1.4); however, there was no general exposure–response pattern. [The Working Group noted that the study population must be partly overlapping with that included in a previous Swedish study by Törnqvist *et al.* (1991). The Working Group also noted that no measurements of cumulative exposure to magnetic fields were available, introducing a high risk for misclassification of the exposure of study subjects, and that no adjustment was made in the risk ratio analysis for mortality among study subjects during follow-up.]

In a linkage study from Norway, Kliukiene *et al.* (1999) used data from the 1960, 1970 and 1980 population censuses to evaluate the incidence of breast cancer in 1 177 129 women who were economically active according to at least one of the censuses. The classification of a job was based on a 3–5-digit industry code and a 3-digit occupation code; the socioeconomic status of the women was defined according to the job title. For a subcohort of women born in 1935 or later, data on age at birth of first child were also available. Exposure to ELF magnetic fields was assessed *a priori* using two different approaches. In the first approach, the number of hours per week during which potential magnetic fields were estimated to be above a background level, defined as 0.1 μT, were classified by an expert panel. In the second approach, measurements of magnetic fields from a previous study (Floderus *et al.*, 1996) of Swedish men were allocated to the women's job titles as reported in the census. In both approaches, exposure was cumulated over years of employment (work hours and μT–years, respectively). Cohort members were linked to the National Cancer Registry for identification of notified cases of breast cancer, and person–years at risk were calculated from the year of entering the study to the date of death or emigration, or to the end of 1992. The SIRs for breast cancer among cohort members were calculated using the rates for the total Norwegian population as a reference. In the two highest categories for number of work hours with exposure to ELF magnetic fields above background, i.e. 1000–1999 h and ≥ 2000 h, the SIRs were 1.05 (95% CI, 1.02–1.07) and 1.08 (95% CI, 1.05–1.12), respectively. The SIRs in the two upper categories for cumulative exposure in μT–years, i.e. 1.5–3.0 and > 3.0 were 1.06 (95% CI, 1.03–1.09) and 1.03 (95% CI, 0.97–1.09), respectively. Using the lowest exposure category as a reference (0 h exposure above background, and cumulative exposure between 0.1 and 0.8 μT–years) and adjusting for socioeconomic status (based on the job title) a Poisson regression analysis showed a risk ratio for breast cancer for the two highest categories for number of work hours with exposure to magnetic fields above background, of 1.08 (95% CI, 1.04–1.12) and 1.14 (95% CI, 1.10–1.19), respectively, and for the highest categories of cumulative exposure of 1.12 (95% CI, 1.07–1.17) and 1.08 (95% CI, 1.01–1.16), respectively. In the subcohort of women born in 1935 or later, the corresponding risk ratio was somewhat lower, and of marginal significance, after adjustment for age at birth of first child.

From employment records, Pira *et al.* (1999) identified a total of 4237 subjects who had worked for at least three months in a geothermal power plant in Italy between 1950 and 1990. After exclusion of all the 225 female workers of 36 men who could not be traced, the remaining 3946 male workers were traced for date of death and cause of death, whenever appropriate, from death certificates and from population files kept by the local municipality. A total of 977 deaths was registered as opposed to 1295 expected on the basis of age- and calendar year-specific national mortality rates applied to the person–years at risk among cohort members, yielding a SMR of 0.75 (95% CI, 0.71–0.80). Eight of the deaths were due to leukaemia and 11 to tumours of the brain and nervous system yielding SMRs of 0.79 (95% CI, 0.34–1.6) and 1.2 (95% CI, 0.57–2.1), respectively. The authors reviewed the working histories of these patients at the power plant and stated that none had worked in activities for which exposure to electric and magnetic fields could be presumed to have occurred.

Mortality from leukaemia was investigated by Harrington *et al.* (2001) in a cohort of 83 977 male and female electricity generation and transmission workers at the former Central Electricity Generating Board of England and Wales for whom computer records were available. All employees were known to have been employed for at least six months with some period of employment in the period 1973–82. Work history records were available until 1993. On the basis of the results from a previous measurement programme on occupational exposure to ELF electric and magnetic fields in parts of the

United Kingdom, exposure of workers in the electricity generation and transmission industry was estimated for 11 different work categories for power-station workers and eight categories for transmission workers. The job history for each worker was classified according to the established job categories and the cumulative occupational lifetime exposure (level multiplied by duration; μT–years) was estimated for each individual. The cumulative exposure in the most recent five-year period was also estimated. The mortality of the total cohort until 1997 was obtained by record linkage with the mortality files of the Central Register of the National Health Service. Compared with mortality rates from England and Wales, the overall SMR from leukaemia among cohort members was 0.84 (95% CI, 0.69–1.01) on the basis of 111 observed cases. Subanalyses by period from date of hire or according to subtype of leukaemia showed no consistent pattern. In the subcohort of 79 972 workers for whom work history data were available, a Poisson regression analysis showed age- and sex-adjusted relative risks of death from leukaemia of 1.5 (95% CI, 0.87–2.5), 0.99 (95% CI, 0.59–1.7), 0.96 (95% CI, 0.55–1.7) and 1.4 (95% CI, 0.68–2.8) among cohort members with a lifetime exposure of 2.5–4.9, 5.0–9.9, 10.0–19.9 and ≥ 20.0 μT–years, respectively, compared with the risk of death from leukaemia among workers with cumulative exposures ≤ 2.4 μT–years. Dose analyses on subtypes showed that only one point estimate, i.e. 'other leukaemias' in the lowest category of exposure, was significantly different from unity (relative risk, 2.0; 95% CI, 1.1–3.7). There was no significant trend of risk for any subtype of leukaemia or for all leukaemias combined with increasing cumulative exposure. A re-analysis using the most recent five years of exposure to ELF electric and magnetic fields did not change the results.

A retrospective cohort mortality study of Swiss Railway employees occupationally exposed to magnetic fields of 16 2/3-Hz and substantial harmonics was conducted by Minder and Pfluger (2001). The cohort comprised all men actively employed as line engineers, shunting-yard engineers, train attendants or stationmasters, or retired from these jobs and alive, identified through several personnel and pension records starting in 1972. The total number of men in the cohort was 18 070, representing 270 155 person–years of observation from 1972–93. Deaths of cohort members from leukaemia or brain tumour identified from death certificates were used as end-points. The assessment of exposure was carried out using a device that measured the magnetic fields in the driver's seat of the engine during complete driving cycles, for different types of train and routes taken. Historical exposure for each five-year calendar period was also assessed based on the number of engines in service and a weighted average of engine-specific exposure. The exposure to magnetic fields of train attendants and stationmasters was assessed from measurements taken at their most frequent places of work. Each cohort member was assigned the exposure associated with his last reported job, which was also generally of the longest duration, due to infrequent job changes. Estimated cumulative exposure in μT–years increased in the period 1930–90 from 9.3 to 25.9 for line engineers, from 2.6 to 13.4 for shunting-yard engineers, from 0.4 to 3.3 for train attendants and from 0.1 to 1.0 for stationmasters. When compared with stationmasters with the lowest exposure, the relative risk for leukaemia was 2.4 (95% CI, 0.97–6.1) for line engineers and 2.0 (95% CI, 0.50–8.1) for shunting-yard engineers. The relative risk for brain tumours (ICD-9 191) was 1.0 (95% CI, 0.23–4.6) for line engineers and 5.1 (95% CI, 1.2–21) for shunting-yard engineers. For cumulative exposure ≥ 75 μT–years compared with exposure 0–4.9 μT–years, the relative risk was 1.6 (95% CI, 0.64–4.2) for leukaemia and 2.4 (95% CI, 0.29–19) for brain tumours. The trend of increasing leukaemia mortality with both cumulative exposure and fraction of time above 10 μT was statistically significant.

(c)   Case–control studies

In the first published studies on occupational exposure to electric and magnetic fields, no measurements of exposure were made; exposure was inferred from the job title, on the assumption that electrical workers were exposed to higher than background fields. Job–exposure matrices, which included scores for exposure probability and exposure intensity in exposed occupations as determined from expert judgement, have also been used. In most studies, no data on exposure to other potential carcinogens were available. Some of the studies presented as case–control studies are based on mortality data collected from death certificates. In these studies, the 'cases' were deaths from the cause of interest (i.e. leukaemia, brain tumour or breast cancer) and the controls were selected from other causes of death; such studies should be seen as mainly exploratory.

More recent studies have included exposure measurements and concerned mostly occupational cohorts analysed using a nested case–control study design. In some instances, the results have been presented according to the type of field measured (i.e. ELF magnetic fields, ELF electric fields). Exposure to potential occupational confounders was generally assessed in these studies.

The results of the case–control studies of ELF magnetic fields are summarized in Table 30 and for ELF electric fields in Table 31.

(i)   Leukaemia

In a case–control study, McDowall (1983) used the deaths recorded in England and Wales for the year 1973. The cases selected were 537 men who had died aged ≥ 15 years from acute myeloid leukaemia. A total of 1074 controls were randomly selected from men who had died aged ≥ 15 years from all causes except leukaemia to match the cases within five-year age groups. The analysis showed an increased odds ratio for acute myeloid leukaemia for all five of the electrical occupations studied; however, this was statistically significant only when all five occupations were analysed combined (odds ratio, 2.1; 95% CI, 1.3–3.6). Further evaluation of the group of 'all electrical occupations and persons of any occupation engaged in an electrical telecommunications industry' gave an odds ratio of 2.3 (95% CI, 1.4–3.7).

In a population-based case–control study, Pearce et al. (1985) identified 546 cases of leukaemia among men aged ≥ 20 years notified to the cancer registry of New Zealand during 1979–83. The 2184 controls were men chosen at random from the

Table 30. Case–control studies of occupational groups with assumed or documented exposure to ELF magnetic fields

Country (reference)	Subjects: cases, controls (recruitment period)	Source of subjects	Source of job information; exposure assessment methods	Estimates of exposure to ELF magnetic fields	No. of cases	Odds ratio (95% CI)	Comments
Leukaemia							
England and Wales (McDowall, 1983)	537 male deaths from AML 1074 male deaths from other causes (controls) (1973)	Death certificates	Death certificates; occupation	All electrical occupations Any occupation in electrical or telecommunications industry	30 36	2.1 2.3	(1.3–3.6) (1.4–3.7)	Matched on age
New Zealand (Pearce et al., 1985)	546 men with leukaemia 2184 men with other cancers (controls) (1979–83)	Cancer registry	Cancer registry; occupation	All electrical occupations	18	1.7 (0.97–3.0)	Matched on age
New Zealand (Pearce et al., 1989) [partly overlapping with Pearce et al. (1985)]	534 men with leukaemia 19 370 men with other cancers (controls) (1980–84) Chronic leukaemia Acute leukaemia	Cancer registry	Cancer registry; occupation	All electrical work age 20–64 age ≥ 65	21 9 12 11 6	1.6 (1.0–2.5) 1.4 (0.71–2.7) 1.9 (1.0–3.3) 2.1 (1.2–3.8) 1.3 (0.62–2.5)	
USA (Loomis & Savitz, 1990)	3400 male deaths from leukaemia 34 000 male deaths from other causes (controls) (1985–86) 903 deaths from AML 414 deaths from CNLL 181 deaths from ALL 800 deaths from CLL	Death certificates; 16 states in the USA	Death certificates	Electrical occupations	76 22 11 6 11	1.0 (0.8–1.2) 1.1 (0.7–1.7) 1.1 (0.8–1.7) 1.5 (0.7–3.4) 0.6 (0.3–1.1)	Adjusted for race and age Not adjusted Not adjusted Not adjusted Not adjusted

Table 30 (contd)

Country (reference)	Subjects: cases, controls (recruitment period)	Source of subjects	Source of job information; exposure assessment methods	Estimates of exposure to ELF magnetic fields		No. of cases	Odds ratio (95% CI)	Comments
France (Richardson et al., 1992)	185 men and women with acute leukaemia 513 men and women with other diseases (controls) (1984–88)	In-patient files; two hospitals	Interview; job–exposure assessment	Any exposure		14	1.7 (0.9–3.5)	Matched on sex, age, ethnic group and place of residence Adjusted for prior chemo-therapy or radiotherapy
Italy (Ciccone et al., 1993)	86 men and women with myeloid leukaemia or MDS 246 hospital and population controls (1989–90)	In-patient files	Personal interview; job-exposure matrix	**Other than from arc welding** Any Moderate/higher		7 3	3.9 (1.2–13) 2.9 (0.6–14)	Matched on sex, age and area of residence
				Possibly and probably exposed Men Women		17 4	1.6 (0.6–4.1) 0.8 (0.2–2.5)	
USA (Sahl et al.,1993; Kheifets et al., 1999)	44 deaths from leukaemia 438 cohort controls (1960–88)	Cohort of electric utility workers	Company personnel records; job-exposure matrix based on measured magnetic fields	< 4 μT–years 4–8 μT–years 8–16 μT–years >16 μT–years		6 3 7 15	1.0 (baseline) 1.0 (0.2–4.8) 1.6 (0.4–6.4) 1.5 (0.4–6.3)	
Sweden (Floderus et al., 1993)	250 men with leukaemia 1121 male population controls (1983–87)	Cancer registry; population registry	Mailed questionnaire and spot measurements; job-exposure matrix	**Mean level** ≤ 0.15 μT 0.16–0.19 μT 0.20–0.28 μT ≥ 0.29 μT ≥ 0.41 μT	(Q1) (Q2) (Q3) (Q4) (90%)	48 50 61 80 32	1.0 (baseline) 0.9 (0.6–1.4) 1.2 (0.8–1.9) 1.6 (1.1–2.4) 1.7 (1.0–2.7)	Matched on age

Table 30 (contd)

Country (reference)	Subjects: cases, controls (recruitment period)	Source of subjects	Source of job information; exposure assessment methods	Estimates of exposure to ELF magnetic fields		No. of cases	Odds ratio (95% CI)		Comments
	112 men with CLL			≤ 0.15 µT	(Q1)	13	1.0	(baseline)	
				0.16–0.19 µT	(Q2)	17	1.1	(0.5–2.3)	
				0.20–0.28 µT	(Q3)	33	2.2	(1.1–4.3)	
				≥ 0.29 µT	(Q4)	41	3.0	(1.6–5.8)	
				≥ 0.41 µT	(90%)	22	3.7	(1.8–7.7)	
	90 men with AML			≤ 0.15 µT	(Q1)	22	1.0	(baseline)	
				0.16–0.19 µT	(Q2)	24	1.0	(0.5–1.8)	
				0.20–0.28 µT	(Q3)	18	0.8	(0.4–1.6)	
				≥ 0.29 µT	(Q4)	23	1.0	(0.6–1.9)	
				≥ 0.41 µT	(90%)	8	0.9	(0.4–2.1)	
USA (London et al., 1994)	2355 men with leukaemia 67 212 men with other cancers (controls) (1972–90)	Los Angeles county cancer registry	Job at diagnosis from medical record; job–exposure matrix based on measured magnetic fields in selected occupations	**Average level**					
				< 0.17 µT		2264	1.0	(baseline)	
				0.18–0.80 µT		61	1.2	(1.0–1.6)	
				≥ 0.81 µT		30	1.4	(1.0–2.0)	
	853 men with ANLL			< 0.17 µT		820	1.0	(baseline)	
				0.18–0.80 µT		23	1.3	(0.9–1.9)	
				≥ 0.81 µT		10	1.3	(0.7–2.3)	
	534 men with CLL			< 0.17 µT		512	1.0	(baseline)	
				0.18–0.80 µT		18	1.6	(1.2–2.3)	
				≥ 0.81 µT		4	0.8	(0.4–1.5)	
	487 men with CML			< 0.17 µT		469	1.0	(baseline)	
				0.18–0.80 µT		8	0.8	(0.5–1.3)	
				≥ 0.81 µT		10	2.3	(1.4–3.8)	

Table 30 (contd)

Country (reference)	Subjects: cases, controls (recruitment period)	Source of subjects	Source of job information; exposure assessment methods	Estimates of exposure to ELF magnetic fields	No. of cases	Odds ratio (95% CI)	Comments
Norway (Tynes et al., 1994a)	52 men with leukaemia 259 cohort controls (1958–90)	Cohort of railway workers	Job history from employment files; job–exposure matrix	Ever worked at an electric line	33	0.7	(0.37–1.4)	Matched on age
				Cumulative exposure				
				None	19	1.0	(baseline)	
				Low (0.1–310 µT–years)	22	1.0	(0.49–2.1)	
				High (311–3600 µT–years)	11	0.49	(0.22–1.1)	
				Very high (1900–3600 µT–years)	4	0.84	(0.25–2.8)	
Canada, France (3 cohorts combined) (Thériault et al., 1994)	140 incident cases of leukaemia 546 cohort controls (Canada, 1970–88; France, 1978–89)	Three cohorts of electric utility workers in Canada (Quebec and Ontario) and France	Company personnel records; job–exposure matrix based on measurements of exposure to magnetic fields	**Cumulative exposure**				
				< 3.1 µT–years	70	1.0	(baseline)	
				> 3.1 µT–years	70	1.5	(0.90–2.6)	
				> 15.7 µT–years	13	1.8	(0.77–4.0)	
	60 incident cases of ANLL 238 cohort controls			< 3.1 µT–years	27	1.0	(baseline)	
				> 3.1 µT–years	33	2.4	(1.1–5.4)	
				> 15.7 µT–years	6	2.5	(0.70–9.1)	
	24 incident cases of CML 93 cohort controls			< 3.1 µT–years	16	1.0	(baseline)	
				> 3.1 µT–years	8	0.61	(0.18–2.1)	
	14 incident cases of ALL 55 cohort controls			< 3.1 µT–years	10	1.0	(baseline)	
				> 3.1 µT–years	4	2.1	(0.12–35)	
	41 incident cases of CLL 157 cohort controls			< 3.1 µT–years	17	1.0	(baseline)	
				> 3.1 µT–years	24	1.5	(0.50–4.4)	
				> 15.7 µT–years	6	1.7	(0.44–6.7)	

Table 30 (contd)

Country (reference)	Subjects: cases, controls (recruitment period)	Source of subjects	Source of job information; exposure assessment methods	Estimates of exposure to ELF magnetic fields	No. of cases	Odds ratio (95% CI)	Comments
Quebec cohort (included in Thériault et al., 1994)	24 men with leukaemia 95 cohort controls (1970–88)			**Cumulative exposure**			
				< 3.1 µT–years	6	1.0	(baseline)
				> 3.1 µT–years	18	0.29	(0.04–1.8)
				> 15.7 µT–years	4	0.45	(0.04–3.8)
	[8 men with ANLL] 32 cohort controls			< 3.1 µT–years	1	1.0	(baseline)
				> 3.1 µT–years	[7]	0.75	(0.00–> 100)
				>15.7 µT–years	2	0.14	(0.00–> 100)
	10 men with CLL 40 cohort controls			< 3.1 µT–years	3	1.0	(baseline)
				> 3.1 µT–years	7	0.25	(0.02–2.6)
				>15.7 µT–years	2	0.27	(0.02–4.2)
France cohort (included in Thériault et al., 1994)	71 incident cases of leukaemia 279 cohort controls (1978–89)			**Cumulative exposure**			
				< 3.1 µT–years	55	1.0	(baseline)
				> 3.1 µT–years	16	1.4	(0.61–3.1)
				> 15.7 µT–years	3	1.9	(0.46–7.8)
	34 men with ANLL 134 cohort controls			< 3.1 µT–years	24	1.0	(baseline)
				> 3.1 µT–years	10	1.8	(0.57–5.4)
				> 15.7 µT–years	1	1.4	(0.03–16.2)
	13 men with CLL 51 cohort controls			< 3.1 µT–years	10	1.0	(baseline)
				> 3.1 µT–years	3	4.8	(0.45–71)
				> 15.7 µT–years	1	2.8	(0.04–68)

Table 30 (contd)

Country (reference)	Subjects: cases, controls (recruitment period)	Source of subjects	Source of job information; exposure assessment methods	Estimates of exposure to ELF magnetic fields	No. of cases	Odds ratio (95% CI)		Comments
Canada, Ontario cohort included in Thériault et al., 1994, updated Miller et al. (1996)[a]	50 men with leukaemia 199 cohort controls 1970–88	Cohort of electric utility workers at Ontario Hydro	Company personnel records; job–exposure matrix for magnetic fields (Positron meter)	**Cumulative exposure** < 3.1 μT–years 3.2–7 μT–years ≥ 7.1 μT–years	10 16 24	1.0 1.7 1.6	(baseline) (0.58–4.8) (0.47–5.1)	Adjustment for potential confounders
	20 men with ANLL [80 cohort controls]			< 3.1 μT–years 3.2–7 μT–years ≥ 7.1 μT–years	3 6 11	1.0 1.9 2.9	(baseline) (0.27–14) (0.42–20)	
	19 men with CLL [76 cohort controls]			< 3.1 μT–years 3.2–7 μT–years ≥ 7.1 μT–years	4 6 9	1.0 0.49 0.25	(baseline) (0.06–4.2) (0.01–4.6)	
Brain tumours								
USA (Lin et al., 1985)	519 male deaths from brain tumours (370 gliomas or glioblastoma multiforme, and 149 astrocytomas) 519 male deaths from other causes (controls)	Maryland state vital records	Job on death certificate; job–exposure matrix	No exposure Possible exposure Probable exposure Definite exposure	323 128 21 27	1.0 1.4 2.0 2.2	(1.1–2.0) (0.94–3.9) (1.1–4.1)	
	432 male deaths from brain tumours of unspecified type 432 male deaths from other causes (controls) (1969–82)			No exposure Possible exposure Probable exposure Definite exposure	286 87 19 15	1.0 0.94 1.3 1.5	(0.68–1.3) (0.60–2.8) (0.68–3.4)	

Table 30 (contd)

Country (reference)	Subjects: cases, controls (recruitment period)	Source of subjects	Source of job information; exposure assessment methods	Estimates of exposure to ELF magnetic fields	No. of cases	Odds ratio (95% CI)		Comments
USA (Speers et al., 1988)	202 male deaths from glioma 238 male deaths from other causes (controls) (1969–78)	Death certificates (East Texas)	Job on death certificate; job-exposure matrix	No exposure Possible exposure Probable exposure Definite exposure	92 68 11 6	1.0 1.2 2.9 infinite	(0.73–1.8) 0.80–10) $p = 0.009$	Adjusted for age
New Zealand (Pearce et al., 1989)	431 men with malignant brain tumours (ICD-9 191) 19 473 men with other cancers (controls) (1980–84)	Cancer registry	Cancer registry occupation	All electrical workers	12	1.0	(0.56–1.8)	
USA (Preston-Martin et al., 1989)	202 men with glioma 202 male neighbourhood controls (1980–84) 70 men with meningioma 70 neighbourhood controls	Los Angeles County Cancer Registry	Work history from questionnaire; electrical occupations	Any exposure duration < 5 years > 5 years Any exposure duration	14/8 16 14 2/3	1.8 1.4 1.8 0.7	(0.7–4.8) (0.7–3.1) (0.8–4.3) (0.1–5.8)	No. of discordant pairs No. of discordant pairs
USA (Loomis & Savitz, 1990)	2173 male deaths from brain cancer (ICD-9 191) 21 730 male deaths from other causes (1985–86)	Death certificates (16 US states)	Death certificate; occupation	Electrical occupations	75	1.4	(1.1–1.7)	Adjusted for race and age

Table 30 (contd)

Country (reference)	Subjects: cases, controls (recruitment period)	Source of subjects	Source of job information; exposure assessment methods	Estimates of exposure to ELF magnetic fields	No. of cases	Odds ratio (95% CI)	Comments
Australia (Ryan et al., 1992)	110 glioma 60 meningioma 417 controls (1987–90)					0.75 (0.30–1.9) 0.90 (0.20–4.1)	Matched on age
Sweden (Floderus et al., 1993)	261 men with brain tumours (astrocytomas and oligodendrogliomas) 1121 male population controls (1983–87)	Cancer registry/ population registry	Postal question-naire and spot measurements; job–exposure matrix	**Mean level** ≤ 0.15 µT (Q1) 0.16–0.19 µT (Q2) 0.20–0.28 µT (Q3) ≥ 0.29 µT (Q4) ≥ 0.41 µT	53 59 72 74 24	1.0 (baseline) 1.0 (0.7–1.6) 1.5 (1.0–2.2) 1.4 (0.9–2.1) 1.2 (0.7–2.1)	
USA (Sahl et al., 1993)	31 deaths from brain cancer (ICD-9 191) 286 cohort controls	Cohort of electric utility workers	Company personnel records; job–exposure matrix based on measured magnetic fields	Treating cumulative mean exposure as a continuous variable. Odds ratio per 25 µT–years of exposure	4	0.81 (0.48–1.4)	
Canada, France (3 cohorts combined) (Thériault et al., 1994)	108 men with brain cancer (ICD-9 191) 415 cohort controls	Three cohorts of electric utility workers in Canada (Québec and Ontario) and France	Company personnel records; job–exposure matrix based on measurements of exposure to magnetic fields (Positron meter)	**Cumulative exposure** < 3.1 µT–years > 3.1 µT–years > 15.7 µT–years	60 48 12	1.0 (baseline) 1.5 (0.85–2.8) 2.0 (0.76–5.0)	

Table 30 (contd)

Country (reference)	Subjects: cases, controls (recruitment period)	Source of subjects	Source of job information; exposure assessment methods	Estimates of exposure to ELF magnetic fields	No. of cases	Odds ratio (95% CI)	Comments
Quebec cohort (Thériault et al., 1994)	24 men with brain cancer 94 cohort controls (1970–88)			**Cumulative exposure**			
				< 3.1 μT–years	6	1.0	(baseline)
				> 3.1 μT–years	18	1.6	(0.38–6.8)
				> 15.7 μT–years	6	1.7	(0.29–9.7)
Ontario cohort (Thériault et al., 1994)	24 men with brain cancer 90 cohort controls (1970–88)			**Cumulative exposure**			
				< 3.1 μT–years	7	1.0	(baseline)
				> 3.1 μT–years	17	1.9	(0.53–6.5)
				> 15.7 μT–years	4	5.5	(0.59–51)
France cohort (Thériault et al., 1994)	60 men with brain cancer 231 cohort controls (1978–89)			**Cumulative exposure**			
				< 3.1 μT–years	47	1.0	(baseline)
				>3.1 μT–years	13	1.4	(0.65–3.1)
				>15.7 μT–years	2	NR	–
Norway (Tynes et al., 1994a)	39 men with brain tumours 194 cohort controls (1958–90)	Cohort of railway workers	Job history from employment files; job–exposure matrix	Ever worked at an electric line	28	0.82	(0.38–1.8)	Matched on age
				Cumulative exposure			
				None	11	1.0	(baseline)
				Low (0.1–310 μT–years)	14	0.81	(0.33–2.0)
				High (311–3600 μT–years)	14	0.94	(0.39–2.3)
				Very high (1900–3600 μT–years)	3	0.97	(0.24–4.0)	Unadjusted
USA (Grayson, 1996)	230 men with brain cancer (ICD-9 191) 920 cohort controls (1970–89)	Cohort of male members of the US Air Force	Work history from personnel records; job–exposure matrix (scores for exposure probability)	Ever exposed	129	1.3	(0.95–1.7)
				1–59[b]	39	1.3	(0.81–2.1)
				60–134	33	0.93	(0.56–1.5)
				135–270	44	1.6	(1.0–2.6)
				271–885	13	1.4	(0.88–2.3)

Table 30 (contd)

Country (reference)	Subjects: cases, controls (recruitment period)	Source of subjects	Source of job information; exposure assessment methods	Estimates of exposure to ELF magnetic fields	No. of cases	Odds ratio (95% CI)		Comments
England & Wales (Harrington et al., 1997)	112 men and women with brain cancer (primary) 654 cohort controls (1972–84)	Cohort of electricity generation and transmission workers	Job history from employment files; job–exposure matrix	≤ 3.0 µT–years 3.1–5.9 µT–years ≥ 6.0 µT–years Unclassifiable	30 37 27 18	1.0 1.3 0.91 1.8	(baseline) (0.75–2.2) (0.51–1.6) (0.93–3.6)	
Sweden (Rodvall et al., 1998)	84 men with glioma 155 population controls (1987–90)	In-patient files and cancer registry/ population registry	Postal question- naire; job–exposure matrix	< 0.20 µT > 0.40 µT		1.0 1.9	(baseline) (0.8–5.0)	Adjusted for socio-economic status and exposure to solvents and plastic materials
	20 men with meningioma 155 population controls			< 0.20 µT > 0.40 µT		1.0 1.6	(baseline) (0.3–10)	
USA (Cocco et al., 1998a)	28 416 deaths from central nervous system cancer (men and women) 113 664 deaths from other causes (controls) (1984–92)	Death certificates (24 US states)	Job on death certificate; job- exposure matrix (exposure yes/no)	White men Black men White women Black women	5271 234 1382 78	1.0 1.0 1.0 1.2	(1.0–1.0) (0.8–1.2) (0.9–1.1) (0.9–1.6)	Adjusted for socioeconomic status and other variables
USA (Cocco et al., 1999)	12 980 female deaths from central nervous system cancer + meningioma 51 920 deaths from other causes (controls) (1984–92)	Death certificates (24 US states)	Job on death certificate; job- exposure matrix: probability and intensity of exposure					

Table 30 (contd)

Country (reference)	Subjects: cases, controls (recruitment period)	Source of subjects	Source of job information; exposure assessment methods	Estimates of exposure to ELF magnetic fields	No. of cases	Odds ratio (95% CI)		Comments
	12 819 deaths from central nervous system cancer			Any exposure level	2901	1.2	(1.1–1.2)	
				Probability				
				Low	2312	1.2	(1.1–1.2)	
				Medium	255	1.2	(1.0–1.4)	
				High	334	1.2	(1.0–1.3)	
				Intensity				
				Low	2200	1.2	(1.1–1.2)	
				Medium	616	1.1	(1.0–1.3)	
				High	85	1.3	(1.0–1.6)	
	161 deaths from meningioma			Any exposure level	34	0.9	(0.6–1.4)	
Breast cancer								
Women								
USA (Loomis *et al.*, 1994b) [included in Cantor *et al.*, 1995]	28 434 deaths from breast cancer (women, excluding homemakers) 113 011 other causes of death (controls) (1985–89)	Death certificates (24 US states)	Job on death certificate; occupation	Electrical occupations	68	1.4	(1.0–1.8)	Adjusted for age, race, social class
USA (Cantor *et al.*, 1995)	33 509 deaths from breast cancer (women, excluding homemakers) 117 794 other causes of death (controls) (1984–89)		Job on death certificate; job–exposure matrix: probability and level of exposure	*White women*				
				Probability				
				Low	8581	0.92	(0.89–0.95)	
				Medium	779	1.1	(1.05–1.3)	
				High	1869	1.1	(1.02–1.2)	
				Level				
				Low	9360	0.94	(0.9–0.96)	
				Medium	1746	1.1	(1.03–1.2)	
				High	123	0.97	(0.8–1.2)	

Table 30 (contd)

Country (reference)	Subjects: cases, controls (recruitment period)	Source of subjects	Source of job information; exposure assessment methods	Estimates of exposure to ELF magnetic fields	No. of cases	Odds ratio (95% CI)	Comments
				Black women			
				Probability			
				Low	1516	0.81 (0.7–0.9)	
				Medium	168	1.3 (1.1–1.6)	
				High	293	1.3 (1.1–1.6)	
				Level			
				Low	1684	0.85 (0.8–0.92)	
				Medium	273	1.3 (1.1–1.5)	
				High	20	1.2 (0.7–2.1)	
USA (Coogan *et al.*, 1996)	6888 women with breast cancer 9529 population controls (1988–91)	4 US states	Usual occupation from telephone interview; job–exposure matrix	**Potential for exposure**			Adjusted for risk factors for breast cancer
				Low	577	1.0 (0.91–1.2)	
				Medium	104	1.1 (0.83–1.4)	
				High	57	1.4 (0.99–2.1)	
	1424 women with premenopausal breast cancer 2675 population controls			Low	91	0.91 (0.69–1.2)	
				Medium	18	0.82 (0.45–1.5)	
				High	20	2.0 (1.0–3.8)	
	5163 women with postmenopausal breast cancer 6421 population controls			Low	462	1.0 (0.89–1.2)	
				Medium	78	1.1 (0.80–1.5)	
				High	35	1.3 (0.82–2.2)	
USA (Coogan & Aschengrau, 1998)	259 women with breast cancer 738 general population controls (1983–86)	5 Upper Cape Cod towns in MA	Work history from questionnaire; job–exposure matrix	Occupational exposure to medium magnetic fields	16	0.9 (0.5–1.7)	Adjusted for risk factors for breast cancer
				Occupational exposure to high magnetic fields	7	1.2 (0.4–3.6)	

Table 30 (contd)

Country (reference)	Subjects: cases, controls (recruitment period)	Source of subjects	Source of job information; exposure assessment methods	Estimates of exposure to ELF magnetic fields	No. of cases	Odds ratio (95% CI)	Comments
Sweden (Forssén et al., 2000)	1767 women with breast cancer 1766 population controls (1960–65)	Cohort of residents near power lines	Occupation from census; job–exposure matrix	**Occupational exposures** < 0.12 µT 0.12–0.19 µT ≥ 0.20 µT **Occupational and residential exposures** < 0.12 µT = 0.12 µT	156 178 62 31 8	1.0 (baseline) 1.0 (0.7–1.4) 1.0 (0.7–1.6) 1.0 (baseline) 0.9 (0.3–2.7)	Matched on age, individual power line and municipality
Men							
USA (Demers et al., 1991)	227 men with incident breast cancer 300 population controls (1983–87)	10 US population-based cancer registries	2 occupations of longest duration from questionnaire; occupation	**Any electrical occupation** Ever exposed < 10 years 10–19 years 20–29 years ≥ 30 years	33 10 6 8 9	1.8 (1.0–3.7) 1.8 (0.7–4.9) 1.8 (0.5–6.2) 1.5 (0.5–4.3) 2.1 (0.7–6.2)	
USA (Rosenbaum et al., 1994)	71 men with incident breast cancer 256 volunteers from cancer screening clinic (1979–88)	Western New York state	Hospital registration cards, city directories; electrical occupations	Electrical occupations	6	0.6 (0.2–1.6)	Adjusted for age, county and heat exposure
Sweden (Stenlund & Floderus, 1997)	63 men with breast cancer 1121 population controls (1985–91)	Cancer registry	Postal questionnaire and spot measurements; job-exposure matrix	**Mean level** ≤ 0.15 µT (Q1) 0.16–0.19 µT (Q2) 0.20–0.28 µT (Q3) ≥ 0.29 µT (Q4) ≥ 0.41 µT	11 17 17 11 4	1.0 (baseline) 1.2 (0.6–2.7) 1.3 (0.6–2.8) 0.7 (0.3–1.9) 0.7 (0.2–2.3)	Adjusted for age, education and exposure to solvents

Table 30 (contd)

Country (reference)	Subjects: cases, controls (recruitment period)	Source of subjects	Source of job information; exposure assessment methods	Estimates of exposure to ELF magnetic fields	No. of cases	Odds ratio (95% CI)		Comments
USA (Cocco et al., 1998b)	178 male deaths from breast cancer 1041 male deaths from other causes (controls) (1985–86)	Death certificates	Occupation of longest duration from questionnaire to next-of-kin; job–exposure matrix	**Probability of exposure**				
				Low	30	1.0	(0.6–1.6)	
				Medium	7	1.2	(0.5–3.1)	
				High	19	1.1	(0.6–1.9)	
				Level of exposure				
				Low	31	1.0	(0.6–1.7)	
				Medium	16	1.1	(0.6–2.0)	
				High	9	1.0	(0.5–2.1)	

ALL, acute lymphoblastic leukaemia; AML, acute myeloid leukaemia; ANLL, acute non-lymphoblastic leukaemia; CLL, chronic lymphocytic leukaemia; CML, chronic myeloid leukaemia; CNLL, chronic non-lymphocytic leukaemia; MDS, myelodysplastic syndrome; NR, not reported; Q, quartile
[a] This study included five cases of leukaemia not included in the initial analysis by Thériault et al. (1994); this explains the different results for Ontario workers reported in the two papers.
[b] Product of potential exposure score and duration in months

Table 31. Case–control studies of occupational groups with assumed or documented exposure to ELF electric fields[a]

Country (reference)	Subjects (recruitment period)	Source of subjects	Source of job information; exposure assessment methods	Estimates of exposure to ELF electric fields	No. of cases	Odds ratio (95% CI)	Comments
Leukaemia							
Canada (Miller et al., 1996) [included in Thériault et al., 1994][b]	50 men with leukaemia 199 cohort controls (1970–88)	Cohort of electric utility workers at Ontario Hydro	Company personnel records; job–exposure matrix based on magnetic fields and measurements of exposure to electric fields (Positron meter)	Electric fields (V/m-years) 0–171 / 172–344 / ≥ 345	11 / 13 / 26	1.0 (baseline) / 2.1 (0.59–7.2) / 4.5 (1.0–20)	Adjusted for socioeconomic status and potential occupational confounders
	20 incident cases of ANLL 80 cohort controls			0–171 / 172–344 / ≥ 345	4 / 6 / 10	1.0 (baseline) / 10 (0.58–172) / 7.9 (0.43–143)	
	19 incident cases of CLL 76 cohort controls			0–171 / 172–344 / ≥ 345	3 / 6 / 10	1.0 (baseline) / 1.3 (0.07–21) / 7.2 (0.31–169)	
France (Guénel et al., 1996) [included in Thériault et al., 1994]	72 men with leukaemia 285 cohort controls (1978–89)	Cohort of electric utility workers at Electricité de France	Company personnel records; job–exposure matrix based on measurements of exposure to electric fields (Positron meter)	Electric fields (V/m-years) < 253 / 253–329 / 330–401 / ≥ 402	38 / 20 / 10 / 4	1.0 (baseline) / 0.96 (0.45–2.0) / 0.71 (0.27–1.9) / 0.37 (0.11–13)	Adjusted for socioeconomic status
	34 men with ANLL 134 cohort controls			percentiles < 50 / ≥ 50–< 75 / ≥ 75–< 90 / ≥ 90	18 / 10 / 4 / 2	1.0 (baseline) / 0.95 (0.45–2.0) / 0.71 (0.26–1.9) / 0.36 (0.10–1.3)	Adjusted for socioeconomic status and exposure to magnetic fields

Table 31 (contd)

Country (reference)	Subjects (recruitment period)	Source of subjects	Source of job information; exposure assessment methods	Estimates of exposure to ELF electric fields	No. of cases	Odds ratio (95% CI)	Comments
USA (Kheifets *et al.*, 1997b) [same study as London *et al.*, 1994]	2355 men with leukaemia 67 212 other cancer cases (1972–90)	Los Angeles county cancer registry	Job at diagnosis from medical record; job–exposure matrix based on measured electric fields in selected occupations	Electric fields (V/m) < 10 10–20 > 20	2296 28 31	1.0 1.2 1.2	(baseline) (0.80–1.9) (0.78–1.7)
	853 men with ANLL			< 10 10–20 > 20	831 11 11	1.0 1.3 1.2	(baseline) (0.68–2.5) (0.59–2.2)
	534 men with CLL			< 10 10–20 > 20	517 9 8	1.0 1.9 1.3	(baseline) (1.1–3.2) (0.72–2.2)
	487 men with CML			< 10 10–20 > 20	478 2 7	1.0 0.39 1.3	(baseline) (0.09–1.6) (0.60–2.8)
Norway (Tynes *et al.*, 1994b)	52 men with leukaemia 259 cohort controls (1958–90)	Cohort of railway workers	Job history from employment files; job-exposure matrix	Electric fields (kV/m-years) None Low (0.1–5) High (> 5–30) Very high (21–30)	19 9 24 3	1.0 0.44 0.98 0.68	(baseline) (0.18–1.1) (0.48–2.0) (0.18–2.6)

Table 31 (contd)

Brain tumours

Country (reference)	Subjects (recruitment period)	Source of subjects	Source of job information; exposure assessment methods	Estimates of exposure to ELF electric fields	No. of cases	Odds ratio (95% CI)		Comments
Norway (Tynes et al., 1994b)	39 men with brain tumours 194 cohort controls (1959–90)	Cohort of railway workers	Job history from employment files; job–exposure matrix	Electric fields (kV/m–years) None Low (0.1–5) High (> 5–30) Very high (21–30)	11 12 16 4	1.0 0.69 1.2 1.2	(baseline) (0.28–1.7) (0.49–2.8) (0.33–4.6)	
Canada (Miller et al., 1996) [included in Thériault et al., 1994]	24 incident cases of malignant brain tumours 96 cohort controls [exact number not given] (1970–88)	Cohort of electric utility workers at Ontario Hydro	Company personnel records; job–exposure matrix based on electric field exposure measurements (Positron meter)	Electric fields (V/m–years) 0–171 172–344 ≥ 345	12 4 8	1.0 0.57 0.99	(baseline) (0.10–3.2) (0.16–6.2)	Adjusted for socioeconomic status and potential occupational confounders

Table 31 (contd)

Country (reference)	Subjects (recruitment period)	Source of subjects	Source of job information; exposure assessment methods	Estimates of exposure to ELF electric fields	No. of cases	Odds ratio (95% CI)		Comments
France (Guénel et al., 1996) [included in Thériault et al., 1994]	69 incident cases of brain tumour (ICD-9 191, 225) 271 cohort controls (1978–89)	Cohort of electric utility workers at Electricité de France	Company personnel records; job–exposure matrix based on electric field exposure measurements (Positron meter)	Electric fields (V/m–years) <238 238–318 319–386 ≥387	29 22 8 10	1.0 2.5 1.4 3.1	(baseline) (0.99–6.2) (0.46–4.5) (1.1–8.7)	Adjusted for socioeconomic status
	59 incident cases of malignant brain tumour (ICD-9 191) 231 cohort controls			Percentiles < 50 ≥ 50–< 75 ≥ 75–< 90 ≥ 90		1.0 2.5 1.6 1.8	(baseline) (0.93–6.8) (0.46–5.4) (0.54–5.7)	Adjusted for socioeconomic status and exposure to magnetic fields

ANLL, acute non-lymphoblastic leukaemia; CLL, chronic lymphocytic leukaemia; CML, chronic myeloid leukaemia

[a] Electric field measurements in occupational studies are made using meters that are worn on the body. The results are therefore difficult to interpret because in this situation the field is distorted and the measurement is sensitive to body position.

[b] This study included five cases of leukaemia not included in the initial analysis by Thériault et al. (1994); this explains the different results for Ontario workers reported in the two papers.

cancer registry, with four controls per case matched on age and year of registration. For the combined group of selected occupations involving potential exposure to electrical and magnetic fields, a marginally significant excess risk for leukaemia was seen (odds ratio, 1.7; 95% CI, 0.97–3.0) on the basis of 18 observed cases. In an extension of this study to cover the registration period 1980–84, Pearce *et al.* (1989) used 19 904 of 24 762 notified cases of cancer among men ≥ 20 years old for whom information on occupation was available (80% of all relevant registry notifications) to evaluate any link between site-specific cancer and 'electrical work'. For each site of cancer under investigation, other sites formed the control group. 'Electrical work' was associated with an increased risk for leukaemia (odds ratio, 1.6; 95% CI, 1.0–2.5) on the basis of 21 observed cases. The odds ratios were generally greater for chronic leukaemia (odds ratio, 2.1; 95% CI, 1.2–3.8) than for acute leukaemia (odds ratio, 1.3; 95% CI, 0.62–2.5) and the risk was generally greater for subjects aged 65 years or more than for those aged 20–64 years.

In a case–control study based on death certificates recorded in 1985 and 1986 in 16 states in the USA (Loomis & Savitz, 1990), 3400 cases of leukaemia among men ≥ 20 years were compared with approximately 34 000 controls matched on year of death and who had died from causes other than brain cancer or leukaemia. Decedents were allocated to the exposed group if the occupation or industry given on their death certificate indicated that they had held a job included in a predefined list of electrical occupations (Milham, 1982). All other jobs were considered as unexposed. There was no association between electrical occupation and leukaemia (odds ratio, 1.0; 95% CI, 0.8–1.2). A slightly increased risk was observed for acute lymphoblastic leukaemia (odds ratio, 1.5; 95% CI, 0.7–3.4).

Richardson *et al.* (1992) conducted a case–control study of men and women ≥ 30 years old, resident in France. The cases had been diagnosed with acute leukaemia in two hospitals in France between 1984 and 1988; the 561 controls were patients in other departments at the same hospitals, matched to cases for sex, age (± 5 years), ethnic group and place and type of dwelling. Information on past medical history including radiotherapy and chemotherapy, drug use, some sources of environmental exposure and exposure related to leisure activities and a full occupational history described by job titles and industrial activities was obtained by personal interview for 204 cases (72% of those eligible) and 561 controls. Case and control subjects for whom the interviewer had recorded poor cooperation (approximately 5%) and case subjects without controls and vice versa were subsequently excluded, leaving 185 (154 acute myeloid leukaemia and 31 acute lymphoblastic leukaemia) cases (50.2% men) and 513 (48.2% men) controls for analysis. Exposure to ELF electric and magnetic fields, benzene, ionizing radiation, exhaust fumes and pesticides were assessed by an industrial hygienist on the basis of the reported occupational history of each study subject. Whenever possible, the exposure to an agent was coded as either low (< 5% of working time), medium (5–50%) or high (> 50%). There were three electronic engineers among cases and none among controls. After adjusting for prior chemotherapy or radiotherapy and taking into account the matching variables in an unconditional logistic regression model, any occupational exposure to electric and magnetic fields (all types of exposure) was shown to be associated with an elevated relative risk for acute leukaemia (odds ratio, 1.7; 95% CI, 0.9–3.5; 14 cases), while occupational exposure to ionizing radiation was not (odds ratio, 0.7; 95% CI, 0.2–2.0). Dividing the group of workers exposed to electric and magnetic fields into arc welders and others gave odds ratios of 1.2 (95% CI, 0.5–3.0) for welders and 3.9 (95% CI, 1.2–13) for others, based on eight and seven cases, respectively. [The Working Group noted that the risk estimation made after the separation of sources of exposure to electric and magnetic fields into arc-welding and non-arc-welding should be regarded as a post-hoc analysis. The Working Group also noted that all seven cases of acute leukaemia in workers exposed to electric and magnetic fields from arc-welding were acute myeloid leukaemia and that there was no information on the subtype distribution among the eight cases who were exposed to electric and magnetic fields from sources other than arc-welding.]

A case–control study within a cohort of telephone linemen at the American Telephone and Telegraph company was conducted by Matanoski et al. (1993). The cases were deaths from leukaemia, except chronic lymphocytic leukaemia, that occurred from 1975–80 among white men who had worked for the company for at least two years. Deaths were identified from company records for all workers who were still employed by the company when they died and for a subset of retired workers. From 177 eligible cases and their matched controls, a complete job history was obtained in 35 sets, each set was composed of a case plus at least one of its matched controls. The assessment of exposure to magnetic fields was made using the EMDEX-C personal monitor to make measurements on 15–61 individuals in each occupational category (204 measurements at 10-second intervals). No assessment of exposure to other potentially leukaemogenic agents was performed. The odds ratio for exposure above median of the mean values was 2.5 (95% CI, 0.7–8.6) compared with exposure below median of the mean values. There was also an indication of a dose–response relationship when subjects were divided into quartiles of peak exposure. [The Working Group noted that little weight should be given to a study in which only 35 of 177 eligible cases were included. It is not listed in Table 30.]

In a hospital-based case–control study conducted in one hospital in northern Italy, 46 men and 40 women aged between 15 and 74 years who had been newly diagnosed during 1989–90 with myeloid leukaemia (acute and chronic) or myelodysplastic syndrome were identified (Ciccone et al., 1993). Two control groups were chosen, one selected from all patients newly diagnosed with other diseases at the same hospital and one selected from the city population in the area of the hospital. Both groups were frequency-matched to the cases on sex, age and area of residence. The response rates were 91% for cases, 99% for the hospital controls and 82% for the population controls, leaving 86 cases (50 patients with acute myeloid leukaemia, 17 with chronic myeloid leukaemia and 19 with myelodysplastic syndrome) and 246 controls for analysis. The occupational history of the study subjects was used by one industrial hygienist to assess the probability of exposure to ELF electric and magnetic fields and to eight other agents or classes of agent known or suspected to increase the risk for myeloid leukaemia, myelodysplastic syndrome or other haematolymphopoietic malignancies. Using logistic regression analysis, male study subjects possibly or probably exposed to electric and magnetic fields had a non-significantly increased odds ratio of 1.6 (95% CI, 0.6–4.1) for myeloid leukaemia or myelodysplastic syndrome combined, compared with subjects not exposed to electric and magnetic fields or any of the other risk factors under study. The equivalent risk estimate for women was 0.8 (95% CI, 0.2–2.5). The estimates were based on 17 men and four women who had been exposed to electric and magnetic fields.

In a cohort study of cancer mortality in 36 221 electricity utility workers who had been employed at the Southern California Edison Company for at least one year between 1960 and 1988, the main analyses used a nested case–control study design, based on 3125 identified causes of death at the end of the follow-up period in 1988 (Sahl *et al.*, 1993). Magnetic fields were measured over 776 person–days in 35 occupational categories using the EMDEX-2 meter. Case–control analyses were presented for 44 cases of leukaemia, but no association with scores for exposure to magnetic fields was observed (mean, median, 99th percentile, fraction above different thresholds). In a re-analysis of these data based on different exposure categories, a modest non-significantly increased risk for leukaemia was apparent (Kheifets *et al.*, 1999).

Within a well-defined population of men who, according to the 1980 census, were employed and living in mid-Sweden, Floderus *et al.* (1993) conducted a study of all men aged 20–64 years notified to the Swedish Cancer Registry with a recent diagnosis of leukaemia (*n* = 426) during 1983–87. For the control group, two subjects per case (*n* = 1700) were chosen from the source population and matched to the case on age. Only acute lymphoblastic leukaemia, acute myeloid leukaemia, chronic myeloid leukaemia and chronic lymphocytic leukaemia were included. A postal questionnaire was used in which a full employment history was requested, including a description of all major work tasks undertaken by the study subject during the 10-year period before the diagnosis (and the equivalent dates for the controls). The questionnaire was completed by 77% of leukaemia patients or their relatives and 72% of the control subjects who received the questionnaire, so that 250 leukaemia cases and 1121 controls were available for analysis. On the basis of the work task held for the longest time by 1015 cases and control participants, a full-day measurement of exposure to ELF electric and magnetic fields at a frequency of 50 Hz was conducted using EMDEX-100 and EMDEX-C meters. Exposure categories were defined on the basis of the quartiles of exposure levels measured among the control subjects. The evaluation of exposure to potential confounders (benzene, other solvents, ionizing radiation and smoking) was based on self-reported information from study subjects and workplace information. On the basis of the job held for the longest time during the 10-year period before diagnosis, the age-adjusted odds ratios for all types of leukaemia combined were 0.9 (95% CI,

0.6–1.4), 1.2 (95% CI, 0.8–1.9) and 1.6 (95% CI, 1.1–2.4) among study subjects with daily mean level of exposure to magnetic fields in the second (0.16–0.19 µT), third (0.20–0.28 µT) and upper (≥ 0.29 µT) exposure quartiles, respectively, when compared with the risk of subjects with exposure in the lower quartile (≤ 0.15 µT). In an extended analysis on leukaemia subtypes, the excess risk seemed to be due exclusively to an increased risk for chronic lymphocytic leukaemia, with an odds ratio for exposure in the upper quartile of 3.0 (95% CI, 1.6–5.8). With exposure above the 90th percentile (≥ 0.41 µT), the odds ratio for chronic lymphocytic leukaemia was 3.7 (95% CI, 1.8–7.7). The results were not changed when potential confounders were taken into consideration; however, no independent risk estimates were given for these potential confounders. [The Working Group noted that the different proportions of postal questionnaires completed by next-of-kin (cases, 67%; controls, 0%) may have affected the odds ratios.]

London et al. (1994) conducted a case–control study based on cancer registry data. The cases were 2355 men aged 20–64 years diagnosed with leukaemia, and reported to the population-based cancer registry for Los Angeles county between 1972 and 1990. The controls were 67 212 men diagnosed with other cancers, excluding malignancies of the central nervous system. Only the occupation recorded in the medical record at the time of diagnosis was available to estimate occupational exposure to electric and magnetic fields. The assessment of exposure was based on measurements of magnetic fields obtained for 278 electrical workers in nine electrical occupations and 105 workers in 18 non-electrical occupations selected at random from the general population. The workers selected from each occupational group wore an EMDEX monitor for one work shift. A task-weighted estimate of exposure to magnetic fields in a given occupation was made. A single exposure index was calculated for all non-electrical occupations for which the mean exposure to magnetic fields was generally lower than that in electrical jobs. Occupational exposure to ionizing radiation, benzene, chlorinated hydrocarbon solvents, other solvents and pesticides was evaluated by an expert panel. Using the magnetic field exposure estimates, the odds ratios were 1.0, 1.2 (95% CI, 1.0–1.6) and 1.4 (95% CI, 1.0–2.0), respectively, for exposure to < 0.17 µT, 0.18–0.80 µT and ≥ 0.81 µT, and the trend was statistically significant. An analysis by leukaemia subtype showed a high odds ratio for chronic myeloid leukaemia (odds ratio, 2.3; 95% CI, 1.4–3.8) for average exposure ≥ 0.81 µT, compared with exposure < 0.17 µT, but there was also some evidence of increased risk for acute non-lymphoblastic leukaemia and chronic lymphocytic leukaemia. According to the authors, these results were not appreciably affected by adjustment for other occupational exposures. Data on electric fields were also collected in this study (Table 31). The measurements of electric fields by occupational group revealed no clear evidence of an association between this exposure and leukaemia, and no exposure–response relationship for any leukaemia sub-type was seen (Kheifets et al., 1997b). [The Working Group noted that the assessment of occupational exposure was based on a single occupation recorded at the time of cancer diagnosis; only a few non-electrical occupations were measured, but they were used as an exposure proxy for all other non-electrical occupations; controls were other cancer cases: if workers in electrical occupations have a lower incidence of other cancers than non-electrical workers, the odds ratio for leukaemia could be spuriously elevated relative to non-electrical workers. It was also noted that the cut-points used for the categorization of exposure correspond to the 97% and 99% percentile in the control population. No clear explanation was given for this apparently unusual choice.]

Tynes *et al.* (1994b) conducted a case–control study of leukaemia nested in a cohort of 13 030 male railway line workers, exposed to 16 2/3-Hz electric and magnetic fields, outdoor station workers and railway electricity workers (railway electricity line workers, installation electricians, radio communication workers and railway power substation workers) selected from the records of all employees working on either electric or non-electric railways in Norway in 1957 and from historical and current databases provided by the railway workers' trade union. The case groups comprised all 52 members (one case was excluded because no work history was available) diagnosed according to the files of the national Norwegian Cancer Registry with leukaemia during the follow-up period 1958–90. Each case was matched on year of birth with four or five controls selected from the cohort (a total of 259 controls). Work histories were combined with a job–exposure matrix for ELF-electric and magnetic fields to provide a simple exposure categorization: i.e. ever exposed versus never exposed to electric railway lines, and a more complex one: i.e. cumulative exposure (μT–year) during a person's entire period of employment (up to the date of diagnosis of cancer or a similar date for the matched controls) with the railways in Norway. Limited information on potential confounders such as exposure to creosote, solvents and herbicides was also collected; information on smoking (ever smokers) was obtained by telephone interviews with the subjects or their work colleagues. Ever exposure to magnetic fields from electric railway lines was associated with an odds ratio of 0.72 (95% CI, 0.37–1.4) for all types of leukaemia combined. An analysis of leukaemia subtypes also showed no association. Using study subjects never exposed to magnetic fields from electric railway lines as the exposure reference category, cumulative exposures of 0.1–310 μT–years (low exposure), 311–3600 μT–years (high) and 1900–3600 μT–years (very high) were associated with odds ratios of 1.0 (95% CI, 0.49–2.1), 0.49 (95% CI, 0.22–1.1) and 0.84 (95% CI, 0.25–2.8), respectively, for leukaemia. Adjustment for smoking habits and potential confounders in a multivariate regression analysis for matched pairs did not change the results. Sub-analyses with inclusion of lag time intervals (5 and 15 years) and exposure windows (5–25 years and 2–12 years) did not reveal any associations. Analysis for electric fields did not show any association with leukaemia (see Table 31).

A large case–control study of exposure to magnetic fields nested within three cohorts of electric utility workers in Quebec and Ontario, Canada, and in France was conducted by Thériault *et al.* (1994). There were small differences in study design and the results were not consistent across the three utilities; each cohort is therefore described separately.

In Quebec, the cohort included all men with at least one year of employment at Hydro-Québec, between January 1970 and December 1988. The observation period ended either at death or December 31, 1988. The cases were 774 men from the cohort newly diagnosed with cancer during this period (24 leukaemia). The controls were 1223 cohort members matched to the cases by year of birth with a case–control ratio of 1:4 for cancer of the haematopoietic system, brain cancer and skin melanoma, and 1:1 for all other cancer sites. Measurements of magnetic field exposure were made with a personal Positron meter, worn for a full working week by 466 workers at Hydro Québec, who had been selected to achieve a representative sample of all workers in 32 occupational groups. The time-weighted average exposure to magnetic fields was calculated from the measurements to construct a job–exposure matrix. Past exposure to magnetic fields was estimated using adjustment factors based on changes in power systems, work techniques and exposure sources. Exposure to other potential occupational carcinogens was evaluated through expert judgement. The odds ratio for cumulative exposure to magnetic fields above median (3.1 µT–years) was 0.29 (95% CI, 0.04–1.8) for all leukaemia and 0.75 (95% CI, 0.00–>100) for acute non-lymphoblastic leukaemia, based on small numbers. No clear association with other leukaemia types was observed.

In Ontario, the cohort comprised men with one full year of employment at Ontario Hydro between 1973 and 1988, as well as men on the pension roll in 1970–73. The observation period ended at death or December 31, 1988. A total of 1472 incident cancer cases (45 leukaemia) were identified from the Ontario Cancer Registry during the study period. The controls were 2080 men selected in the same way as those from Hydro Québec. Measurements of exposure to magnetic fields were made for 771 workers with 260 job titles. The occupations were then combined in 17 broad categories, based on mean exposure, occupational profiles and consideration of past changes in these factors. These 17 occupational categories were used as the rows of the job–exposure matrix on magnetic fields. Exposure to other occupational agents ((2,4-dichlorophenoxy)acetic acid, (2,4,5-trichlorophenoxy)acetic acid and benzene) was assessed from consultation with experts (Miller *et al.*, 1996). The odds ratio for cumulative exposure to magnetic fields above 3.1 µT–years was 3.1 (95% CI, 1.1–9.7) for all leukaemia and 6.2 (95% CI, 0.95–78) for acute non-lymphoblastic leukaemia. Non-significant increases in odds ratios were also observed for chronic lymphocytic leukaemia.

In France, the cohort included men with at least one year of employment at Électricité de France–Gaz de France during 1978–89. The cases were 1905 men identified from company medical records who were newly diagnosed with cancer (71 leukaemia) during the same period. This group of cases included all workers diagnosed with cancer while they were active in the company. Since the identification of cases was not possible for cancer diagnosed after retirement, the observation period ended at termination of employment or December 1989. The controls were 2803 subjects matched to the cancer cases by year of birth in the same way as in the cohort of Hydro Québec workers. The method of assessment of exposure to magnetic fields was similar to that used in Quebec.

Measurements were made using a Positron meter worn by 829 workers for a full working week, selected from 37 occupational groups defined *a priori*. Past exposure to magnetic fields was assessed using adjustment factors. Estimates of exposure to other potential occupational carcinogens were also evaluated using expert judgement in a separate job–exposure matrix. The odds ratios for cumulative exposure to magnetic fields above 3.1 μT–years were 1.4 (95% CI, 0.61–3.1) for all leukaemia and 1.8 (95% CI, 0.57–5.4) for acute non-lymphoblastic leukaemia.

For the three cohorts combined, the odds ratios for all leukaemia were 1.5 (95% CI, 0.90–2.6) for cumulative exposure to magnetic fields above median (3.1 μT–years) and 1.8 (95% CI, 0.77–4.0) for exposure above the 90th percentile (15.7 μT–years). The odds ratios for acute non-lymphoblastic leukaemia were 2.4 (95% CI, 1.1–5.4) and 2.5 (95% CI, 0.70–9.1), respectively. However, there was no clear trend of increased risk with increasing exposure. Elevated odds ratios were also observed for chronic lymphocytic leukaemia for cumulative exposure during the 20 years prior to diagnosis (Thériault *et al.*, 1994).

Data from the Ontario Hydro cohort were re-analysed (Miller *et al.*, 1996). This re-analysis included five additional cases of leukaemia not considered in the initial analysis by Thériault *et al.* (1994) (a total of 50 cases). A refined assessment of exposure to potential occupational confounders was also used. The odds ratio for all leukaemia decreased from 2.0 to 1.7 (95% CI, 0.58–4.8) for cumulative exposure between 3.2 and 7 μT–years and from 2.8 to 1.6 (95% CI, 0.47–5.1) for cumulative exposure ≥ 7.1 μT–years, after adjustment for potential occupational confounders. For acute non-lymphoblastic leukaemia, the corresponding odds ratios were reduced from 3.0 to 1.9 (95% CI, 0.27–14) and from 5.0 to 2.9 (95% CI, 0.42–20), respectively. This report also described the risk for leukaemia in relation to exposure to electric fields which were also measured by the Positron meter (Table 31). For leukaemia, the odds ratios for cumulative exposure to electric fields between 172 and 344 V/m–years and for exposure ≥ 345 V/m–years, as compared with exposure below 172 V/m–years (median), were 2.1 (95% CI, 0.59–7.2) and 4.5 (95% CI, 1.0–20), respectively, after adjustment for potential occupational confounders. For acute non-lymphoblastic leukaemia, and the main component, acute myeloid leukaemia, the odds ratios associated with electric fields were elevated but did not reach statistical significance. Analysis of the combined effects of electric and magnetic fields showed that exposure to electric fields carried a greater risk for leukaemia than exposure to magnetic fields. It was shown that risk for leukaemia was more particularly associated with duration of exposure above the exposure threshold (Villeneuve *et al.*, 2000).

The effects of electric fields were also investigated among electric utility workers from France. These workers were part of the Canada–France study (Thériault *et al.*, 1994). Electric fields were recorded by a Positron meter at the same time as magnetic fields and were used to assess the exposure to electric fields by occupation in a job–exposure matrix (Guénel *et al.*, 1996). No association between cumulative expo-sure to electric fields and leukaemia was observed in this study (Table 31).

Feychting *et al.* (1997) looked at combined residential and occupational exposure (see section 2.3.1).

In an Italian study, Pulsoni *et al.* (1998) compared selected characteristics of 335 patients with acute promyelocytic leukaemia aged > 15 years with those of 2894 patients aged > 15 years diagnosed with other acute myeloid leukaemia. Patients were identified from the files of a clinical database, initiated in 1992, until 1997. A significant association was found between working as an electrician and development of acute promyelocytic leukaemia with an age-adjusted odds ratio of 4.4 (95% CI, 2.0–9.7). [The Working Group noted that the occupational group considered (i.e. electricians) comprised less than 1% of the comparison group of other acute myeloid leukaemia patients making interpretation difficult.]

(ii)   *Brain tumours* (see Table 30)

Brain tumours without further histological classification represent a heterogeneous group of lesions. Studies based on death certificates only may include deaths from secondary tumours that have metastasized from an unknown primary cancer, or tumours that are histologically benign (Percy *et al.*, 1981). Where possible the results reported here are specifically for malignant tumours or for known histological types.

Death certificates for white men in Maryland who died between 1969 and 1982 were used to conduct a case–control study on brain tumours (Lin *et al.*, 1985). A total of 951 men aged ≥ 20 years who had died from a tumour of the brain (519 gliomas, glioblastoma multiforme, or astrocytomas) were matched by age and date of death with controls who had died from non-malignant diseases. The occupation recorded on the death certificate was used to classify the subjects according to a predefined category of exposure to electric and magnetic fields (definite, probable, possible or no exposure). Jobs were classified according to a list of 'electrical occupations' revised from that of Milham (1982). Using the no-exposure group as referent, the odds ratios for primary brain tumours increased with increasing probability of exposure to electric and magnetic fields.

Speers *et al.* (1988) conducted a study based on mortality data in East Texas, USA during the period 1969–78. The cases were 202 white male decedents between 35 and 79 years of age who had been diagnosed with glioma. The controls were 238 men selected from among white residents of the East Texas study area who had died from a cause other than brain tumour. Information abstracted from the death certificate included the usual occupation for which exposure to electric and magnetic fields was classified using the system proposed by Lin *et al.* (1985). The analysis by level of exposure to electric and magnetic fields yielded an increase in risk with increasing probability of exposure with a significant linear trend.

In a cancer registry-based study in New Zealand, Pearce *et al.* (1989) used 19 904 cases of cancer notified from 1980–84 among men ≥ 20 years old, for whom information on occupation was available (80% of all relevant registry notifications) to evaluate any link between site-specific cancer and 'electrical work'. For each site of cancer studied, patients with cancer at other sites formed the control group. Among 481 patients with brain cancer (ICD-9 191), 12 had been employed in electrical work, giving an odds ratio for brain tumours of 1.0 (95% CI, 0.56–1.8) on the basis of the 12 observed cases.

Preston-Martin *et al.* (1989) conducted a case-control study on brain tumours in Los Angeles county, USA. The cases were men 25–69 years of age for whom a first diagnosis of glioma or meningioma had been made during 1980–84. Two hundred and seventy-two of 478 eligible cases (202 gliomas and 70 meningiomas) and 272 controls were available for analysis. A complete work history was obtained for each subject, together with information on previous brain diseases, head traumas, alcohol and tobacco habits and diet. Work in an occupation with suspected exposure to electric and magnetic fields, according to Milham's definition (Milham, 1982), was associated with an increased odds ratio for glioma (1.8; 95% CI, 0.7–4.8), and the risk increased with increase in the number of years spent working in these occupations. The association was strongest for astrocytoma (odds ratio, 4.3; 95% CI, 1.2–16) for > 5 years). The authors noted that confounding from occupational exposure to other harmful agents (e.g. solvents) may be an alternative explanation for this finding. [The Working Group noted that selection bias may have occurred because only living patients could be interviewed.]

The study by Loomis and Savitz (1990) based on death certificates in 16 states in the USA, described in the section on leukaemia, also presented results for brain cancer. The cases were 2173 deaths in men from brain cancer (ICD-9 191). The controls were selected from among men who had died from other causes with a 10:1 ratio. Men who were reported to have been electrical workers on their death certificates had an odds ratio of 1.4 (95% CI, 1.1–1.7) when compared with non-electrical occupations.

Ryan *et al.* (1992) also looked at exposure to ELF electric and magnetic fields in the electrical and electronics industries and found no increase in glioma and meningioma.

Floderus *et al.* (1993) (described in the section on leukaemia) conducted a case–control study of all individuals with a recent diagnosis of brain tumour (*n* = 424) during 1983–87. Only patients with histologically confirmed astrocytoma (type I-IV) or oligodendroglioma were included. A questionnaire was completed by 76% of patients or their relatives and 72% of control subjects, leaving 261 cases of brain tumour and 1121 controls for analysis. Exposure was defined as in the section on leukaemia (p. 215). On the basis of the 10-year period before diagnosis, the age-adjusted odds ratios for all types of brain tumour combined were 1.0 (95% CI, 0.7–1.6), 1.5 (95% CI, 1.0–2.2) and 1.4 (95% CI, 0.9–2.1) among study subjects with a daily mean exposure to electric and magnetic fields in the second (0.16–0.19 μT), third (0.20–0.28 μT) and upper (≥ 0.29 μT) exposure quartiles, respectively, when compared with the lower quartile (≤ 0.15 μT). When exposure was above the 90th percentile (≥ 0.41 μT), the odds ratio was 1.2 (95% CI, 0.7–2.1). The results were unchanged when the potential confounders were taken into consideration. [The Working Group noted that the different proportions of postal questionnaires completed by next-of-kin; for cases (85%) and for controls (0%) may have affected the odds ratios.]

In a study on cancer mortality among employees at the Southern California Edison Company described in the section on leukaemia (Sahl *et al.*, 1993), brain cancer (ICD-9 191) (31 cases) was also investigated and the results are summarized in Table 30. No association between brain cancer mortality and scores of exposure to magnetic fields was apparent.

The Canada–France study on electric utility workers, described in the section on leukaemia (Thériault *et al.*, 1994), also presented results for brain cancer (ICD-9 191); they are shown in Table 30. For the three cohorts combined, the odds ratios for all brain cancers were 1.5 (95% CI, 0.85–2.8) for cumulative exposure to magnetic fields above median (3.1 μT–years) and 2.0 (95% CI, 0.76–5.0) for exposure above the 90th percentile (15.7 μT–years). In the analysis by histological subtype, the risk for astrocytoma was particularly elevated in the highest exposure category (odds ratio, 12; 95% CI, 1.1–144), but this result was based on only five exposed cases, and according to the authors was dependent on the statistical method used.

The association between malignant brain cancer and electric fields among workers at Ontario Hydro was investigated by Miller *et al.* (1996) (Table 31). No association between exposure to electric fields and brain cancer was apparent.

The relationship between exposure to electric fields and risk of brain tumour (ICD-9 191, 225) (59 malignant and 10 benign cancers) was also investigated in workers at Électricité de France (Table 31) (Guénel *et al.*, 1996), who were part of the Canada–France study described above. Using the arithmetic mean of electric field measurements obtained with the Positron meter, the odds ratio in the highest exposure category was 3.1 (95% CI, 1.1–8.7), but the risk did not increase monotonically with exposure. There was no clear indication of an increased risk when exposure was assessed using the geometric mean of electric fields.

In parallel to the leukaemia study described above, Tynes *et al.* (1994b) conducted a case–control study of brain tumours (unspecified) nested in a cohort of 13 030 male railway workers. The case group comprised all 39 cohort members diagnosed according to the files of the national Norwegian Cancer Registry with brain tumour, during the follow-up period 1958–90. Each case was matched on year of birth with four or five controls selected from the cohort (a total of 194). Ever exposure to electric railway lines was associated with an odds ratio of 0.82 (95% CI, 0.38–1.8) for brain tumours. Using study subjects who had never been exposed to electric railway lines as the exposure reference category, cumulative exposure to magnetic fields of 0.1–310 μT–years (low exposure), 311–3600 μT–years (high) and 1900–3600 μT–years (very high) were associated with odds ratios for brain tumours of 0.81 (95% CI, 0.33–2.0), 0.94 (95% CI, 0.39–2.3) and 0.97 (95% CI, 0.24–4.0), respectively. Subanalyses with inclusion of lag time intervals (5 and 15 years) and exposure windows (5–25 years and 2–12 years) did not reveal any associations. No association between brain tumours and exposure to electric fields was apparent (see Table 31).

In a case–control study nested within a cohort of male members of the US Air Force with at least one year of service in the period 1970–89, 230 cases of brain tumour (ICD-9 191) were matched on year of birth and race with 920 controls (Grayson, 1996). Complete job histories were linked to a job–exposure matrix that assessed the probability of exposure to ELF electric and magnetic fields (definite, probable, possible or no exposure) by job title and time of employment. The odds ratio for workers ever exposed to ELF electric and magnetic fields was 1.3 (95% CI, 0.95–1.7). However, no clear trend relating risk for brain tumour to cumulative exposure was apparent.

In a substudy from Sweden, Feychting et al. (1997) estimated the separate and combined effects of occupational and residential exposure to ELF magnetic fields on the risk for tumours of the central nervous system (see section 2.3.1 on residential exposure).

Mortality from brain cancer was investigated by Harrington et al. (1997) in a case–control study nested in a cohort of 84 018 men and women employed for at least six months between 1972 and 1984 as electricity generation and transmission workers at the Central Electricity Generating Board of England and Wales. Computerized work histories were available for a part of the cohort from 1972 and for all cohort members from 1979. Follow-up of cohort members until the end of 1991 in the national mortality files revealed a total of 176 deaths from brain cancer of which 112 were confirmed through the national cancer registry as primary brain cancers (case group). Approximately six controls per case were chosen from the cohort and matched to the corresponding case on sex and date of birth, giving a total of 654 controls who were all alive at the date of diagnosis of the corresponding case. Exposure assessment was based on an earlier set of measurements of exposure to ELF electric and magnetic fields (50 Hz) in the electricity supply industry made for 675 person–work shifts (Merchant et al., 1994); the cumulative exposure was categorized into tertiles on the basis of the distribution among all study subjects. Using the study subjects in the lower tertile of cumulative exposure to electric and magnetic fields as the exposure reference category (≤ 3.0 μT–years), study subjects in the middle and upper tertiles had odds ratios for primary brain cancer of 1.3 (95% CI, 0.75–2.2) and 0.91 (95% CI, 0.51–1.6), respectively. Subjects who could not be classified according to their cumulative exposure had an odds ratio of 1.8 (95% CI, 0.93–3.6). There was no significant association between the risk for brain cancer and any of the potential confounders included in the study.

In an update of this cohort study, Sorahan et al. (2001) analysed brain tumour mortality until 1997 for the subset of 79 972 study subjects for whom computerized work histories were available for the period 1971–93. A Poisson regression analysis showed age- and sex-adjusted relative risks for death from brain tumour of 0.88 (95% CI, 0.53–1.5), 0.65 (95% CI, 0.41–1.0), 0.68 (95% CI, 0.42–1.1) and 0.68 (95% CI, 0.33–1.4) among cohort members with a lifetime exposure to magnetic fields of 2.5–4.9, 5.0–9.9, 10.0–19.9 and ≥ 20.00 μT–years, respectively, compared with the risk for death from brain tumour among workers with cumulative exposure ≤ 2.4 μT–years.

A re-analysis using the most recent five years of exposure to ELF magnetic fields did not change the results substantially.

In a small study from Sweden, Rodvall *et al.* (1998) identified 105 histologically confirmed cases of intracranial glioma (ICD-9 191) and 26 of meningioma (ICD-9 192.1), newly diagnosed in men aged 25–74 years during 1987–90. The cases were identified from the files of a large university hospital and the regional cancer registry. A total of 155 controls was selected from population listings and matched to the cases on date of birth and parish. Only controls who were alive at the time of diagnosis of the corresponding case were included. A postal questionnaire requesting data on the occupational history of the study subjects was completed for 84 (80%) of the glioma cases (71 by the patient and 13 by a close relative), 20 (77%) of the meningioma cases (19 by the patient and one by a close relative) and, after the inclusion of a number of replacement controls, by 155 [response rate unknown] of the control subjects themselves. The analyses used multiple logistic regression models with adjustment for socioeconomic status and self-reported occupational exposure to solvents and plastic materials. Ever having worked in an electrical occupation was associated with odds ratios of 1.0 (95% CI, 0.4–2.4) for glioma and 1.8 (95% CI, 0.3–3.6) for meningioma. Employment in a job classified by an electrical engineer as probably highly exposed to magnetic fields showed odds ratios of 1.6 (95% CI, 0.6–4.0) for glioma and 2.1 (95% CI, 0.4–10) for meningioma. Risks were also analysed according to the exposure to electric and magnetic fields classified using a previously constructed job–exposure matrix for ELF electric and magnetic fields (Floderus *et al.*, 1993, 1996), applied to the job history of the study subjects. Ever having been in an occupation with exposure to magnetic fields > 0.4 μT was associated with odds ratios of 1.9 (95% CI, 0.8–5.0) and 1.6 (95% CI, 0.3–10) for glioma and meningioma, respectively.

Mortality data including all death certificates from 24 US states for the period 1984–92 were used to explore the association of industry and occupation with risk for brain cancer (Cocco *et al.*, 1998a). The cases were 28 416 subjects $\geq$ 25 years old who had died from cancer of the brain (ICD-9 191) and other parts of the central nervous system (ICD-9 192), and the controls were 113 664 subjects who had died from non-malignant diseases other than those affecting the central nervous system, frequency-matched to cases by state, race, sex and age. The subjects were classified as having been exposed or unexposed to electric and magnetic fields and other potential risk factors for brain cancer (herbicides, other pesticides, solvents, lead, contact with animals, contact with the public) using an a-priori job–exposure matrix. Brain cancer showed a consistent association with high socioeconomic status. Exposure to electric and magnetic fields was not associated with risk in any sex–race strata, although an odds ratio of 1.2 (95% CI, 0.9–1.6) was observed among African-American women.

In a re-analysis of 12 980 women based on death certificates, and a refined job–exposure matrix using exposure scores for probability and intensity of exposure, the odds ratio for tumours of the central nervous system was 1.2 (95% CI, 1.1–1.2) for women with any exposure to electric and magnetic fields. Slightly increased odds ratios of 1.2–1.3 were observed for high exposure probability or high exposure intensity (Cocco *et al.*, 1999).

(iii) *Pooled analysis* (leukaemia and brain tumours)

A pooled analysis of the data from three studies of electric utility workers (Sahl *et al.*, 1993, California, USA; Thériault *et al.*, 1994, France, Ontario, Quebec; Savitz & Loomis, 1995, USA) including four companies and five utilities where quantitative measurements of magnetic fields had been carried out, was conducted to examine the relation between cumulative exposure to magnetic fields and risk of leukaemia and brain tumours (Kheifets *et al.*, 1999). Overall, excluding the data for Ontario, the results indicated a small increase in risk for both brain cancer (relative risk, 1.8; 95% CI, 1.1–2.9) and leukaemia (relative risk, 1.4; 95% CI, 0.85–2.1) for exposure > 16 µT–years as compared to exposure < 4 µT–years. For a 10 µT–year increase in exposure, the relative risks were 1.12 (95% CI, 0.98–1.3) and 1.09 (95% CI, 0.98–1.2) for brain cancer and leukaemia, respectively. There was some consistency of the results across the utility companies.

(iv) *Female breast cancer*

Breast cancer was analysed using the mortality database of 24 US states for the period 1985–89 in a case–control study (Loomis *et al.*, 1994b). After exclusion of 'homemakers', the cases were 28 434 women > 19 years old whose underlying cause of death had been breast cancer and the controls were 113 011 women who had died from other causes, excluding brain cancer and leukaemia. The usual occupation as recorded on their death certificates was used to classify women according to the likelihood of having been exposed to electric and magnetic fields, using an extended list of 'electrical occupations'. The odds ratio for the association between electrical occupation and cancer, adjusted on race and social class, was 1.4 (95% CI, 1.0–1.8).

The same US mortality database was analysed by Cantor *et al.* (1995) using an alternative method for exposure assessment. The study included 33 509 women who had died from breast cancer and 117 794 controls selected from women who had died from causes other than cancer. Exposure scores were determined for each occupation using different indices for levels of exposure to ELF electric and magnetic fields and exposure probability. The results were presented separately for black and white women. There was no consistent excess risk with increasing level or probability of exposure to ELF electric and magnetic fields.

Coogan *et al.* (1996) conducted a case–control study of 6888 (81%) respondent cases out of 8532 eligible women ≤ 74 years of age with breast cancer diagnosed between April 1988 and December 1991 in Maine, Massachusetts, New Hampshire and Wisconsin. The controls were 9529 (84%) respondents out of 11 329 eligible women, frequency-matched on age and state of residence, and identified from driver's license and the lists of the Health Care Financing Administration. Because all subjects were interviewed by telephone, a listed telephone number was required for eligibility by both cases and controls. The women were asked about their usual occupation, which was classified into one of four categories of potential exposure to 60-Hz magnetic fields (high, medium, low and background exposure), as defined by an industrial hygienist. Compared to women with background exposure, an odds ratio of 1.4 (95% CI, 1.0–2.1) was found for women whose usual occupation was in the high exposure category. The odds ratios for the medium and low exposure categories did not differ appreciably from unity. No significant difference was seen between pre- and post-menopausal women (odds ratio, 2.0 (95% CI, 1.0–3.8) and odds ratio, 1.3 (95% CI, 0.82–2.2), respectively) in the highest exposure groups.

Coogan and Aschengrau (1998) carried out a case–control study on 259 of the 334 women residing in the Upper Cape Cod area who were diagnosed with breast cancer in 1983–86. They selected 738 controls by random-digit dialling, from lists of Medicare beneficiaries or from the death certificates of women who had lived in the same area. Complete work histories were obtained for each subject, and jobs were classified according to their potential for higher than background exposure to magnetic fields (high, medium or no exposure). Residential exposure to magnetic fields from power lines and substations was also considered, as well as exposure to magnetic fields from electrical appliances in the home. Suspected or established risk factors for breast cancer were included in the analyses as potential confounders. There was no association between breast cancer risk and occupational exposure to magnetic fields, nor with any other source of magnetic fields. The adjusted odds ratios for jobs with potential exposure to high electric and magnetic fields and jobs with potential exposure to medium electric and magnetic fields were 1.2 (95% CI, 0.4–3.6) and 0.9 (95% CI, 0.5–1.7), respectively. No association was observed with duration of employment in these occupational groups.

In a study from Sweden, Forssén et al. (2000) estimated the separate and combined effects of occupational and residential exposure to ELF magnetic fields on the risk for female breast cancer (see section 2.3.1). Occupational exposure data were available for 744 cases and 764 controls and both contemporary residential and occupational exposure data were available for 197 cases and 200 controls. No increased risk in breast cancer was associated with occupational exposure to ELF magnetic fields.

(v)   *Male breast cancer*

Demers et al. (1991) investigated occupational exposure to ELF electric and magnetic fields in 227 incident cases of breast cancer in males, 22–90 years old, identified in 1983–87 in 10 population-based cancer registries in the USA (320 cases were eligible). Three hundred controls matched on age and study area (out of 499 eligible controls) were selected by random-digit dialling for controls aged under 65 years and from Medicare lists for older controls. Personal interviews using a standardized questionnaire were used to obtain a partial work history (information on the two longest-held occupations). The estimates of exposure to electric and magnetic fields were based on job titles. Data on several suspected risk factors for male breast cancer were also collected. The odds ratio associated with jobs entailing exposure to electric and magnetic fields was 1.8 (95% CI, 1.0–3.7). No significant trend with increasing duration of employment in an exposed occupation was observed. No confounding by the non-occupational risk factors investigated in the study was observed. [The Working Group noted that the participation rate was low, especially among controls.]

In a case–control study based on mortality data from 24 states in the USA in the period 1985–88, Loomis (1992) analysed 250 males aged < 19 years who had died from breast cancer and approximately 2500 controls, selected from men who had died from other causes and matched by year of death. Four of the cases had an electrical occupation listed on their death certificate (odds ratio, 0.9). [The Working Group noted the limited number of exposed cases.]

Rosenbaum et al. (1994) studied 71 incident cases of male breast cancer diagnosed in western New York between 1979 and 1988, and 256 controls selected from men who had been screened for cancer in the Prevention–Detection Clinic (voluntary cancer screening). The cases and controls were resident in the same areas and were matched by race, year of diagnosis or screening and age. Occupational exposure to electric and magnetic fields was evaluated using a job–exposure matrix, based on the assumption that workers in 'electrical occupations' were exposed to higher than background electric and magnetic fields. Exposure to electric and magnetic fields was associated with an odds ratio of 0.6 (95% CI, 0.2–1.6). [The Working Group noted that selection bias may have occurred for controls and information bias for occupation.]

To study the relationship between exposure to ELF magnetic fields and male breast cancer, Stenlund and Floderus (1997) re-used the study design and the control group established five years earlier by Floderus et al. (1993) in the previously described study from mid-Sweden of leukaemias and brain tumours. The new study included 92 men who had been diagnosed with breast cancer during 1985–91 in any part of Sweden. Fifteen patients with cancer of the breast were ineligible for study because permission to contact the patient or a relative was not obtained, or because no relatives were identified. The postal questionnaire was completed by the patient or a close relative for 63 of the cases (69%). The response rate for the controls was 72% giving 1121 controls for analysis. Using the same job–exposure matrix as in the original study and following the same strategy of analysis, the authors found odds ratios for breast cancer of 1.2 (95% CI, 0.6–2.7), 1.3 (95% CI, 0.6–2.8) and 0.7 (95% CI, 0.3–1.9) associated with exposure levels to electric and magnetic fields in the second, third and upper quartile, respectively. [The Working Group noted, as did the authors, that the breast cancer cases were not selected from the same study base as that defined for control subjects, implying a possibility for bias in the selection of study subjects.]

In a case–control mortality-based study on male breast cancer, Cocco et al. (1998b) used the data of the US national mortality follow-back survey. The cases were 178 men who had died from breast cancer in a sample of 1% of all adult deaths that occurred in the USA in 1986 (excluding Oregon) and all men who had died from breast cancer in

1985 among black and white adults 25–74 years old. The controls were 1041 male decedents selected from men who had died from all other causes of death, after the exclusion of smoking- and alcohol-related causes, and matched on race, age and region of death. Questionnaires were sent to the next-of-kin of the decedents to obtain information on sociodemographic variables; the longest-held occupation and industry, and non-occupational data on consumption of selected dietary items; alcohol consumption, tobacco smoking, and medical history. Occupational exposure to electric and magnetic fields, solvents, herbicides, other pesticides, high temperatures and polycyclic aromatic hydrocarbons was estimated from a job–exposure matrix that included scores for exposure intensity and probability of exposure. Although an increased odds ratio was observed in men with high socioeconomic status, and in certain occupations or industries not classified as having high exposure to electric and magnetic fields, no association was observed with the probability or the intensity of exposure to electric and magnetic fields, organic solvents, polycyclic aromatic hydrocarbons, herbicides and other pesticides.

(vi)   *Other cancer sites*

Several other sites of cancer have been investigated in relation to ELF electric and magnetic fields, particularly in case–control studies nested within occupational cohorts of electric utility workers, since data on all sites of cancer were collected in these studies. However, there was generally no indication of an increased risk for any site of cancer other than those already described for leukaemia and brain tumours. The results obtained for the sites of cancer that were considered of interest *a priori* in studies of electric utility workers, namely lymphomas (Sahl *et al.*, 1993; Thériault *et al.*, 1994) and malignant melanoma (Thériault *et al.*, 1994) showed no association with exposure to magnetic fields.

In a population-based case–control study in the four northern counties of Sweden, Hallquist *et al.* (1993) identified a total of 188 surviving patients with thyroid cancer who were aged 20–70 years at the time of diagnosis during 1980–89. The cases were drawn from the national Swedish Cancer Registry and represented 81% of all cases of thyroid cancer notified in persons of the same age group during the period of interest in the four counties (44 patients had other diseases and were excluded). The original histopathological diagnoses were re-evaluated by a pathologist, resulting in the exclusion of seven patients who were reclassified as having diseases other than thyroid cancer. One subject died shortly before the interview, leaving 180 histologically verified cases for study. For each case, two living controls close in age and of the same sex, and resident in one of the four northern counties, were drawn from the national population registry. Information on the occupational history of the study subjects and on known and suspected risk factors for thyroid cancer was obtained through a postal questionnaire that was completed by 171 case subjects (response rate, 95%) (123 women and 48 men), and 325 control subjects (240 women and 85 men) (response rate, 90%). Five male cases had worked as linemen while no controls had reported that occupation. The authors stated that this occupation might entail exposure to ELF electric and magnetic fields; however, no exposure estimates were given. Three of these linemen were exposed to impregnating agents, i.e. chlorophenols and creosote, which were found in the analysis to be significantly associated with thyroid cancer (odds ratio, 2.8; 95% CI, 1.0–8.6). Employment as an electrical worker was associated with an odds ratio of 1.9 (95% CI, 0.6–6.1) on the basis of eight exposed cases.

In order to study the relationship between exposure to ELF magnetic fields and testicular cancer, Stenlund and Floderus (1997) re-used the study design and the control group established five years earlier by Floderus *et al.* (1993) in the previously described study from mid-Sweden of leukaemias and brain tumours. The extended study included 214 men diagnosed with testicular cancer during 1983–87 and living in the original catchment area. The postal questionnaire was completed by the patient [proportion not given] or a relative [proportion not given] for 144 of the 185 eligible subjects with testicular cancer (78%). The response rate among the controls was 72% leaving 1121 controls. Using the same job–exposure matrix as in the original study and following the same strategy of analysis, the authors found odds ratios for testicular cancer of 1.3 (95% CI, 0.7–2.4), 1.4 (95% CI, 0.8–2.7) and 1.3 (95% CI, 0.7–2.5) associated with exposure to magnetic fields in the second, third and upper quartile (0.16–0.19 µT, 0.20–0.28 µT and ≥ 0.29 µT), respectively. Among the 13% of study subjects who were exposed to the highest estimated levels (≥ 0.41 µT), the odds ratio for testicular cancer was 2.1 (95% CI, 1.0–4.3). In a subsequent analysis on subtypes of testicular cancer, the authors observed an increased risk for non-seminomas particularly in subjects less than 40 years of age (odds ratio, 7.1 (95% CI, 1.4–36), odds ratio, 7.1 (95% CI, 1.3–38), odds ratio, 8.1 (95% CI, 1.7–39) and odds ratio, 16 (95% CI, 2.7–95) in the four exposure quartiles, respectively).

3. Studies of Carcinogenicity in Experimental Animals

3.1 Chronic exposure studies

The results of one- and two-year rodent bioassays are summarized in Table 32.

3.1.1 *Mouse*

Groups of 100 male and 100 female B6C3F$_1$ mice, six to seven weeks of age, were exposed for 18.5 h per day for two years to linearly polarized 60-Hz magnetic fields that were nearly pure, transient-free and had less than 3% total harmonic distortion. [The group-size used in this study was twice that commonly used in chronic rodent bioassays; this enlargement was purposely chosen to increase the statistical power of the experimental design, and thereby increase its ability to identify potentially weak carcinogenic effects (Portier, 1986).] Groups of animals were continuously exposed to field strengths of 2 μT, 200 μT or 1000 μT, or were intermittently exposed (1 h on/1 h off for 18.5 h per day) to a field strength of 1000 μT. Parallel sham control groups were housed within an identical exposure apparatus, but were exposed only to ambient magnetic fields. The exposure system has been described by Gauger *et al.* (1999). The design of these studies allowed for continuous monitoring of magnetic field strength and waveform throughout the two-year exposure period. At termination, all animals received a full necropsy and complete histopathological evaluations were performed on all gross lesions collected from all study animals, in addition to examinations of 43 tissues per animal in all study groups. Body weights were comparable in all groups, but a statistically significant reduction in survival time was observed in male mice subjected to continuous exposure to field strengths of 1000 μT. (This effect was not seen in female mice.) When compared to controls, no increases in the incidence of neoplasms at any site were observed in males or females from any treated group. In fact, a statistically significant reduction in the incidence of malignant lymphoma was observed in female mice exposed intermittently to 1000 μT (32/100 in controls versus 20/100; $p = 0.035$) and significant decreases in the combined incidence of lung tumours were observed in both male (30/100 in controls versus 19/100; $p = 0.04$) and female mice (11/95 in controls versus 2/99; $p = 0.008$) exposed to field strengths of 200 μT. In addition, statistically significant reductions in the total incidence of malignant neoplasms (all sites) were seen in female mice continuously exposed to field strengths of 200 μT (55/100 in controls versus 39/100;

Table 32. Summary of statistically significant findings ($p < 0.05$) from one- and two-year rodent bioassays on carcinogenicity of magnetic fields

Reference	Animal model (species/strain/sex)	Exposure (frequency: field strength)	Statistically significant differences from sham controls (tumour type, trend, field strength)	Comment
McCormick et al. (1999); National Toxicology Program (1999a)	Mouse/B6C3F$_1$/ male	60 Hz: 2 µT, 200 µT, 1000 µT or 1000 µT-intermittent	Lung tumours, ↓, 200 µT	
	Mouse/B6C3F$_1$/ female		Malignant lymphoma, ↓, 1000 µT-intermittent; Lung tumours, ↓, 200 µT; Malignant neoplasms, all sites, ↓, 200 µT; Malignant neoplasms, all sites, ↓, 1000 µT	
Mandeville et al. (1997)	Rat/Fischer 344/ female (F$_0$, F$_1$)	60 Hz: 2 µT, 20 µT, 200 µT, 2000 µT	None	
Yasui et al. (1997)	Rat/Fischer 344/ male	50 Hz: 500 µT, 5000 µT	Fibroma of subcutis, ↑, 5000 µT; Invasive neoplasms (all), ↓, 500 µT	Histopathology limited to gross lesions identified at necropsy
	Rat/Fischer 344/ female		None	Histopathology limited to gross lesions identified at necropsy
Boorman et al. (1999a); National Toxicology Program (1999a)	Rat/Fischer 344/ male	60 Hz: 2 µT, 200 µT, 1000 µT or 1000 µT-intermittent	Leukaemia, ↓, 1000 µT-intermittent; Preputial gland carcinoma, ↑, 200 µT; Skin trichoepithelioma, ↑, 1000 µT; Thyroid C-cell tumours, ↑, 2 µT; Thyroid C-cell tumours, ↑, 200 µT; Thyroid C-cell carcinoma, ↑, 2 µT	
	Rat/Fischer 344/ female		Adrenal cortex-adenoma, ↓, 1000 µT-intermittent	

↑, increase; ↓, decrease

$p = 0.015$) and 1000 μT (55/100 in controls versus 40/100; $p = 0.024$) (McCormick et al., 1999; National Toxicology Program, 1999a).

In a site-specific chronic bioassay, a group of 380 female C57BL/6 mice was exposed to a circularly polarized 60-Hz magnetic field at a field strength of 1420 μT for up to 852 days. The incidence of haematopoietic neoplasms in these mice was compared with that observed in a negative (untreated) control group of 380 female C57BL/6 mice and in a sham-treated control group of 190 female C57BL/6 mice. Chronic exposure to magnetic fields had no statistically significant effects on animal survival or on the incidence or latency of haematopoietic neoplasms in this study. At study termination, the final incidence of lymphomas in mice exposed to magnetic fields was 36.8% (140/380) compared with an incidence of 34.7% (66/190) in sham controls. The incidence of histiocytic sarcomas was 23.7% (90/380) in mice exposed to magnetic fields versus 22.1% (42/190) in sham controls, yielding a total incidence of haematopoietic neoplasia of 56.3% (107/190) in sham controls versus a total incidence of 59.2% (225/380) in the group exposed to circularly polarized magnetic fields. No statistically significant differences were observed (Babbitt et al., 2000). In addition to the investigation of the effects of magnetic fields on haematopoietic tumours, a post-hoc histopathological analysis of brain tissues from animals in this study was performed to investigate the possibility that magnetic fields are a causative agent for primary brain tumours. Consistent with the results for haematopoietic neoplasms, histopathological examination of brains from this study provided no support for the hypothesis that exposure to magnetic fields is a significant risk factor for induction of brain tumour since no brain tumours were identified in any of the three groups described above (Kharazi et al., 1999). [The Working Group noted that the primary strengths of this study are that it evaluated the potential carcinogenicity of a previously unstudied type of exposure (circularly polarized rather than linearly polarized magnetic fields), and used very large experimental groups thus increasing its statistical power and its ability to identify effects of modest magnitude.]

Using a unique study design in which three consecutive generations of CFW mice were exposed to extremely high flux densities (25 mT) of 60-Hz magnetic fields, an increased incidence of malignant lymphoma in the second (no statistical analysis given) and third generations (2/41 versus 37/92; $p < 0.001$) of exposed animals was reported (Fam & Mikhail, 1996). [The Working Group noted several deficiencies in the design and conduct of this experiment. These include small and variable group sizes, a very small number of observed malignant lesions in the F_2 generation and weaknesses in exposure assessment. Of particular concern is the inadequate control of environmental factors, including the heat, noise and vibration generated by the exposure system, and the noise and vibration made by the ventilation equipment. Because non-specific stressors have been demonstrated to increase the growth of transplantable lymphomas and other tumours in mice and to decrease survival in several other animal models, inadequate control of environmental conditions may confound the study results. These possible confounders render this study difficult to interpret.]

3.1.2   *Rat*

Groups of 100 male and 100 female Fischer 344 rats, six to seven weeks of age, were exposed for 18.5 h per day for two years to linearly polarized 60-Hz magnetic fields that were nearly pure, transient–free and had less than 3% total harmonic distortion. Groups of animals were continuously exposed to field strengths of 2 μT, 200 μT or 1000 μT, or were intermittently exposed (1 h on/1 h off) to a field strength of 1000 μT. Parallel sham control groups were housed within an identical exposure array, but were exposed to low ambient magnetic fields. The exposure system has been described by Gauger *et al.* (1999). The design of these studies allowed for continuous monitoring of magnetic field strength and waveform throughout the two-year exposure period. At termination, all animals received a full necropsy and complete histo-pathological evaluations were performed on all gross lesions collected from all study animals, in addition to examinations of 45 tissues per animal in all study groups. Body weight and survival were comparable in all groups. Significant differences from tumour incidences in controls observed in this study included a statistically significant decrease (50/100 in controls versus 36/100; $p < 0.045$) in the incidence of leukaemia in male rats exposed intermittently to field strengths of 1000 μT, a statistically significant increase in the incidence of preputial gland carcinomas in male rats exposed to magnetic field strengths of 200 μT (but not to 1000 μT) (0/100 in controls versus 5/100; $p = 0.032$), a statistically significant increase in the incidence of trichoepitheliomas of the skin in male rats exposed continuously to 1000 μT (0/100 in controls versus 5/100; $p = 0.029$), a statistically significant decrease in the incidence of adenomas of the adrenal cortex in female rats exposed intermittently to 1000 μT (6/100 in controls versus 0/100; $p = 0.02$), and statistically significant increases in the incidence of thyroid C-cell tumours (adenomas + carcinomas) in male rats exposed to field strengths of 2 μT (16/99 in controls versus 31/100; $p = 0.005$) and 200 μT (16/99 in controls versus 30/100; $p = 0.009$). There was also a marginal increase in thyroid C-cell tumours in male rats exposed to a continuous field strength of 1000 μT (16/99 in controls versus 25/100; $p = 0.055$), but no increase in animals intermittently exposed to 1000 μT (16/99 in controls versus 22/100; $p = 0.147$). In male rats, there was also a statistically significant increase in thyroid C-cell carcinomas at 2 μT (1/99 in controls versus 7/100; $p = 0.03$) and a non-significant increase in this rare tumour in animals intermittently exposed to 1000 μT (1/99 in controls versus 5/100; $p = 0.1$) (Boorman *et al.*, 1999a; National Toxicology Program, 1999a). [An examination by the Working Group of the historical controls used by the National Toxicology Program for the 10 most recent bioassays conducted using an identical diet (NTP-2000 diet) showed a historical incidence for thyroid C-cell tumours of 17% (102/603) with a range from 2% (1/50) to 28% (14/50), indicating no discernible problem with the controls for this tumour in this study. In the same historical database, the incidence of thyroid C-cell carcinomas is 1.7% (10/603) varying from 0% (0/50) observed for four of the 10 control datasets to 4% observed for two of the datasets (2/50 and 4/100).] The original authors (Boorman *et al.*, 1999a) concluded [without the benefit of adequate historical controls] that their finding was equivocal and the peer-review committee of the National Toxicology Program (1999a) reached the same conclusion. [The Working Group noted that the lack of a dose–response relationship and the as yet unknown mechanism of thyroid C-cell carcinogenesis prevent a clear interpretation of this finding. For these reasons, because the results cannot be interpreted as clearly negative and because the data are insufficient to be listed as clearly positive, the Working Group concluded that the evaluation should remain equivocal.]

In another study, groups of gestating Fischer 344 rats were exposed to similar, linearly polarized 60-Hz magnetic fields at field strengths of 2 µT, 20 µT, 200 µT and 2000 µT for 20 h per day from day 20 of gestation. At weaning, groups of 50 female offspring were exposed to the same intensities and magnetic fields as the dams had been for 20 h per day for two years. The experimental design included a group of 50 female rats as cage controls and 50 female sham-exposed controls. During the lifetime of the animals, the study was conducted using a blinded design in which investigators were unaware of group identities. Toxicological end-points included a standard battery of evaluations of the living animals, followed after death by the histopathological evaluation of gross lesions and 50 tissues per animal. The authors noted no statistically significant increases in the incidence of any tumour at any site evaluated (Mandeville *et al.*, 1997). [The Working Group noted that this study included exposure during the perinatal and juvenile periods, and its design addressed the possibility of enhanced sensitivity of younger animals to the effects of magnetic fields. The blinded design of the bioassay precluded any possible influence of investigator bias on study results.]

In a third chronic bioassay in rats, groups of 48 male and 48 female Fischer 344 rats, five weeks of age, were exposed for an average of 22.6 h per day to 50-Hz magnetic fields at strengths of 500 µT or 5000 µT for two years; control groups of 48 males and 48 females received sham exposure for the same period. At termination of the study, all animals received a complete necropsy. Histopathological evaluations were performed on all gross lesions, and on all sites suspected of tumoral lesions. Survival was comparable in all study groups, and differential white blood cell counts performed after 52, 78 and 104 weeks of exposure failed to identify any effects of exposure to magnetic fields. The authors reported that exposure to magnetic fields had no effect on survival of animals of either sex or on the total incidence or number of neoplasms. The only histopathological findings that were statistically significant were an increase in the incidence of a benign lesion (fibroma) in the subcutis of male rats exposed to field strengths of 5000 µT (2/48 in controls versus 9/48; $p < 0.05$) and a decrease in the total incidence of invasive neoplasms in male rats in the group exposed to 500 µT (6/48 in controls versus 1/48; $p < 0.05$). When compared with incidence in sham controls (8/48), no increases in the incidence of thyroid C-cell tumours (benign and malignant) were observed in male rats exposed to 500 µT (10/48) or 5000 µT (6/48). The incidence of thyroid C-cell carcinomas was less than 2% in all groups.

Although the increased incidence of fibromas in male rats exposed to 5000 µT was statistically significant when compared with concurrent male sham controls, the incidence of lesions was stated to be comparable to that observed in historical controls in the same laboratory [range not given]. On this basis, the authors concluded that the increase in fibroma incidence observed in rats exposed to 5000 µT was not exposure-related. Because no differences in the incidence of metastatic neoplasms were seen in the study, the observed decrease in the incidence of invasive malignancy appears to be without biological significance (Yasui *et al.*, 1997). [The Working Group noted that this study provided only limited information because of the incomplete histopathological evaluation. Because thyroid C-cell carcinomas frequently involve the entire lobe of the thyroid gland in Fischer rats, this level of histopathological evaluation should be adequate for this tumour. However, since thyroid C-cell adenomas are generally small and difficult to separate from hyperplasia, this evaluation cannot be easily compared with the National Toxicology Program (1999a) study.]

3.2   Exposure in association with known carcinogens

3.2.1   *Multistage studies of mammary cancer*

A number of studies have investigated the effect of exposure to magnetic fields on the incidence of mammary cancer in rodents. All of these have been multistage studies in which female rats were treated with a chemical carcinogen, *N*-methyl-*N*-nitrosourea (MNU) or 7,12-dimethylbenz[*a*]anthracene (DMBA), followed by exposure to magnetic fields of different strengths and for different time intervals.

(*a*)   *Multistage studies with* N-*methyl*-N-*nitrosourea*

Groups of 50 female outbred rats (obtained from the Oncology Research Center, Republic of Georgia) between 55 and 60 days of age, were intravenously injected with 50 mg/kg bw MNU. Starting two days after this treatment, the animals were exposed to a 50-Hz, 20-µT magnetic field daily for 0.5 h (group 1) or 3 h (group 2), or to a 20-µT static magnetic field, daily for 0.5 h (group 3) or 3 h (group 4). A fifth group received MNU only. Four control groups each of 25 rats received no MNU, but were exposed to the same magnetic fields as described above for groups 1–4. A 10th group of 50 rats received no treatment at all. The rats were followed for a period of two years after injection of the carcinogen. In comparison with a 59% (27/46) incidence of mammary tumours in rats receiving MNU only, a significant increase was observed in the groups exposed daily for 3 h to a 50-Hz magnetic field (93%, 43/46; $p < 0.05$) or to a static magnetic field (87%, 39/45; $p < 0.05$). These increases were not seen in rats exposed to a 50-Hz field or a static field for 0.5 h per day. In comparison with a 0% (0/48) incidence of mammary tumours in rats that received no MNU and no exposure to magnetic fields, a statistically significant increase was observed ($p < 0.05$) in the incidence of mammary tumours in non-MNU-treated rats exposed for 3 h per day to a

50-Hz magnetic field (30%, 7/23) [$p < 0.01$; χ^2 test]. The authors also noted that exposure to either type of magnetic field for 3 h per day shortened the latent period of tumour development [no statistical analysis given] and led to a change in the morphological spectrum of the mammary tumours, with more adenocarcinomas than fibroadenomas (Beniashvili et al., 1991).

Groups of 40 outbred female rats (obtained from the Oncology Research Center, Republic of Georgia), one month of age, were kept on a 12-h/12-h light–dark cycle, and intravenously injected with MNU (50 mg/kg bw) three times per week. Starting two days after the first dose of MNU, the animals were exposed daily for 3 h to either a 20-μT static magnetic field or a 50-Hz, 20-μT magnetic field. A group of 30 rats received MNU only. Mammary adenocarcinomas were found in 7/22 (32%) of the MNU-treated controls and in 12/30 (40%) and 15/33 (45%) of the animals exposed to the static and 50-Hz magnetic fields, respectively. These differences were not statistically significant. The mean latent period for development of mammary tumours was significantly decreased ($p < 0.05$) in the rats exposed to the 50-Hz fields (125 ± 7 days), but not to the static fields (162 ± 11 days), compared with the latent period observed in the MNU-treated controls (166 ± 4 days). In the same series of studies, experiments were carried out with animals that were kept in constant darkness or in constant light. Incidences of mammary tumours decreased to 1/38 (2.6%), 1/48 (2.1%) and 2/45 (4.4%), respectively, for the MNU, MNU plus static field and MNU plus 50-Hz field groups that had been kept in the dark. Conversely, under conditions of constant light, the tumour incidences were increased: 20/35 (57%), 25/41 (61%) and 34/42 (81%), respectively, for the three groups (Beniashvili et al., 1993; Anisimov et al., 1996).

(b)   Multistage studies with 7,12-dimethylbenz[a]anthracene

The results of these studies are summarized in Table 33.

Female Sprague-Dawley rats, 52 days of age, were given 5 mg DMBA by gavage. Administration of DMBA (5 mg by gavage) was repeated at weekly intervals up to a total dose of 20 mg/animal. Beginning immediately after the first dose of DMBA, the treatment groups were exposed 24 h per day for 13 weeks to either a static magnetic field of 15 mT (18 rats), a 50-Hz magnetic field of 30 mT (1 group of 15 and 1 group of 18 rats) or a non-uniform 50-Hz magnetic field ranging from 0.3–1 μT (36 rats). Control groups equal in size to those of the treated rats were sham-exposed. Rats were palpated weekly to assess development of mammary tumours. After 13 weeks, all rats were necropsied and the number and weight or size of the tumours were determined. The exposure system was adequately described and consisted of six identical solenoidal coils and six sham coils of the same dimensions. The DMBA-treated age-matched reference control groups were kept in a separate room (ambient field, 0.05–0.15 μT). In sham-exposed animals and reference controls, the tumour incidence varied between 50 and 78% in the different experiments. The average number of mammary tumours per tumour-bearing animal varied between 1.6 and 2.9. In none of the experiments did exposure to magnetic fields significantly alter tumour incidence

Table 33. Multistage studies of mammary cancer in female Sprague-Dawley rats treated with 7,12-dimethylbenz[a]anthracene (4 weekly gavage doses of 5 mg/animal, unless otherwise stated) and exposed to magnetic fields for 13 weeks (unless otherwise indicated)

Reference	Exposure conditions	Exposure and control groups (no. of animals)	No. of animals with tumours	Tumour incidence (%)	Total no. of tumours	No. of tumours per tumour-bearing animal	Remarks
Mevissen et al. (1993)	15 mT, static	Exposed (18)	10	56	17	1.7 ± 0.31	
		Sham-exposed (18)	14	78	30	2.1 ± 0.28	
		Reference control (8)	6	75	16	2.7 ± 0.35	
	30 mT, 50 Hz	Exposed (18)	14	78	40	2.8 ± 0.63 *	* $p < 0.05$ compared with sham-exposed animals
		Sham-exposed (18)	10	56	22	2.2 ± 0.47	
		Reference control (18)	12	67	20	1.7 ± 0.23	
	30 mT, 50 Hz	Exposed (15)	6	40	11	1.8 ± 0.34	
		Sham-exposed (18)	9	50	14	1.6 ± 0.17	
		Reference control (9)	5	55	12	2.4 ± 0.37	
	0.3–1 µT, 50 Hz	Exposed (36)	21	58	47	2.2 ± 0.3	
		Sham-exposed (36)	21	58	60	2.9 ± 0.45	
Löscher et al. (1993)	100 µT, 50 Hz	Exposed (99)	51	52 *	82	~1.6	All figures calculated from curves * $p < 0.05$ compared with sham-exposed animals
		Sham-exposed (99)	34	34	62	~1.8	
Baum et al. (1995)[a]	100 µT, 50 Hz	Exposed (99)	65	66	134	~2.1	No. of animals with adenocarcinoma: exposed, 62; sham-exposed, 49 ($p < 0.05$) Tumour volume in exposed animals significantly larger than in sham-exposed animals ($p < 0.05$)
		Sham-exposed (99)	57	58	113	~2.0	
		Exposed, no DMBA (9)	0	0	–	–	
		Sham-exposed, no DMBA (9)	0	0	–	–	
Löscher et al. (1994)	0.3–1 µT, 50 Hz	Exposed (36)	24	67	77	3.2 ± 0.54	
		Sham-exposed (36)	22	61	95	4.3 ± 0.83	

Table 33 (contd)

Reference	Exposure conditions	Exposure and control groups (no. of animals)	No. of animals with tumours	Tumour incidence (%)	Total no. of tumours	No. of tumours per tumour-bearing animal	Remarks
Mevissen et al. (1996a)	10 μT, 50 Hz	Exposed (99)	66	67	151	~ 2.5	
		Sham-exposed (99)	60	61	129	~ 2.5	
		Exposed, no DMBA (9)	0	0	–	–	
		Sham-exposed, no DMBA (9)	0	0	–	–	
Mevissen et al. (1996b)	50 μT, 50 Hz	Exposed (99)	69	70 *	193	~ 2.7	* p < 0.05 compared with sham-exposed animals
		Sham-exposed (99)	55	56	139	~ 2.5	
		Exposed, no DMBA (9)	0	0	–	–	
		Sham-exposed, no DMBA (9)	0	0	–	–	
Mevissen et al. (1998a)	100 μT, 50 Hz	Exposed (99)	83	84 *	297	~ 3.8	* p < 0.05 compared with sham-exposed animals
		Sham-exposed (99)	62	63	230	~ 3.8	
Ekström et al. (1998)	250 μT, 50 Hz	Exposed (60)	42	70	102	2.4	A single intragastric dose (7 mg/animal) of DMBA was given one week before exposure to the magnetic field for 21 weeks, 15 s on/15s off
	500 μT, 50 Hz	Exposed (60)	42	70	90	2.1	
		Sham-exposed (60)	43	72	111	2.6	
Anderson et al. (1999); National Toxicology Program (1999b)	100 μT, 50 Hz	Exposed (100)	86	86 (carc.)	528 * (carc.)	5.3 ± 4.4 *	* p < 0.05, decrease
	500 μT, 50 Hz	Exposed (100)	96	96	561	6.5 ± 4.9	
	100 μT, 60 Hz	Exposed (100)	96	96	692	6.9 ± 4.8	
		Sham-exposed (100)	92	92	691	6.9 ± 4.8	

Table 33 (contd)

Reference	Exposure conditions	Exposure and control groups (no. of animals)	No. of animals with tumours	Tumour incidence (%)	Total no. of tumours	No. of tumours per tumour bearing animal	Remarks
Anderson et al. (1999); National Toxicology Program (1999b)	100 µT, 50 Hz	Exposed (100)	48	48	90	0.9 ± 1.3	DMBA-treatment with 4×2 mg/animal at weekly intervals
	500 µT, 50 Hz	Exposed (100)	38	(carc.) 38	(carc.) 79	0.8 ± 1.3	
		Sham-exposed (100)	43	43	102	1.0 ± 1.9	
Boorman et al. (1999b); National Toxicology Program (1999b)	100 µT, 50 Hz	Exposed (100)	90	90	494 *	4.9 ± 4.2 *	* $p < 0.05$, decrease. A single intra-gastric dose of 10 mg/rat followed by exposure to magnetic field for 26 weeks
	500 µT, 50 Hz	Exposed (100)	95	(carc.) 95	(carc.) 547	5.5 ± 3.9	
	100 µT, 60 Hz	Exposed (100)	85	85 *	433 *	4.3 ± 3.9 *	
		Sham-exposed (100)	96	96	649	6.5 ± 4.8	
Thun-Battersby et al. (1999)	100 µT, 50 Hz	Exposed (99)	64	65 *	166	~ 2.3	* $p = 0.044$ compared with sham-exposed animals DMBA-treatment with a single oral dose (10 mg/rat), at one week after the start of exposure to magnetic field for 27 weeks
		Sham-exposed (99)	50	51	116	~ 2.6	

carc., carcinoma; DMBA, 7,12-dimethylbenz[a]anthracene

[a] This study provided histological confirmation of the results reported in Löscher et al. (1993).

although in one of the groups exposed to the 50-Hz magnetic field of 30 mT, the number of tumours per tumour-bearing animal was significantly increased ($p < 0.05$). This increase was not seen in the other experiment at 30 mT or in the combined analysis. Furthermore, exposure to the static magnetic field of 15 mT significantly enhanced the tumour weight. Exposure to the non-uniform magnetic field (50 Hz; 0.3–1 µT) had no significant effects on tumour multiplicity or tumour sites. The authors concluded that these experiments suggest that magnetic fields at high flux densities may act as a promoter or co-promoter of mammary cancer. However, they considered this interpretation to be tentative because of the limitations of this study, particularly the small sample size used to study exposure to a magnetic field. Confirmation would require further experiments with larger groups of animals (Mevissen *et al.*, 1993).

A group of 99 female Sprague-Dawley rats, 52 days of age, was exposed to a homogeneous 50-Hz magnetic field of 100 µT for 24 h per day for a period of 13 weeks; another group of 99 rats was sham-exposed. The exposure chambers (Merritt coils, adequately described) were identical for both groups of animals. DMBA (5 mg) was administered by gavage to both groups on the first day of exposure to the magnetic field and at weekly intervals thereafter up to a total dose of 20 mg/rat. The animals were palpated once weekly to assess the development of mammary tumours. Eight weeks after the first dose of DMBA, the incidence of palpable mammary tumours in rats exposed to the magnetic field was significantly higher than in sham-exposed animals ($p < 0.05$). This difference was observed throughout the period of exposure, at the end of which the tumour incidence in exposed rats was 52% (51/99) versus 34% (34/99) in the sham-exposed controls ($p < 0.05$). The tumour size ($p = 0.013$) and the number of tumours per animal ($p < 0.05$) were also increased in the exposed group (Löscher *et al.*, 1993). To provide histopathological confirmation, data on palpable tumours were examined separately for the study described above. Interrupted step sections were prepared from all mammary glands from all animals, yielding 50–60 sections/rat. The incidence of mammary tumours (all types) was 66% (65/99) in magnetic field-exposed and 58% (57/99) in sham-exposed rats ($p > 0.05$). The percentage of animals with mammary adenocarcinomas was significantly higher in the group exposed to DMBA plus a magnetic field than in the sham-exposed controls treated with DMBA (62/99 versus 49/99, $p < 0.05$). Forty-five other organs or tissues were also examined and no significant changes in tumour incidence were noted (Baum *et al.*, 1995).

Using the same protocol (exposure to DMBA and magnetic fields) as Löscher *et al.* (1993) with the addition of a full histopathological review, one group of 36 female Sprague-Dawley rats was exposed to a magnetic field of 0.3–1 µT at 50 Hz and a second group was sham-exposed. In this study, 67% (24/36) of the animals exposed to the magnetic field versus 61% (22/36) of the sham-exposed controls had mammary tumours ($p > 0.05$) and there were also no differences observed in the number of tumours per animal or in average tumour size (Löscher *et al.*, 1994).

Using the same experimental protocol (DMBA, magnetic fields) as Löscher *et al.* (1994), one group of 99 female Sprague-Dawley rats was exposed continuously to

50-Hz, 10-μT magnetic fields for 13 weeks and a second group of 99 rats was sham-exposed. At autopsy, 61% (60/99) of the sham-exposed and 67% (66/99) of the magnetic field-exposed rats had developed macroscopically visible mammary tumours ($p > 0.05$). The average size of the individual tumours and the average sum of all tumours per tumour-bearing rat were similar in both groups (Mevissen *et al.*, 1996a).

The experimental design described above was used to study the effects of a higher field strength of 50 μT. Within eight weeks after the first DMBA administration, the group of rats exposed to a 50-Hz, 50-μT magnetic field exhibited significantly more ($p = 0.028$) palpable mammary tumours than sham-exposed animals. Autopsy revealed significantly more ($p < 0.05$) macroscopically visible mammary tumours (69/99) in rats exposed to magnetic fields than did controls treated with DMBA alone (55/99). No differences in the numbers of tumours per tumour-bearing animal or tumour size were seen (Mevissen *et al.*, 1996b).

Löscher and Mevissen (1995) published a regression analysis of the four studies described above (Löscher *et al.*, 1993, 1994; Baum *et al.*, 1995; Mevissen *et al.*, 1996a,b) in which exposure level was compared with the percentage increase in incidence over controls of palpable mammary tumours after 13 weeks of treatment. This analysis demonstrated a highly significant ($p < 0.01$) trend.

A previous study (Löscher *et al.*, 1993; Baum *et al.*, 1995) was replicated in the same laboratory under the same experimental conditions (exposure to DMBA and a 50-Hz, 100-μT magnetic field). After nine weeks of treatment, the incidence of palpable mammary tumours in the group exposed to a magnetic field was significantly higher than that in the sham-exposed group ($p < 0.05$). This difference was maintained throughout the remainder of the period of exposure. At 13 weeks, the incidence of macroscopically visible mammary tumours was 63% (62/99) in controls and 84% (83/99) in exposed rats ($p < 0.05$). No differences were observed in the number of tumours per tumour-bearing rat or in the average tumour size. The addition of this data point to the previous regression analysis by Löscher & Mevissen (1995) did not markedly alter the significant trend ($p < 0.05$) (Mevissen *et al.*, 1998a).

In a study performed in another laboratory, female Sprague-Dawley rats, 52 days of age, were randomly allocated to one of three groups of 60 animals each. All rats received a single gavage dose of 7 mg DMBA on day 1 of the experiment. Beginning one week later, groups were exposed to intermittent (15 s on/15 s off) transient-associated 50-Hz magnetic fields at a field strength of 250 μT or 500 μT and another group was sham-exposed. The exposure treatment was continued for 24 h per day for 21 weeks, with intermissions for animal care and observations. Animals were palpated twice weekly to identify mammary tumours, but no histological analysis was carried out. The tumour incidence in the two groups exposed to the magnetic fields was 70% (42/60 in both groups), and the incidence in the DMBA-treated controls was 72% (43/60). The total numbers of tumours were 102 (250 μT exposure group), 90 (500 μT exposure group) and 111 (sham-exposed). These values were not statistically different.

Total tumour weight and total tumour volume were not statistically different between groups (Ekström *et al.*, 1998).

Groups of 100 female Sprague-Dawley rats, 50 days of age, received four weekly gavage doses of 5 mg DMBA per animal. After the first DMBA dose, exposure to ambient fields (sham exposure), 50-Hz magnetic fields (field strength of 100 or 500 μT) or 60-Hz fields (field strength of 100 μT) was initiated. The animals were exposed to magnetic or sham fields for 18.5 h per day, seven days per week for 13 weeks. In a second study, groups of 100 female Sprague-Dawley rats received lower doses of DMBA (2 mg/animal per week for four weeks). After the first dose of DMBA, rats were exposed to ambient fields (sham exposure) or 50-Hz magnetic fields at a field strength of 100 or 500 μT for 18.5 h per day, seven days per week for 13 weeks. Rats were weighed and palpated weekly for the presence of mammary tumours. Palpable mammary tumours were examined histologically. Exposure to magnetic fields had no effect on body weight gains or on the time of appearance of mammary tumours in either study. In the first study, the mammary cancer incidences were 92% (92/100), 86% (86/100), 96% (96/100) and 96% (96/100) for the DMBA-treated control, 50-Hz, 100-μT, 50-Hz, 500-μT and 60-Hz, 100-μT groups, respectively. The average numbers of mammary carcinomas per animal were 6.9, 5.3 ($p < 0.05$, decrease), 6.5 and 6.9 for the same groups, respectively. In the second study, the mammary cancer incidences were 43% (43/100), 48% (48/100) and 38% (38/100) for the DMBA-treated control, 50-Hz, 100-μT and 50-Hz, 500-μT groups, respectively. There was no effect of exposure to magnetic fields on the number of tumours per rat or tumour size (Anderson *et al.*, 1999; National Toxicology Program, 1999b). [The Working Group noted that the high tumour incidence observed in the first study effectively precluded the identification of increases in tumour incidence at the end of the study. The decrease in average number of tumours per animal exposed to magnetic fields of 100 μT coincided with a 7% decrease in survival which could have affected this finding.]

Groups of 100 female Sprague-Dawley rats, 50 days of age, received a single dose of 10 mg DMBA by gavage, followed by exposure to ambient fields (strength < 0.1 μT; sham exposure), 50-Hz magnetic fields (strength, 100 or 500 μT) or 60-Hz magnetic fields (strength, 100 μT). The animals were exposed for 18.5 h per day, seven days per week, for 26 weeks. Rats were palpated weekly for mammary tumours. After 26 weeks of exposure to magnetic fields or sham exposure, the animals were killed and the mammary tumours counted and measured; mammary tumours were confirmed histologically. Exposure to magnetic fields had no effect on body weight gain or the time of appearance of mammary tumours. The incidence of mammary cancer was 96% (96/100), 90% (90/100; $p = 0.07$)), 95% (95/100; $p = 0.52$) and 85% (85/100; $p = 0.009$) for the DMBA-treated control, 50-Hz, 100-μT, 50-Hz, 500-μT, and 60-Hz, 100-μT groups, respectively. The total numbers of carcinomas were 649, 494 ($p < 0.05$), 547 and 433 ($p < 0.05$) for the same groups, respectively. The number of fibroadenomas varied from 276 to 319 per group, with the lowest number in the 60-Hz, 100-μT exposure group. Measurement of the tumours revealed no difference in tumour size between groups (Boorman *et al.*, 1999b; National Toxicology Program, 1999b). [The Working Group noted that the high tumour incidence observed in sham control animals effectively precluded the identification of increases in tumour incidence at the end of the study. However, unlike the first National Toxicology Program experiment (Anderson *et al.*, 1999; National Toxicology Program, 1999b), the decreases in average number of tumours per animal were probably not associated with differences in survival.]

Using the protocol of Mevissen *et al.* (1998b) the treatment period was extended to 27 weeks. Groups of 99 female Sprague-Dawley rats, 45–49 days of age, were exposed either to sham fields or to 50-Hz, 100-μT magnetic fields for 24 h per day, seven days per week. A single dose of 10 mg DMBA/rat by gavage was administered one week after study start to both sham-exposed rats and those exposed to magnetic fields (rather than four weekly doses of 5 mg). The animals were palpated once weekly from week 6 onwards to assess the development of mammary tumours. The incidence of palpable mammary tumours in the group exposed to DMBA plus magnetic fields was increased by week 13 ($p = 0.029$) and continued to be elevated throughout the study in comparison with the incidence in the DMBA-treated, sham-exposed group. At study termination, the incidence of histologically verified mammary tumours was 50.5% (50/99) in controls and 64.7% (64/99) in exposed rats, the difference being statistically significant ($p = 0.044$). The incidence of adenocarcinomas was not significantly different between the two groups (42.4%, 42/99 in controls versus 52.5%, 52/99 in exposed rats). When tumour incidence was evaluated separately for each of the six mammary complexes, the most pronounced effect of exposure to the magnetic field was seen in the L/R1 glands, where the overall tumour incidences were 18.2% (18/99) and 30.3% (30/99) for control and exposed rats, respectively ($p < 0.05$). No differences in the size of mammary tumours or number of tumours per animal were noted (Thun-Battersby *et al.*, 1999). [The Working Group questioned the feasibility of attributing the site of origin of a mammary tumour to a specific mammary gland, but agreed that increases in L/R1-3 as a complex appear to be exposure-related.]

3.2.2   *Multistage studies of skin cancer*

(*a*)   *Mouse (conventional)*

Groups of 32 female SENCAR mice, six to seven weeks of age, were sham-exposed or exposed to 60-Hz, 2000-μT continuous magnetic fields (Merritt exposure system described in Stuchly *et al.* (1991) [geomagnetic field not given]) for 6 h per day, five days per week for 21 weeks with or without weekly co-promotion with 1 μg 12-*O*-tetradecanoylphorbol 13-acetate (TPA) [four groups in total]. All animals received a single topical initiation with DMBA on the dorsal skin at a dose of 10 nmol (2.56 μg) dissolved in 200 μL acetone one week prior to exposure to the magnetic field. Any macroscopically visible tumours and other tissue abnormalities such as enlarged spleens and lymph nodes were examined histopathologically. [Minimum size of papillomas was not reported.] The development of papillomas in the magnetic field-exposed mice treated with TPA was no earlier than in sham-exposed animals treated with TPA ($p = 0.898$). At termination of the experiment, the total number of animals with papillomas was 28/31 in the sham-exposed group treated with TPA versus 29/32 animals in the matching group of animals exposed to the magnetic field. The average number of papillomas per tumour-bearing animal was 10 in both sham-exposed and magnetic field-exposed animals treated with TPA. All lesions were benign but two mice from the group exposed to the magnetic field and treated with TPA were found to have one papilloma each with very mild invasion of the squamous epithelium and another mouse was diagnosed as having a lymphoma. The investigators concluded that exposure to a magnetic field did not act as a promoter since none of the DMBA-initiated mice developed papillomas in the absence of treatment with TPA regardless whether or not they were exposed to a magnetic field. A positive control group (21 mice that received 2 µg TPA twice a week) showed no increase in tumour incidence compared with the control group (mice that received 1 µg TPA once a week) [p-value not given]; the author suggested a saturation of the system allowing no opportunity to detect a co-promotional effect (McLean et al., 1991).

In a second study, two groups of 48 female SENCAR mice, six weeks of age, were sham-exposed or exposed to a 60-Hz, 2000-µT magnetic field (Merritt exposure system described in Stuchly et al. (1991)) for 23 weeks; all animals received a weekly application of 0.3 µg TPA. All animals received a single topical initiation with DMBA (10 nmol dissolved in 200 µL acetone) on the dorsal skin one week prior to exposure to the magnetic field. An increased rate of papilloma development in animals exposed to magnetic fields compared with the sham-exposed controls was reported to be significant at weeks 16 ($p = 0.01$, Fisher exact test), 17 ($p = 0.02$) and 18 ($p = 0.02$), but was not significant at the end of the study ($p = 0.16$). The number of tumours per animal was not statistically different from the controls at the end of the study ($p = 0.21$, Wilcoxon test), but a significant difference was also reported at weeks 16 ($p = 0.01$), 17 ($p = 0.03$) and 18 ($p = 0.03$) (Stuchly et al., 1992). [The authors did not report any significant tests on papillomas per tumour-bearing animal although it is clear that the average number of tumours per tumour-bearing animal was lower than that seen in the first study of MacLean et al. (1991)]. The same study was prolonged to 52 weeks; TPA treatment was discontinued after week 24. The authors reported no overall increase in total tumours or papillomas alone in animals exposed to magnetic fields, but more animals exposed to magnetic fields developed squamous-cell carcinoma (8/48) than did sham-exposed animals (1/48) [$p < 0.03$, Fisher exact test two-sided]. [The number of tumours per animal was not reported.] The conclusion of the authors was that exposure to magnetic fields may accelerate progression to malignancy (McLean et al., 1995).

Two further studies replicating the 23-week study described by Stuchly et al. (1992) and using the same experimental design were performed. Two groups of 47 or 48 female SENCAR mice were used in each replicate. At week 23, tumour incidence in the first replicate was identical in animals exposed to a 2000-µT magnetic field and sham-exposed controls. In the second replicate, tumour incidence in mice exposed to a 2000-μT magnetic field was significantly reduced compared to the incidence in sham-exposed controls ($p = 0.04$, Fisher's exact test). The total number of tumours seen by Stuchly et al. (1992) was higher in the group exposed to the magnetic field (86) than in the sham-exposed group (48) ($p = 0.15$), whereas the opposite effect was seen in the two replicates (33 in the group exposed to the magnetic field versus 50 in the sham-exposed group [$p = 0.26$] and 27 in the group exposed to the magnetic field versus 86 in the sham-exposed group [$p = 0.01$]). The study does not support a role for exposure to a magnetic field as a strong co-promoter in the mouse skin-tumour model (McLean et al., 1997).

In another study of skin tumour promotion, groups of 30 female NMRI/HAN mice, seven weeks old, initiated with 25.6 μg DMBA in 200 μL acetone on the shaved dorsal skin or uninitiated (acetone only), were exposed to a 50-Hz magnetic field of strength of 50 or 500 μT or to sham-exposure conditions for 19 h per day on weekdays and for 21 h per day during weekends for a period of 104 weeks [six groups in total]. The exposure system has been described by Rannug et al. (1993a). In addition, TPA (given topically at a dose of 3.08 μg in 200 μL acetone twice per week), was used as a positive control in groups treated with either acetone alone or DMBA in acetone. The detection limit for the size of papilloma was 2 mm. After 104 weeks, the animals were assessed by complete necropsy and histopathological evaluation of skin tumour types. The survival of uninitiated mice exposed to the 500-μT magnetic field was significantly reduced in comparison to uninitiated controls ($p = 0.029$; 10% versus 23.3%). Exposure to the magnetic field had no significant effect on survival in any other group. Skin samples taken from members of each group at different times and analysed for hyperplasia showed a hyperplastic response in animals treated with DMBA plus TPA and those treated with TPA alone; no increased hyperplastic response was observed in any of the groups exposed to the magnetic field. No skin tumours were found in uninitiated mice in either the sham-exposed group or the group exposed to the magnetic field but one uninitiated TPA-treated animal had two skin tumours. No statistically significant differences in the number of animals with skin tumours or in the total number of skin tumours were reported in the initiated animals. The authors concluded that exposure to magnetic fields did not promote skin tumour incidence, nor act as a complete skin carcinogen. Full necropsy at death showed no significant change in incidence of other tumours associated with exposure to magnetic fields (Rannug et al., 1993b).

In a second study in female SENCAR mice, the tumour-promoting effects of continuous and intermittent exposure to 50-Hz magnetic fields for up to 105 weeks were examined. Starting one week after initiation with 2.56 μg DMBA in 200 μL acetone, groups of 40 mice were exposed to continuous magnetic fields at strengths of 50 or 500 μT or to intermittent fields (15 s on/15 s off) at the same field strengths. Untreated (acetone), sham-treated (DMBA alone) and TPA-treated (initiated and uninitiated) groups were also included. No skin tumours were reported in animals treated with DMBA and then exposed to continuous magnetic fields at either 50 μT or 500 μT. Four skin tumours were found in 4/40 animals in the group intermittently exposed to a field strength of 50 µT and 13 skin tumours were found in 5/40 animals animals in the group intermittently exposed to 500 µT compared with two skin tumours in 2/40 DMBA-treated sham-exposed animals. These increases were not significantly different when the two intermittently exposed groups were combined and compared with the sham-exposed group [p-value not given], but a significant difference was obtained when combined comparisons of continuous versus intermittent exposure were performed [$p = 0.014$, analysis according to Peto (1974)]. [The Working Group noted that it is not clear whether the pooling of data across groups is appropriate.] Similar findings were reported for the total number of skin tumours [no significant difference except when the pooled continuous exposure group was compared to the pooled intermittent exposure group; $p < 0.01$.] A linear regression analysis comparing the cumulative number of skin tumours in skin tumour-bearing animals exposed intermittently to different field strengths with the DMBA-treated controls gave a one-sided p-value of 0.045 which the authors noted as suggesting a dose dependency. No hyperplastic response was reported in animals exposed to magnetic fields. Histopathological investigation showed that carcinomas (squamous, spindle or basal) occurred in both the intermittently exposed groups and in the positive control group (DMBA plus TPA), but not in the DMBA-treated control group (where both lesions observed were papillomas). The investigators concluded that intermittent exposure could be weakly promoting, but interpretation is uncertain because of the possibility of induced electric fields due to mechanical switching and associated occurrence of transients (Rannug et al., 1994).

Groups of 56 female SENCAR mice, four to six weeks of age, were initiated with 10 nmol DMBA in 200 µL acetone before being exposed to ambient magnetic fields (mean field strength, 0.11 µT) or continuous magnetic fields of 2000 µT, 60 Hz for 23 weeks. Exposure to magnetic fields was combined with topical application of TPA once a week at doses of 0 (acetone only), 0.85 nmol (1.04 µg), 1.7 nmol (2 µg) or 3.4 nmol (4.2 µg) [a total of eight groups]. An additional group of 40 uninitiated mice was given a weekly dose of 3.4 nmol (4.2 µg) TPA and exposed to an ambient-strength magnetic field. The incidence of skin lesions [minimum size of tumour detected was not reported] was monitored and tumours were histologically evaluated at the end of the study. Tumour incidence [reported only as plots] did not differ between animals exposed to magnetic fields and ambient controls. [The Working Group noted that data on tumour multiplicity, total tumours, time to first tumour and survival of animals were not reported although some were shown in the figures, and p-values were given for comparisons of tumour multiplicity and incidence. The Working Group also noted several inconsistencies between the p-values and data reported; for example, a p-value of 0.32 was given in the Table for the group in which 3.4 (4.2 µg) nmol TPA was used as a promoter compared with the ambient control but a Figure illustrating the same results reported that 55/56 animals in the ambient group had tumours and that 56/56 animals in the group exposed to magnetic fields developed tumours. It was also noted that many of the p-values were repeated in both tumour incidence and tumour multiplicity tables.] The conclusion of the authors was that, within the sensitivity limits of this animal model and the exposure parameters employed, no promotional or co-promotional effects of exposure to a 2-mT magnetic field were observed in the two-stage skin cancer model (Sasser *et al.*, 1998).

(b)   Mouse (genetically modified)

Groups of 21–22 female mice of a transgenic hybrid strain (K2) that overexpresses the human ornithine decarboxylase gene and groups of 21–22 female non-transgenic littermates, seven to nine months of age, were exposed to ultraviolet light (200 J/m², 35 min/day, three times per week) for 10.5 months in order to induce skin cancer. The investigation of possible promotional or co-promotional effects of ELF magnetic fields was performed in groups of sham-exposed mice and mice exposed for 24 h to intermittent magnetic fields (field strengths of 1.3, 13 and 130 µT, each applied in succession for 20 min, followed by a 2-h pause) or continuous magnetic fields (50 Hz, 100 µT). The field generator used for this study was well described. Skin tumour incidence and tumour multiplicity were monitored, and the tumours were histologically evaluated at the end of the study. An increase in the rate of onset of macroscopically detectable tumours, with a minimum size of 2 mm, was reported in both transgenic and non-transgenic animals exposed to magnetic fields. This effect was statistically significant in the combined analysis, i.e. when all animals exposed to a magnetic field were compared with all controls ($p < 0.015$), and also when transgenic animals exposed to intermittent and continuous magnetic fields were compared to the ultraviolet light-treated controls ($p < 0.025$), but this effect was not significant in non-transgenic mice ($p < 0.15$). No significant differences were seen in the individual comparisons between the separate groups exposed to a magnetic field and the controls. Measurements of human ornithine decarboxylase activity did not show any significant changes as a function of exposure to magnetic fields ($p < 0.10$). The authors concluded that exposure to magnetic fields accelerated tumour growth (Kumlin *et al.*, 1998a). [The Working Group considered that the evaluation of these findings is difficult for two reasons. Firstly, a new skin cancer model using transgenic mice was used, but the investigators did not include a positive control, which would have been helpful in the interpretation of the findings, and, secondly, the results on tumour incidence as a function of time were presented only as values summed over the groups continuously and intermittently exposed to magnetic fields.]

3.2.3   Multistage studies of liver cancer

(a)   Mouse

Groups of 50 female CBA/S mice, 3–5 weeks of age, were exposed to ionizing radiation (using a 4- or 6-MV linear accelerator). The total body dose was 4 Gy delivered as three equal fractions of 1.33 Gy at one-week intervals. Simultaneously, the animals were exposed either to a 50-Hz vertical magnetic field of field strength 1.26, 12.6 and 126 µT (each applied in succession for 20 min) or were sham-exposed. A third group served as cage controls. The mice were exposed to the magnetic field or sham-exposed for 24 h per day for 1.5 years. An increase in the incidence of basophilic liver foci in mice exposed to magnetic fields was statistically significant when compared to sham-exposed mice (p = 0.002; Poly-3 test). The incidence of liver carcinomas showed a slight, but not statistically significant (p = 0.147) increase in the mice exposed to a magnetic field (14/48, 28%) compared to the sham-exposed group (7/50, 14%) (Heikkinen *et al.*, 2001).

(b)   *Rat*

Two studies in rats have been conducted to evaluate the possible tumour promoting and/or co-promoting effects of a 50-Hz magnetic field on chemically initiated liver tumours. The rat liver foci assay was used for the experiments (Rannug *et al.*, 1993b,c). Preneoplastic lesions were initiated in male Sprague-Dawley rats weighing approximately 200 g [age not specified] by intraperitoneal administration of 30 mg/kg bw *N*-nitrosodiethylamine (NDEA) 24 h after a 70% partial hepatectomy. These studies reported the number, the volume percentage and the mean area of altered hepatic foci in the liver, as determined by use of two enzyme markers, the placental glutathione-S-transferase (GST-P) and γ-glutamyltranspeptidase (γGT). Markers were measured by immunohistochemical methods using sections from the right lateral lobe of the liver.

In the first study, rats were exposed to either a 50-Hz magnetic field (4 groups of 10 rats) or to phenobarbital (2 groups of 10 rats) given in the diet at a concentration of 300 ppm for 12 weeks, beginning one week after administration of NDEA. Two separate experiments were conducted. In the first experiment, the flux densities to which the animals were exposed were 0.5 and 50 μT and for the second experiment 5 and 500 μT were used. The fields were switched on for 19–21 h per day, seven days per week. The exposure system is described in Rannug *et al.* (1993a). The experiments were performed in a blind fashion and sham controls (2 groups of 10 rats) were used as matching controls. The body weight gains and liver weights did not differ between animals exposed to magnetic fields and controls whereas the liver weight was significantly increased in rats used as positive controls (phenobarbital). The number of γGT-positive foci per cm^2, the mean focus area (cm$^2 \times 10^{-4}$) and the number of foci per cm^3 in groups exposed to magnetic fields did not differ significantly from the number seen in the controls. A significant difference was seen when the positive control (NDEA + phenobarbital) was compared to the control rats (NDEA) ($p < 0.01$). A significant increase in the percentage volume occupied by γGT-positive foci was observed at a field strength of 50 μT compared with the control rats ($p < 0.05$). In the second experiment, no such increase in γGT-positive foci was seen in association with exposure to magnetic fields. However, the percentage volume occupied by foci was lower in the first experiment (experiment 1: NDEA, 0.002 ± 0.001; experiment 2: NDEA, 0.006 ± 0.003), and the magnitude was similar to that seen in the first experiment in animals treated with NDEA and exposed to a magnetic field of 0.5 μT (0.005 ± 0.002) and in animals treated with NDEA and exposed to a magnetic field of

50 μT (0.005 ± 0.001). No effects on GST-P-positive foci were observed for any para-meters measured in either experiment when compared with the NDEA control (Rannug *et al.*, 1993c).

A second study, undertaken to investigate the possible interaction between exposure to a 50-Hz magnetic field, partial hepatectomy, initiation with NDEA and promotion with phenobarbital, was performed as described above, but the exposure to a magnetic field began immediately after partial hepatectomy. Four groups (9–10 rats each) were included in the study and treated as follows: one group was treated with NDEA alone, one group with NDEA followed by phenobarbital and two groups were given NDEA and phenobarbital and were exposed to 50-Hz magnetic fields, at field strengths of 0.5 and 500 μT. No effects on the liver:body weight ratio or body weight gain in the treated animals were observed after the 12-week exposure period when compared with controls. In the group exposed to magnetic fields of 0.5 μT, the number of γGT-positive foci per cm^3 (931 ± 131) was significantly decreased compared with the matching control group (1413 ± 181) ($p < 0.05$), and a reduction in GST-P-positive foci was associated with the higher field strength (500 μT) for the mean focus area and the percentage volume occupied by foci ($p < 0.05$) (Rannug *et al.*, 1993a).

3.2.4   *Multistage studies of leukaemia or lymphoma*

(a)   *Mouse (conventional)*

Groups of female CBA/S mice, 25 ± 5 days of age, were exposed either to X-rays (four fractions of 1.31 Gy at a dose rate of 0.45 Gy/min at four-day intervals; 64 mice in total), to pulsed magnetic field (vertical 20-kHz field with a field strength of 15 μT; 53 mice in total) for the lifetime of the animal or to both X-rays and pulsed magnetic field (exposure to magnetic field started immediately after each X-ray exposure; 63 mice in total) or received no treatment (stray-field, field strength < 0.7 μT; 47 mice in total). The mice were killed when moribund and autopsied; haemoglobin concentration, leukocyte counts and differential leukocyte counts were determined and 10 tissues were examined histologically. The diagnosis of lymphoma by microscopy was confined to thymic and non-thymic types. The difference in mean survival time between mice exposed to X-rays alone versus those exposed to both X-rays and magnetic field was not significant. However, mean survival time in mice exposed to magnetic field alone was significantly reduced in comparison to untreated controls ($p = 0.002$). Lymphoma incidence was increased in animals exposed to X-rays; however exposure to magnetic fields had no effect: the incidence of lymphomas was 42/64 (65.6%) in mice exposed to X-rays alone, 45/63 (71.4%) in mice exposed to both X-rays and magnetic field, 3/53 (5.7%) in mice exposed to magnetic field alone and 3/47 (6.4%) in untreated controls. The body weights of mice exposed to X-rays alone or both X-rays and magnetic field were not significantly different from those of controls, but the body weights of mice exposed to magnetic field alone were significantly greater than those of controls. Under the environmental conditions used, pulsed magnetic fields did not affect the frequency of spontaneous lymphoma, or the frequency of lymphomas induced by exposure to X-rays (Svedenstål & Holmberg, 1993). [The Working Group noted that animals were introduced to the experimental conditions at different time points, in some cases over a period of nearly one year.]

Groups of newborn Swiss Webster mice were subcutaneously injected with 0 or 35 μg DMBA in 1% gelatin, and, two weeks later, exposed to a 50-Hz magnetic field at field strength of 1 mT for 3 h per day, six days a week for 16 weeks, or were sham-exposed. All surviving mice were killed and autopsied at 32 weeks of age. No thymic lymphoma/leukaemia was found in the 84 sham-exposed mice that received 1% gelatin alone. There was no significant difference in survival times. The incidences of pre-malignant changes in thymus, early thymic lymphoma and advanced lymphoma were: sham-exposed males treated with DMBA, 18/80 (22.5%); males exposed to a magnetic field and treated with DMBA, 24/89 (27.0%); sham-exposed females treated with DMBA, 28/75 (37.3%); and females exposed to a magnetic field and treated with DMBA, 26/76 (34.2%); and demonstrated no statistically significant differences within a sex. The incidence of 'dense' metastatic infiltrations in the livers of the mice treated with DMBA and exposure to a magnetic field was significantly greater ($p < 0.01$) than that in the sham-exposed group treated with DMBA. The number of granulocytic leukaemia-bearing mice in the former group was 5/165 and that in the latter group was 4/155. The authors concluded that the study had found no evidence for a 'striking promotion effect' of this magnetic field on the incidence of lymphoma or leukaemia induced by DMBA (Shen et al., 1997).

Groups of female C57BL/6 mice, 28–32 days of age, were either sham-exposed (190 mice) or exposed to ionizing radiation (^{60}Co γ-rays) at 3.0, 4.0 or 5.1 Gy (dose rate, ~ 0.2 Gy/min; 3 groups of 190 mice) or to a 60-Hz magnetic field at a strength of 1.4 mT (380 mice) or to both ionizing radiation and magnetic fields for 18 h per day (3 groups of 380 mice). A group of 380 negative controls was exposed to ambient magnetic field only. All animals were kept until natural death or were killed at 852 days (the mean lifespan of the negative controls). All animals were necropsied and sections of thymus, spleen, lymph nodes, lungs, mainstem bronchi, sternum, kidneys, liver and brain as well as gross lesions in these tissues were examined. Lymphomas were classified as lympho-blastic, lymphocytic, immunoblastic, plasma cell and follicular centre cell. Non-lymphoid tumours included myelogenous leukaemia and histiocytic sarcoma. Because of considerable overlap, the follicular-centre-cell, immunoblastic and plasma-cell lymphomas were combined. There were no significant differences in mortality between the groups. The relative frequencies and general occurrence of haematopoietic neoplasia were similar for both animals exposed to magnetic fields and sham-exposed mice that had received the same ionizing radiation treatment. An exception was the reduced inci-dence rate of lymphoblastic lymphoma at death in the group exposed to 5.1 Gy and the magnetic field compared to the group exposed to 5.1 Gy only ($p = 0.05$). The total tumour incidences for groups exposed to magnetic fields were not significantly different from those of unexposed animals ($p = 0.55$, χ^2 test). The authors concluded that the results establish a lack of any overall effect of treatment with a single high level of exposure to magnetic field on the incidence of haematopoietic tumours (Babbitt *et al.*, 2000).

Groups of 50 female CBA/S mice, 3–5 weeks of age, were exposed to ionizing radiation (using a 4- or 6-MV linear accelerator), The total body dose was 4 Gy delivered as three equal fractions of 1.33 Gy at one-week intervals. Simultaneously, the animals were exposed either to a 50-Hz vertical magnetic field of strength 1.26, 12.6 and 126 μT (each applied in succession for 20 min) or were sham-exposed. A third group served as cage controls. The mice were exposed to a magnetic field or sham-exposed for 24 h per day for 1.5 years. In this study, survival until the end of the study in the group exposed to ionizing radiation plus the time-varying magnetic field was 66%, in comparison to a survival until the end of the study of 54% in the group that received ionizing radiation plus sham exposure. The incidence of lymphoma in the group exposed to radiation plus a magnetic field was 22% in comparison to an incidence of lymphoma of 30% in the group treated with radiation plus sham exposure. The authors concluded that the data from this study did not support a role for magnetic fields as a tumour promoter (Heikkinen *et al.*, 2001).

(b)   Mouse (genetically modified)

Groups of 103–111 female (C57BL/LiA × CBA × C57BL/6)fBR-(TG)pim-1 transgenic Eμ-*Pim*-1 (PIM) mice, which carry the *pim*-1 oncogene and are highly sensitive to lymphoma induction by *N*-ethyl-*N*-nitrosourea (ENU), 38–52 days of age, were sham-exposed or exposed to a 50-Hz magnetic field at a field strength of 1, 100, 1000 μT (continuous) or 1000 μT (intermittent 15 min on, 15 min off) for 20 h per day for up to 18 months; a group of 97 female wild-type C57BL/6Ntac mice was sham-exposed. No differences in body weights were observed throughout the experiment. Thirty Eμ-*pim*-1 mice and 30 wild-type mice each received an intraperitoneal injection of 50 mg/kg bw ENU. After nine months of treatment, the incidence of lymphoblastic lymphoma in these controls was 60% in the Eμ-*pim*-1 mice and 5% in the wild-type mice. The incidence of lymphoma (lymphoblastic and non-lymphoblastic) was 2/97 in wild-type mice and 32/111, 31/105, 27/103, 32/105 and 36/104 in the sham-exposed mice and in those exposed to continuous magnetic fields of 1 μT, 100 μT and 1000 μT and those exposed intermittently to 1000 μT, respectively. The incidence of a renal glomerular disease, not associated with lymphoma incidence, varied from 9% to 19% between the groups. All of 12 representative cases of lymphoma expressed the T-cell marker Thy 1, none carried B-cell markers. The authors concluded that, compared to the results of the sham exposure, there was no statistically significant increase in the incidence of lymphoma or its subtypes or in the time to appearance of the tumours in any of the groups exposed to magnetic fields (Harris *et al.*, 1998). [The Working Group noted that the 235 'healthy' survivors from the six groups were not autopsied. This is unlikely to have had an effect on the results obtained using this model since the lymphomas are likely to have been rapidly lethal.]

Groups of 30 male and 30 female Eµ-*pim*-1 mice (see definition above; 'high-inci-dence' model) [age unspecified] received a single intraperitoneal injection of 25 mg/kg bw ENU followed one day later by exposure to a 60-Hz continuous magnetic field at a field strength of 0 (sham controls), 2, 200 or 1000 µT or to an intermittent field of 1000 µT for 18.5 h per day for 23 weeks. The exposure system has been described by Gauger *et al.* (1999). Groups of 30 male and 30 female TSG-*p53* mice (knock-out; 'low-incidence' model) [age unspecified], not pre-treated with ENU, were either sham-exposed or exposed to a continuous magnetic field of 1000 µT for 18.5 h per day for 23 weeks. All animals underwent a limited gross necropsy. Survival until the end of the study was similar for both strains in all treatment groups, except in Eµ-*pim*-1 males exposed continuously to a magnetic field of 1000 µT (77% versus 60% in sham controls). The incidence of lymphomas in male Eµ-*pim*-1 mice in the sham-exposed group was 59% compared with 47%, 43% and 57% in the groups exposed to fields of 2 or 200 µT, or to the intermittent field of 1000 µT, respectively (not significant). Conti-nuous exposure to the magnetic field at 1000 µT resulted in a decreased lymphoma incidence (23%), which was statistically significant ($p = 0.041$, Fisher's exact test; $p = 0.054$, life-table test). There was no significant difference in the incidence of lymphomas between groups of female Eµ-*pim*-1 mice. The incidence was 49% in sham controls and 45%, 45%, 47% and 53% in the four groups exposed to magnetic fields. In the TSG-*p53* mice, the incidence of lymphoma was 3% in male controls, 0% in exposed males, 3% in female controls and 7% in exposed females (not significant). The authors concluded that their study demonstrated no increased risk of lymphoma in either Eµ-*pim*-1 or TSG-*p53* (knock-out) mice exposed to 60-Hz magnetic fields (McCormick *et al.*, 1998).

(c)   Other studies

Although studies with fully transformed cells are outside the immediate scope of this monograph, the Working Group briefly considered a series of in-vivo studies conducted to determine the influence of magnetic fields on the growth and proli-feration of transplantable tumour cells. In these studies, mice or rats received injections of leukaemia cells and were subsequently exposed to magnetic fields. The results of these studies were uniformly negative: no effects of exposure to magnetic fields were identified in any study reported (Thomson *et al.*, 1988; Sasser *et al.*, 1996; Morris *et al.*, 1999; Devevey *et al.*, 2000).

3.2.5   Multistage studies of neurogenic cancer

One study has been conducted to determine the influence of 60-Hz magnetic fields on the induction of neurogenic tumours by transplacental exposure to ENU. On day 18 of gestation, female Fischer 344 rats received a single intravenous dose of ENU (5 mg/kg bw) and were randomized into groups of 32 dams which were either sham-exposed or exposed to magnetic fields at strengths of 2 µT, 20 µT, 200 µT and

2000 µT. The magnetic field exposure system has been described in detail by Mandeville *et al.* (1997). After parturition, dams were exposed to magnetic fields together with their pups until weaning. At weaning, female pups were selected into groups of 50, and exposure to the magnetic fields was continued until study termination at age 65 weeks. A blinded histopathological analysis was performed on the brain (three levels) and spinal cord (three levels) of all animals. Exposure to magnetic fields had no influence on survival in ENU-treated animals. No significant differences between the sham-exposed group and the groups exposed to magnetic fields were seen in the incidence of all neurogenic tumours, glial tumours in the central nervous system or schwannomas in the peripheral nervous system. When compared with a total incidence of neurogenic tumours of 61% (30/49) in sham controls, the incidences of all neurogenic tumours in groups exposed to 60-Hz magnetic fields were 53% (26/49), 56% (28/50), 48% (24/50) and 46% (23/50) in the groups exposed to 2, 20, 200 and 2000 µT, respectively. None of these differences was statistically significant. The authors concluded that their study provided no evidence that 60-Hz magnetic fields have a promoting effect on neurogenic tumours (Mandeville *et al.*, 2000).

4. Other Data Relevant to an Evaluation of Carcinogenicity and its Mechanisms

4.1 Adverse effects other than cancer in humans

4.1.1 *Reproductive and developmental effects*

The effects of two kinds of exposure to magnetic fields are addressed in this section: those associated with power-frequency fields (ELF fields of 50 or 60 Hz) and those associated with video display terminals. The fields associated with video display terminals are typically of varying frequencies in the ELF range as well as higher frequencies (300 Hz–100 kHz). Studies of the possible effects of exposure to such fields on reproductive outcome are discussed below, including some that have examined mutations in maternal or paternal germ cells, investigations of exposure during prenatal development and other possible magnetic field-induced changes in fetal or maternal physiology. The information has been obtained primarily from epidemiological studies and a number of extensive reviews that have been published recently (Chernoff *et al.*, 1992; Brent *et al.*, 1993; Shaw & Croen, 1993; Huuskonen *et al.*, 1998a; Shaw, 2001).

(a) *Exposure to ELF electric and magnetic fields during pregnancy*

Studies of maternal exposure to power-frequency fields have focused principally on the use of electric blankets and electrically heated beds, other sources of residential exposure and a few occupational studies.

The use of electric blankets (exclusively older models) and electrically heated beds can add appreciably to total exposure to ELF electric and magnetic fields. It has been estimated that use of electric blankets increases overall exposure to electric fields by 36% over that of non-users (Preston-Martin *et al.*, 1988; see also section 1). Although these appliances are frequently used by pregnant women, the available studies present little evidence to support an association of exposure to ELF electric and magnetic fields with adverse reproductive outcomes (Wertheimer & Leeper, 1986; Dlugosz *et al.*, 1992; Juutilainen *et al.*, 1993; Bracken *et al.*, 1995; Li *et al.*, 1995; Belanger *et al.*, 1998). The first suggestion of potential adverse effects came from the study by Wertheimer and Leeper (1986), who reported an increase in the number of spontaneous abortions and of infants who showed below-average fetal growth associated with winter conception, and hence with increased use of electrically heated beds and blankets. However, a number of methodological inadequacies call into question the validity of these results (Hatch, 1992). A subsequent study showed that the use of heated water beds or electric blankets during pregnancy was not associated with intrauterine growth retardation or reduced birth weight (Bracken *et al.*, 1995). The same group also examined the occurrence of spontaneous abortion in pregnant women who used electric blankets or heated water beds: the use of electric blankets did result in an elevated risk ratio (1.8; 95% CI, 1.1–3.1) for spontaneous abortions whereas the use of water beds in homes with wire codes associated with elevated ELF electric and magnetic fields did not increase the risk ratios (Belanger *et al.*, 1998).

In a case–control study that examined specific congenital malformations, cleft palate ($n = 121$), cleft lip ($n = 197$), anencephalus and spina bifida ($n = 224$), no effects on odds ratios were evident in the offspring of women who had been exposed to ELF electric and magnetic fields through the use of electric blankets or electrically heated beds (Dlugosz *et al.*, 1992). The possible association of neural tube defects with the use of heated beds and blankets by pregnant women was assessed in the offspring of a cohort of 23 491 women (Milunsky *et al.*, 1992). No evidence of an association was observed between the use of electric blankets and adverse pregnancy outcomes.

Three studies examining maternal exposure to ELF fields and the risk for cancer in the children exposed *in utero* are reviewed in section 2.2.

In the study of Li *et al.* (1995), the use of electric blankets was not associated with increased anomalies of the urinary tract (odds ratio, 1.1; 95% CI, 0.5–2.3). In a subgroup of women ($n = 5$) with a history of reduced fertility an odds ratio of 4.4 (95% CI, 0.9–23) was observed.

[The Working Group noted the problem in the interpretation of the data on heated beds arising from the potentially large variability in exposure. This is due both to recall bias on duration and frequency of use and power setting of the appliance and to the large variation in the strengths of the fields produced by the different appliances.]

The studies on the effects of residential exposure (other than from heated beds) on pregnancy outcome have focused primarily on spontaneous abortion and birth weight of the offspring. The first such study (Wertheimer & Leeper, 1989) reported a positive correlation between elevated exposure to ELF magnetic fields at ~ 1 µT in electrically heated homes (from electric heating elements in the ceiling) and fetal loss. In another study, in which ELF magnetic fields were measured in the home, an association was also observed between higher fields (average > 0.2 µT; > 0.6 µT at the front door) and spontaneous abortion in early pregnancy (Juutilainen *et al.*, 1993). In contrast, a subsequent study showed no association between exposure to measured magnetic fields above 0.2 µT, or to high wire-codes, and adverse pregnancy outcome, including miscarriage, low birth weight or pre-term delivery (Savitz & Ananth, 1994). Another study that found no relationship between exposure to magnetic fields > 0.2 µT and reduced birth weight or growth retardation was reported by Bracken *et al.* (1995).

Few studies of occupational exposure of women to ELF magnetic fields in relation to pregnancy outcome have been made. In a study of women involved in the manufacturing of semiconductors, no increase in the risk for spontaneous abortion was observed even for workers exposed to the strongest fields (time-weighted average (TWA) ≥ 0.9 μT) (Swan *et al.*, 1995).

A prospective cohort study was conducted to assess the effects of exposure to ELF magnetic fields on spontaneous abortion. Exposure was measured by a dosimeter worn on the body. A weak, non-significant association was observed with exposure to fields above 0.2 μT when the TWA was used as the exposure metric. However, a significantly increased risk (approximately threefold) was found when the exposure metric used was maximum exposure above 1.6 μT. The risk was limited to women who indicated that measurements had been taken during a 'typical' day and was further increased if the subjects had a history of difficulties during pregnancy (Li *et al.*, 2001).

(b) *Paternal exposure to ELF electric and magnetic fields*

The investigations of the relationship between paternal exposure to ELF electric and magnetic fields and potentially adverse reproductive outcomes have been almost exclusively conducted in occupational settings. The available studies have largely focused on cancer in the offspring; however, a number of other end-points have also been investigated, including male infertility, perinatal death, spontaneous abortion, congenital anomalies, number of offspring, male:female birth ratio and low birth weight.

The studies examining risk for childhood cancer associated with paternal exposure to ELF electric and magnetic fields are reviewed in section 2.2.

The frequency of abnormal pregnancy outcome (described as congenital malformations and fertility difficulties) was reported to be significantly increased among the wives of workers in high-voltage switchyards (Nordström *et al.*, 1983). Buiatti *et al.* (1984) found more cases of male infertility in radioelectrical workers than in controls (odds ratio, 5.9; 95% CI, 0.9–40). No association was seen between semen abnormalities and electrical occupations (Lundsberg *et al.*, 1995) and no significant increase in abnormal birth outcome was reported for offspring of power-industry workers (Törnqvist, 1998). Baroncelli *et al.* (1986) reported no effect on the number of children per family when fathers worked in a high-voltage substation. For male workers in industries associated with exposure to ELF fields, the proportion of male offspring was slightly reduced, while the number of offspring of female workers was significantly reduced (Irgens *et al.*, 1997). [The Working Group noted that 'number of children' is a particularily weak end-point with respect to developmental toxicology.]

(c) *Exposure to mixed ELF and higher-frequency electric and magnetic fields*

Many of the studies on the relation between exposure to ELF fields and reproductive effects in humans have addressed the question of whether pregnant women are at risk when exposed to the ELF fields associated with video display terminals. Most of these studies have investigated spontaneous abortion and congenital abnormalities in the offspring. There is little evidence for an association between exposure to fields from video display terminals and spontaneous abortion. Of ten studies (McDonald *et al.*, 1986; Ericson & Källén, 1986a,b; Westerholm & Ericson, 1987; Goldhaber *et al.*, 1988; Bryant & Love, 1989; Windham *et al.*, 1990; Nielsen & Brandt, 1990; Schnorr *et al.*, 1991; Lindbohm *et al.*, 1992), only two showed a significantly increased risk. The study by Goldhaber *et al.* (1988) showed an odds ratio of 1.8 (95% CI, 1.2–2.8), although later analyses suggest that differential reporting of exposure was the source of the association (Hertz-Picciotto *et al.*, 1992). Lindbohm *et al.* (1992) observed increased risk ratios among women in Finland who used video display terminals with high-intensity fields (peak-to-peak value, > 0.9 µT) for more than 10 h per week (risk ratio, 3.4; 95% CI, 1.4–8.6). In contrast, a large study conducted in the USA showed no dose–response relationship and no increased risk (Schnorr *et al.*, 1991). The strongest fields to which subjects were exposed in this study were weaker than those in the Lindbohm study. Most of the studies reported risk ratios from 1.1 to 1.2 but with 95% confidence intervals that include 1.0. Furthermore, in studies that assessed duration of exposure per day, there was generally no evidence for an increase in risk in association with longer exposure times.

Studies of reproductive health outcomes other than spontaneous abortion have also been made. Low birth weight, pre-term delivery, intrauterine growth retardation and perinatal mortality have been considered when evaluating exposure to fields from video display terminals, but these other end-points have rarely shown any indication of an effect of exposure to ELF fields from video display terminals. Over half a dozen studies (for a review, see Shaw, 2001), including two large studies with more than 1500 cases and 21 000 controls by McDonald *et al.* (1986) and Windham *et al.* (1990), have shown no significant reduction in birth weight associated with use of video display terminals, although intrauterine growth retardation was somewhat elevated (odds ratio, 1.6; 95% CI, 0.92–2.9) with greater use of video display terminals. Elevated risks (odds ratio > 1.5) have occasionally been observed for perinatal death (Bjerkedal & Egenaes, 1987) and congenital abnormalities (Ericson & Källén, 1986a,b); however, the risks were not significantly different from those in controls, and other studies did not confirm the results.

4.1.2   *Immunological effects*

The effects of exposure to magnetic fields on various markers of immune function were studied in two groups of workers: one group comprised 10 hospital personnel operating magnetic resonance tomographs and the other group was composed of 10 industrial workers operating induction heaters. A group of 23 workers served as non-exposed controls. Operaters of magnetic resonance tomographs exposed to static magnetic fields at ≥ 0.5 mT for an indeterminate time showed no significant reductions in their concentration of interleukin-2 or the number of monocytes in their blood. The operaters of induction heaters had been exposed to magnetic fields at either 50–600 Hz (up to 2 mT) or to 2.8–21 kHz (0.13–2 mT) for at least two years, and often longer than five years. The numbers of natural killer cells and monocytes were significantly increased in the exposed group while monocytes had significantly reduced phagocytic activity compared with those from unexposed personnel. For the two subjects with the highest exposure, the natural killer cell counts were > 700 cells/μL blood compared with 276 ± 124 cells/μL for the controls. Blood samples drawn from these two subjects eight months later still showed elevated counts of natural killer cells (671 and 1202 cells/μL, respectively) while controls had 281 ± 115 cells/μL, indicating that the elevated readings had not been due to an unknown confounding factor at the time of the first blood sampling (Tuschl *et al.*, 2000).

A group of 16 young men aged 20–30 years were exposed to 50-Hz, 10-μT magnetic fields from 23.00 to 08.00. In the first experiment, exposure was continuous for one night, in a second experiment exposure was intermittent, i.e. 1 h 'off' and 1 h during which the field was switched between 'on' and 'off' every 15 s. Sixteen other men were the sham-exposed controls. Blood samples were collected at 3-hourly intervals from 11.00 to 20.00 and hourly from 22.00 to 08.00. No significant differences were observed between exposed and sham-exposed men in haemoglobin concentration, haematocrit, or counts of erythrocytes, platelets, total leukocytes, monocytes, lymphocytes, eosinophils or neutrophils. The numbers of CD3, CD4, CD8, natural killer cells and B cells were also comparable between the two groups (Selmaoui *et al.*, 1996a).

4.1.3   *Haematological effects*

A survey of neurovegetative disorders and haematological effects was conducted in a group of three men and 10 women who had worked near electrical transformers, high-tension cabling (13 kV) and a power generator. In one room the 50-Hz field was 1.2–6.6 μT at floor level and 0.3–1.2 μT at 1.5 m above floor level. The magnetic fields in an adjacent room also used by the group were 0.2–0.3 μT and 0.09–0.12 μT, respectively. The subjects had worked on the premises for at least 8 h per day for one to five years. The occurrence of neurovegetative disorders was assessed from self-rating questionnaires completed by the exposed workers and matched control groups. A comparative analysis of the questionnaires showed that the exposed group suffered a significant increase in physical fatigue, psychological asthenia, lipothymia, decreased libido, melancholy, depressive tendency and irritability. [The Working Group noted the possibility of subjective bias in self-reporting questionnaires.] There was also a significant decrease in total lymphocytes and CD2, CD3 and CD4 lymphocytes, as well as an increase in the number of natural killer cells. Leukopenia and neutropenia were seen in two subjects who were chronically exposed to a field strength of 1.2–6.6 μT. The effects disappeared when exposure stopped, and reappeared when exposure was resumed (Bonhomme-Faivre *et al.*, 1998).

4.1.4   *Neuroendocrine effects*

Melatonin, a hormone produced in the mammalian pineal gland, is secreted in a circadian pattern to give high concentrations at night and low concentrations during the day. The circadian release of melatonin is known to influence certain physiological functions and to modulate the release of other hormones. Although relatively little is known about the mechanism by which changes in melatonin in humans may affect health and well-being, plausible hypotheses exist which suggest that alterations in this hormone may influence the risk for cancer.

On the basis of experimental studies that showed reductions in melatonin concentrations in animals exposed to ELF electric or magnetic fields, Stevens (1987) proposed what has been described as the 'melatonin hypothesis' (Stevens, 1987; Stevens & Davis, 1996). Exposure-related suppression of the night-time rise in melatonin concentration may explain the adverse health outcomes, including cancer — specifically breast or prostate cancer — reproductive problems and neurodegenerative diseases (Wilson *et al.*, 1989). A prerequisite to establishing the relevance and validity of the melatonin hypothesis in humans exposed to ELF electric and magnetic fields would be to determine whether melatonin is indeed suppressed during or following exposure.

(*a*)   *Exposure under laboratory conditions*

Studies of endocrine function in humans exposed to 50- or 60-Hz magnetic fields under laboratory conditions have been conducted in four laboratories. As shown in Table 34, the results have been principally negative with respect to the demonstration of exposure-related effects. Night-time exposure of human volunteers to magnetic fields under controlled exposure and lighting conditions had no apparent effect on nocturnal blood concentrations of melatonin when compared with sham-exposed subjects. Endocrine parameters, other than melatonin, have not been shown to be affected by exposure to 50- or 60-Hz electric or magnetic fields.

In the first of a series of double-blind studies, 33 young male volunteers, aged 18–35 years, were exposed to intermittent, circularly polarized magnetic fields of 1 µT or 20 µT, or sham-exposed (11 subjects per group) between 23.00 and 07.00 under controlled environmental and exposure conditions (Graham *et al.*, 1996). Overall, exposure had no effect on melatonin concentrations in serum, as measured by radio-immunoassay. However, men with a pre-existing low melatonin production showed significantly reduced melatonin concentrations when exposed to the 20-µT field. In a second experiment, 40 men were identified who had low melatonin concentrations in their serum. Each of these volunteers slept in the exposure facility for two nights. On one night the men were sham-exposed and on the other they were exposed to 60-Hz, 20-µT magnetic fields. In this experiment, exposure had no effect on melatonin concentrations and the original finding was not replicated in these low-melatonin subjects (Graham *et al.*, 1996). In a third study, using a cross-over design in which each subject served as his own control, 40 young men were sham-exposed for one

Table 34. Melatonin levels in human volunteers

Reference	Assay	Exposure	Response	Comment
Wilson et al. (1990)	Early morning excretion of urinary metabolite of melatonin	60-Hz EMFs generated by pulsed alternating or direct current supply to electric blankets at night for 7–10 weeks	No overall effect; transient increases in 7/28 users of one type of blanket	Realistic, but concomitant lack of control over lifestyle
Graham et al. (1996)	Night-time serum melatonin concentrations	60-Hz, intermittent fields of 1 or 20 µT for 8 h at night	No effect; possible effect on low-melatonin subjects not replicated in larger study	
Graham et al. (1997)	Night-time serum melatonin concentrations	60-Hz, continuous fields of 1 or 20 µT for 8 h at night	No effect	
Selmaoui et al. (1996b)	Night-time serum melatonin concentrations and excretion of its major urinary metabolite	50-Hz, continuous or intermittent fields of 10 µT for 9 h at night	No effect	
Wood et al. (1998)	Night-time serum melatonin concentrations	50-Hz intermittent sinusoidal or square-wave fields of 20 µT for 1.5–4 h at night	Possible delay and reduction of night-time melatonin concentrations in subgroup	Inconsistent, variable data; incomplete volunteer participation
Graham et al. (2000a)	Early morning excretion of urinary melatonin and its metabolite	60-Hz circularly polarized magnetic field of 28.3 µT overnight for 4 consecutive nights	No effect on night-time concentrations	Exposed subjects showed less intra-individual consistency on night 4

From NRPB (2001)
EMF, electric and magnetic field

night and exposed continuously through the second night to a 60-Hz, 20-μT magnetic field. Again, no overall effects on melatonin were found (Graham *et al.*, 1997).

Another study conducted in the same laboratory examined 30 healthy young male volunteers who were exposed to 60-Hz, 28.3-μT magnetic fields for four consecutive nights (Graham *et al.*, 2000a). Melatonin concentrations were determined by measurement of 6-hydroxymelatonin sulfate (a major melatonin metabolite) in the morning urine samples each day. Again, no overall effect of exposure on melatonin concentrations was observed. The consistency of intra-individual urinary measurements over the four test nights was much higher in the control samples than in those of the exposed subjects.

In a study of the effects of exposure to a 50-Hz, 1-μT magnetic field on sleep patterns in eight women and 10 men, Åkerstedt *et al.* (1999) measured concentrations of melatonin, growth hormone, prolactin, testosterone and cortisol in peripheral blood during a night of sham exposure or a night of exposure to the magnetic field. Exposure to the magnetic field showed effects on sleep patterns, but no significant endocrine or neuroendocrine effects due to exposure were observed.

The other major laboratory study on neuroendocrine effects measured not only melatonin but also pituitary, thyroid and adrenocortical hormones in 32 young men aged 20–30 years who were divided into three groups that were sham-exposed or exposed to continuous or intermittent linearly polarized 10-μT magnetic fields. The exposure-treatments and sampling of serum and urine were conducted over two 24-h periods. No significant differences in serum melatonin and urinary 6-hydroxymelatonin sulfate were found between the test and control groups (Selmaoui *et al.*, 1996b). There were also no differences observed in the concentrations or circadian variations of thyroid-stimulating hormone, follicle-stimulating hormone, luteinizing hormone, triiodothyronine, thyroxine, thyroxine-binding globulin, cortisol, 17-hydroxycorticosteroids or on thyroxine-binding index (Selmaoui *et al.*, 1997).

In another study (Wood *et al.*, 1998), the effect of exposure to magnetic fields on melatonin patterns was evaluated in 30 male volunteers. Once the nocturnal melatonin curve had been determined for each individual, the subjects were divided into groups that were either sham-exposed or exposed to magnetic fields (50 Hz, 20 μT) before, during or after the time of peak concentration of melatonin. Exposure preceding the rising segment of the curve significantly delayed the peak in one exposed individual and showed a similar trend in others. The authors noted the suggestive nature of their results but considered them to be preliminary. [The Working Group noted that this is the only laboratory study in humans which sugggested an apparent effect of exposure to magnetic fields on neuroendocrine parameters, including melatonin. Because of the preliminary nature of this study, a number of concerns have been raised over experimental design and statistical analysis.]

(b)   *Exposure in occupational and residential environments*

A heterogeneous group of epidemiological studies has also evaluated endocrine function in humans exposed to ELF magnetic fields in the relatively uncontrolled environment of occupational and residential settings. In contrast to the negative results of laboratory studies, all of these studies noted some perturbation in the excretion of 6-hydroxymelatonin sulfate in exposed groups. The perturbations were not, however, consistent across studies and the exposure parameters also differed from one study to the next; they included use of electric blankets, exposure to 16 $^2/_3$-Hz fields (for railway engineers) as well as exposure to 60 Hz in residential settings and to 50 Hz and 60 Hz in occupational settings.

One of the earliest studies to measure melatonin via the urinary metabolite hydroxymelatonin sulfate (6-OHMS) focused on users of electric blankets. Forty-two volunteers used standard or modified continuous-polymer-wire blankets for eight weeks. The continuous-polymer-wire blankets produced fields that were 50% stronger than those of conventional blankets (0.66 versus 0.44 µT) and they switched on and off twice as often. The subjects in this study served as their own controls. Seven of 28 volunteers using the continuous-polymer-wire blankets were found to have a statistically significant increase in mean night-time urinary excretion of 6-OHMS at the cessation of exposure. This is the only study to report an increase in melatonin or its metabolite in association with higher-than-background exposure to ELF electric or magnetic fields. No changes were seen in the users of conventional electric blankets (Wilson et al., 1990).

A study of Swiss railway workers compared 66 engineers (average exposure of most-exposed workers, 20 µT) with 42 other employees (train attendants and station managers; average exposure of least-exposed workers, 1 µT). Each volunteer served as his own control. Morning and evening urinary concentrations of 6-OHMS were determined during leisure periods and on the day following resumption of work. Evening concentrations of 6-OHMS appeared to be lower by a factor of 0.81 (95% CI, 0.73–0.90) during workdays compared with leisure days in engine drivers, but not in controls. The evening concentrations recovered significantly during leisure periods, which suggests that the effects were reversible. In contrast, morning concentrations of 6-OHMS from engineers and controls did not differ much between workdays and leisure days (Pfluger & Minder, 1996). [The Working Group noted that the interpretation of this study is hampered by the difficulties in accounting for the effects of shift work in some of the subjects and a crude assessment of exposure. It should also be noted that the predominant exposure is to 16 $^2/_3$-Hz fields, not the 50- or 60-Hz fields considered in most of the other studies.]

The effects of exposure to 60-Hz magnetic fields (and ambient light) were studied in 142 male electric utility workers. Melatonin was measured as 6-OHMS in post-workshift urinary samples over a three-day sampling period. The groups compared were 29 power generation workers (mean exposure, 0.32 µT), 56 linemen and sub-station operators (mean exposure, 0.23 µT) and 57 utility maintenance and administrative staff (mean exposure, 0.15 µT). In addition to field intensity measured as TWA fields, the temporal stability of the fields was also determined. Exposure intensity, as measured by the geometric mean magnetic field, was not associated with 6-OHMS excretion. However, evaluation of the temporal stability parameter showed that men in the highest quartile had lower 6-OHMS concentrations on the second and third days of the sampling period when compared with the men in the lowest quartile (Burch et al., 1998, 1999).

The same research team studied workers in substations in three-phase environments compared with workers in one-phase environments. Over three consecutive workdays, an apparent field-dependent reduction in mean nocturnal and post-work concentrations of 6-OHMS was reported for men who worked more than 2 h per day in a substation with a three-phase environment. No difference was observed among those men who worked 2 h or less or those who worked in one-phase environments (Burch et al., 2000).

A third occupational study was carried out among female garment-industry workers including 39 production workers and 21 office workers, who served as controls. Exposure assessment varied with the type of machine used and was based on magnetic field measurements made around each type of machine. Exposure to 50-Hz magnetic fields was quite high (> 1 μT) for one group, approximately 0.3–1 μT for a second exposure group and about 0.15 μT for the control group. Morning void urine samples were collected on the Friday and Monday for three consecutive weeks. The average 6-OHMS concentration in the urine on Fridays was lower in the factory workers than in the control group, but no monotonic dose–response pattern was observed. The 6-OHMS concentrations measured in urine samples taken on Monday and Friday were not different for test subjects and controls. Multivariate analysis identified exposure to a magnetic field, smoking and age as significant explanatory variables associated with decreased 6-OHMS excretion (Juutilainen et al., 2000).

The most comprehensive study of the effects of residential exposure to magnetic fields on neuroendocrine response was conducted in Seattle, Washington. The 203 participants were women selected from a group that participated as controls in a case–control study of breast cancer and exposure to electric and magnetic fields. Magnetic fields were measured in the participants' bedrooms, and personal field measurements were made during the same 72 h. Total night-time urine samples collected during the three consecutive nights of the measurement period were used to assess the concentration of 6-OHMS. The results showed that decreasing concentrations of 6-OHMS in the night-time urine were associated with increasing magnetic field strength, as measured in the women's bedrooms at night. The magnetic field effect was seen primarily in women who used medication (e.g. beta blockers, calcium-channel blockers and psychotropic drugs), and was strongest during the times of the year with the shortest nights. These findings were particularly marked when the exposure measure was the proportion of night-time magnetic field measurements ≥ 0.2 μT. The reduction in the concentration of 6-OHMS in urine was not correlated with personal field measurements > 0.2 μT, variability in field measurements, use of electric blankets or wire codes (Davis et al., 2001). [The Working Group noted that

extensive field assessment data were collected and analysed in this study and considerable effort had been made to consider and control potential confounding.]

4.1.5   *Behavioural and physiological effects*

A number of volunteer studies have investigated the effects of static or ELF electric
and/or magnetic fields on perception, the electrical activity of the brain, memory and
reasoning, mood, hypersensitivity and heart rate.

(*a*)   *Static fields*

Prior to the development of magnetic resonance imaging (MRI) techniques, few
studies of the effects of static magnetic fields on volunteers had been documented
(summarized by WHO, 1987), although various anecdotal reports from laboratories
using large accelerators existed. However, with the advent of superconducting magnet
technology and MRI in the late 1970s, volunteers could be routinely exposed to static
fields $\geq$ 1.5 T. Most of the reported effects of acute exposure are consistent with
known mechanisms of action.

(i)   *Perception of electric fields*

The electric charge induced on the surface of a person exposed to a static electric
field can be perceived by its interaction with body hair, particularly on the head.
Clairmont *et al.* (1989) reported that volunteers had a threshold of perception around
20 kV/m, and that fields above about 25 kV/m produced annoying sensations.

(ii)   *Perception of magnetic fields*

Schenck *et al.* (1992) reported dose-dependent sensations of vertigo, nausea and a
metallic taste in the mouth in volunteers exposed in MRI systems to static magnetic
fields of 1.5 or 4 T; however, gradient and higher-frequency magnetic fields also seem
to have contributed to the total exposure. The sensations reported occurred only
during movement of the head. In addition, magnetic phosphenes (described below)
could sometimes be seen during eye movement in a static magnetic field of at least
2 T.

(iii)   *Cognition*

In a static field, Lorentz forces will be exerted on the ion flow through nerve
membranes, although these may not be of biological significance at field strengths < 2 T
(Tenforde, 1992). The possible cognitive effects immediately after volunteers had been
exposed for one hour to a static magnetic field of 8 T were investigated by Kangarlu
et al. (1999). The written and oral tests comprised a standard 'mini-mental' status examination of cognitive function and other standard tests of cognition and motor function.
The performance in these tests after exposure did not differ from that in the tests conducted before exposure. An earlier study of a large number of volunteers had reported a
lack of effect of exposure to static magnetic fields from magnetic resonance imaging

equipment of 0.15 T in a variety of cognitive tests, although anxiety was increased in the exposed group following exposure (Sweetland *et al.*, 1987).

(iv)   *Cardiac effects*

On theoretical grounds, Kinouchi *et al.* (1996) noted that the Lorentz force exerted on the blood flow generates an electrical potential across the blood vessel. In practice, so-called 'flow' potentials are readily demonstrated in large animal species, such as dogs, baboons and other monkeys exposed to static fields stronger than ~ 0.1 T. Generally, the largest flow potentials occur across the aorta after ventricular contraction and appear superimposed on the T-wave of the electrocardiogram (Tenforde, 1992).

In addition, a 5–10% reduction in blood flow in the aorta is predicted to occur in static fields of 10–15 T, due to magneto–hydrodynamic interactions (Kinouchi *et al.*, 1996). However, Kangarlu *et al.* (1999) noted that following exposure to an 8-T static field, volunteers showed no change in heart rate or diastolic or systolic blood pressure, compared with values measured before exposure; the values recorded during exposure were also reported as unchanged.

(b)   *ELF electric and magnetic fields*

The nervous system functions by virtue of electrical signals and may be thought particularly vulnerable to ELF electric and magnetic fields. Various studies have been carried out on the effects of ELF electric and magnetic fields on perception, electrical activity of the brain, cognitive processes (i.e. thinking and memory), mood, hyper-sensitivity, sleep and heart rate.

(i)   *Perception of electric fields*

It is well established that ELF electric fields can be perceived due to the field-induced vibration of body hair. The threshold for perception by hair vibration shows wide individual variation: 10% of exposed subjects were found to have detection thresholds of 10–15 kV/m at 50–60 Hz, and 5% of subjects could detect fields as weak as 3–5 kV/m. Although these effects are not considered to be a hazard, hair vibration and tingling became an annoyance to test subjects at field strengths > 20 kV/m (Deno & Zaffanella, 1975). Of greater biological significance may be the occurrence of capacitive spark discharges or microshocks, generated when two objects of different potential come into close proximity and the electric breakdown field strength of the air is exceeded. The threshold for the perception of spark discharges by a small proportion (10%) of a group of volunteers close to a grounded object has been reported to be 0.6–1.5 kV/m at 50 or 60 Hz, while the threshold was 2.5–6 kV/m for the rest of the group (Bernhardt, 1988).

(ii)   *Magnetic phosphenes*

Exposure to power-frequency magnetic fields < 1 mT is generally regarded as imperceptible. In contrast, exposure of the head to magnetic flux densities at 20 Hz above about 5 mT up to about 50 Hz, 15 mT, will reliably induce faint, flickering, visual sensations called magnetic phosphenes (Lövsund *et al.*, 1979, 1980a,b). Similar sensations can be induced by electric currents applied directly via electrodes attached to the head (Lövsund *et al.*, 1980b). It is generally agreed that phosphenes result from the interaction of the induced electric current with electrically sensitive cells in the retina. The maximum current density in the retina associated with the generation of magnetically induced phosphenes has been estimated to be about 11 mA/m² at 20 Hz (Wake *et al.*, 1998), based on calculations using a realistic, electrically heterogeneous model of the human head.

(iii)   *Electroencephalograms and event-related brain potentials*

The electrical activity of the brain, recorded as an electroencephalogram, conveys information of a general nature that characterizes the mental state of a person. The electroencephalogram is used in the diagnosis of a variety of pathological conditions. Event-related brain potentials, which are also recorded using electrodes placed on the scalp, convey more specific information concerning brain activity evoked by a sensory stimulus (evoked potentials) and, after about 100 min, by subsequent cognitive processes.

Two double-blind studies on the effects of exposure to 45-Hz, 1000 Amps/m [1.26 mT], magnetic fields on electroencephalograms have been reported. Changes were observed in the alpha, delta and beta frequency bands and in auditory evoked potentials (Lyskov *et al.*, 1993a,b). A phase reversal and a slower decrease in the amplitude of the major components of visual evoked potentials have been reported during exposure to very intense (60 mT), pulsed magnetic fields, although they had no effect on visual acuity (Silny, 1984, 1985, 1986). In contrast, no marked effect on visual, auditory or somatosensory evoked potentials was reported during exposure to fields of weaker intensity (28 μT) (Graham *et al.*, 1999).

The possible effects of electromagnetic fields on event-related potentials have been investigated mostly in conjunction with various cognitive tests. In general, few effects have been found; those that were noted have tended to be subtle and transitory (Crasson *et al.*, 1999). For example, small changes in latency and amplitude of a late component (P300) of the event-related potential associated with cognitive function were observed when subjects exposed to combined electric (9-kV/m) and magnetic (20-μT) fields were asked to discriminate between frequent and infrequent stimuli (Cook *et al.*, 1992; Graham *et al.*, 1994). When subjects were exposed to a magnetic field of 50 Hz, 100 μT, changes were observed in event-related brain potentials during performance of a listening task, in which auditory discrimination is tested (Crasson *et al.*, 1999).

(iv)   *Cognition*

A number of studies have looked for evidence of changes in cognitive ability during or after exposure to power-frequency electromagnetic fields. Reaction time, vigilance or sustained attention, memory function, and tasks involving time perception and information processing have all been tested. Some changes have been reported, but the effects were not consistent between studies. For example, studies have reported both increases (Cook *et al.*, 1992; Kazantzis *et al.*, 1996) and decreases (Graham *et al.*, 1994; Preece *et al.*, 1998) in the accuracy of task performance. Similarly, several studies have reported decreased reaction time (Graham *et al.*, 1994; Whittington *et al.*, 1996), or no effect (Podd *et al.*, 1995; Preece *et al.*, 1998).

(v)   *Mood*

The possibility that environmental exposure to power-frequency electromagnetic fields might be associated with a variety of negative mood states has been assessed in several double-blind laboratory studies in which volunteers have completed mood assessment checklists before and after exposure. None of these studies (Stollery, 1985; Cook *et al.*, 1992; Graham *et al.*, 1994; Crasson *et al.*, 1999) reported any effects although Stollery (1985) reported decreased arousal in one of two participating groups.

(vi)   *Hypersensitivity*

It has been reported that some people are sensitive to electric and magnetic fields. The symptoms of sensitivity include sleep disturbance, general fatigue, difficulty in concentrating, dizziness, eye strain, facial skin problems such as eczema and sensations of itching, burning or stinging. Several double-blind laboratory provocation studies have been carried out. Generally, the patients and volunteers who participated in these studies were not reliably able to identify the presence of electric or magnetic fields, and neither subjective symptoms nor biochemical measures were significantly related to the exposure conditions (Andersson *et al.*, 1996; Lonne-Rahm *et al.*, 2000).

(vii)   *Sleep electrophysiology*

Several studies have examined the effect of exposure to electric and magnetic fields on sleep, monitored using electroencephalograms and self-assessment. One study reported a reduction of 'slow-wave' sleep, total sleep time and depth of sleep in subjects exposed to a relatively weak power-frequency magnetic field (50 Hz, 1 µT) (Åkerstedt *et al,*. 1999). In contrast, another study reported that intermittent exposure to 60-Hz, 28-µT magnetic fields was associated with a poor and irregular pattern of sleep (Graham *et al.*, 1999).

(viii)   *Heart rate*

A statistically significant slowing of heart rate, recorded as the interbeat interval, during exposure to 60-Hz electric and magnetic fields (9 kV/m and 20 µT) has been reported in several studies (Cook *et al.*, 1992; Graham *et al.*, 1994). However, these effects were not observed at higher or lower field strengths. No effect on heart rate or blood pressure was seen during acute exposure to 50-Hz, 100-µT magnetic fields (Whittington *et al.*, 1996).

A more recent study reported an altered heart rate variability during exposure to an intermittent 60-Hz magnetic field at night (Sastre *et al.*, 1998). However, in a pooled analysis of several studies conducted at the same institute, Graham *et al.* (2000b) later reported that this effect was observed only in studies where hourly blood sampling had taken place as part of a different experiment. The authors hypothesized that blood sampling may have altered the arousal of the subjects, allowing interaction with the magnetic field to affect heart-rate variability.

(*c*)   *Epidemiological studies*

Several epidemiological studies have been carried out over the past 20–30 years on the incidence of neurodegenerative diseases, suicide and depression, and cardio-vascular disease in relation to occupational or residential exposure to ELF electric and magnetic fields (reviewed in Portier & Wolfe, 1998; International Commission on Non-Ionizing Radiation Protection, 1998).

(i)   *Neurodegenerative diseases*

Many studies have focused on amyotrophic lateral sclerosis, a progressive degene-rative motor neuron disease, and Alzheimer disease, a progressive irreversible degene-rative disease of the brain, in groups of people occupationally exposed to ELF electric and magnetic fields.

Several studies on amyotrophic lateral sclerosis have been published (Deapen & Henderson, 1986; Gunnarsson *et al.*, 1992; Davanipour *et al.*, 1997; Johansen & Olsen, 1998; Savitz *et al.*, 1998a,b). The combined results from the two studies of uti-lity workers (Johansen & Olsen, 1998; Savitz *et al.*, 1998a,b) show a clear increase in mortality from amyotrophic lateral sclerosis in association with exposure to ELF magnetic fields. This increase is unlikely to be due to chance but may be confounded by exposure to electric shocks.

Five studies have been conducted on Alzheimer disease in relation to exposure to ELF electric and magnetic fields (Sobel *et al.*, 1995, 1996; Feychting *et al.*, 1998b; Savitz *et al.*, 1998a,b). When all studies are considered together, there appears to be an association between the occurrence of the disease and estimated exposure to ELF electric and magnetic fields. However, since this result is mainly confined to studies with weaker designs, support for the hypothesis of a link between Alzheimer disease and exposure to ELF electric and magnetic fields is weak (International Commission on Non-Ionizing Radiation Protection, 1998).

(ii)   *Suicide and depression*

A number of studies have examined possible associations between the incidence of suicide and residential or occupational exposure to ELF electric and magnetic fields (Reichmanis *et al.*, 1979; Perry *et al.*, 1981; Baris & Armstrong, 1990; Baris *et al.*, 1996a,b; Johansen & Olsen, 1998; van Wijngaarden *et al.*, 2000). Only the most recent study provides some support for the original findings of Reichmanis *et al.* (1979) and Perry *et al.* (1981) suggesting a relation between suicide and exposure to magnetic fields from overhead power lines (International Commission on Non-Ionizing Radiation Protection, 1998).

The relationship between the prevalence of depressive symptoms and residential or occupational exposure to ELF electric and magnetic fields has been investigated in several studies (Dowson *et al.*, 1988; Poole *et al.*, 1993; McMahan *et al.*, 1994; Savitz *et al.*, 1994; Verkasalo *et al.*, 1997). Overall, the findings are inconsistent and difficult to interpret (International Commission on Non-Ionizing Radiation Protection, 1998).

(iii)   *Cardiovascular disease*

Reduced heart rate variability after exposure to 60-Hz magnetic fields has been reported (Sastre *et al.*, 1998). Although inconsistent with the findings of others (Graham *et al.*, 1999), the results suggested that such exposure might be associated with an increased incidence of cardiovascular disease and death. Two studies have examined mortality from cardiovascular disease among electric utility workers (Baris *et al.*, 1996b; Savitz *et al.*, 1999). The overall mortality from cardiovascular and ischaemic disease was generally lower in the study cohorts than in the general population, although the most recent study (Savitz *et al.*, 1999) found that longer duration of employment in jobs with elevated exposure to ELF magnetic fields was associated with an increased risk for death from arrhythmia-related conditions and acute myocardial infarction. Nevertheless, the International Commission on Non-Ionizing Radiation Protection (1998) considered the evidence relating cardiovascular effects to elevated exposure to magnetic fields as weak, and the possible association between exposure and altered autonomic control of the heart is speculative.

4.2   Adverse effects other than cancer in experimental systems

4.2.1   *Reproductive and developmental effects*

(a)   *Static magnetic fields*

(i)   *Homogeneous fields*

The results obtained in studies on reproduction and development and exposure to relatively homogeneous static magnetic fields (fields without strong gradients) consistently fail to indicate any strong, easily detectable adverse effects. No effects have been seen on frog embryos exposed to a static magnetic field of 2.5 kG (0.25 T) (Hansson Mild *et al.*, 1981); on prenatal development, based on standard teratological and several postnatal evaluations in gestating mice exposed at 1 T (Konermann & Mönig, 1986); on the development of the testis and epididymis in mice after exposure to 0.5–0.7 T *in utero* (Tablado *et al.*, 2000); on reproductive performance in mice exposed to 0.49 T (Grzesik *et al.*, 1988), or on spermatogenesis in male mice exposed to 0.3 T (Withers *et al.*, 1985). When male and female mice were mated in a 3.5-T magnetic field, the number of gestating mice decreased to 21% compared with 68% after matings under sham-exposure conditions (Zimmermann & Hentschel, 1987). The effect was not seen if mating occurred after removal from the magnetic field, suggesting that only mating behaviour in the strong magnetic field was affected. No adverse effects on fetal development were observed. In contrast, Strand *et al.* (1983) observed a significant enhancement of fertilization when ova or sperm of rainbow trout (*Salmo gairdneri*) were exposed to a 1-T static magnetic field.

Only two studies have reported significant effects of static magnetic fields on embryonal development.

Batches of fertilized eggs from two species of sea urchin (*Lytechinus pictus* and *Strongylocentrotus purpuratus*) were exposed to fields produced by permanent magnets. Static fields delayed the onset of mitosis in both species by a length of time that was dependent on the time interval between exposure and fertilization. Fields of 30 mT, but not 15 mT, caused an eightfold increase in the incidence of exogastrulation in *L. pictus*, whereas neither of these fields produced exogastrulation in *S. purpuratus* (Levin & Ernst, 1997).

Light microscopy and electron microscopy showed changes in chick embryo cerebella when the embryos were exposed to a 20-mT static field either on day 6 of development or during the first 13 days of development (Espinar *et al.*, 1997).

(ii)   *Static fields with strong gradients*

Two studies have reported effects of static magnetic fields with high spatial gradients on the embryonic development of frogs. The early embryonic growth of *Rana pipiens* was strongly inhibited in a 1-T field with a gradient of 0.84 T/cm (Neurath, 1968). The rate of malformation was increased in *Xenopus laevis* embryos grown in 1-T magnetic fields with gradients from 10–1000 T/m (Ueno *et al.*, 1984). The authors of both studies discussed possible mechanisms related to the effects of magnetic forces on iron-containing molecules and oxygen molecules.

(b)   *Strong static magnetic fields combined with weaker time-varying fields*

Magnetic resonance imaging produces a combination of strong static magnetic fields, radiofrequency fields and time-varying ELF and very low-frequency gradient fields. Few studies have addressed possible developmental effects of the combined fields typical of magnetic resonance imaging.

No effects were seen on embryonal development of frogs when frog spermatozoa, fertilized eggs or embryos were exposed to a combination of a 7.05-kG (0.705-T) static magnetic field and a 30-MHz radiofrequency field (Prasad *et al.*, 1982).

Groups of 15 gestating C57BL/6J mice were subjected to magnetic resonance imaging conditions on day 7 of gestation for 36 min, using a 1.5-T static magnetic field combined with a radiofrequency field of 64 MHz (Tyndall & Sulik, 1991). The incidence of eye malformations — towards which this mouse strain is genetically predisposed — was significantly increased in exposed animals compared with a sham-exposed group. A similar exposure to magnetic resonance imaging conditions also produced statistically significant effects on crown–rump length and craniofacial perimeter, which are less sensitive teratological parameters in these mice (Tyndall, 1993).

Chick embryos were simultaneously exposed to a static magnetic field of 1.5 T for 6 h and to 64-MHz radiofrequency field pulses and switched magnetic field gradients for 4 h. Hatching time and the migration, proliferation and death of motoneurons in the lateral motor column in the chick were unaffected by exposure under conditions of magnetic resonance imaging. Embryo development proceeded normally. There were no obvious adverse effects of exposure to magnetic resonance on differentiation of the major organs, no increase in the incidence of gross abnormalities and no evidence of lesions and malformations (Yip et al., 1994).

(c)   ELF electric fields

Several studies have addressed possible effects of 60-Hz electric fields on reproduction and development in rats (Charles River CD and Sprague-Dawley), using field strengths of 80 kV/m (Seto et al., 1984), 100 kV/m (Sikov et al., 1984; Rommereim et al., 1987), 112–150 kV/m (Rommereim et al., 1989) or 10–130 kV/m (Rommereim et al., 1990). The studies involved large group sizes and exposure over several generations. Overall, the studies did not reveal any consistent adverse effects. Malformations were increased and fertility was decreased in one study (Rommereim et al., 1987). These effects were not confirmed in a companion replicate experiment or in further studies by the same group.

Exposure to 50-Hz electric fields at 50 kV/m did not have any significant effects on the growth and development of eight-week-old male rats exposed for 8 h per day for four weeks. Negative results were also obtained in rabbits exposed for 16 h per day from the last two weeks of gestation to six weeks after parturition (Portet & Cabanes, 1988).

A three-generation study was conducted on Hanford Miniature swine kept in a 60-Hz, 30-kV/m electric field for 20 h per day, seven days per week. Two teratological evaluations were performed on the offspring of the F_0 generation. The incidence of malformations was decreased in the first teratological evaluation after four months of exposure (significant only if analysed by fetus), but was increased in the second evaluation, after 18 months of exposure. An increased number of malformations was also found in one group of offspring of the F_1 generation at 18 months (exposed in utero and from birth), but not in another group of F_1 offspring 10 months later. A complete teratological evaluation was performed only for the latter group of offspring. The inconsistency of these results precludes any conclusion that there is a causal relationship between exposure to an electric field and developmental effects (Sikov et al., 1987).

(*d*)   *ELF magnetic fields*

(i)   *Mammalian teratological studies*

Mouse

In the experiments of Rivas *et al.* (1985), 25–27 gestating Swiss mice per group were exposed to 50-Hz pulsed magnetic fields at 83 μT or 2.3 mT. The number of live births per litter and the mean birth weight were slightly lower in the exposed animals, but the differences from the controls were not statistically significant.

Gestating CBA/Ca mice were exposed from day 0 to day 18 of gestation to 50-Hz or 20-kHz magnetic fields in two independent experiments. In the first experiment, 55 females were exposed to a field of 50 Hz, 13 μT (sinusoidal) or a field of 20 kHz, 15 μT (peak-to-peak). A group of 45 sham-exposed animals served as controls. The second experiment involved 33 females exposed to a 50-Hz, 130-μT field and 34 controls. In addition to standard teratological evaluation, micronuclei were determined in erythrocytes from maternal bone marrow. The numbers of skeletal variations were increased consistently in all exposed groups. The variations were similar in all exposure groups, and suggestive of decreased ossification. The incidence of fetuses with at least three skeletal variations showed a statistically significant increase in all exposed groups compared with corresponding controls. No other significant differences were found in any other maternal or fetal parameters (Huuskonen *et al.*, 1998b).

Gestating CD-1 mice were either exposed from day 0 to day 17 of gestation to a 50-Hz sinusoidal magnetic field at 20 mT or sham-exposed, and the development of the fetuses was evaluated. A total of 90 exposed and 86 sham-exposed control females were analysed. Exposure to magnetic fields was associated with longer and heavier fetuses at term, even when adjusted for litter size, and the fetuses had fewer external abnormalities. The incidence of fetuses with one or more cervical ribs was signifi-cantly increased, but the finding was no longer significant when analysed using methods accounting for possible litter effects. The incidences of external and internal abnormalities and resorptions, and of other parameters measured were unaffected (Kowalczuk *et al.*, 1994).

In Hebrew University mouse pre-implantation embryos (94 to 303 per group) exposed to 1-Hz, 20-Hz or 50-Hz magnetic fields, a significant increase in the percentage of embryos with arrested development was seen after 72 h of exposure at 20 Hz or 50 Hz. Inhibition of hatching and further development was seen in more than 50% of blastocysts. No exposure-related differences were noted in the rate of deve-lopment in those embryos that continued to develop (Zusman *et al.*, 1990).

No consistent effects were seen in preimplantation CBA/S mouse embryos exposed to 50-Hz magnetic fields at 13 μT. The vitality of the embryos was not affected by the exposure, and the timing of the development up to the blastocyst stage was similar to that in controls (Huuskonen *et al.*, 2001b).

Rat

Zecca *et al.* (1985) exposed groups of 10 gestating Sprague-Dawley rats to a 50-Hz, 5.8-mT magnetic field for 3 h per day during the period of organogenesis (days 6–15). No malformations were observed, and the numbers of visceral or skeletal variations were not increased. Resorptions and total post-implantation losses were doubled in the exposed group, but these differences were not statistically significant. [The Working Group noted that the small group sizes meant that the study had very little statistical power to show any effects.]

Huuskonen *et al.* (1993) exposed gestating Wistar rats (70–72 per group) to a 50-Hz, 35.6-µT sinusoidal magnetic field or to a 20-kHz, 15-µT (peak-to-peak) saw-tooth magnetic field on days 0–20 of gestation for 24 h per day. The number of fetuses with minor skeletal anomalies was significantly higher in both exposed groups compared with controls. The number of implants and living fetuses per litter showed a statistically significant increase after exposure to the 50-Hz fields. No effects on the incidence of external or visceral malformations or resorptions were found.

The effects of 50-Hz sinusoidal magnetic fields on embryo implantation, maternal serum estradiol, progesterone, testosterone and melatonin concentrations, and on estrogen and progesterone receptor densities in the uterus were studied during pre-implantation and implantation periods in rats (Huuskonen *et al.*, 2001a). Groups of 60 gestating Wistar rats were exposed to the magnetic fields at 10 or 100 A/m (13 or 130 µT) or sham-exposed for 24 h per day from day 0 of gestation, and killed at regular intervals between 70 h and 176 h after ovulation. No effects on the total number of implantations were seen, although there were statistically significant differences in the estrogen-receptor and progesterone-receptor densities at some time points.

The incidence of minor skeletal anomalies in fetuses was significantly increased when Wistar rats (12 dams per group) were exposed continuously to a 50-Hz magnetic field with a flux density of 30 mT from day 1 to day 20 of gestation. Increased skeletal ossification was noted, possibly indicating accelerated prenatal development (extra-thoracic ribs, particularly comma-shaped). Compared with controls a significantly lower number of fetuses with reduced ossification of pelvic bones was also observed, indicating that ossification was accelerated by exposure to a magnetic field (Mevissen *et al.*, 1994).

No effects were reported in 175 gestating Sprague-Dawley rats exposed throughout gestation for 20 h per day to a 60-Hz magnetic field at 1000 µT, or in a second group of 174 animals exposed to an average field of 0.6 µT (0.33–1.2 µT) as a result of leakage from the system used to expose the first group. The 170 control animals were exposed to an ambient field of 0.1 µT. A decrease in the number of fetuses per litter was found in the group exposed to 1000 µT in the first study, but this decrease was not repeated in a replicate group in that study. Fetal body weight and incidences of external, visceral and skeletal malformations and variations were similar in all groups, and there were no signs of maternal toxicity (Rommereim *et al.*, 1996).

Continuous or intermittent exposure to 60-Hz magnetic fields during the period of major organogenesis had no adverse effects on fetal development or maternal toxicity in Sprague-Dawley rats. In this study, 46–55 gestating females per group were either sham-exposed or exposed for 18.5 h per day to linearly polarized, sinusoidal 60-Hz magnetic fields at flux densities of 2, 200 or 1000 µT, or to intermittent fields (1 h on/ 1 h off) at 1000 µT from gestation day 6 to day 19. Some statistically significant differences between the exposed and sham-exposed animals were seen among the many parameters measured, but no dose–response relationships or any other consistent patterns suggestive of adverse effects were observed. In contrast, a clear response to a positive control (ethylenethiourea) was reported (Ryan *et al.*, 1996).

The possible developmental effects of 180-Hz magnetic fields (third harmonic of 60 Hz) alone or in combination with 60-Hz fields were evaluated in groups of 18–20 Sprague-Dawley rats exposed for 18.5 h per day from gestation day 6 to day 19 to a 60-Hz field or a 180-Hz field at 0.2 mT, or to a 60 + 180-Hz field (10% third harmonic; total field, 0.2 mT). Exposure to a magnetic field had no effects on maternal health, litter size, litter weight or fetal development. The incidence of fetal anomalies was comparable in all groups, with the exception of rib variants, which were increased in the exposed groups, with a statistically significant increase in the group exposed to 60 + 180 Hz. The increase in the number of rib anomalies was within the variation observed in historical controls, and the authors concluded that the effect was not biologically significant (Ryan *et al.*, 2000).

In an in-vitro study, Hebrew University Sabra strain rat embryos (10.5 days old) were exposed for 48 h to pulsed magnetic fields at frequencies of 20, 50 or 70 Hz. The number of embryos was 32–40 in the treated groups and 60 in the control group. [The field intensities were not reported.] Exposure to the magnetic fields resulted in retarded development, and an increased incidence of malformed embryos was seen after exposure to 50 and 70 Hz. The main malformations observed were absence of telencephalic, optic and otic vesicles and of forelimb buds (Zusman *et al.*, 1990).

(ii)   *Mammalian perinatal exposure and behavioural effects*

One of the most sensitive systems for investigating the impact of putative toxic agents employs the perinatal exposure of animals and the assessment of anomalies in the subsequent adult expression of neural and behavioural responses (Lovely, 1988). However, few studies have been conducted on the potential neurobehavioural teratological effects of in-utero exposure to ELF electric and magnetic fields.

Mouse

In a study of postnatal development and behaviour after prenatal exposure, 21 CD1 mice were exposed throughout gestation to a sinusoidal 50-Hz, 20-mT magnetic field. Three possibly field-dependent effects were noted: exposed animals performed the air-righting reflex about two days earlier than controls; exposed males weighed significantly less than controls at 30 days of age; and exposed animals remained on a Rota-rod for less time as juveniles than sham-exposed control mice ($n = 23$). No field-dependent effect on the surface-righting reflex or eye opening was reported, in contrast to the findings of Zusman *et al.* (1990) (see above). There was a suggestion that exposed animals took slightly longer to avoid a cliff edge, but this difference was of borderline significance. In the activity wheel, a slightly increased activity of exposed females and a slightly decreased activity of exposed males was noted compared with control mice, but these effects were not considered by the authors to be of any biological significance. The reduction in running time on a Rota-rod, observed in juvenile mice, may represent an impairment of motor coordination during adolescence induced by the magnetic field. No gross impairments of postnatal development or behaviour were seen in the exposed mice (Sienkiewicz *et al.*, 1994).

Seven gestating CD1 mice were exposed for the whole gestation period to a vertical, sinusoidal, 50-Hz magnetic field at 5 mT. Eight control animals were sham-exposed. The male offspring were raised without exposure to magnetic fields, and 10 males per group (no more than two from each litter) were tested at 82–84 days of age for deficits in spatial learning and memory in a radial arm maze. No effects on performance were observed (Sienkiewicz *et al.*, 1996).

Rat

In early studies, rats were exposed to 60-Hz electric fields during gestation. The offspring were tested using operant-avoidance behavioural methods at 80 days of age. Perinatally exposed rats performed the task more slowly, but were able to avoid shocks during these tests equally as well as control animals (Persinger & Pear, 1972). Two related studies on postnatal development were reported in which rats were exposed *in utero* to 60-Hz electric fields. In rats exposed *in utero* from gestation day 0 until day 8 after parturition, movement, standing and grooming were increased when compared to controls at 14 days of age. There was a significant decrease in the percentage of exposed offspring displaying the righting reflex. A negative geotropism was seen in exposed offspring in a parallel study where exposure began at day 17 of gestation and was terminated 4 days after weaning. All differences were transient and were no longer evident when the animals were tested at 21 days of age (Sikov *et al.*, 1984).

After continuous exposure of dams to a pulsed 20-Hz electromagnetic field throughout gestation, the weight of Sprague-Dawley rat offspring at day 1 of age was reduced but it was increased after exposure to a 100-Hz field. The weights of the offspring of dams exposed to a 50-Hz field were decreased only from 21–28 days of age. When combined into one group, exposed animals showed a statistically signifi-cant delay, compared to controls, in eye opening. No effect was seen on the surface-righting reflex (Zusman *et al.*, 1990). [The field intensities were not reported.]

Increased male accessory sex-organ weights were noted in Sprague-Dawley rats prenatally exposed to a 15-Hz pulsed magnetic field with 0.3-ms pulse duration and a peak intensity of 0.8 mT. Gestating animals (6 per group) were exposed for two 15-min periods on days 15–20 of gestation, a period critical for the sexual differentiation of the male rat brain. At parturition, no exposure-related effects on number of live fetuses, average weight or anogenital distance were noted. At day 120 postpartum, the male offspring of the exposed dams exhibited diminished territorial scent-marking behaviour and increased accessory sex-organ weights. Concentrations of circulating testosterone, luteinizing hormone and follicle-stimulating hormone were unchanged, as were epididymal sperm counts. The authors concluded that in-utero exposure to magnetic fields had caused incomplete masculinization (McGivern *et al.*, 1990).

The developmental increase in the activity of choline acetyltransferase was examined in the brains of fetuses and offspring from Sprague-Dawley rats exposed to a 60-Hz, 500-mG (50-µT), sinusoidal, circularly polarized magnetic field for one month before gestation and during gestation and lactation. Choline acetyltransferase activity in the brain was assessed at four time points during fetal development and at five and 10 days after parturition. Six animals per group were examined at each time point. No differences were observed between the exposed and control rats (Sakamoto *et al.*, 1993).

Female Sprague-Dawley rats were exposed to 60-Hz combined electric and magnetic fields for 23 h per day, from day 5 to day 19 post-conception, to study the effects of exposure on somatic growth and cortical development, as well as biochemical and morphological maturation of the neopallium. The animals were exposed to fields of 1 kV/m and 1 mT, 100 kV/m and 0.1 mT, and 100 kV/m and 1 mT. Pups were killed at birth or on postnatal days 5, 12 or 19 for biochemical and morphological studies. No macroscopic or microscopic changes were observed. A small but significant reduction in cortical weight was observed in rats exposed to 1 kV/m and 1 mT, and a small but significant increase of cortical weight after exposure to 100 kV/m and 0.1 mT. The concentrations of DNA, RNA, protein and cerebroside were measured in the neopallium. Slight but significant reductions in RNA and protein concentrations were measured during the first days of exposure to 100 kV/m and 0.1 mT, and a small reduction in RNA concentration in animals exposed to 100 kV/m and 1 mT. The authors concluded that the exposure had either no effect or else caused minimal changes in somatic growth and cerebral development (Yu *et al.*, 1993).

The effects of postnatal exposure to combined electric and magnetic fields (in the same combinations as above) on the development of the cerebellum were studied in newborn Sprague-Dawley rats. The pups were exposed for 7–8 h per day from the day of birth and killed after one, two or three weeks. No morphological changes were observed in the exposed group. There was a small but statistically significant decrease in brain weight in the group exposed to 1 kV/m and 1 mT. The concentrations of DNA and RNA in the cerebellum showed some statistically significant differences after exposure to 1 kV/m and 1 mT and 100 kV/m and 0.1 mT, but not at 100 kV/m and 1 mT. In animals exposed to 1 kV/m and 1 mT, DNA and RNA concentrations were elevated at six and 13 days, but not at 20 days. In animals exposed to 100 kV/m and 0.1 mT, DNA and RNA concentrations were initially (day 8) lower than in the control animals; concentrations in the exposed and control groups were approximately the same at 14 days and were higher in exposed animals than in controls at 22 days. Protein concentrations were lower in the exposed animals than in controls at eight days, but higher at 14 and 22 days (Gona *et al.*, 1993).

Altered behaviour after perinatal exposure to an electric and magnetic field has been reported. Groups of rats were either sham-exposed or exposed for 20 h per day to a combination of a 60-Hz (30-kV/m) electric and 100-μT magnetic field for 22 days *in utero* and during the first eight days *post partum*. As adults, male rats were trained to perform a multiple, random-interval operant task. The responses of the rats that had been exposed *in utero* to the electric and magnetic field gradually became significantly slower than those of the sham-exposed controls. Once the difference in response rate was established, it was found to persist even after experimental extinction of the response followed by reconditioning. The exposed rats did not differ from sham-exposed controls in terms of body mass, physical appearance, grossly observed activity level or incidence of disease (Salzinger *et al.*, 1990).

(iii)   *Mammalian multi-generation studies*

Swiss mice were exposed to 50-Hz pulsed (5 ms) magnetic fields at either 2.3 mT or 83 mT from day 0 to day 120 *post partum* for the first generation, and from conception throughout embryological development and up to day 120 *post partum* for the second generation. In the first generation, no changes were observed in body weights or serum glucose, protein, cholesterol or triglyceride concentrations. In the second generation, the body weights and serum glucose concentrations of the exposed mice were significantly lower at 60 and 120 days *post partum* and the triglyceride concentration was decreased at 120 days, compared to sham-exposed control mice (Rivas *et al.*, 1987).

A study aimed at reproductive assessment by continuous breeding investigated reproductive performance in rats over several generations. Groups of Sprague-Dawley rats (40 breeding pairs per group) were sham-exposed or exposed continuously for 18.5 h per day to linearly polarized, sinusoidal 60-Hz magnetic fields at field strengths of 2, 200 or 1000 μT or to an intermittent (1 hour on, 1 hour off) magnetic field of 1000 μT. No exposure-related toxicity was observed in any of the three generations examined. Fetal viability and body weight were similar in all groups, and there were no differences between test and control groups in any measure of reproductive performance (number of litters per breeding pair, percentage of fertile pairs, latency to parturition, litter size or sex ratio). Teratological examinations were not performed (Ryan *et al.*, 1999).

(iv)   *Effects of paternal exposure on mammalian reproduction*

Male OF1 mice were exposed from the age of 6 weeks until adulthood to a sinusoidal 50-Hz, 15-μT magnetic field to study possible alterations in testis histology and its endocrine function. Female mice that were exposed chronically to the same field from the age of 6 weeks were mated at 20 weeks with the exposed males. The offspring were kept under the same experimental conditions. When the offspring reached sexual maturity, the testes of 30 exposed and 30 control males were analysed. A significant increase in testis size and weight was observed. This increase was associated with increased testosterone concentrations in the interstitial tissue, as was shown by histological analysis. Complete spermatogenesis occurred in both control and exposed animals (Picazo et al., 1995).

A flow cytometric study was performed to monitor the effects of a 50-Hz sinusoidal magnetic field on mouse spermatogenesis. Groups of five male hybrid (C57BL/Cne × C3H/Cne)F_1 mice, aged 8–10 weeks, were exposed to a field strength of 1.7 mT for 2 or 4 h. Flow cytometry measurements to distinguish various cell types were performed 7, 14, 21, 28, 35 and 42 days after exposure. No effects were observed in animals exposed for 2 h. In groups exposed for 4 h, a statistically significant decrease in the number of elongated spermatids was observed 28 days after the treatment, suggesting a possible cytotoxic and/or cytostatic effect of the exposure on differentiating spermatogonia (De Vita et al., 1995).

Six weeks of continuous exposure to circularly polarized 50-Hz magnetic fields at 1, 5 or 50 μT did not change the plasma testosterone concentration in groups of 48 male Wistar-King rats (Kato et al., 1994a).

The possible effects of 50-Hz magnetic fields on the fertility of male rats were investigated in Sprague-Dawley rats, aged 20 weeks, exposed to a sinusoidal, 50-Hz magnetic field at 25 μT for 90 days before they were mated with unexposed females. Ten males per group were used (13 in the control group), and each male was mated with two females. The number of conceptions was significantly decreased from 24/26 (92%) in the control group to 10/20 (50%) in the exposed group. The effect persisted in a second mating at 45 days after cessation of exposure (12 conceptions; 60%), but not at 90 days (16 conceptions; 80%). There was also a significant increase in the total number of resorptions, from two in the female controls to six in the females mated with the exposed males. [As only the total number of resorptions was reported, possible litter effects could not be evaluated.] The numbers of implantations and viable fetuses per litter were not significantly affected. The effect of exposure on the fertility of females (10 animals per group) was also evaluated in this study. The 90-day exposure was carried out under the same conditions as used for the males and resulted in a statistically significant decrease in the number of conceptions, from 100% in the controls to 60% in the exposed females. The mean number of implantations per litter decreased from 9.9 to 4.7 and the mean number of viable fetuses per litter from 9.6 to 4.3. These differences were statistically significant. The total number of resorptions was similar in exposed and control females (Al-Akhras et al., 2001).

(v)   *Chick and quail embryos exposed to magnetic fields* in vitro

The initial report of Delgado *et al.* (1982) stated that pulsed magnetic fields at frequencies of 10, 100 and 1000 Hz (pulse duration, 5 ms, peak flux density, 0.12, 1.2 or 12 µT) resulted in a large increase in the percentage of abnormalities noted in chick embryos incubated for two days. In later experiments by the same group, the teratogenic effect seemed to depend on the waveform used (Ubeda *et al.*, 1983, 1985). In still later experiments, the effect seemed to depend on the orientation of the chick embryo relative to the geomagnetic field (Ubeda *et al.*, 1987). The effect of 100-Hz, 0.4-µT or 1-µT pulsed fields on chick embryos was not always reproducible. In the combined data from 13 experiments (40–50 eggs per experiment), 35% of the exposed and 30% of the control embryos were abnormal. However, there was a significant correlation between the variations in the results and extremely small time-dependent changes in the local geomagnetic field. This finding was interpreted by the authors as suggesting that the effect might occur only at some specific values of the geomagnetic field (Leal *et al.*, 1989).

In all the experiments described above, the chick embryos were examined directly after incubation for 48 h in the magnetic field. Ubeda *et al.* (1994) incubated the eggs for an additional nine days after exposure for 48 h, and the embryos were then inspected in a blinded manner. The group sizes of the exposed and sham-exposed embryos ranged from 72 to 92, and an additional 276 embryos were used as background controls. The 100-Hz fields were similar to those used in previous studies, with 1 µT peak amplitude and 5 ms pulse duration, but two different pulse waveforms were used with rise times of 1.2 and 85 µs. The number of developmental anomalies was increased in the exposed groups, indicating that the abnormalities seen in the previous studies are irreversible. The increase was significant ($p = 0.007$) only for the waveform with the shorter rise time.

Two studies failed to reproduce the results of Delgado *et al.* (1982) (Maffeo *et al.*, 1984, 1988). A large well-designed international study ('Henhouse project') aimed at replicating Delgado's results (summarized in Berman *et al.*, 1990) was carried out in six separate laboratories using identical equipment and standardized experimental procedures. The eggs, however, came from different sources, and the local geomagnetic fields were different. While the results were not uniform, the combined data showed a significant ($p < 0.001$) increase in abnormal embryos in the exposed group. [The Working Group noted that these results may be compromised by the different field strengths used in different laboratories.]

The experiments of Juutilainen *et al.* (1986) showed a higher percentage of abnormalities compared to controls in chick embryos exposed during the first two days of development to a 100-Hz magnetic field with a pulsed waveform similar to that used by Delgado, but also with sinusoidal and rectangular waveforms. In another series of experiments with sinusoidal waveforms, similar effects were found upon exposure to a wide range of frequencies (Juutilainen & Saali, 1986). The effects of 100-Hz sinusoidal fields with a field strength of 1 A/m (1.3 µT) were confirmed in experiments with a large number of eggs (Juutilainen, 1986). Further experiments showed similar effects from exposure to 50-Hz sinusoidal fields, and the results suggested that the field strength–response curve has a sharp threshold at 1 A/m (1.3 µT) (Juutilainen *et al.*, 1987).

Apart from the extensive series of experiments by Juutilainen and colleagues, there have been few other studies on sinusoidal fields. Cox *et al.* (1993) attempted to replicate in part the findings of Juutilainen *et al.* Two hundred White Leghorn chick eggs were exposed to a 50-Hz, 10-µT magnetic field for 52 h and a second group of 200 eggs was incubated in a background field of 0.2 µT. The incubation was continued for 68 h after removal of the eggs from the magnetic field after which the embryos were examined. No difference in malformation rate was observed between the exposed and control embryos. Most of the experimental conditions in the laboratories of Cox and Juutilainen were similar. However, the static (geomagnetic) field was only 17 µT in Cox's laboratory compared with 44–50 µT in Juutilainen's laboratory.

An extensive series of experiments was conducted to study the effects of pulsed and sinusoidal magnetic fields on chick embryo development, involving over 2500 White Leghorn chick embryos. The experiments were performed over five years in five separate studies. In four of these, a pulsed 100-Hz field with a peak amplitude of 1 µT was used (similar to the field used in the Henhouse study). In the last study, the embryos were exposed for 48 h to a 60-Hz, 4-µT sinusoidal magnetic field. The number of abnormalities was always higher in exposed embryos, but in one of the pulsed-field studies, the difference was small and not statistically significant. Overall, the number of abnormalities was approximately doubled in embryos exposed to the pulsed 100-Hz magnetic fields and approximately tripled by exposure to the sinusoidal 60-Hz magnetic fields. Both effects were highly significant. According to the authors, the lack of response in one of their studies could have been due to a change in the genetic composition of the breeding stock before the start of that study. The authors proposed that genetic differences in susceptibility to magnetic fields may explain the inconsistent results between laboratories (Farrell *et al.*, 1997).

The effects of 50-Hz and 100-Hz magnetic fields on the development of quail embryos were investigated in eggs produced by 10 females. Data were reported separately for each female. In each experiment, two eggs from each female were used: one exposed and one control. Sham experiments conducted with 240 eggs showed that there was no difference between the exposure and control locations in the incubator. The eggs were exposed to 50-Hz or 100-Hz magnetic fields with rectangular wave-form and intensities of 0.2, 1.2, 3.3 and 3.2 µT. The embryos were exposed for 48 h and then inspected in a blind manner. The number of abnormalities was higher in the exposed embryos than in the controls. However, the increase did not reach statistical significance for the embryos exposed to 50 Hz. Comparison of the data for the individual females suggests that there might be genetic differences in sensitivity (Terol & Panchon, 1995).

(vi)   *Other non-mammalian embryos*

Ramirez *et al.* (1983) reported reduced egg laying in fruit flies (*Drosophila melano-gaster*) and reduced survival during development after exposure to pulsed 100-Hz, 1.8-mT and sinusoidal 50-Hz, 1-mT magnetic fields. No effects were seen in a similar study using 60-Hz, 1-mT fields (Walters & Carstensen, 1987).

Graham *et al.* (2000c) studied the effects of magnetic fields on 'developmental stability', which describes the ability of an organism to maintain a consistent pheno-type under given genetic and environmental conditions. *Drosophila melanogaster* were exposed for their entire lives (egg to adult) to 60-Hz magnetic fields at 1.5 or 80 μT. The exposed flies in both groups showed a significant reduction in body weight, compared to controls. The flies exposed to the 80-μT field showed reduced developmental stability measured both by fluctuating asymmetry (asymmetrical wing veins) and frequency of phenodeviants (fused abdominal segments). [The Working Group noted that developmental stability is a new concept, that could potentially be a very useful tool for detecting relatively weak environmental effects.]

Exposure to sinusoidal magnetic fields of 60 Hz, 0.1 mT has been reported to delay the development of Medaka fish embryos (*Oryzias latipes*) (Cameron *et al.*, 1985), sea urchin embryos (*Strongylocentrotus purpuratus*) at 60 Hz, 0.1 mT (Zimmerman *et al.*, 1990) and zebrafish embryos (*Dario rerio*) at 50 Hz, 1 mT (Skauli *et al.*, 2000). No malformations were found in these studies.

(vii)   *Interactions with known teratogens*

Cultures of embryonic cells of *Drosophila melanogaster* were used to assess the potential developmental toxicity of exposure to a 60-Hz, 100-μT field for 16–18 h. Exposure to the magnetic field alone was not teratogenic and exposure did not enhance the effects of retinoic acid, hydroxyurea or cadmium, which were all clearly teratogenic in this model. Additional experiments, in which embryos were exposed at 10 or 100 μT for their entire development up to the adult stage, did not produce a significant increase in developmental abnormalities (Nguyen *et al.*, 1995).

Exposure to 50-Hz, 10-mT magnetic fields modified the embryotoxic effect of ionizing radiation on chick embryos, but no effects of exposure to magnetic fields alone were observed. In this study, several experiments were performed with X-ray doses of 4 or 5 Gy given on days 3 or 4 of development. The magnetic field was applied either during the first 2–40 h of embryonic development (before X-ray treatment), or during the 12 h immediately after the X-ray treatment. The embryos were examined at day 9 of development. Embryotoxicity was expressed as the sum of embryonic deaths and malformations. Exposure to the magnetic field before the X-ray treatment seemed to protect the embryos from X-ray-induced toxicity, while an enhancement of the embryo-toxicity was seen when exposure to the magnetic field followed the X-ray irradiation. Both the protective effect and the enhancing effect were seen consistently in several experiments and were statistically significant (Pafková & Jerábek, 1994). A similar protective effect against subsequent exposure to the chemical teratogens insulin and tetracyclin was described for 50-Hz, 10-mT magnetic fields. The authors sought to explain the interactions of magnetic fields with X-rays and chemical teratogens on the basis of magnetic-field-induced oxidative stress (Pafková et al., 1996).

4.2.2   Immunological effects

   (a)   In-vivo studies

      (i)   Static fields

The humoral and cell-mediated immune responses were studied in mice (LAF1/J) following exposure to 1.5-T static magnetic fields for six days. The immune response of spleen lymphocytes to sheep erythrocytes was tested by assaying the number of Jerne plaques formed by spleen lymphocytes, and by measuring the concentration of IgM in the serum. The mitogen-stimulated proliferation index of the spleen lymphocytes was also tested using concanavalin A, phytohaemagglutinin and lipopolysaccharide. In no case did lymphocytes from exposed mice respond differently from those from control animals (Tenforde & Shifrine, 1984).

A series of studies investigated the influence of 60-mT magnets implanted over several brain regions on the immune response in male and female Wistar rats. After implantation, the animals were challenged with sheep erythrocytes or bovine serum albumin and tested 14, 24 and 34 days later. Control rats were implanted with iron beads (and sham-operated, when appropriate, to conform to the treatment of exposed animals). The rats were tested for the plaque-forming cell response, local hyper-sensitivity skin reactions and experimental allergic encephalomyelitis. The authors reported that placing magnets over each of three regions of the brain could have effects on the immune system not seen in controls (Jankovic et al., 1991). Furthermore, while surgical induction of lesions in the brain in the nucleus locus ceruleus or pinealectomy caused a reduced immune response, implantation of the magnets reversed these effects (Jankovic et al., 1993a,b). Old rats (aged 22 months) that underwent pinealectomy and magnet implantation also showed recovery of immune responses as did the younger animals in the earlier study (Jankovic et al., 1994).

      (ii)   ELF electric and magnetic fields

In a study of the effects of electric fields alone, male and female Swiss-Webster mice were exposed for 30 or 60 days (21 h per day) to 60-Hz electric fields of 100 kV/m. No significant differences in serum immunoglobins (IgG and IgM), complement levels or distribution of T or B lymphocytes were found in comparison with sham-exposed control mice. A statistically significant decrease in leukocyte and lymphocyte counts was found after exposure for 60 days but these counts were elevated compared to controls in a subsequent experiment (Morris & Ragan, 1979).

Male C57BL/6 mice were exposed to 60-Hz, 100-µT magnetic fields for 1, 5, 10, 21, 49 and 105 days. For each exposure period, three replicates were evaluated using a battery of 20 immune assays. When these data were analysed using linear statistical methods, no significant difference in any immune parameter was found (Marino *et al.*, 2000). [Although the authors noted the increased variance in their data for the two longest exposure times, the Working Group considered this analysis too speculative to include in its evaluation.]

The 7,12-dimethylbenz[*a*]anthracene (DMBA) model for breast cancer was used to study immunological effects in rats exposed to horizontal, 50-Hz, 50-μT magnetic fields for 91 days (24 h per day, seven days per week). The geomagnetic field produced a static component of 16 μT parallel to, and 36 μT perpendicular to, the 50-Hz field. After 13 weeks, a marked suppression in T-cell proliferation capacity under concanavalin A challenge was observed. No change in nocturnal concentrations of serum melatonin was seen in exposed animals after 9 or 12 weeks of exposure (Mevissen *et al.*, 1996b).

Female Sprague-Dawley rats were exposed to magnetic fields of 50 Hz, 0.1 mT (24 h per day, seven days per week) for 2, 4, 8 and 13 weeks. Proliferative capacity and production of interleukin-2 were investigated in primary splenic cultures of T and B lymphocytes. Significantly fewer viable splenic lymphocytes were observed at all times in the exposed animals compared with sham-exposed controls. The proliferation rate of B cells, either unstimulated or stimulated with pokeweed mitogen, was comparable in exposed rats and sham-treated animals. In contrast, the proliferation rate of T cells stimulated with concanavalin A from exposed animals showed a statistically significant increase at two weeks, a slight reduction at four weeks, no statistically significant change at eight weeks and a statistically significant reduction at 13 weeks, compared with controls. Non-stimulated T-cell proliferation was unchanged at these treatment times. The addition of melatonin at 1, 10 and 100 nM did not change the T-cell proliferation rate in concanavalin-A-treated cultures from animals exposed to magnetic fields for two and four weeks nor did the addition of melatonin to these cultures change the T-cell proliferation from that observed in cultures treated with concanavalin A alone. The same was true for sham controls except for one experiment with 100 nM melatonin where the response to concanavalin A and melatonin was significantly higher than the reponse to concanavalin A alone. No changes in production of interleukin-2 were observed after any treatment of B lymphocytes. The authors noted that the triphasic alteration in T-cell function during the 13-week treatment period resembled the responses that are seen during prolonged administration of a chronic mild stress, i.e. activation, tolerance and suppression. They concluded that long-term exposure to magnetic fields may lead to impaired immune surveillance in female rats (Mevissen *et al.*, 1998b).

The activation of interleukins by activated T and B lymphocytes was studied in female Sprague-Dawley rats exposed to 50-Hz, 0.1-mT magnetic fields, under conditions described previously by Mevissen *et al.* (1996b). In the first experiment, DMBA-treated rats were exposed for 14 weeks under conditions that suppressed the T cell-stimulated proliferative response, following exposure to magnetic fields for 13 weeks. This experiment failed to show any difference in production of interleukin-1 by mitogen-activated B cells between exposed and sham-exposed animals. In the second experiment, non-DMBA-treated rats were exposed for 1, 7 and 14 days to 50-Hz, 0.1-mT magnetic fields. No significant changes were observed in the production of interleukin-1 or interleukin-2 by stimulated B or T cells (Haussler *et al.*, 1999).

Natural and adaptive immunity were studied in rats born and raised for six weeks in a 60-Hz magnetic field for 20 h per day. Twenty days after mating, gestating Fisher F344/N rats were exposed to intensities of 2, 20, 200 μT and 2 mT, or kept under control conditions (< 0.02 μT). At weaning, the offspring were separated from their dams and kept under the same field conditions. The following immunological para-meters were examined: total T and B cells; CD5+, CD4+ and CD8+ sub-population pattern and natural killer cell activity in splenic lymphocytes; hydrogen peroxide, nitric oxide and tumour necrosis factor production by peritoneal macrophages. In comparison with the control group, there was a significant reduction in the number of CD5+, CD4+ and CD8+ populations, with the greatest reduction occurring at 2 mT. A smaller but nevertheless significant reduction was observed in CD5+ cells at 200 μT. Linear regression analysis showed a dose–response effect with increasing magnetic field intensity. Furthermore, B lymphocyte populations (Ig+ cells) showed a significant reduction ($p < 0.05$) at 20 and 200 μT, but these results did not show a dose-related response. Natural killer cell activity decreased by 50% ($p < 0.05$) at 2 mT. In peritoneal macrophages, no significant changes were observed in the activity of tumour necrosis factor or secretion of nitric oxide, but background and phorbol ester-stimulated production of hydrogen peroxide increased ($p < 0.05$ and $p < 0.001$, respectively) (Tremblay *et al.*, 1996). [The Working Group noted that the significance of these effects was greatly reduced when a comparison was made with sham-exposed animals; only two end-points remained statistically significant.]

Male and female B6C3F$_1$ mice and female Fischer 344 rats were exposed continuously (18.5 h per day) to 60-Hz, 0.002-, 0.2- and 1-mT magnetic fields; one additional group was exposed to an intermittent (1 h on, 1 h off) 1-mT magnetic field. After exposure periods of 21, 28 or 90 days the parameters examined included spleen and thymus weights and cellularity, antibody-forming cells, delayed-type hyper-sensitivity, splenic lymphocyte subsets, susceptibility to infection with *Listeria mono-cytogenes* and natural killer cell activity. No statistically significant differences were found between exposed and sham-exposed mice, except in the activity of natural killer cells and occasional differences in delayed-type hypersensitivity for which there was no clear dose-related pattern. After 28 days of exposure, the activity of natural killer cells in female mice showed a statistically significant increase only at 1 mT, whereas non-significant increases occurred in a dose-dependent manner. Isolated changes in the activity of natural killer cells were seen in three groups of mice exposed for 42 days. A dose-related reduction in activity of natural killer cells was observed in female mice exposed for 90 days, but this was not consistently reproducible (House *et al.*, 1996).

The influence of 60-Hz magnetic fields on the clinical progression of leukaemia was investigated in male Fischer 344 rats. Large granular lymphocytic leukaemia cells from spleens of leukaemic rats were injected intraperitoneally into young rats, which were subsequently exposed to magnetic fields at either 2 µT or 1 mT, for 20 h per day, seven days per week for five, six, seven, eight, nine or 11 weeks. Changes in growth of the spleen and infiltration of large granular lymphocytic leukaemia cells into the spleen and liver were monitored. No significant and consistent differences were observed between groups exposed to a magnetic field and the control group, whereas progression was enhanced in a positive control group exposed to 5 Gy γ-rays (Morris *et al.*, 1999).

A model for transplantable acute myeloid leukaemia in rats was applied to examine the influence of 50-Hz, 0.1-mT magnetic fields. Leukaemic cells were injected intravenously into the lateral tail vein of Norway rats, which were sub-sequently exposed to the magnetic field for 18 h per day (14.00 to 08.00), seven days a week, for up to 27 days. The geomagnetic field was 47 µT, with projection onto the horizontal axis of 8 µT. The investigators measured survival time, body weight, haematological parameters and infiltration of blood, bone marrow, spleen and liver by leukaemic cells. The results showed no significant changes in rats treated with leukaemic cells and exposed to a magnetic field compared with those injected with leukaemic cells, but not exposed to a magnetic field, for any of the parameters involved in leukaemia progression (Devevey *et al.*, 2000).

The effects of 7-Hz and 40-Hz square wave magnetic fields on the immunological response were studied in female Lewis rats. Immediately before exposure, the test rats were injected with emulsified spinal column prepared from female Lewis rats. Groups of rats were exposed to magnetic fields of 0.05 µT or 0.5 µT (peak intensity) of either 7 or 40 Hz for 6 minutes every hour for 8 h per day for 15 days. The field patterns were designed to be similar to those that might occur during geomagnetic storms. After exposure, the brains of the rats were examined for infiltration of lymphocytes (mononuclear cells) and mast cells. In rats exposed to 7-Hz, 0.05-µT fields, there were fewer infiltration foci than in any of the other groups. Rats exposed to 40-Hz, 0.5-µT fields had more foci in the right thalamus while those exposed to 7 Hz at the same intensity showed more foci in the left thalamus. The total number mast cells within the thalamus was also increased by the treatments (Cook *et al.*, 2000).

Eight adult male baboons (*Papio cynocephalus*) were exposed for six weeks to 60-Hz vertical electric fields of 6 kV/m and horizontal magnetic fields of 50 µT. Blood samples taken before exposure, at the end of the period of exposure and six weeks after exposure ended were examined by standard immunological methods to determine total leukocyte count and total T cell (CD3+), T helper cell (CD4+), cytotoxic T cell (CD8+), B cell and natural killer cell numbers. A second experiment was conducted in which six of the eight animals used previously were exposed to higher field intensities: 30-kV/m electric fields and 0.1-mT magnetic fields. In the first experiment, exposure-related reductions ($p < 0.05$) in CD3+ and CD4+ counts, interleukin-2 receptor expression and T-cell proliferation in response to pokeweed mitogen were observed. In the second experiment, there was a similar but reduced response in these parameters compared with the pre-exposure control values. A comparison of the results in the two experiments showed group × period interactions (indication of significance) for total leukocyte count and CD4+ and CD8+ ratios, but the higher exposures did not show greater effects (Murthy *et al.*, 1995).

(b)   In-vitro studies

(i)   Static fields

Static magnetic fields generated by a 0.5-T magnetic resonance imaging unit were used to study activation markers and interleukin release in mononuclear cells from human peripheral blood. The cells were exposed to the fields for 2 h at 24 °C, then cultured for 24 h at 37 °C with or without PHA stimulation. The cells were assayed for expression of CD25, CD69 and CD71 by immunofluorescence microscopy, and the concentrations of interferon-γ, tumour necrosis factor α and interleukin-4 were measured in the medium using an enzyme-linked immunosorbent assay. Exposure to the magnetic field caused a reduction in CD69 expression, which was enhanced under PHA stimulation compared with controls. An increased release of interferon-γ and interleukin-4 occurred in unstimulated cells, but a reduced release was seen under PHA stimulation compared to controls. The release of tumour necrosis factor α, interleukin-6 and interleukin-10 was unchanged (Salerno *et al.*, 1999).

Apoptosis, intracellular calcium concentrations and lymphocyte and macrophage functions were measured in C57BL/6 mouse macrophages, splenic lymphocytes and thymic cells exposed for 24 h to static magnetic fields ranging from 25 to 150 mT. Cytofluorometric analysis showed a decreased phagocytic uptake of fluorescent latex microspheres, with a concomitant increase in the concentration of intracellular calcium ions in the macrophages. Exposure to the magnetic fields also decreased the concanavalin A-induced mitogenic response in lymphocytes; this was also associated with increases in calcium ion influx. In addition, exposure gave rise to increased apoptosis in thymic cells, as determined by flow cytometry (Flipo *et al.*, 1998).

(ii)   ELF electric and magnetic fields

Natural killer cell activity in human peripheral blood was examined following in-vitro exposure to 50-Hz magnetic fields. Phytohaemagglutinin or interleukin-2-stimulated lymphocytes or unstimulated control cells were exposed to 50-Hz magnetic fields before or during cytotoxicity tests, and then mixed with different target cancer cell lines (Daudi, Raji, U937, H14, IGROV, SW626, K562, HL60). Exposure to magnetic fields of 0.1, 0.035, 1.8 and 10 mT, with exposure times between 4 h and 7 days, took place in two independent laboratories. The results from both laboratories showed that exposure to 50-Hz magnetic fields with strengths of up to 10 mT did not affect the cytotoxic activity of human natural killer cells (Ramoni *et al.*, 1995).

Peritoneal mast cells from Sprague-Dawley rats were tested for function and histamine release in response to 48/80 (a standard mast cell-stimulating compound).

The cells were exposed to a 60-Hz, 5-mT magnetic field for periods of 30 min to 2 h, either before or during the test. No significant degranulation occurred during exposure to the magnetic field and the cells showed no reduction in sensitivity to the degranulating agent, 48/80 (Price & Strattan, 1998).

The effect of magnetic fields on intracellular free calcium was studied in thymocytes from Sprague-Dawley rats. The cells were exposed to 50-Hz, 0.1-mT horizontal or vertical magnetic fields or to 50-Hz, 0.14-mT circularly polarized fields for 30 min; the effects of consecutive 20-min periods of exposure to vertical and horizontal magnetic fields were also examined. In addition, control or lectin-activated thymocytes, splenocytes and peripheral blood lymphocytes were exposed to a 50-Hz, 5-mT vertical magnetic field for 30 min. No changes in intracellular free calcium concentration were observed in any of these experiments (Nishimura et al., 1999). Intracellular calcium was also monitored in the Jurkat lymphocyte T-cell line. The cells were pre-incubated for 8 min to establish a baseline, and subsequently exposed for 8 min to a 50-Hz, 0.15-mT magnetic field, or sham-exposed in a blinded fashion. No effects on the concentration of intracellular free calcium were found (Wey et al., 2000).

Mononuclear cells from human peripheral blood were exposed to either static or pulsed, 50-Hz, 3-mT magnetic fields and assayed for proliferative responses and production of the cytokines interleukin-2, interferon-γ and tumour necrosis factor α. Pulsed 50-Hz fields with a 120-ns rise time, a 100-μs fall time and a duty cycle of 2/5 were applied for 15 min every 2 h for 6 h, for a total exposure time of 45 min. Proliferative response was measured with and without PHA stimulation, and cytokine concentrations were determined with biological immunoenzymatic assays. There was no effect of the static or pulsed fields on cell proliferation, and the cytokine concentrations, and transcriptional or translational processes in the exposed cells did not differ from those in the controls (Pessina & Aldinucci, 1997).

In a later study, mononuclear cells from human peripheral blood were exposed for 12 h to pulsed magnetic fields of 50-Hz, 3-mT square waves with a rise time of 120 ns, a fall time of 2 ms and a duty cycle of 1/2. In unstimulated cells, a reduction in tumour necrosis factor α was seen immediately after exposure, but no changes were observed in either interleukin-1β or interleukin-2. In contrast, cells stimulated with PHA immediately before exposure to the fields showed progressive increases ($p < 0.05$) in the concentrations of interleukin-1β and tumour necrosis factor α at 24 and 48 h after treatment. The concentration of interleukin-2 was also increased, but only at the end of exposure; proliferation indices were also significantly increased 48 h after treatment (Pessina & Aldinucci, 1998).

Murine cytotoxic T-lymphocytes were used to target B-lymphocytic tumour cells in a standard assay to test chromium release from the B-lymphocytic tumour cells into the medium as an indicator of cell disruption by the murine cytotoxic T-lymphocytes. The latter were exposed through agar bridges to 60-Hz, sinusoidal electric fields for 48 h at intensities of 0.1, 1 and 10 mV/cm in the medium. Following exposure, a 4-h cytotoxicity assay was performed. The results showed a non-significant reduction (7%) in cytotoxicity after exposure to 0.1 mV/cm, and significant reductions of 19% ($p < 0.0005$) and 25% ($p < 0.005$) after exposure to 1 and 10 mV/cm, respectively (Lyle et al., 1998).

Human leukocytes exposed at 20 °C to 0.2-, 1- or 3-ms pulses from a spark discharge with electric fields of 2.6 kV/cm and higher — as single pulses and with up to 10 pulses at 5-s intervals — showed that breakdown of the membrane barrier was field intensity-dependent (Hansson Mild et al., 1982).

Mononuclear cells obtained from the peripheral blood of 25 healthy donors were exposed for 12 h to a 3-Hz pulsed magnetic field of 4.5 mT, with a 1/2 duty cycle and rise and fall times less than 100 μs (the field was somewhat smoothed due to coil inductance). Each pulse had a maximum value of 13 T/s, which produced a maximum induced electric field in the medium of 23 mV/m. Cell proliferation, as measured by the uptake of tritiated thymidine, was inhibited in mononuclear cells from 24/25 donors, by up to 60%. There were no differences in blastogenic response between exposed and control cultures (Mooney et al., 1986).

4.2.3   Haematological effects

(a)   Static fields

Female NMRI mice were exposed to a static magnetic field of 3.5 T. The haematological parameters were generally unaffected. Necropsy and histopathological investigations revealed no pathological alterations (Zimmermann & Hentschel, 1987).

The potential adverse effects of subchronic exposure to a strong static magnetic field were evaluated in adult Fischer rats and their offspring. A battery of clinical tests in adult male and female rats and their offspring detected no adverse biological effects that could be attributed to a 10-week period of exposure to a 9.4-T static magnetic field (High et al., 2000).

(b)   ELF electric and magnetic fields

The acute effects of power-frequency magnetic fields on haematopoiesis were studied in 10-week-old CBA/H mice known to be susceptible to the induction of acute myeloid leukaemia after exposure to ionizing radiation. Mice were exposed to 50-Hz, 20-mT magnetic fields for seven days. Samples of blood and bone marrow were taken from three mice in each group immediately after exposure (day 0) and at about the same time on days 2, 4, 7, 10 and 18. Up to 19 days after exposure, no significant effects on peripheral blood characteristics were observed. Assays of bone-marrow stem cells and myelomonocytic progenitor cells also failed to show significant differences between exposed and control mice (Lorimore et al., 1990).

Spleen colony formation was examined in male CBA mice exposed to 50-Hz, 0.022-mT magnetic fields for 1 h, at the same time of day, for five successive days (5 h per five days). The number of colony-forming units was higher in the exposed animals than in the untreated controls but was not higher than that counted in sham-exposed animals. Significant changes were seen in the thymus weight and thymus index of exposed animals when compared with both control and sham-exposed animals. In a second study, mice were given a sublethal dose of X-rays (6 Gy) followed 2 h later with the same magnetic field treatment as above, i.e. 5 h every five days. The number of colonies per spleen showed a consistent, significant increase with exposure to the magnetic field and the number of colony forming units per femur was decreased. In a third study, bone marrow was taken from mice that had been exposed to 50-Hz, 0.022-mT magnetic fields for 5 h per five days, and injected into mice that had been exposed to a lethal dose of X-rays (9 Gy). The number of colony forming units per femur in the recipient mice was significantly reduced at days 1 and 4 after injection (Korneva *et al.*, 1999).

Two hundred and forty adult male Sprague-Dawley rats were exposed for 8 h per day to 50-Hz, 25-kV/m or 100-kV/m vertical electric fields for 35, 55 and 155 days and the animals were killed after 140, 164 and 315 days, respectively. Irrespective of the duration of exposure, the mode of grounding and the field strength, no statistical differences in body weight, morphology or histology of the liver, heart, mesenteric lymph nodes or blood parameters were found in the exposed animals compared to controls (Margonato *et al.*, 1993).

Adult male Sprague-Dawley rats were exposed to a 50-Hz magnetic field with a flux density of 5 μT for 22 h per day for 32 weeks. Haematological variables were measured in blood samples taken before exposure to the field and at 12-week intervals during exposure. No differences were detected between the exposed and control groups in haematology and haematochemistry, or in the concentrations of the neurotransmitters dopamine and serotonin in the brain (Margonato *et al.*, 1995).

Adult male Sprague-Dawley rats were exposed for 8 h per day on five days per week for eight months to 50-Hz electric and magnetic fields of two different field strength combinations: 5 μT and 1 kV/m and 100 μT and 5 kV/m. The animals were kept under constant controlled illumination for 24 h per day. Blood samples were collected for determination of haematological variables before exposure and at 12-week intervals during exposure. No pathological changes were observed at either field strength combination in animal growth rate, in morphology and histology of the tissue specimens collected from the liver, heart, mesenteric lymph nodes, testes and bone marrow, or in serum chemistry (Zecca *et al.*, 1998).

Haematological and serum chemistry variables were examined in groups of female Sprague-Dawley rats exposed to uniform, vertical 60-Hz electric fields at 100 kV/m for 15, 30, 60 or 120 days. Another group of rats was sham-exposed. Blood samples were collected from all animals within 3 h after exposure and analysed for leukocyte and erythrocyte counts, haemoglobin concentration, reticulocyte and thrombocyte counts, bone marrow cellularity and prothrombin times, serum iron and serum alkaline phosphatase concentrations and serum triglyceride values. Significant differences between exposed and sham-exposed rats were seldom seen. Statistical evaluation of these data did not detect any consistent effect of the electric field (Ragan *et al.*, 1983).

Groups of Fischer 344/N rats and B6C3F$_1$ mice were exposed to 60-Hz magnetic fields or sham-exposed for eight weeks. Magnetic field strengths were 0.002, 0.2 and 1 mT. The whole-body exposure was continuous for 18.5 h per day, seven days per week. An additional group of rats and mice was exposed intermittently (1 h on/1 h off) to 1-mT magnetic fields for the same period of time. The animals were kept on a 12 h/12 h light/dark cycle. There were no gross, histological, haematological or bio-chemical lesions that could be attributed to magnetic field exposure (Boorman et al., 1997).

4.2.4   Neuroendocrine effects

The hypothesis that reduced pineal function may promote the development of breast cancer in humans was initially suggested by Cohen et al. (1978). Several mechanisms whereby changes in pineal or serum melatonin concentrations may affect the risk for breast cancer have been proposed. These include the observation, at least in some animal species, that decreased melatonin concentrations cause elevations in circulating levels of estrogen and progesterone which increase cell proliferation in the stem cell population of the breast and may thus increase the risk for cancer in this tissue (Cohen et al., 1978). Other authors have suggested that melatonin may directly suppress the growth of human mammary tumour cells (Blask & Hill, 1986) and of cells of other cancer types, particularly melanoma, prostate cancer, ovarian cancer, bladder cancer and leukaemia (Stevens, 1993) and that melatonin may act as a scavenger of free radicals, preventing oxidative damage to DNA (Reiter et al., 1995), at least at pharmacological levels (Cridland et al., 1996). It has also been suggested that melatonin may modulate immune responsiveness (Maestroni et al., 1986).

Studies of the effects of electric and magnetic fields on melatonin concentrations have mostly been carried out in rats but have also used mice, Djungarian hamsters (Phodopus sungorus), sheep and primates, including humans. The experimental details of the animal studies are given in Table 35. Human data are discussed in section 4.1.4 and summarized in Table 34.

(a)   Electric fields

Several studies by one group of authors (Wilson et al., 1981, 1986; Reiter et al. 1988), reported that exposure to electric fields significantly suppressed pineal melatonin and the activity in the pineal gland of the enzyme N-acetyltransferase, which is important in the synthesis of melatonin. These effects appeared within three weeks of exposure, but disappeared within three days after cessation of exposure. A similar suppression of pineal melatonin was reported by these authors to occur after the prenatal and neonatal exposure of rats to power-frequency electric fields; no simple dose–response relationship was apparent. Grota et al. (1994) reported that exposure of rats to power-frequency electric fields had no effect on pineal melatonin concentrations

Table 35. Studies in animals of melatonin concentrations in response to exposure to ELF electric and magnetic fields

Reference	Assay	Exposure	Response	Comment
ELF electric fields				
Rats				
Wilson *et al.* (1981, 1983)	Night-time pineal melatonin concentrations and NAT enzyme activity in adult rats	60 Hz, 1.7–1.9 kV/m (not 65 kV/m) due to equipment failure; 20 h per day for 30 days	Reduced pineal melatonin and SNAT activity	Data combined in one experiment because of variability
Wilson *et al.* (1986)	Night-time pineal melatonin concentrations and SNAT enzyme activity in adult rats	60 Hz, 65 kV/m (39 kV/m 'effective') for up to 4 weeks	Pineal melatonin and SNAT activity reduced after 3 weeks of exposure; recovered 3 days after exposure ceased	
Reiter *et al.* (1988)	Night-time pineal melatonin concentrations in immature rats	60 Hz, 10, 65 or 130 kV/m during gestation and 23 days postnatally	Night-time peak reduced and delayed in exposed animals	No simple dose–response relationship
Grota *et al.* (1994)	Night-time pineal melatonin concentrations and NAT enzyme activity and serum melatonin in adult rats	60 Hz, 65 kV/m for 20 h per day for 30 days	No effect on night-time melatonin and NAT; serum melatonin decreased	
ELF magnetic fields				
Mice				
Picazo *et al.* (1998)	Serum melatonin concentrations in fourth generation of male mice	50 Hz, 15 µT for four generations	Reduced night-time concentrations	Experimental procedures not fully described
de Bruyn *et al.* (2001)	Night-time plasma melatonin concentrations in mice	50 Hz, 0.5–77 µT (2.75 µT average) 24 h/day from conception to adulthood	No effect	

Table 35 (contd)

Reference	Assay	Exposure	Response	Comment
Rats				
Martínez-Soriano *et al.* (1992)	Serum melatonin concentrations in adult rats	50 Hz, 5 mT for 30 min during the morning for 1, 3, 7, 15 and 21 days	Serum melatonin reduced on day 15 [no values for days 1, 7 or 21]	Technical difficulties; brief description of method
Kato *et al.* (1993)	Pineal and serum melatonin concentrations in adult rats	50 Hz circularly polarized, 1, 5, 50 or 250 µT for 6 weeks	Night-time and some daytime reductions in serum and pineal melatonin	Questionable comparisons with historical controls
Kato *et al.* (1994b)	Serum melatonin concentrations in adult rats	50 Hz, circularly polarized, 1 µT for 6 weeks	Night-time melatonin concentrations reduced, returning to normal within one week	Comparison with sham-exposure
Kato *et al.* (1994c)	Pineal and serum melatonin concentrations in adult, pigmented rats	50 Hz, circularly polarized, 1 µT for 6 weeks	Night-time pineal and serum melatonin concentrations reduced	Comparison with sham-exposed and historical controls
Kato *et al.* (1994d)	Serum melatonin concentrations in adult rats	50 Hz, horizontally or vertically polarized, 1 µT for 6 weeks	No effect	Comparison with sham-exposed and historical controls
Kato *et al.* (1994a)	'Antigonadotrophic' effect of melatonin on serum testosterone in adult rats	50 Hz, circularly polarized, 1, 5 or 50 µT for 6 weeks	No effect	Comparison with sham-exposure
Selmaoui & Touitou (1995)	Night-time serum melatonin concentrations and pineal NAT activity in adult rats	50 Hz, 1, 10 or 100 µT for 12 h once or for 18 h per day for 30 days	Reduced melatonin and NAT activity after 100 µT (acute) and 10 and 100 µT (chronic)	
Bakos *et al.* (1995, 1997, 1999)	Night-time excretion of melatonin urinary metabolite in adult rats	50 Hz, 1, 5, 100 or 500 µT for 24 h	No significant effects compared with baseline pre-exposure controls	

Table 35 (contd)

Reference	Assay	Exposure	Response	Comment
Mevissen et al. (1996b)	Night-time pineal melatonin concentrations in non-DMBA-treated adult rats	50 Hz, 10 µT for 13 weeks	No effect	A small part of a larger, well-planned mammary tumour study
Löscher et al. (1998)	Night-time serum melatonin concentrations in adult rats	50 Hz, 100 µT for 1 day or 1, 2, 4, 8 or 13 weeks	No consistent efects on melatonin	The few positive effects could not be replicated
John et al. (1998)	Night-time excretion of melatonin urinary metabolite in adult rats	60 Hz, 1 mT for 20 h/day for 10 days or 6 weeks; 1 mT inter-mittent for 1 h or for 20 h/day for 2 days	No effect	
Djungarian hamsters				
Wilson et al. (1999)	Night-time pineal melatonin concentrations	60 Hz, 100 µT for 15 min, 2 h before dark	Suppression of night-time peak	
Yellon (1994)	Night-time pineal and serum melatonin concentrations	60 Hz, 100 µT for 15 min, 2 h before dark	Reduced and delayed night-time peak; less effect and absent in 2nd and 3rd replicates, respectively	Considerable variation between replicate studies
Yellon (1996a)	Night-time pineal and serum melatonin concentrations	60 Hz, 100 µT for 15 min, 2 h before dark	Reduced and delayed night-time peak; less effect in second replicate	Considerable variation between replicate studies
Yellon (1996b)	Night-time pineal and serum melatonin concentrations; adult male reproductive status	60 Hz, 100 µT for 15 min, 2 h before dark for 3 weeks	No effect on pineal or serum melatonin; no effect on melatonin-induced sexual atrophy	Second part of above study

Table 35 (contd)

Reference	Assay	Exposure	Response	Comment
Truong *et al.* (1996)	Night-time pineal and serum melatonin concentrations; male puberty, assessed by testes weight	60 Hz, 100 μT for 15 min, 2 h before dark from 16 to 25 days of age	Reduced and delayed night-time peak; this effect absent in second replicate; no effect on development of male puberty	Considerable variability in melatonin concentrations between replicate studies
Truong & Yellon (1997)	Night-time pineal and serum melatonin concentrations	60 Hz, 10 or 100 μT (continuous) or 100 μT (intermittent) for 15 or 60 min before or after onset of dark period	No effect	
Yellon & Truong (1998)	Night-time rise in pineal and serum melatonin concentrations; testes weight	60 Hz, 100 μT in complete darkness; 15 min/day for up to 21 days	No effect, even in the absence of photoperiodic cue	
Niehaus *et al.* (1997)	Night-time pineal and serum melatonin concentrations; testis cell numbers	50 Hz, 450 μT (peak) sinusoidal or 360 μT (peak) rectangular fields, 56 days	Increased cell number and night-time serum melatonin concentrations after exposure to rectangular field	Animals on 'long-day' schedule; difficult to interpret
Wilson *et al.* (1999)	Night-time pineal melatonin concentrations and testis and seminal vesicle weights in short-day (regressed) animals	60 Hz, 100 or 500 μT; continuous and/or intermittent, starting 30 min or 2 h before onset of darkness; for up to 3 h on up to 42 days	Reduced pineal melatonin after acute (15 min) exposure; reduced gonad weight but not melatonin after 42-day exposure	Authors suggest a stress-like effect

Table 35 (contd)

Reference	Assay	Exposure	Response	Comment
ELF electric and magnetic fields				
Suffolk sheep				
Lee *et al.* (1993, 1995)	Night-time serum melatonin concentrations and female puberty, detected by rise in serum progesterone	60-Hz, 6-kV/m and 4-µT fields generated by overhead power lines; 10 months	No effect of electric and magnetic fields; strong seasonal effects	Two replicate studies; open-air conditions
Non-human primates				
Rogers *et al.* (1995a)	Night-time serum melatonin concentration in baboons	60-Hz, 6-kV/m and 50-µT fields (6 weeks), 30-kV/m and 100-µT fields (3 weeks)	No effect	
Rogers *et al.* (1995b)	Night-time serum melatonin concentration in baboons	60-Hz, irregular and intermittent sequence of 6-kV/m and 50-µT fields or 30-kV/m and 100-µT fields accompanied by 'transients'	Reduced serum melatonin levels	Preliminary study on two animals

Data from National Radiological Protection Board (1992, 2001)
DMBA, 7,12-dimethylbenz[*a*]anthracene; NAT, *N*-acetyltransferase; SNAT, serotonin-*N*-acetyltransferase

or activity of *N*-acetyltransferase, although concentrations of serum melatonin were significantly depressed.

The difficulties in reproducing some of the earlier findings on the effects of power-frequency electric fields on melatonin have been discussed by Brady and Reiter (1992).

(b)   Magnetic fields

(i)   Studies in mice

One large-scale study (National Toxicology Program, 1996) reported that exposure to continuous or intermittent 60-Hz magnetic fields (up to 18.5 mT) had no effect on serum or pineal melatonin concentrations in mice. Chronic exposure to 50-Hz magnetic fields of varying intensity (1–130 µT), as part of a tumour promotion study, was reported to have no effect on the night-time excretion of a urinary metabolite of melatonin in mice that had been pre-exposed to ionizing radiation (4 Gy) (Heikkinen *et al.*, 1999).

(ii)   Studies in rats

An extensive series of studies was conducted in male rats to assess the effects of exposure to circularly or linearly polarized power-frequency magnetic fields of up to 250 µT for up to six weeks on pineal and serum melatonin concentrations. [The Working Group noted that a major difficulty with the interpretation of the results of many studies by this group was that the sham-exposed rats were sometimes treated as 'low-dose' groups because they were exposed to stray magnetic fields (< 2%) generated by the exposure system; thus, statistical comparison was sometimes made with historical controls. Such procedures fail to allow for the interexperimental variability (Kato *et al.*, 1993, 1994b,c,d).]

In the first study, night-time pineal and serum melatonin concentrations were shown to be significantly reduced after exposure for 6 weeks to circularly polarized power-frequency magnetic fields of up to 250 µT, compared with melatonin concentrations in historical controls; in contrast, there was no difference between the concentrations measured in the exposed and concurrent, sham-exposed groups (Kato *et al.*, 1993). The effect had disappeared one week after cessation of exposure (Kato *et al.*, 1994b). A further study with a different, pigmented, rat strain showed a night-time suppression of serum and pineal melatonin in exposed animals compared with both sham-exposed animals and historical controls (Kato *et al.*, 1994c). In contrast to these results, exposure to horizontally or vertically polarized power-frequency magnetic fields for six weeks had no effect on pineal or serum melatonin when compared to sham-exposed animals and historical controls. The reason for this difference between the effects of circularly polarized and horizontally or vertically polarized fields was not clear (Kato *et al.*, 1994d). The last study of this series tested the hypothesis that a reduction in serum melatonin might be correlated with an increase in serum testosterone. However, animals exposed to circularly polarized 50-Hz magnetic fields were found to have serum testosterone levels similar to those in their sham-exposed counterparts (Kato *et al.*, 1994a).

Four other groups investigated the effects of magnetic fields on serum and pineal melatonin concentrations in rats and came to inconsistent, but generally negative, conclusions. One study reported that acute or chronic exposure of rats to horizontally polarized power-frequency magnetic fields significantly depressed night-time serum melatonin concentrations and activity of *N*-acetyltransferase in the pineal gland (Selmaoui & Touitou, 1995). In another study, exposure to a vertical or horizontal power-frequency magnetic field (50 Hz, 100 μT) had no effect on the circadian excretion of 6-sulphatoxymelatonin, the major urinary metabolite of melatonin (Bakos *et al.*, 1995, 1997). As part of a larger study on the effects of electromagnetic fields on mammary tumours induced by 7,12-dimethylbenz[*a*]anthracene and pineal function in rats, no effect of exposure to a magnetic field of 50 Hz, 10 μT, continuously for 13 weeks, was found on pineal melatonin concentrations in animals treated with 7,12-dimethylbenz[*a*]anthracene (Mevissen *et al.*, 1996b; see also Löscher *et al.*, 1998). Exposure of rats to power-frequency magnetic fields (60 Hz, 1 mT) for up to six weeks under a variety of conditions intended to maximize magnetic field sensitivity had no effect on the circadian excretion of the major urinary metabolite of melatonin (John *et al.*, 1998).

(iii)   *Studies in seasonal breeders*

Four laboratories have investigated the effects of exposure to ELF electric and magnetic fields on pineal activity, serum melatonin concentrations and reproductive development in animals that breed seasonally. Three research groups examined these effects in Djungarian hamsters in which the duration of the night-time rise in melatonin secretion during the shortening days of autumn and winter inhibits reproductive activity.

The most complete data come from a series of studies by Yellon and colleagues. Acute exposure to a power-frequency magnetic field (60 Hz) two hours before the onset of darkness reduced and delayed the night-time rise in serum and pineal melatonin concentration, but this effect was diminished in a second replicate study and absent in the third (Yellon, 1994). Similarly variable results on pineal and serum melatonin concentrations were reported by Yellon (1996a,b) and Truong *et al.* (1996). In addition, both studies found that exposure to a magnetic field had no effect on reproductive development, even in reproductively repressed hamsters kept on 'short-day' (winter) light/dark schedules, which might be thought to be sensitive to reduced and delayed night-time melatonin elevation. A fourth study, under experimental conditions that were different from those in the previous studies found no effect of exposure to magnetic fields on the night-time melatonin concentrations (Truong & Yellon, 1997). Finally, a brief exposure to power-frequency magnetic fields before the night-time rise in pineal and serum melatonin concentrations had no effect even in complete darkness, i.e. in the absence of a strong photoperiodic cue (Yellon & Truong, 1998).

In studies from a different laboratory, chronic exposure of Djungarian hamsters, which were kept on 'long-day' (summer) light/dark schedules, to 'rectangular' power-frequency magnetic fields (50 Hz; 360 or 450 µT) was reported to increase testis cell numbers and night-time concentrations of melatonin in serum, whereas exposure to sinusoidal power-frequency magnetic fields had little effect. The authors concluded that the in-vivo effects of magnetic fields may have been dependent on their waveform, and that the rapidly changing waveform of the rectangular fields was a more effective biological stimulus (Niehaus et al., 1997). [The Working Group noted that the results are not easy to interpret: increased melatonin concentrations in the Djungarian hamster are usually accompanied by decreased testicular activity.]

More recently, the effect of exposure to power-frequency (60 Hz) magnetic fields on pineal melatonin concentration, serum prolactin concentration and testicular and seminal vesicle weights have also been studied in Djungarian hamsters that had been shifted to a short-day light/dark regime in order to induce sexual regression. Night-time pineal melatonin concentrations were reduced after acute exposure, but this effect diminished with prolonged exposure. In contrast, induced sexual regression, as indicated by the reduction in testicular and seminal vesicle weights, seemed to be enhanced rather than diminished by prolonged exposure to the magnetic field, suggesting a possible stress response (Wilson et al., 1999).

The fourth set of studies of the effects of electric and magnetic fields on seasonal breeders was conducted with Suffolk sheep, which have a long gestational period and become reproductively active in the autumn, as the day-length shortens. In two replicate studies, Suffolk lambs were exposed outdoors to the magnetic fields generated by overhead transmission lines for about ten months. No effect of exposure was observed on serum melatonin concentrations or on the onset of puberty (Lee et al., 1993, 1995).

(iv)   Studies in non-human primates

Chronic exposure of three male baboons (Papio cynocephalus) to a combination of 60-Hz electric and magnetic fields (6 kV/m, 50 µT and 30 kV/m, 100 µT) for 6 weeks had no effect on night-time serum melatonin concentrations (Rogers et al., 1995a). A preliminary study, based on data from two baboons, showed a marked suppression of the night-time rise in melatonin after exposure of the animals for three weeks to an irregular, intermittent sequence of combined electric and magnetic fields in which switching transients were generated (Rogers et al., 1995b).

(v)   Cellular effects

The effects of magnetic fields on the function of the serotonin receptor, 5-HT1B, were studied in tissue samples of rat and guinea-pig brain and in Chinese hamster ovary cells transfected with the human form of the receptor. The tissue and cell samples were exposed to 50-Hz magnetic fields (0.01–10 mT) for 30 or 60 min before specific assays were performed. The authors observed an effect of the field on the affinity constant of 5-HT1B receptors, which decreased (in a sigmoidal fashion at field intensities between 0.05 and 2 mT, with a threshold at 0.6 mT) when the response saturated. Functionally, the magnetic fields inhibited the action of a 5-HT1B agonist to produce cyclic adenosine monophosphate (cAMP) and also caused a change in the cellular activity of the receptors, as demonstrated by the inhibition of synaptosomal release of 5-HT1B receptors from rats, guinea-pigs and humans (Massot *et al.*, 2000). [The Working Group noted that it is unclear how these results relate to changes *in vivo*.]

4.2.5   *Behavioural effects*

Studies on the effects of ELF electric and magnetic fields on behaviour have included tests of:

— perception and detection;
— arousal and aversive activity responses; and
— learning and memory.

The studies of specialized response systems that operate in various animal species and are associated with exposure to electric and magnetic fields, such as communication of food location (in honeybees), electroreception systems (as found in certain fish species) and homing and navigation (found in several species of birds), are not discussed here.

(*a*)   *Static fields*

Davis *et al.* (1984) observed that neither exposure of mice to a 60-Hz, 1.65-mT nor a 1.5-T static magnetic field resulted in any behavioural alterations in a passive avoidance learning test.

Exposure to a strong static magnetic field (600 mT) for 16 h per day for 14 weeks inhibited avoidance behaviour in rats (Nakagawa & Matsuda, 1988). A taste-aversion study was conducted in rats exposed to very high-intensity static magnetic fields (9.4-T magnet) for 30 min. The exposure was significantly associated with taste aversion (Nolte *et al.*, 1998) and the effect lasted throughout the eight days of post-exposure follow-up.

The effects of exposure of rats to static magnetic fields were also investigated in a maze test to assess alterations in learning ability. Newborn male and female rats exposed to a field of 0.5 T for 14 days postnatally showed no significant change in learning ability compared with sham-exposed controls when tested one month after exposure (Hong *et al.*, 1988).

If given a choice, rats preferred to stay out of a high-voltage static electric field when the field strength was ≥ 55 kV/m, whereas they showed no such aversion at field strengths ≤ 42.5 kV/m. Changes in the air concentrations of either positive or negative ions had no effect on aversive or non-aversive behaviour (Creim *et al.*, 1993).

(*b*)   *ELF electric and magnetic fields*

(i)   *Behavioural effects related to perception of fields*

Behavioural studies in several animal species provide evidence that the animals perceive the presence of electric and magnetic fields and suggest that electric fields may directly alter behaviour. A number of investigations have reported the threshold of detection of a vertical 60-Hz electric field to be in the range of 4–10 kV/m in rats (Stern *et al.*, 1983). Male rats were trained to press a lever in the presence of the field and not to press in its absence. Control procedures showed that the behaviour required the rat to be in the electric field and that the behaviour was not controlled by any of several potentially confounding variables. Female rats, evaluated in a similar experimental signal-detection system showed comparable detection thresholds of 3–10 kV/m (Stern & Laties, 1985).

The thresholds for the perception of electric fields by animals other than rats have also been evaluated: they ranged from an average of 12 kV/m in baboons (with one animal perceiving a field as weak as 5 kV/m) (Orr *et al.*, 1995b) to 35 kV/m in miniature swine, as determined by use of preference aversion measurements (Kaune *et al.*, 1978). The perception threshold for mice was 25 kV/m, using arousal as the response indicator (Rosenberg *et al.*, 1983), and that for chickens and pigeons was approximately the same (Graves *et al.*, 1978; Graves, 1981). Human volunteers in certain postures were able to perceive a 9-kV/m electric field (Graham *et al.*, 1987). It appears that changes in various environmental factors, such as relative humidity, can alter perception thresholds (Weigel & Lundstrom, 1987). Cutaneous sensory receptors that respond to 60-Hz electric fields have been identified in the cat paw (Weigel *et al.*, 1987).

Detection of magnetic fields by animals is presumed to be quite different from that for perceiving electric fields and a wide divergence of results has been reported in various studies designed to evaluate detection of and response to magnetic fields by animals. By use of conditional suppression techniques to measure the response, rats were shown to be able to perceive the presence of a magnetic field as weak as 0.2 mT, with a 7–65-Hz frequency range (Smith *et al.*, 1994).

(ii)   *Activity, aversion responses*

The arousal response of animals exposed to a stimulus is a less precise index of perception than the responses discussed above. Arousal and preference or avoidance responses have been determined at several field strengths for ELF electric and magnetic fields. Such responses were observed in mice exposed to 60-Hz electric fields. At 25, 50 and 100 kV/m the responses were transient and not sustained with prolonged or repeated exposure (Hackman & Graves, 1981; Rosenberg *et al.*, 1981). When exposed to field strengths of 25 kV/m, rats preferred to spend their inactive period in the field (60 Hz). At 75–100 kV/m, they avoided exposure (Hjeresen *et al.*, 1980). To determine the strength of the aversion in rats, Creim *et al.* (1984) examined taste-aversion associated with exposure to electric fields. The animals showed no taste-aversion behaviour when exposed to electric fields up to 133 kV/m. Static fields (approximately 75 kV/m) were also ineffective in producing taste aversion in rats (Creim *et al.*, 1995).

There was no indication of aversive behaviour in mice exposed for 72 h to 1.65-mT static or 60-Hz magnetic fields (Davis *et al.*, 1984). These results were confirmed in other experiments showing a lack of aversive behaviour in rats exposed for 1 h to a 60-Hz, 3.03-mT magnetic field (Lovely *et al.*, 1992). The internal body currents induced by this level of exposure were comparable to those from strong electric fields. Because aversion was not demonstrated with magnetic fields, but was observed with electric fields, these results suggest that the aversive behaviour is not due to internal body currents. One study, using special combinations of parallel static and alternating magnetic fields (at cyclotron resonance frequencies), reported a reduction in exploratory behaviour at much lower intensities of static fields (500 and 50 µT) (Zhadin *et al.*, 1999).

At higher field intensities, both static (490 mT) and 50-Hz, 18-mT magnetic fields caused a decrease in irritability of rats after extended (2 h per day, 20 days) exposure (Trzeciak *et al.*, 1993). No other significant behavioural effects were observed.

Swine exposed to 30-kV/m electric fields were reported to prefer the field during the day and to avoid exposure during the night (Hjeresen *et al.*, 1982). At comparable exposures to 60-Hz, 30-kV/m electric fields, minor behavioural changes were observed in baboons, which appeared to be related to the animals' perception of the fields (Rogers *et al.*, 1995c). Even at field strengths up to 65 kV/m, no aversive behaviour was noted in non-human primates, although some increase in social stress was induced in groups of baboons exposed to 60-kV/m electric fields (Easley *et al.*, 1991). [The Working Group noted that only a few other behavioural changes have been reported after exposure of animals to electric fields up to 100 kV/m. Furthermore, the alterations that were seen in most studies were not persistent. Indeed, the animals quickly habituated to the presence of the electric field.]

(iii)   *Neurobehavioural teratology*

Neurobehavioural teratology studies are reviewed in section 4.2.1.

(iv)   *Learning, performance and memory*

As early as 1970, studies were conducted to examine the effects of ELF electric and magnetic fields on learning and performance. Macaques (*Macaca nemestrina*) exposed to weak electric fields of 4–10 Hz showed a disruption of the timing behaviour of an operant schedule response (Gavalas *et al.*, 1970; Gavalas-Medici & Day-Magdaleno, 1976). Operant behaviour in baboons was studied after exposure to 30- and 60-kV/m electric fields, but other than an initial work stoppage in exposed animals, no effect on operant behaviour was observed (Rogers *et al.*, 1995c). No effects were seen on response rate, number of errors or extinction of a simple task motivated by appetite (Rogers *et al.*, 1995d). Social behaviour of baboons was not affected by exposure to a 30-kV/m electric field for 12 h per day for 6 weeks (Coelho *et al.*, 1991).

The spatial learning task is considered to reflect the 'working' or 'short-term' memory. Tests of the performance of rodents in a maze are the usual methods of assessing spatial learning during exposure. In adult male mice exposed for 45 min to a 50-Hz, 0.75-mT magnetic field immediately before testing in a radial arm maze, the rate of learning the task was significantly reduced, compared with unexposed controls, although the overall accuracy of the memory was not affected (Sienkiewicz *et al.*, 1998). A similar test performed by rats exposed to 60-Hz, 0.75-mT magnetic fields showed a significant retardation of learning compared with sham-exposed controls. Treatment of the animals with the cholinergic agonist, physostigmine, before the exposure to the magnetic field reversed the field effect (Lai, 1996). Rats were also tested in a water maze to evaluate performance after exposure to a 1-mT magnetic field for 1 h. No differences between exposed and sham-exposed animals were observed in learning (ability to locate the platform); however, the swimming speed of the exposed rats was significantly lower than that of controls (Lai *et al.*, 1998).

Thomas *et al.* (1986) and Liboff *et al.* (1989) reported that timing discrimination in rats was disrupted even by a very weak 60-Hz magnetic field of 26 μT. These results, however, could not be replicated in later studies conducted under the same exposure conditions (Stern et al., 1996).

Rats exposed overnight to a 7-Hz, ~ 50-nT magnetic field coupled with a known geomagnetic activity showed no significant differences in number of errors made or speed of acquisition of the learning task when compared with sham-exposed controls (McKay & Persinger, 1999).

Studies have also been conducted to evaluate learning, performance and memory in animals given operant tasks after exposure to combined electric and magnetic fields.

In a study in rats, Salzinger *et al.* (1990) exposed animals to 60-Hz fields of 30 kV/m and 100 μT. When the performance of complex operant tasks was tested at various times during the light/dark cycle, a slightly slower response was observed in exposed animals at one point in the cycle.

Groups of eight baboons were sham-exposed or exposed to 6-kV/m, 50-μT, and subsequently to 30-kV/m, 100-μT fields at 60 Hz (Orr *et al.*, 1995a). In contrast to the effect seen with exposure to the electric field (30 kV/m) alone, there was no decrement of performance in the animals exposed to the combined fields. The authors suggest that the 100-μT magnetic field may have blocked the transient loss of performance. Other behavioural end-points were also unaffected in exposed baboons at these combinations of exposure to electric and magnetic fields (Coelho *et al.*, 1995). Macaques (*Macaca nemestrina*) were exposed to combined 60-Hz fields of 3–30 kV/m and 10–90 μT for 18 h per day for three weeks. No changes were observed in the performance of the exposed animals in a food-motivated operant task compared with sham-exposed controls (Wolpaw *et al.*, 1989).

There is clear evidence that animals can perceive electric fields at field strengths in the range of 3–10 kV/m and above. The ability to perceive ELF magnetic fields at low intensities is less well established, with both positive and negative indications of perception.

[The Working Group noted that both static and ELF magnetic fields have been shown to influence animals in learning and memory tasks. The data indicate that, for exposure to electric fields, aversive behaviour occurs above approximately 30–50 kV/m, depending on the animal species. For ELF magnetic fields, intensities of up to at least 18 mT do not appear to influence aversion. However, there have also been studies that showed no effects of exposure on either learning or memory acquisition. The results of a number of studies suggest that observation of field-related effects requires that exposure is closely coupled to testing in time, and may be more related to acquisition of the task or even to the state of arousal of the animal than to an effect on the memory itself. Because studies in this area show variable results, demonstrating both decreases in apparent learning ability and memory, and lack of any effect, it is difficult to draw firm conclusions as to the robustness of the effects of exposure to magnetic fields on learning.]

4.3   Effects of ELF electric and magnetic fields on bone healing[1]

In some situations the interaction of electric and magnetic fields with biological systems may be beneficial and lead to new medical applications. Examples include nuclear magnetic resonance imaging (MRI), electromagnetically induced hyperthermia, and bone healing with pulsed electromagnetic fields. This section will briefly discuss the evidence for the beneficial effects of exposure to electric and magnetic fields on bone healing in humans. Similar studies in animals and in-vitro experiments dealing with this topic have been reviewed in detail elsewhere (Portier & Wolfe, 1998) and will not be discussed here.

The therapeutic effects of specific low-energy, time-varying magnetic fields (pulsed electromagnetic fields, PEMF) on bone healing were first documented in 1973. These effects are based on the generation of electrical currents within the bone tissue by magnetic induction. These currents enhance the activity of bone-forming cells, the osteo-blasts (Cane *et al.*, 1993). Initially, this form of treatment with athermal energy was used mainly for patients with juvenile or adult non-unions, i.e. bone fractures showing no sign of union nine months after injury, despite the usual forms of surgical treatment, inclu-ding bone grafting. The biological effectiveness of PEMF therapy in augmenting bone healing has been confirmed by several double-blind and placebo-controlled prospective studies, and supported by laboratory studies (for an early review, see Bassett 1989).

---

[1] For the sake of completeness, this section was added by the IARC secretariat after the Working Group Meeting.

A group of 125 patients with non-united fractures of the tibial diaphysis were treated for >10h/day with PEMF (15-Hz pulse burst; 1.5-mT peak flux density; 25-µs decay time; 4-kHz pulse repeat rate; 5-ms burst duration). This treatment resulted in an estimated electric field pulse of 0.1–0.2 V/m in the bone tissue. Healing of the fracture was observed in 87% of the patients who required an average treatment duration of 5.2 months (range, 2–22 months). Failure of the therapy was attributed by the authors to inadequate immobilization of the fractured bone, separation of opposing fractured bone surfaces by more than 1 cm, exposure for less than 10 h/day, or incorrect positioning of the induction coils (Bassett *et al.*, 1981).

A double-blind, randomized, placebo-controlled trial of magnetic field therapy was conducted in 16 patients with tibia fractures that had not healed for at least one year. All patients received full leg plasters and were divided into a treatment group (9 patients; average age, 38 years) and a placebo group (7 patients; average age, 29 years). The patients were instructed to use PEMF stimulators — active or inactive — for 12–16 h/day, with coils designed to fit around the cast of each patient. The active devices produced a 1.5-mT peak intensity, 5-ms burst waveform repeated at 15 Hz. At 24 weeks, the fractures in 5/7 patients in the placebo group and 5/9 patients in the treatment group had healed. The authors concluded that magnetic field therapy is not more effective than the traditional treatment of these fractures (Barker *et al.*, 1984).

A group of 45 patients with tibial shaft fractures were included, over a period of 6 years, in a double-blind multi-centre trial to assess the effect of exposure to PEMF on bone fracture healing. The fractures had not shown union for 16–32 weeks, and some were characterized by severe displacement, angulation or the presence of injury to soft tissue and skin. Plaster immobilization was used in all patients (mean age, 35 years), 20 of whom received active electromagnetic stimulation (15-Hz pulse burst; 200-µs pulse duration; 25-µs decay time; peak flux density not reported). The other 25 patients (mean age, 45 years; significantly different from that of the treatment group) served as controls and used non-functioning stimulators. Treatment continued for 12 weeks, for 12 h/day. In the treatment group, union was observed in five cases, progress to union in five patients, but no progress was reported in 10 patients. In the control group the numbers of patients were one, one and 23, respectively ($p = 0.002$) (Sharrard, 1990).

In two randomized placebo-controlled clinical trials with 32 and 37 patients, respectively, treatment with pulsed magnetic fields (1.8 mT; repeat frequency, 75 Hz; rise time, 1.3 ms; estimated peak intensity of the induced electric field, 50 mV/m) for 8 h/day for up to 90 days induced a significant increase in trabecular bridging after surgical bone transection (Borsalino *et al.*, 1988; Mammi *et al.*, 1993).

A randomized double-blind prospective study was conducted in 195 patients who underwent lumbar interbody fusion, a surgical procedure aimed at connecting adjacent vertebrae. The study comprised 98 patients exposed to 15-Hz pulse-burst signals of 1.5 mT (rise time, 25 µs; repeat rate, 4 kHz; burst duration, 5 ms) and 97 patients in the placebo group. Both groups were asked to wear a brace for 8 h/day. Fusion was observed in 60 of 65 patients (92%) in the treatment group and in 63 of 97 (65%) patients carrying a placebo device ($p < 0.005$). The success rate in 33 patients who used the active brace for less than 4 h/day was similar to that of the placebo group (Mooney, 1990).

Another double-blind clinical trial was conducted in patients undergoing limb-lengthening surgery, which involves transection of the bone, distraction of the bone ends, and regeneration of bone tissue in the distraction gap. Patients were asked to wear the induction devices — active or placebo — for 4 h/day for up to 12 months. Exposure conditions to the pulsed magnetic fields were similar to those described in the previous study (Mooney, 1990). Bone densities were measured by X-ray analysis at the mid-point of the gap and at the proximal and distal ends. No difference between the treatment and placebo groups was observed either in limb-lengthening rate or in bone density at the distraction gap. However, there was a significant increase of bone density in the proximal segment in the field-exposed group, and a significant reduction of bone loss at the distal side (Eyres et al., 1996).

A recent clinical trial used low-amplitude PEMF on 19 patients with non-union or delayed union of the long bones. The stimulator device produced 0.3-ms pulses repeated at 80 Hz, with maximum magnetic fields of 0.01–0.1 mT, i.e. considerably weaker than in previous studies. Stimulation was applied for 9–12 h/day until mobility at the fracture site had disappeared. Among the 13 patients (age range, 9–90 years; period since injury 8–108 weeks, average 41.3 weeks) who completed the treatment, 11 had successfully healed bones, after treatment periods of 4–27 weeks. The two unsuccessful cases had bone gaps greater than 1 cm after removal of dead bone after infection. The authors concluded that weak magnetic fields may be effective in stimulating bone healing (Satter-Syed et al., 1999).

A recent review discussed studies in which magnetic fields were applied to promote bone-healing, to treat osteoarthritis and inflammatory diseases of the musculoskeletal system, to alleviate pain, to enhance healing of ulcers and to reduce spasticity. The action of magnetic fields on bone healing and pain alleviation was confirmed in most of the trials. In the treatment of other disorders the results have been contradictory. Application times varied between 15 minutes and 24 hours per day for three weeks up to 18 months. There seems to be a relationship between longer daily application time and positive effects, particularly in bone-healing. Of the 12 well-controlled studies dealing with the application of pulsed magnetic fields on bone healing, 11 reported a more rapid and improved healing process, compared to placebo controls. In all these studies, a treatment duration of 8–12 hours per day appeared to be required to produce the beneficial effect. It was noted that optimal dosimetry for this type of therapy has yet to be established (Quittan et al., 2000).

4.4   Genetic and related effects

4.4.1   *Genotoxic effects*

(*a*)   *Studies in humans*

(i)   *Static magnetic fields*

No data were available to the Working Group.

(ii)   *ELF electric and magnetic fields*

Several studies were carried out to investigate the clastogenic effects of exposure to power-frequency electric and magnetic fields and transient electric currents.

Chromosome analyses were performed on lymphocytes from 32 workers occupationally exposed for more than 20 years to 50-Hz electric and magnetic fields in 380-kV switchyards. Comparison with a control group of 22 workers of similar age and occupation, who had not been exposed to electric or magnetic fields, showed that neither the numbers of structural chromosomal changes nor the frequencies of sister chromatid exchange were increased (Bauchinger *et al.*, 1981).

Chromosomal aberrations in lymphocytes from three groups of welders were examined. The technology used by one group of welders gave rise to elevated concentrations of nickel in their serum and urine. Although all three groups were exposed to essentially the same electrical discharges, only the welders with the higher concentrations of nickel had increased chromosomal aberrations in their lymphocytes. There were no correlations between the number of aberrations and the concentration of nickel in the serum or the duration of occupational exposure, but correlations were found with the number of cigarettes smoked. It was concluded that certain welding processes produce fumes that seem to have effects on chromosomes, but that fields from welding *per se* do not seem to cause increased aberrations (Elias *et al.*, 1989).

Lymphocytes from the peripheral blood of 20 switchyard workers (9 smokers, 11 non-smokers) were assayed for chromosomal anomalies. The rates of chromatid and chromosome breaks were found to be significantly increased compared to those in lymphocytes from 17 control subjects (7 smokers, 10 non-smokers) (Nordenson *et al.*, 1984).

In a follow-up to the previous study, data were reported on 38 employees of electric power companies, 19 of whom worked on the repair and maintenance of circuit breakers and disconnectors in 400-kV substations. The other 19 served as controls and were exposed only to normal environmental electric and magnetic fields. Coded blood samples were analysed for the presence of chromosomal aberrations, sister chromatid exchanges (SCE), and cells with micronuclei. Compared with the control group, the exposed subjects showed a statistically significant increase in chromosomal aberrations and cells with micronuclei, but not in the frequency of SCE. Because similar results were obtained in studies of lymphocytes exposed *in vitro* to transient electric currents (spark discharges), the increase in chromosomal damage in substation workers may be associated with exposure to transient electric currents during work (Nordenson *et al.*, 1988).

Chromosomal aberrations, SCEs, replication indices and micronuclei were analysed in lymphocytes from the peripheral blood of 27 non-smoking power linemen who had considerable long-term exposure to 50-Hz electric and magnetic fields. An equal number of non-smoking telephone linemen were matched pair-wise with the exposed workers for age and geographical region, and served as controls. No differences between the groups were observed with respect to SCEs, replication indices or micronuclei. The overall frequency of chromosomal aberrations was higher in the exposed workers than in the controls, but the difference was not significant. However, the mean rate of lymphocytes with chromatid-type breaks was significantly higher in the power linemen than in the reference group. The excess of aberrant cells was observed mainly in lymphocytes from power linemen who had smoked earlier in their life. Although the interpretation is complicated by the confounding effect of having been a smoker, these results suggest that exposure to 50-Hz electric and magnetic fields is associated with a slight increase in chromatid breaks (Valjus *et al.*, 1993).

Thirteen workers in a high-voltage laboratory and 20 controls participated in a cross-sectional, matched-pairs study of cytogenetic damage. During cable testing the workers were exposed to static, alternating (50 Hz), or pulsed electric and magnetic fields. The magnetic field strength was normally 5–10 µT, but was occasionally much higher. Chromosomal aberrations, SCE and aneuploidy were studied in lymphocytes from the peripheral blood of exposed workers and controls. In addition, chromosomal aberrations were investigated in lymphocyte cultures treated with hydroxyurea and caffeine, to inhibit DNA synthesis and repair. Among seven laboratory workers (all smokers) the mean number of chromosome breaks/200 cells was 2.3, as compared with 0.7 for controls matched for job, age and smoking habits. The comparable figures for the cultures treated with hydroxyurea and caffeine were 12.0 and 6.0, respectively. No field-related increase was detected in non-smokers by either method. The other genetic parameters did not differ between the exposed workers and the controls (Skyberg *et al.*, 1993).

A cytogenetic analysis was carried out on cultured (48 h) peripheral lymphocytes of Swedish train drivers exposed to relatively strong magnetic fields up to > 100 µT. A pilot study with lymphocytes from 18 train drivers indicated a significant difference in the frequency of cells with chromosomal aberrations (gaps included or excluded) in comparison with seven concurrent controls (train dispatchers) and a control group of 16 office workers. The frequencies of cells with chromosome-type aberrations (excluding gaps) were about four times higher in train drivers than in office workers ($p < 0.01$) and dispatchers ($p < 0.05$). Seventy-eight percent of the train drivers had at least one cell with chromosome-type aberrations per 100 compared with 29% for the dispatchers and 31% for the office workers. In a follow-up study on a group of 30 train drivers, half of the cytogenetic slides from each subject were examined in one laboratory and the remainder was analysed in another; a statistical analysis showed no difference in results between laboratories, so the data were pooled. The results showed an increase ($p < 0.05$) in the frequency of cells with chromosome-type aberrations (gaps excluded) in the train drivers compared with that in a control group of 30 policemen. Sixty percent of the train drivers had one or more cells per 100 cells with chromosome-type aberrations compared with 30% among the policemen. These results support the hypothesis that exposure to magnetic fields at mean intensities of 2–15 µT can induce chromosomal damage *in vivo* (Nordenson *et al.*, 2001).

A cross-sectional study was carried out in a Norwegian transformer factory of 24 workers exposed to electric and magnetic fields and mineral oil, and 24 matched controls. The exposure group included employees in the high-voltage laboratory and in the generator-welding department. Electric and magnetic fields and oil mist and vapour were measured. Blood lymphocytes were cultured and analysed for chromosomal aberrations. In addition to conventional cultures, the lymphocytes were also treated with hydroxyurea and caffeine to inhibit subsequent DNA synthesis and repair. The results of the conventional lymphocyte cultures were similar in the exposure group and the controls for all cytogenetic parameters. In the cultures in which DNA synthesis and repair were inhibited, the cytogenetic parameters of the lymphocytes from generator welders did not differ from those of the controls, whereas in lymphocytes from workers in the high-voltage laboratory, the numbers of chromatid breaks, chromosome breaks and aberrant cells were significantly increased compared with control values. More years of exposure and smoking increased the risk of aberrations. No increase in cytogenetic damage in exposed workers compared to controls was detected with the conventional lymphocyte assay. In repair-inhibited cultures, however, there were indications that electric and magnetic fields in combination with exposure to mineral oil may produce chromosomal aberrations (Skyberg *et al.*, 2001).

(*b*)   *Studies in animals*

(i)   *Static magnetic fields*

Wing spot tests were performed in *Drosophila melanogaster* to examine the possible mutagenic activity of a static magnetic field. A DNA repair-defective mutation was introduced into the conventional test system to enhance the frequency of the mutant spots. Third instar larvae were exposed to a horizontal 5-T static magnetic field for 24 h. After moulting, wings were examined under a microscope to detect hair spots (large single and twin spots) with mutant morphology, indicative of somatic recombination. The exposure caused a statistically significant enhancement of somatic recombination compared with the unexposed control. This enhancement was suppressed to the control level by treatment with vitamin E, a non-specific antioxidant. Enhancement of non-disjunction, terminal deletions and gene mutations was not detected (Koana *et al.*, 1997). [The Working Group noted that there is limited information on the genetic effects of static magnetic fields.]

(ii)   *ELF electric and magnetic fields*

Cytogenetic effects, DNA breaks, DNA cross-links

The effect of long-term exposure to a magnetic field on subsequent cell proliferation and the frequency of SCE was examined in the peripheral lymphocytes of female Wistar rats following in-vivo exposure to a 50-Hz, 30-mT magnetic field for 7 or 28 days. As a positive control, another group of rats was treated with cyclophosphamide. The magnetic field influenced neither the frequency of SCE nor the proliferation characteristics of cultured peripheral lymphocytes (measured as mitotic indices and proliferation index) (Zwingelberg *et al.*, 1993).

The acute effect of magnetic fields on DNA integrity was examined in male Sprague-Dawley rats (age, 2–3 months; weight, 250–300 g), which were exposed in the Helmholtz coil system to a 60-Hz magnetic field at flux densities of 0.1, 0.25 and 0.5 mT for 2 h. Four hours after exposure, the rats were killed and DNA single-strand and double-strand breaks were assayed in brain cells by single-cell gel electrophoresis ('comet' assay) at neutral and alkaline pH. A significant increase in DNA single-strand breaks was observed in all cases, and the effect was dose-dependent. No significant effect on DNA double-strand breaks was observed after exposure to the 0.1-mT magnetic field, but a significant increase was seen at flux densities of 0.25 and 0.5 mT (Lai & Singh, 1997a).

Studies were conducted to determine whether treatment with melatonin and the spin-trap compound *N-t*-butyl-α-phenylnitrone could block the effect of magnetic fields on brain-cell DNA. Rats were injected with melatonin (1 mg/kg bw, subcutaneously) or *N-t*-butyl-α-phenylnitrone (100 mg/kg bw, intraperitoneally) immediately before and after two hours of exposure to a 60-Hz, 0.5-mT magnetic field. Brain cells were assayed by single-cell gel electrophoresis and both treatments were found to block induction by the magnetic field of DNA single- and double-strand breaks. Since melatonin and *N-t*-butyl-α-phenylnitrone are efficient scavengers of free radicals, the authors inferred that free radicals may play a role in DNA damage induced by magnetic fields (Lai & Singh, 1997b).

In the same laboratory, DNA–protein and DNA–DNA cross-links were studied by use of the single-cell gel electrophoresis assay. Male Sprague-Dawley rats (age, 2–3 months; weight, 250–300 g) were exposed in the Helmholtz coil system to a sinusoidal 60-Hz, 0.5-mT magnetic field for 2 h. Rats were killed 4 h after exposure. Most of the increase in DNA migration induced by the magnetic field was observed only after treatment with proteinase-K, suggesting the presence of DNA–protein cross-links. In addition, when brain cells from control rats were exposed to X-rays, an increase in DNA migration was observed, the extent of which was independent of treatment with proteinase-K. However, the X-ray-induced increase in DNA migration was retarded in cells from animals exposed to magnetic fields even after treatment with proteinase-K, suggesting that DNA–DNA cross-links were also induced by the field. The effects of magnetic fields were also compared with those of mitomycin C, a known inducer of

DNA cross-links. The pattern of effects was similar with the two agents (Singh & Lai, 1998).

Degradation of DNA was measured by single-cell gel electrophoresis in brain cells of CBA mice exposed *in vivo* continuously to 50-Hz, 0.5-mT magnetic fields for 2 h, 5 days or 14 days. No differences between the groups exposed for 2 h and 5 days and controls were observed. However, in the group exposed to the magnetic fields for 14 days, a significantly extended brain cell DNA migration was observed ($p < 0.05$) (Svedenstål *et al*, 1999a).

CBA mice were exposed outdoors to 50-Hz electromagnetic fields, with a flux density of about 8 μT, generated by a 220-kV transmission line. Possible genotoxic effects as well as effects on body weight, leukocyte and erythrocyte counts, and the level of ornithine decarboxylase activity in spleen and testis were determined after 11, 20 and 32 days of exposure. Ornithine decarboxylase is an enzyme involved in the synthesis of putrescine from ornithine, which is one of the polyamine synthesis pathways. DNA degradation was studied in brain cells by single-cell gel electrophoresis. After 32 days of exposure, a highly significant increase of the tail:head ratio of the comets was observed ($p < 0.001$), indicative of DNA damage. A decreased number of mononuclear leukocytes ($p < 0.05$) was observed in mice exposed for 20 days (Svedenstål *et al*, 1999b). [The Working Group noted that the single-cell gel electrophoresis assay following in-vivo exposure is particularly protocol-dependent, specifically with respect to the method of killing the animals and the treatment of tissue samples between exposure of the animal and analysis of the tissues.]

Dominant lethal mutations

Sexually mature, male C3H/He mice, aged 8–10 weeks at the beginning of a study to determine the induction of dominant lethal mutations, were exposed continuously for two weeks to a vertical, 50-Hz, 20-kV/m electric field or sham-exposed. Current densities induced in the testes were estimated to be approximately 100 μA/m². After exposure, each male was mated weekly with two different female mice for eight weeks. In this way, female mice were inseminated with sperm that had been exposed to the electric field at different stages of the spermatogenic cycle. Another group of male mice was exposed to 169-keV X-rays for about 150 min and served as a positive control. In this group, the estimated dose to the testis was 1.5 Gy. Whereas the positive controls gave clear evidence of mutagenesis, no significant changes related to exposure to an electric field were observed in fertilization rates or in survival of embryos before or after implantation (Kowalczuk & Saunders, 1990).

In a second assay of dominant lethal mutation a total of 42 male mice were exposed for eight weeks to a 50-Hz, 10-mT sinusoidal magnetic field. A group of 47 males was used as simultaneous cage controls. Each male was subsequently mated with two females on weeks 1, 3, 5, 7 and 9 post-exposure. The numbers of gestating females, corpora lutea, and live and dead implants were recorded. Multiple logistic regression analyses were used to examine the effects of exposure on fertilization rate, pre-implan-tation survival and post-implantation survival. There were no differences in overall response between the exposed and control groups, nor was any significant effect of exposure seen at any number of weeks after exposure. Thus, exposure to power-frequency magnetic fields at 10 mT for the approximate period of spermatogenesis did not appear to induce dominant lethal mutation in the germ cells of male mice (Kowalczuk et al., 1995). [The Working Group noted that the dominant lethal mutation assay is not sufficiently sensitive to allow detection of weak mutagens.]

(c)   In-vitro studies

(i)   Static magnetic fields

Micronucleus formation

The formation of micronuclei occurs when a chromosome or a chromosome fragment is released from the nucleus as a result of strand breakage or disturbance of spindle function during mitosis.

The effects of exposure to a 4.7-T homogeneous static magnetic field on the frequency of micronuclei induced in cultured CHL/IU cells by mitomycin C were studied. Simultaneous exposure to the magnetic field and mitomycin C for 6 h significantly decreased the frequency of mitomycin C-induced micronucleated cells analysed after post-treatment culture periods of 18, 42, 54 and 66 h. In both field-exposed and control groups, the highest frequency of mitomycin C-induced micro-nucleated cells was observed 42 h after treatment; the frequency decreased gradually after this time (Okonogi et al., 1996).

Mutation

The possible mutagenic and co-mutagenic effects of strong static magnetic fields were examined in a bacterial mutagenicity test. A super-conducting magnet was used to generate a homogeneous static magnetic field with a flux density of up to 5 T. Expo-sure to this field produced no mutations in four strains of *Salmonella typhimurium* (TA100, TA1535, TA1537 and TA98) or in *Escherichia coli WP2 uvrA* either using the pre-incubation method or in the plate-incorporation assay. In the co-mutagenicity test, *E. coli WP2 uvrA* cells were treated with various chemical mutagens and simulta-neously exposed to a 2-T or a 5-T static magnetic field. The mutation rate in the group exposed to the magnetic field was significantly higher than in the unexposed group when cells were treated with six different alkylating agents. Exposure to the magnetic field did not affect the mutagenicity of 2-aminoanthracene, 9-aminoacridine, N^4-aminocytidine or 2-acetoamidofluorene (Ikehata et al., 1999).

(ii)   *ELF electric and magnetic fields*
Chromosomal aberrations and sister chromatid exchange

Many researchers have determined the frequency of chromosomal aberrations and sister chromatid exchange in response to exposure to ELF electric and magnetic fields. Several studies with peripheral human blood lymphocytes have been reported in which exposure *in vitro* to 50- or 60-Hz magnetic fields (30 μT–7.5 mT) caused no increase in the frequency of chromosomal aberrations or sister chromatid exchanges (Cohen *et al.*, 1986a,b; Rosenthal & Obe, 1989; Antonopoulos *et al.*, 1995; Paile *et al.*, 1995). However, 72 hours of continuous in-vitro exposure to 10-ms pulses of a magnetic field (50 Hz, 1.05 mT) caused a significant increase in the frequency of chromosomal aberrations and sister chromatid exchanges in the lymphocytes from three male donors (Khalil & Qassem, 1991).

In-vitro exposure of human peripheral lymphocytes to a 50-Hz electric current with a current density of 1 mA/cm^2 did not induce any chromosome damage. Exposure to 10 spark discharge pulses (duration, 3 μs) with a peak field strength in the samples of 3.5 kV/cm, however, resulted in chromosome breaks at a frequency similar to that induced in lymphocytes *in vitro* by 0.75 Gy ionizing radiation (Nordenson *et al.*, 1984).

Lymphocytes from human peripheral blood were exposed for 48 h to 50-Hz magnetic fields (62.8, 80, 88.4, 504, 1061, 1750 and 2500 μT) and examined for cytogenetic effects. No significant changes in chromosomal aberration or sister chromatid exchange frequencies were observed. Combined treatments with mutagens (mitomycin C or X-rays) and 50-Hz magnetic fields did not reveal any significant synergistic, potentiating or antagonistic effects between magnetic fields and these mutagens (Maes *et al.*, 2000).

Exposure of Chinese hamster V-79 cells to 25-μs pulses of a magnetic field (0.18–2.5 mT) repeated at 100 Hz for 24 h did not increase sister chromatid exchanges (Takahashi *et al.*, 1987).

A significant increase in sister chromatid exchanges was found in mouse m5S cells exposed for 42 h to a strong ELF magnetic field (50 Hz, 400 mT) (Yaguchi *et al.*, 1999). In contrast, exposure to magnetic fields $\leq$ 50 mT did not cause any increase in the frequency of sister chromatid exchange in these cells. Chromosomal aberration analysis revealed an increased frequency of chromatid-type aberrations such as gaps in response to exposure to magnetic fields with flux densities of 50 mT and higher (Yaguchi *et al.*, 2000). Intermittent (15 s on, 15 s off) exposure of human amniotic cells to a 50-Hz, 30-μT magnetic field over 72 h doubled the frequency of chromosomal aberrations including gaps ($p < 0.05$). However, increased frequencies of chromosomal aberrations were not observed after continuous exposure to a 300-μT magnetic field (Nordenson *et al.*, 1994). When mouse m5S cells were exposed to a 60-Hz, 50-mT or a 50-Hz, 400-mT magnetic field after pre-exposure to X-rays (3 Gy) or mitomycin C (1 μM), chromatid-type aberrations were enhanced by the ELF magnetic field (Yaguchi *et al.*, 2000). Combined exposure of lymphocytes from human peripheral blood to an ELF magnetic field (60 Hz, 1.4 mT) and ionizing radiation (3 Gy) resulted in a higher frequency of tetraploid cells than that produced by ionizing radiation alone (Hintenlang, 1993).

DNA strand breaks

Various studies in which cultured mammalian cells were exposed to ELF magnetic fields (0.2 μT–5 mT) found no induction of DNA single-strand breaks (Reese *et al.*, 1988; Fiorani *et al.*, 1992; Fairbairn & O'Neill, 1994; Cantoni *et al.*, 1996). In contrast to the results of an in-vivo exposure study in rats (Lai & Singh, 1997a), there was no significant increase in DNA strand breaks, as measured by single-cell gel electro-phoresis, in cultured MO54 human brain tumour cells exposed for 30 min to strong 50- or 60-Hz fields (5–400 mT) (Miyakoshi *et al.*, 2000a).

In a study examining the combined effects of ELF electric and magnetic fields and oxidative stress, human Raji cells were treated with hydrogen peroxide and simulta-neously exposed to a pulsed field with a peak amplitude of 5 mT (pulse duration, 3 ms; pulse frequency, 50 Hz). Analysis of these cells by the single-cell gel electrophoresis assay showed no effect of the electric and magnetic field on the number of DNA single-strand breaks induced by hydrogen peroxide (Fairbairn & O'Neill, 1994). Similarly, exposure to 50-Hz fields (20-kV/m electric, 0.2-mT magnetic, or a combi-nation of these) had no influence on DNA single-strand breaks (measured by alkaline elution) in Chinese hamster ovary, CCRF-CEM and McCoy's cells pre-treated with methylmethane sulfonate, potassium chromate, ultraviolet radiation or hydrogen peroxide (Cantoni *et al.*, 1995, 1996).

Three studies reported on the modifying effects of exposure to ELF electric and magnetic fields on the repair of DNA damage induced in human lymphocytes by ionizing radiation. Two of the studies found no inhibition of the repair of DNA damage induced by ionizing radiation (100 Gy or 5 Gy) after post-irradiation exposure of the cells to pulsed magnetic fields (repetition rate, 50 Hz; peak intensity, 2.5 mT) or to 60-Hz magnetic or electric fields, as judged by indices of DNA rejoining and unscheduled DNA synthesis (Cossarizza *et al.*, 1989a; Frazier *et al.*, 1990). However, when human glioma MO54 cells were exposed to 50- and 400-mT magnetic fields for 30 min after X-ray irradiation at 4 °C (to inhibit enzymatic strand rejoining), a slight but significant increase in the number of DNA strand breaks was observed (Miyakoshi *et al.*, 2000a).

Lymphocytes from male Wistar rats were exposed for 3 h to either static or 50-Hz magnetic fields at 7 mT. In some cases, hydrogen peroxide or $FeCl_2$ was added to the medium. DNA damage (single-strand breaks and alkali-labile sites) were detected using the single-cell gel electrophoresis assay. Exposure to the static or 50-Hz fields did not produce any detectable DNA damage, nor did hydrogen peroxide or $FeCl_2$ alone. However, when lymphocytes were incubated with $FeCl_2$ and simultaneously exposed to 7-mT magnetic fields, the number of damaged cells was significantly increased and reached about 20% after exposure to static and 15% after exposure to 50-Hz magnetic fields (Zmyslony et al., 2000).

Micronucleus formation

Micronucleus formation was examined in human lymphocytes exposed to 50-Hz sinusoidal electric fields at 0.5, 2, 5 and 10 kV/m in air. No difference was found between the frequency of micronuclei in cultures exposed to the electric fields at any of the intensities tested and that in unexposed control cultures (Scarfi et al., 1993). When mitomycin C was added to the cultures, the frequency of micronuclei increased significantly, but no difference was found between field-exposed and unexposed cultures. Many studies have found no effect of sinusoidal and pulsed power-frequency fields of 30 μT to 2.5 mT on micronucleus formation (Saalman et al., 1991; Scarfi et al., 1991, 1994; Lagroye & Poncy, 1997; Scarfi et al., 1999). In contrast, two studies have shown positive effects. A statistically significant increase in the frequency of micronuclei in human squamous-cell carcinoma SCL II cells was observed after continuous exposure to 50-Hz magnetic fields (0.8 and 1.0 mT) for 48 h and 72 h (Simkó et al., 1998a). However, in a non-transformed cell line cultured from human amniotic fluid used in the same study, there was no increase in the number of micronuclei induced by similar exposure to an ELF magnetic field. Another group reported an increase in the number of micronuclei in the same cell line after horizontal exposure to a 50-Hz, 1-mT magnetic field (Simkó et al., 1998b). Exposure to a 50-Hz, 100-μT magnetic field for 24 h after 6 Gy γ-radiation caused a significant increase in the number of binucleated cells with micronuclei compared to exposure to γ-radiation alone (Lagroye & Poncy, 1997).

Scarfi et al. (1997) reported an increase in the number of micronuclei in human lymphocytes from donors with Turner syndrome when the cells were exposed for 72 h to magnetic fields pulsed at 50 Hz (peak flux density was 2.5 mT, rise time 1.2 ms, pulse width ~ 2 ms, rate of change of 1.0 T/s, and the induced electric field was estimated to be 0.05 V/m). However, they observed no change in the number of micronuclei seen in lymphocytes from either normal donors or those with Turner syndrome when these cells were exposed to sinusoidal 50-Hz magnetic fields at 1 mT for 72 h (Scarfi et al., 1996).

Mutation

The Ames assay using different strains of *Salmonella typhimurium* (TA100, TA98, TA97a and TA1102) revealed no effect of exposure to ELF magnetic fields (60, 600 and 6000 Hz; 0.3 mT, for 48 h) on mutation frequency (Morandi et al., 1996). Juutilainen and Liimatainen (1986) found no increase of mutation in *S. typhimurium* strains TA100 and TA98 exposed to 100-Hz magnetic fields (0.13, 1.3 or 130 μT), alone or in combination with the chemical mutagens 4-nitro-*ortho*-phenylenediamine or sodium azide. Exposure of Chinese hamster cells to ELF magnetic fields for seven days (1 μT; 50 Hz) did not cause a significant increase in the mutation frequency of the *Hprt* gene encoding the enzyme hypoxanthine-guanine phosphoribosyl transferase (Nafziger *et al.*, 1993). However, exposure of human melanoma MeWo cells to a strong magnetic field (50 Hz, 400 mT) resulted in an increase in the number of mutations of this gene. When the MeWo cells were exposed to the magnetic field in an annular culture plate (diameter, 15 cm), the frequency of *HPRT* mutations increased from the centre of the plate towards the edge, indicating increased mutation frequency with increasing current density. Under conditions of inhibited DNA synthesis, no induction of mutation was observed. Specifically, there was increased mutation induction during the S phase of the cell cycle (Miyakoshi *et al.*, 1996a, 1997). In a study of exposure of *Drosophila melanogaster* larvae to an ELF magnetic field (20 mT) using an annular plate, the frequency of somatic mutations increased as a function of induced current (Koana *et al.*, 2001). In a direct examination of the effects of electric fields, a 10-h exposure to a 60-Hz, 10-V/m electric field induced about twice as many *Hprt* gene mutations as in sham-exposed larvae (Ding *et al.*, 2001).

Exposure to a strong magnetic field (50 Hz, 400 mT) induced mutations in the *HPRT* gene of *p53*-deficient human osteosarcoma cells. These mutations were suppressed by expression of the wild-type *p53* gene introduced on a plasmid (Miyakoshi *et al.*, 1998a).

In a study of the mutagenic effects of ELF fields, continuous exposure to a 60-Hz, 5-mT magnetic field for six weeks did not significantly increase the frequency of *Hprt* mutations in CHO-K1 cells (Miyakoshi *et al.*, 1999).

Concomitant exposure to an ELF magnetic field (60 Hz, 3 mT) and menadione, a compound that induces the formation of free radicals, or *N*-methyl-*N*-nitrosourea, an alkylating agent, did not influence the mutation rate in the *E. coli lacI* target gene in *lacI*-transgenic rat embryo fibroblasts (Suri *et al.*, 1996).

Two studies on mutation induction by combined exposure to ionizing radiation and EMF fields showed that the frequency of mutations in the *Hprt* gene induced by X-rays (3 Gy) in CHO-K1 cells was significantly increased by exposure to 5-mT ELF magnetic fields during 1–6 weeks following X-ray irradiation (Miyakoshi *et al.*, 1999; Walleczek *et al.*, 1999). In a third study, human glioma cells were exposed for eight days to a 60-Hz, 5-mT magnetic field following X-ray irradiation (4 Gy). The frequency of *HPRT* gene mutation was increased approximately fourfold, compared to that induced by X-rays alone (Ding *et al.*, 2001). These results show that ELF electric and magnetic fields can modulate the effects of ionizing radiation.

4.4.2   *Effects relevant to non-genotoxic carcinogenesis*

(*a*)   *In-vivo studies*

(i)   *ELF electric and magnetic fields*

The influence of a 50-Hz magnetic field and simulated solar radiation on ornithine decarboxylase (ODC) activity and polyamines was studied in mouse epidermis. Chronic exposure of mice to combined magnetic fields and simulated solar radiation had no persistent effects on ODC activity or polyamines in comparison with animals exposed to ultraviolet radiation alone, although the same magnetic field treatment had previously been found to accelerate skin tumour development (Kumlin *et al*, 1998b). In an acute 24-h experiment, elevation of putrescine and down-regulation of ODC activity were observed in animals exposed to a 50-Hz, 100-μT magnetic field. No effect was seen 24 h after exposure to simulated solar radiation alone (Kumlin *et al*, 1998a).

A biomarker study was conducted during an ongoing initiation–promotion assay with SENCAR mice (see section 3, Sasser *et al.*, 1998). The study focused on early biochemical changes in epidermal cells associated with skin tumour promotion, including labelling index, ODC activity, protein kinase C activity and epidermal thickness, which were obtained from animals that had been DMBA-initiated and (12-*O*-tetradecanoylphorbol 13-acetate)-promoted and subsequently exposed to a 60-Hz, 2-mT magnetic field for 6 h per day for 5 days per week. No differences were reported for ODC activity or epidermal thickness, but a significant increase in down-regulation of the activity of protein kinase C was seen in the field-exposed mice at certain times during the promotion phase (DiGiovanni *et al.*, 1999). [The Working Group noted that ambient levels of protein kinase C activity in the pilot experiment were substantially lower (by a factor of 10).] The authors concluded that the inability of the 60-Hz, 2-mT magnetic field to alter early biomarkers provides further evidence for the lack of a promotional or co-promotional effect of magnetic fields seen in the tumour development assay (Sasser *et al.*, 1998).

Female rats were exposed to a 50-Hz, 50-μT magnetic field for a period of six weeks, in combination with oral administration of the chemical carcinogen DMBA. In control rats, exposure to the magnetic field alone resulted in an approximate doubling of ODC activity in mammary tissue, and a significant increase in ODC activity in the spleen, but not in the liver, small intestine, bone marrow or ear skin. Combined treatment with the magnetic field and DMBA was not more effective in increasing ODC activity than treatment with DMBA alone, except in liver tissue (Mevissen *et al.*, 1995).

      (*b*)   *In-vitro studies*

          (i)   *Static magnetic fields*

Cell proliferation

Fibroblasts from fetal human lung were exposed to static magnetic fields of 0.2 T, 1.0 T and 1.5 T for 1 h per day on five consecutive days. Cell cycle analyses of synchronously and non-synchronously growing cells were conducted and population doublings (parameters used to describe cell growth) were calculated. The proliferation kinetics of the cells were analysed for 21 days to rule out mid-term effects. No statistically significant differences between exposed and sham-exposed cells were observed. The calculations of population doublings did not reveal any modulation of cell growth during exposure. Proliferation kinetics did not provide evidence of any mid-term growth modulation effects of repeated exposure to magnetic fields (Wiskirchen *et al.*, 2000).

Fibroblasts from fetal human lung were exposed to a static 1.5-T magnetic field for 1 h three times a week for three weeks. Population doublings and cumulative population doublings were calculated weekly to detect treatment-related differences in overall cell growth. No significant differences between groups were found. Clonogenic activity, DNA synthesis, cell cycle and cell proliferation kinetics were not altered by exposure to the magnetic field (Wiskirchen *et al.*, 1999).

MCF-7 human breast cancer cells were exposed for different lengths of time (5–180 min) to the static magnetic field generated by a 0.2-T magnetic resonance tomograph. This treatment significantly decreased the incorporation of [^{3}H]thymidine into DNA in these cells (Pacini *et al.*, 1999a).

Gene expression

Exposure to a static magnetic field of 0.18–0.2 T for 1–6 days did not affect the growth of HeLaS3 cells. The effects of X-rays or heat treatment, which caused a transient delay in cell growth were not enhanced by subsequent exposure to the static magnetic field. Expression of the *c-fos* oncogene was measured in the HeLaS3 cells after exposure to the magnetic field for 2–24 h. No *c-fos* mRNA was detectable in unexposed cells, but it was expressed following incubation at a temperature of 45 °C for 10 and 15 min, and the expression was further enhanced by subsequent exposure of the cells to the magnetic field for 4 h (Hiraoka *et al.,* 1992).

Signal transduction

The effects of a static magnetic field generated by a 0.2-T magnetic resonance tomograph on cultured human neuronal cells (FNC-B4) were examined. Examination of the cells by scanning electron microscopy immediately after 15 min of exposure showed a significant change in cell morphology. At the same time, thymidine incorporation and inositol lipid signalling were significantly reduced. Sham-exposed control cells or non-neuronal cells (mouse leukaemia cells, human breast carcinoma cells) were unaltered. The release of endothelin-1 from FNC-B4 cells was much reduced after exposure to the magnetic field for only 5 min. However, no field-related alterations were found in 12 different DNA microsatellite sequences selected as indicators of genome instability (Pacini *et al.*, 1999b).

Apoptosis

Static magnetic fields generated by permanent magnetic discs showed an inhibitory effect on apoptotic cell death induced by various agents such as hydrogen peroxide, heat shock, ageing in culture and dexamethasone treatment in mitogen-stimulated human lymphocytes and certain human cell lines (U937, CEM and Burkitt lymphoma cells). For U937 cells, the reduction in apoptosis was first evident at flux densities of 0.6 mT and increased in a dose-dependent fashion until 6 mT, at which a plateau was reached that extended to 66 mT. Similar results were seen for CEM cells, but there was no anti-apoptotic effect of exposure to static magnetic fields on human lymphocytes or Burkitt lymphoma cells. Studies to test the involvement of calcium ion influx suggested that the inhibitory effect of magnetic fields on apoptosis is mediated by an enhanced influx of calcium ions from the extracellular medium. Thus, this effect would be limited to cells in which calcium influx has an anti-apoptotic effect (Fanelli *et al.*, 1999).

Exposure of the human leukaemic cell line HL60 to a 50-Hz, 45-mT magnetic field for a minimum of 1 h induced an increase in the number of apoptotic cells, but this effect was not observed in lymphocytes from human peripheral blood (Hisamitsu *et al.*, 1997; Narita *et al.*, 1997). In mouse haematopoietic progenitor cells (FDCP-mix (A4)), no alteration in the frequency of apoptosis was detected after exposure to 50-Hz magnetic fields at 6 µT, 1 mT or 2 mT for various lengths of time up to seven days (Reipert *et al.*, 1997). Rat tendon fibroblasts and rat bone-marrow osteoprogenitor cells were exposed to static magnetic fields and 60- or 1000-Hz alternating fields at flux densities of up to 0.25 mT. Various combinations of field strengths and frequencies resulted in increased apoptosis and detachment of the cells from the substratum, or in failure to attach (Blumenthal *et al.*, 1997). An increased frequency of apoptotic cells was found in a transformed human squamous-cell carcinoma line (SCL-II), but not in a non-transformed human amniotic fluid cell line after exposure to a 50-Hz field with a flux density of about 1.0 mT (Simkó *et al.*, 1998b).

Ismael *et al.* (1998) examined spontaneous and dexamethasone-induced apoptosis in thymocytes and spleen cells from mice exposed to 60-Hz magnetic fields at 0.4–1 µT or static magnetic fields of 8–20 µT. The animals were exposed continuously (24 h per day) for 12 months. The relative weights of the thymus and the spleen did not differ between control and exposed groups. Cells were isolated from both these organs, incubated with or without dexamethasone (10^{-7} M) and examined for apoptosis. Spontaneous apoptosis was not different between groups. Statistically significant increases were observed in dexamethasone-induced apoptosis only in thymocytes from animals exposed to 60-Hz, 0.4–1.0-µT magnetic fields.

The results reviewed above may suggest an increase in apoptosis in some cell types under certain experimental conditions of exposure to electric or magnetic fields, but further studies would be useful.

(ii)   *ELF electric and magnetic fields*

Cell proliferation

No significant change was observed in the proliferation characteristics of Chinese hamster ovary cells exposed to magnetic fields at either 220 µT or 5 mT (Livingston *et al.*, 1991; Miyakoshi *et al.*, 1996b). Oscillatory, time-dependent changes in cell proliferation, however, have been found in SV40-3T3 mouse fibroblasts after a single

1 h exposure to a 50-Hz, 2-mT magnetic field (Schimmelpfeng & Dertinger, 1997). In another study, a 30-min exposure to magnetic fields (50 Hz, 80 μT) caused an increase in the rate of proliferation of human epithelial amnion cells, but no effects were seen with other combinations of flux density and exposure duration (Kwee & Raskmark, 1995). SV40-transformed cells derived from a healthy donor and from an ataxia telangiectasia patient were exposed to an ELF magnetic field (50 Hz, 400 mT) for 2 h after pretreatment with 6 and 4 Gy X-rays, respectively. The magnetic field had no effect on the X-ray-induced reduction of survival of either cell line (Miyakoshi et al., 1994). When human myelogenous HL60 cells were exposed to an ELF magnetic field (45 mT) for 1 h, apoptotic cells were observed, but the same study found no apoptosis in human peripheral lymphocytes exposed under the same conditions (Narita et al., 1997).

The effects of ELF pulsed fields on cell proliferation were studied in cultured human lymphocytes from 24 young and 24 old donors (mean ages, 24 and 86 years, respectively). The pulse duration of the fields was about 2 ms and the repetition rate was 50 Hz, yielding a duty cycle of 1/10. The intensity of the magnetic field was 2.5 mT, and its average time variation of the order of 1 T/s. The maximum induced electric field was estimated to be 0.02 mV/cm. The cultures were exposed for 24–66 h, and then incubated for a further 6 h to allow incorporation of [³H]thymidine. The exposure to the pulsed fields had no effect on control lymphocytes but increased the phytohaemagglutinin-induced proliferation of the lymphocytes from the two donor groups. The effect was stronger in lymphocytes from old people. These cells normally show a reduced proliferative capability, but after exposure to the pulsed magnetic field, the incorporation of [³H]thymidine was similar to that observed in lymphocytes from young subjects (Cossarizza et al., 1989b).

The effects of rapidly changing magnetic gradient fields were examined in fetal human lung fibroblasts exposed for 2–24 h to trapezoid-shaped waveforms of 500- and 75-Hz base frequency and an amplitude of 2 mT. Proliferation of the cells was monitored for three weeks after exposure. Cell cycle analysis was carried out until 24 h after cessation of exposure to detect alterations in cell division. No differences in proliferation or cell cycle distribution between exposed and unexposed cell cultures were observed (Rodegerdts et al., 2000).

A study using human breast cancer MCF-7 cells (provided by D. Blask, Cooperstown, NY) reported the effects of exposure to ELF electric and magnetic fields on cell proliferation. When these cells were exposed to a sinusoidal electric and magnetic field (60 Hz, 1.2 μT) with concomitant melatonin treatment (10^{-9} M) the proliferation-inhibiting effect of this compound was reduced (Liburdy et al., 1993). The same series of studies demonstrated that exposure to ELF sinusoidal, but not fullwave rectified, electric and magnetic fields reduced the ability of tamoxifen, an agent used clinically for the treatment of breast cancer, to inhibit cell proliferation (Harland & Liburdy, 1997; Harland et al., 1999). These results, with respect to both melatonin and tamoxifen, have been independently replicated by Blackman et al. (2001) using the same cells, provided by the Liburdy laboratory. In a study with melatonin-insensitive MCF-7 cells, obtained from the Japanese cell bank, no effect of exposure to fields of 60 Hz, 5 mT was observed on cell growth either in the presence or absence of melatonin (Tachiiri *et al.*, 1999). [The Working Group noted that some but not all MCF-7 cell lines are responsive to melatonin; responsiveness may depend on the presence of estrogen receptors in these cells.]

Exposure of the yeast *Saccharomyces cerevisiae* to a 50-Hz, 120-µT magnetic field delayed recovery from the growth inhibition induced by ultraviolet B (UVB) radiation. The progression of the cell cycle after UVB exposure was also modified by the magnetic field (Markkanen *et al.*, 2001).

The effects of exposure to pulsed electric and magnetic fields were studied in lymphocytes from the peripheral blood of 25 patients with Down syndrome, a disorder in which premature ageing is characterized by precocious immune system derangement, including age-related defective proliferative capability of lymphocytes. After exposure to the pulsed fields, a significant increase in phytohaemagglutinin-induced cell proliferation was observed in cells from children and young adults with Down syndrome, but this phenomenon was much more evident in lymphocytes from older Down syndrome patients (Cossarizza *et al.*, 1991).

In studies that have used incorporation of [^{3}H]thymidine into nuclear DNA as an indicator of cell proliferation, the effects of exposure to electric and magnetic fields on cell growth *in vitro* have been mixed. Human fibroblasts exposed to various frequencies (15–4000 Hz) of magnetic fields (2.3–560 µT) showed enhanced DNA synthesis (Liboff *et al.*, 1984). Exposures to 0.18–2.5-mT pulsed electromagnetic fields, at specific ranges of pulse width, pulse height and pulse repetition rate, also stimulated DNA synthesis in Chinese hamster V79 cells (Takahashi *et al.*, 1986). In contrast, exposure to ELF electric and magnetic fields (1–200 Hz, 230–650 µT) inhibited DNA synthesis in phyto-haemagglutinin-induced human lymphocytes (Conti *et al.*, 1983; Mooney *et al.*, 1986), whereas no significant effect on DNA synthesis was seen in HL-19 normal human fibro-blasts exposed for 30 h to 50-Hz magnetic fields over a wide range of flux densities (20 µT–20 mT) (Cridland *et al.*, 1996).

Gene expression

The transcription of the oncogene *c-myc* in human leukaemia HL60 cells was enhanced by a 20-min exposure to an ELF magnetic field (150–15 Hz: maximum expression at 45 Hz, 200 µT–2.3 mT) (Goodman & Shirley-Henderson, 1991; Goodman *et al*, 1994). It was also reported that chloramphenicol transferase expression was enhanced when a specific DNA region upstream of the *c-myc* gene was transfected into human HeLa cells as a chloramphenicol transferase construct, followed by a 20-min exposure to a 60-Hz, 8-µT magnetic field (Lin *et al.*, 1994). Other studies, however, have failed to reproduce the enhancement of *c-myc* expression by exposure to ELF electric and magnetic fields (Lacy-Hulbert *et al.*, 1995; Saffer & Thurston, 1995; Balcer-Kubiczek *et al.*, 1996; Miyakoshi *et al.,* 1996b). Exposure to an ELF magnetic field

(60 Hz, 1 mT, for 75 min) caused no change in transcription rate of *c-myc* or the β-actin gene in HL60 cells, but it enhanced transcription of 45S ribosomal-RNA (Greene *et al.*, 1993). [The Working Group noted that HL60 cells display a high endogenous level of *c-myc* expression.]

In human T-lymphoblastoid cells (CEM-CM3), the transcriptional activities of *c-fos*, *c-myc* and the protein kinase C gene, which are associated with signal transduction, were increased by exposure to a 60-Hz, 100-μT magnetic field although this effect was strongly dependent on the duration of exposure and the cell density during the assay (Phillips *et al.*, 1992).

A chloramphenicol transferase gene construct containing the upstream regulatory region of the *c-fos* gene (from base pair –700 to +42) was transfected into HeLa cells. The cells were then exposed to a 60-Hz, 6-μT magnetic field for up to 40 min. An approximately 1.2-fold increase in chloramphenicol transferase-protein expression was seen 60 min after a 20-min exposure (Rao & Henderson, 1996). [The Working Group noted that the chloramphenicol transferase activity appeared to be very low, suggesting that the response of the construct was poor under the conditions tested.]

Primary and immortalized rat tracheal epithelial cells exposed for 30 min to a 50-Hz, 100-μT magnetic field displayed an approximately 3-fold enhancement of *c-jun* protein expression. In the same study, however, *c-fos* expression was decreased to approximately 70% of control levels following a 5-h exposure to the magnetic field (Lagroye & Poncy, 1998). Exposure to 60-Hz magnetic fields at 5.7 or 570 μT for 10–40 min caused no change in *c-fos* mRNA expression in HL60 cells (Balcer-Kubiczek *et al.*, 1996).

Transcription of the gene encoding the heat shock protein hsp70 was increased by about 1.8-fold in HL60 cells exposed for 20 min to a 60-Hz, 8-μT magnetic field. Increased transcription of the heat shock gene *SSA1* in the yeast *Saccharomyces cerevisiae* was observed under the same conditions (Goodman *et al.*, 1994). By means of the chloramphenicol transferase assay, the increased expression of hsp70 in HL60 cells, induced by exposure to ELF magnetic fields (60 Hz, 8 μT), was shown to be caused by enhanced binding of the *c-myc* protein to sites within the heat shock protein promoter region (Lin *et al.*, 1998a,b). However, in C3H mouse mammary carcinoma-derived 34i cells, a 20-min exposure to magnetic fields (50 Hz, 1.5 and 3 mT) had no effect on the expression of hsp70 or hsp90 (Kang *et al.*, 1998). Likewise, 2–20 h of exposure to a 50-Hz, 50-mT magnetic field had no influence on the expression of hsp70 protein in HL60RG cells. In this study, however, hsp70 expression induced by mild heat treatment (40 °C or 42 °C) could be suppressed by simultaneous application of the magnetic field (Miyakoshi *et al.*, 2000b).

The exposure of Chinese hamster ovary cells to a 50-Hz, 400-mT magnetic field caused a transient increase (maximum approximately 6 h) in the expression of the gene encoding the neuron-derived orphan receptor 1 (NOR-1); exposure to a 5-mT field had no effect (Miyakoshi *et al.*, 1998b).

Signal transduction

The focus of several investigations on the effect of electric and magnetic fields on cellular signal transduction has been the role of calcium, since it is intimately involved in the regulation of many signal transduction pathways.

Mononuclear blood cells from healthy adult volunteers were stimulated with phyto-haemagglutinin and exposed to a squared waveform field (3 Hz, 6 mT). The uptake of Ca^{2+} was lower than in cells treated with phytohaemagglutinin alone (Conti *et al.*, 1985). Conversely, exposure of rat thymocytes for 60 min to an induced 60-Hz electric field of 1.0 mV/cm produced an average 2.7-fold increase in concanavalin A-dependent Ca^{2+}-uptake compared to that in unexposed, isothermal control cells (Walleczek & Liburdy, 1990). Oscillatory increases in the concentration of intracellular calcium were induced in human Jurkat cells exposed to a 50-Hz, 0.1-mT magnetic field (Lindström *et al.*, 1993). In a further study, the same cells displayed oscillations in intracellular Ca^{2+} when exposed to magnetic fields with a wide frequency range (5–100 Hz), the strongest effect being seen at 50 Hz. At this frequency, the response showed a no-effect threshold at 0.04 mT, and a plateau at 0.15 mT (Lindström *et al.*, 1995a, 1996).

Oscillations of free intracellular calcium were seen in individual Jurkat cells in response to exposure to a 50-Hz, 0.15-mT magnetic field. In contrast, a CD45-deficient Jurkat cell line did not respond to stimulation by a magnetic field. The phosphatase activity of CD45 may regulate the activity of p56lck tyrosine kinase by removing an inhibitory phosphate. By using Jurkat cells that expressed a chimeric molecule, comprising the cytoplasmic phosphatase domain of CD45, the field-induced calcium response could be restored (Lindström *et al.*, 1995b).

Exposure to magnetic fields (50 Hz, 0.1 mT) also resulted in a significant increase in the concentration of inositol 1,4,5-trisphosphate in Jurkat cells. This effect was not inhibited by chelation of intracellular calcium ions, which implies that the oscillations in calcium concentration induced by the magnetic fields were not due to direct stimulation of the calcium-dependent phospholipase C-γ1, an enzyme involved in the formation of inositol 1,4,5-triphosphate (Korzh-Sleptsova *et al.*, 1995).

A later study reported that exposure to magnetic fields (60 Hz, 0.15 mT) had no effect on intracellular calcium signalling in Jurkat E6-1 cells (Lyle *et al.*, 1997).

Exposure of HL60 cells to an electric field (60 Hz, 10–100 V/m) for 1 h signifi-cantly decreased the activity of cytosolic protein kinase C. However, no concomitant rise in membrane-bound protein kinase C activity was observed, indicating that the electric field promotes down-regulation of cytosolic protein kinase C activity (Holian *et al.*, 1996).

Changes in signal transduction events as a result of exposure to magnetic fields have been described in a number of studies with human B-lineage lymphoid cells and chicken lymphoma B-cells (DT40). Other investigators, however, have failed to replicate these findings. Some of the studies are summarized below.

Exposure of human B lymphoid cells to a magnetic field (60 Hz, 0.1 mT) stimu-lated various tyrosine kinases, which resulted in tyrosine phosphorylation of many proteins and subsequent activation of phosphokinase C in a time-dependent manner. Analysis of various steps in the signal transduction pathway led the authors to conclude that the growth regulation of B lymphoid cells may be altered by the activation of a specific tyrosine kinase (Lyn) by the magnetic field (Uckun et al., 1995).

Exposure of DT40 chicken lymphoma B cells to a vertical magnetic field (60 Hz, 0.1 mT) resulted in the activation of phospholipase C-γ2, leading to increased turnover of inositol phospholipids. This activation is mediated by Bruton's tyrosine kinase (BTK), which was shown to be the responsive target for interaction with the magnetic field (Kristupaitis et al., 1998).

In an attempt to replicate the findings described above, Miller & Furniss (1998) examined the effects of magnetic fields on wildtype DT40 cells, on BTK-deficient DT40 cells and on BTK-deficient cells that had been reconstituted with the human *BTK* gene. The cells were all obtained from the Uckun laboratory. No effects were seen on production of inositol-1,4,5-trisphosphate, BTK-activation or tyrosine phosphorylation after exposure of these cells to 60-Hz, 0.1-mT magnetic fields. The authors suggest that the conflicting results may be due to some critical parameter in the exposure environ-ment that is different between laboratories.

In a further study aimed at replication of previous findings, Woods et al. (2000) exposed human B lymphoid cells (obtained from Uckun's laboratory) and chicken lymphoma DT40 cells (obtained from Miller's laboratory, but originally from Uckun) to a 60-Hz, 0.1-mT magnetic field, with or without a parallel, static magnetic field of 0.046 mT. No significant changes were detected in tyrosine phosphorylation or in acti-vation of Lyn and Syk tyrosine kinases in either cell line.

Exposure of β-galactosidase-transfected PC12-VG cells stimulated by forskolin to a 400-mT magnetic field for 4 h enhanced β-galactosidase expression. This enhanced expression was significantly inhibited by calcium entry blockers and almost completely suppressed by concomitant treatment with calphostin C, a protein kinase C inhibitor (Ohtsu et al., 1995). The induction of expression of the neuron-derived orphan receptor (NOR-1) gene by exposure to a 50-Hz, 400-mT magnetic field was also inhibited by treatment with various Ca^{2+} influx inhibitors (Miyakoshi et al., 1998b).

Ornithine decarboxylase activity is controlled by a signal transduction pathway associated with cell proliferation. In a study using human lymphoblastoid cells (CEM), mouse myeloma cells (P3) and rat hepatoma cells (Reuber H35), exposure to 60-Hz electric fields (10–1000 V/m) caused a transient, several-fold increase in the activity of ornithine decarboxylase (Byus et al., 1987). A twofold increase in the acti-vity of this enzyme was also seen in mouse L929 cells exposed to a 60-Hz magnetic field (1–100 μT) (Litovitz et al., 1991). The same group also showed a dose–response relationship with a consistently elevated activity of ornithine decarboxylase at flux densities > 4 μT (Mullins et al., 1999). However, two studies designed to replicate this result with L929 cells from the same or a different source, failed to find a significant change in the activity of ornithine decarboxylase as a result of exposure to electric or magnetic fields (Azadniv *et al.*, 1995; Cress *et al.*, 1999).

Three mammalian tumour cell lines (human promyelocytic leukaemia HL60 cells, mouse ascites tumour ELD cells and mouse teratocarcinoma F9 cells) were used to determine the effects of exposure to magnetic fields on ornithine decarboxylase gene expression. All cell lines showed elevated levels of activity of the enzyme when exposed during culture to a 50-Hz, 30-µT vertical sinusoidal magnetic field for 24, 48 or 72 h. The increase ranged from about 20% in HL60 cells to up to five- to sixfold in ELD cells compared to the controls. The effect was stronger at later stages of growth, when the inherent activity of ornithine decarboxylase is lower (Mattsson & Rehnholm, 1993).

Two lymphoblastic leukaemia cell lines of human origin, Jurkat cells and CEM-CM3 cells, were exposed to horizontal or vertical magnetic fields (50 Hz, 0.10 mT). Exposure to the vertical magnetic field for 3 h or 3 days increased the activity of ornithine decarboxylase in the Jurkat cells by 77% and 47%, respectively. Only a small effect of exposure to the horizontal magnetic field was seen, perhaps due to the lower intensity of the induced electric field. However, the CEM-CM3 cells did not respond to either type of exposure (Valtersson *et al.*, 1997).

Growth factors and differentiation

The effects of pulsed electric and magnetic fields on mitogen-stimulated lymphocytes from aged human volunteers (mean age, 88 years) were studied by measuring the production of interleukin-2 and the expression of interleukin-2 receptor in these cells. The pulse duration of the magnetic field was about 2 ms, the repetition rate 50 Hz, the intensity 2.5 mT, and the average time variation was of the order of 1 T/s. The maximum induced electric field was estimated to be 0.02 mV/cm. Control cultures were maintained in the same incubator in a position where no electric or magnetic field was detectable. [The Working Group noted the close proximity of the control and the exposed samples.] Cultures were exposed for 18 h for evaluation of interleukin-2 receptor-positive cells and percentage of T-activated lymphocytes, and for 24 and 48 h for examining the production of interleukin-2. In exposed cultures that showed increased [^{3}H]thymidine incorporation compared with unexposed controls, the production of interleukin-2 was lower, but the percentages of interleukin-2 receptor-positive cells and of T-activated lymphocytes were increased (Cossarizza *et al.*, 1989c).

A study of the ability of nerve growth factor-stimulated PC-12 cells, derived from the rat adrenal medulla, to produce neurites under a variety of conditions of exposure to magnetic fields, was designed to establish those field parameters critical for production of biological effects. Twenty-three hours of exposure both to sub-optimal concentrations of nerve growth factor and to a flux density series of vertical, 45-Hz magnetic fields demonstrated reduced neurite outgrowth at flux densities between 5 and 10 µT, where the inhibition reached a plateau. The cell response at the periphery of culture dishes of different diameter was identical to that at the centre of the dishes indicating that the induced electric current was not responsible for the effect (Blackman et al., 1993). The frequency response, which was tested from 15 Hz–70 Hz for each of six flux densities (3.5–9.0 µT), displayed frequency-specific profiles of inhibition of neurite outgrowth (Blackman et al., 1995). Trillo et al. (1996) showed that different specific flux densities of static and alternating magnetic fields (at 30 Hz, 0.79–2.05 µT alternating and 1.97 µT static; at 45 Hz, 0.29–4.11 µT alternating and 2.96 µT static) could produce a characteristic, but slightly different inhibition response, in which a narrow flux density region around the value that had produced maximal inhibition, displayed no inhibition. The effects observed using 45-Hz fields were not seen when the static field was reduced to 1.97 µT. Blackman et al. (1999) tested the frequency dependence of the findings of Trillo et al. (1996) using flux densities for maximum effects at 45 Hz and observed a maximal inhibition at 45 Hz with lesser inhibition at 42.5 and 47.5 Hz, and no inhibition at 40 and 50 Hz. Blackman et al. (1996) showed that the neurite outgrowth response changed from field-induced inhibition to enhancement when the static magnetic field was changed through a series of angles from parallel to perpendicular to the alternating magnetic field. In a study based on the work of Blackman and colleagues, McFarlane et al. (2000) observed a field-induced (50 Hz, 4–8 µT) inhibition (~ 22%) of neurite outgrowth in PC-12 cells cultured in 15% serum (weakly differentiating conditions) and enhancement (~ 17%) of the outgrowth in cells cultured in 4% serum (strongly differentiating conditions). No significant changes were observed at higher or lower flux densities.

A Friend erythroleukaemia cell line that can be chemically induced to differentiate was used to determine whether magnetic fields could alter cell proliferation and differentiation in a manner similar to that of a chemical tumour promoter. Exposure of this cell line to 60-Hz fields resulted in a dose-dependent inhibition of differentiation, with a maximal inhibition of 40% at 4 µT. Exposure at 2.5 µT caused a 20% inhibition while a 1-µT field was ineffective. At flux densities in the range of 0.1–1 mT, cell proliferation was stimulated up to 50% above that of sham-treated cells. The activity of telomerase, a marker of undifferentiated cells, decreased 100-fold when the cells were induced to differentiate under sham conditions, but only 10-fold when the cells were exposed to a 50-µT magnetic field. In summary, exposure to ELF electric and magnetic fields appears to partially block the differentiation of Friend erythroleukaemia cells, and this results in a larger population of cells remaining in the undifferentiated, proliferative state, which is similar to results obtained with chemical tumour promoters (Chen et al., 2000).

Intercellular communication

Ubeda et al. (1995) observed that the increased gap-junctional intercellular communication induced in C3H10T1/2 mouse embryo cells by physiological concentrations of melatonin, could be completely eliminated when the cells were exposed for 1 h to vertical, 50-Hz, sinusoidal magnetic fields at 1.6 mT.

Li *et al.* (1999) exposed Chinese hamster lung cells to the tumour promoter TPA alone or in combination with a 50-Hz magnetic field. Combined treatment with 5 ng/mL TPA for the last hour of a 24-h period of exposure to magnetic flux densities of 0.2, 0.4 or 0.8 mT, significantly inhibited gap-junctional intercellular communication compared with TPA treatment alone. The inhibition was dependent on the flux density.

Gap-junctional intercellular communication was also studied in clone 9 cells treated with 2.5 mM chloral hydrate for 24 h prior to exposure to a 45-Hz, 23.8-μT magnetic field, in parallel with a 36.6-μT static magnetic field for 40–45 min. There was no statistically significant effect of exposure to the magnetic field on gap-junctional intercellular communication (Griffin *et al.*, 2000).

Cell transformation

In a study using anchorage-independent growth as an index, mouse epidermal JB6 cells (clone 41) were exposed to magnetic fields (60 Hz, 1, 10 and 100 μT) for eight or 14 days, resulting in a 1.2–3.2-fold increase in colony-forming efficiency of trans-formants (West *et al.*, 1996). In contrast, in a co-culture of C3H10T1/2 mouse fibro-blasts and mutant daughter 10e cells, intermittent exposure to an ELF magnetic field (60 Hz, 100 μT, 1 h, four times a day) for 28 days caused no increase in focus formation. In the same culture system, however, concomitant exposure to the magnetic field and treatment with TPA (10–100 ng/mL) caused a significant increase in focus formation (by an average of 150%) compared with that in cell cultures treated with TPA alone (Cain *et al.*, 1993).

There are reports suggesting that ELF electric and magnetic fields have no effect on cell transformation. In a soft-agar assay, 60-Hz magnetic fields of 0.01, 0.1, 1.0 or 1.1 mT flux density did not induce anchorage-independent growth of mouse epidermal JB6 cells, enhance TPA-induced transformation, increase the maximum number of transformed colonies or produce a shift in the dose–response curve (Saffer *et al.*, 1997). Similarly, in another study, continuous exposure to a magnetic field of 60 Hz, 200 μT for 24 h showed no effect in two transformation systems (Syrian hamster embryo cells and CH310T1/2 clone 8) with or without post-treatment with TPA (Balcer-Kubiczek *et al.*, 1996).

Cultures of primary Syrian hamster dermal cells were continuously exposed to power-line frequency magnetic fields of 10, 100 and 1000 μT for 60 h, with or without prior exposure to an immortalizing dose (1.5 Gy) or a non-immortalizing dose (0.5 Gy) of ionizing radiation. Exposure to the magnetic field alone did not immortalize these cells at a detectable frequency (1×10^{-7} or higher) or enhance the frequency of immor-talization induced by ionizing radiation (Gamble *et al*, 1999). The lack of cell-trans-forming activity of pulsed electric and electric and magnetic fields had previously been shown in a BALB/3T3 cell transformation assay (Jacobson-Kram *et al.*, 1997).

CH310T1/2 clone 8 cells were exposed for 24 h to strong magnetic fields (5–400 mT, 60 Hz) to investigate a change in transformation frequency as analysed by focus formation. No significant increase in transformation frequency was seen after exposure to the magnetic fields alone, but exposure to 3 Gy X-rays followed by exposure to the magnetic fields for 24 h decreased the transformation frequency in comparison with exposure to X-rays alone. In addition, long-term exposure for six weeks at 60 Hz, 5 mT significantly suppressed both spontaneous and X-ray-induced transformation (Miyakoshi *et al.*, 2000c).

4.5   Mechanistic considerations

Limited data are available on the effects of static fields alone. Therefore, the following considerations of a possible mechanism will address primarily ELF electric and magnetic fields.

It is widely agreed that certain alterations in the genetic structure of the cell are causally related to cancer. There is little experimental or theoretical evidence that mutations could be directly caused by ELF magnetic fields. The results of most genetic toxicology studies of ELF magnetic fields have been negative. However, a single laboratory has reported that exposure of human cells to extremely high ELF flux densities (≥ 400 mT), which far exceeds the field intensities encountered in residential or occupational environments, induces sister chromatid exchange, chromatid-type aberrations and mutation in the *HPRT* gene.

It is also relevant to ask whether ELF electric and magnetic fields have effects similar to those of known 'non-genotoxic' carcinogens, 'tumour promoters' or 'cocarcinogens', i.e. agents that seem to enhance cancer by a mechanism other than that of direct DNA damage.

There is little evidence that ELF electric or magnetic fields can cause malignant transformation of cells in culture. There have been relatively few studies of the effects of ELF electric and magnetic fields on DNA repair or genomic stability in mammalian cells, and the results are inconclusive. There is some evidence for an effect of magnetic fields on cellular kinetics: few studies using in-vitro systems have shown enhancement of apoptosis. The results of studies on cell proliferation using a variety of exposure conditions and cell types have varied from inhibition to enhancement. The cell-proliferation response to physical and chemical factors has also been reported to be altered by exposure to ELF magnetic fields. The available experimental evidence suggests that ELF electric or magnetic fields are not cytotoxic.

The effects of ELF electric and magnetic fields on signal transduction have been reported to include changes in intracellular calcium levels and protein phosphorylation, but a number of studies have reported negative findings. These results cannot be used to identify plausible cancer-related pathways.

Several research groups have reported changes in gene expression resulting from exposure to ELF magnetic fields. However, other studies have failed to replicate many of these results.

In relation to both the genotoxic and non-genotoxic cellular and molecular end-points that have been studied, many of the data concern changes evoked following a combined exposure: that is, experiments involving (electric or) magnetic fields together with other agents. There is only weak evidence that ELF magnetic fields potentiate the effects of chemical agents, or ionizing or ultraviolet radiation.

The risk for cancer can also be enhanced through systemic effects in humans or animals. For example, it has been suggested that the hormone melatonin may suppress mammary cancer through hormonal mechanisms; anticarcinogenic effects through free-radical scavenging have also been proposed. Several human and animal studies have investigated possible suppressing effects of ELF magnetic fields on melatonin, but the results are equivocal. Although most experimental studies of the possible immunotoxicity of exposure to magnetic fields have yielded negative results, effects on T-cell proliferation capacity in animals have been reported. However, the effects are inconsistent.

5. Summary of Data Reported and Evaluation

5.1   Exposure data

Static electric and magnetic fields arise from both natural and man-made sources, whereas electric and magnetic fields in the extremely low-frequency (ELF) range (3–3000 Hz) are mostly associated with man-made sources. These are numerous and include electric power systems, electric and electronic appliances and industrial devices. Environmental levels of ELF fields are very low. Exposure levels for the general population are typically 5–50 V/m for electric fields and 0.01–0.2 µT for magnetic fields. Considerably higher exposure occurs for shorter durations and in some occupational settings.

It should be noted that the earth's magnetic field (25–65 µT, from equator to poles) is a static field to which everyone is exposed.

Measurements of electric and magnetic fields are used to characterize sources and levels of exposure to humans. The capabilities of instruments to measure such fields have advanced in recent years, particularly for magnetic fields. In addition to simple, easy-to-use hand-held survey instruments, there are now portable personal exposure meters capable of recording and describing the statistical, threshold, frequency and waveform characteristics of magnetic field exposure. The limiting factor in exposure assessment is not instrumentation but the lack of a consensus as to what exposure characteristics should be measured that are biologically relevant.

Computational methods are available to calculate fields and their parameters for instrument calibration, laboratory exposure systems and certain categories of indoor and outdoor sources. The difficulties in the use of computation methods to characterize exposure to magnetic fields include the lack of complete knowledge as to the magnitude, direction and location of all relevant current flows on conductors. Such difficulties pose special challenges to the use of calculations of ELF magnetic fields to estimate historical exposure from power lines. Where computational methods are used to calculate human exposure in epidemiological studies, it is desirable to understand the overall uncertainty in the calculated values.

In order to understand the effects of electric and magnetic fields on animals and humans, their electrical properties have to be considered. Static magnetic fields, which are not attenuated by the organism, can exert forces on moving charges, orient magnetic structures and affect the energy levels of some molecules. Static and ELF electric fields are greatly attenuated inside the body.

Exposure to ELF electric and magnetic fields results in induction of electric fields and associated currents in tissues. The magnitudes and spatial patterns of these fields depend on whether the external field is electric or magnetic, its characteristics (e.g. frequency, magnitude, orientation and waveform) and the size, shape and electrical properties of the exposed body. This is a basic physical mechanism for interaction of ELF magnetic fields with tissues. The induced electric field increases with the frequency of the external field and the size of the object. A well-established effect of induced fields above a threshold level is the stimulation of excitable cells. Typical residential exposure results in very small induced electric fields, while some occupational exposure and exposure directly under very high-voltage power lines may result in electric fields of the order of 1 mV/m in some tissues. Non-perceptible contact currents under some conditions are calculated to produce electric fields exceeding 1 mV/m in the bone marrow of a child. Residential levels of ELF electric and magnetic fields produce much lower fields in tissues.

Beyond this well-established interaction mechanism, a number of hypotheses have been advanced: radical pair mechanisms, ion charge-to-mass resonance mechanisms, stochastic resonance, action on biogenic magnetite, etc. Theoretical and experimental evidence for the relevance of these mechanisms is being sought actively.

There are well established in-vivo and in-vitro exposure systems that can provide electric fields of up to the order of 150 kV/m and ELF magnetic fields up to 2 mT. Magnetostatic fields up to 5.0 T can be produced in the laboratory.

5.2   Human carcinogenicity data

Effects in children

Since the first report suggesting an association between residential ELF electric and magnetic fields and childhood leukaemia was published in 1979, dozens of increasingly sophisticated studies have examined this association. In addition, there have been numerous comprehensive reviews, meta-analyses, and two recent pooled analyses. In one pooled analysis based on nine well conducted studies, no excess risk was seen for exposure to ELF magnetic fields below 0.4 μT and a twofold excess risk was seen for exposure above 0.4 μT. The other pooled analysis included 15 studies based on less restrictive inclusion criteria and used 0.3 μT as the highest cut-point. A relative risk of 1.7 for exposure above 0.3 μT was reported. The two studies are closely consistent. In contrast to these results for ELF magnetic fields, evidence that electric fields are associated with childhood leukaemia is inadequate for evaluation.

No consistent relationship has been seen in studies of childhood brain tumours or cancers at other sites and residential ELF electric and magnetic fields. However, these studies have generally been smaller and of lower quality.

The association between childhood leukaemia and high levels of magnetic fields is unlikely to be due to chance, but it may be affected by bias. In particular, selection bias may account for part of the association. Case-control studies which relied on in-home measurements are especially vulnerable to this bias, because of the low response rates in many studies. Studies conducted in the Nordic countries which relied on historical calculated magnetic fields are not subject to selection bias, but suffer from very low numbers of exposed subjects. There have been dramatic improvements in the assessment of exposure to electric and magnetic fields over time, yet all of the studies are subject to misclassification. Non-differential misclassification of exposure (similar degrees of misclassification in cases and controls) is likely to result in bias towards the null. Bias due to unknown confounding factors is very unlikely to explain the entire observed effect. However, some bias due to confounding is quite possible, which could operate in either direction. It cannot be excluded that a combination of selection bias, some degree of confounding and chance could explain the results. If the observed relationship were causal, the exposure-associated risk could also be greater than what is reported.

Numerous studies of the relationship between electrical appliance use and various childhood cancers have been published. In general, these studies provide no discernable pattern of increased risks associated with increased duration and frequency of use of appliances. Since many of the studies collected information from interviews that took place many years after the time period of etiological interest, recall bias is likely to be a major problem.

Studies on parental occupational exposure to ELF electric and magnetic fields in the preconceptional period or during gestation are methodologically weak and the results are not consistent.

Effects in adults

Residential exposure

While a number of studies are available, reliable data on adult cancer and residential exposure to ELF electric and magnetic fields, including the use of appliances, are sparse and methodologically limited. None of the studies reported so far has included long-term or personal measurements. Although there have been a considerable number of reports, a consistent association between residential exposure and adult leukaemia and brain cancer has not been established.

For breast cancer and other cancers, the existing data are not adequate to test for an association with exposure to electric or magnetic fields.

Occupational exposure

Studies conducted in the 1980s and early 1990s pointed to a possible increased risk of leukaemia, brain tumours and male breast cancer in jobs with presumed exposure to ELF electric and magnetic fields above average levels. The interpretation of these studies was difficult mainly due to methodological limitations and lack of appropriate exposure measurements. Also, a bias towards publication of positive findings could not be excluded.

Several large studies conducted in the 1990s of both leukaemia and brain cancer made use of improved methods for individual assessment of occupational exposure to magnetic fields, and to potential occupational confounders, mainly through the combined use of systematic workplace measurements, individual job history descriptions, and the development of associated job–exposure matrices. However, because the exposure within occupational groups is highly variable, job–exposure matrices do not eliminate all uncertainties regarding the workers' exposure levels. Some of these studies reported increased cancer risk for intermediate or high magnetic field exposure categories. There was no consistent finding across studies of an exposure–response relationship and no consistency in the association with specific sub-types of leukaemia or brain tumour. Evidence for cancers at other sites was not adequate for evaluation.

Although the assessment of exposure to electric fields is difficult, these fields have been measured occasionally in populations of workers using individual exposure meters. Across the studies, no consistent association of electric field strengths with any particular malignancy was noted.

5.3   Animal carcinogenicity data

Four long-term bioassays have been published in which the potential oncogenicity in experimental animals of exposure to ELF magnetic fields was evaluated in over 40 different tissues using standard chronic toxicity testing designs. Three of the studies were conducted in rats (two in both sexes including one with restricted histopathological evaluation, and one in females only) and one in mice (males and females). Three of the four studies (two rat studies and one mouse study) provide no evidence that exposure to ELF magnetic fields causes cancer in any target organ. The fourth found an increased incidence of thyroid C-cell tumours (adenomas plus carcinomas) in male rats exposed to ELF magnetic fields at two intermediate flux densities, which did not demonstrate a dose–response relationship, and a marginal increase at the highest flux density. In the lowest-exposure group, thyroid C-cell carcinomas significantly exceeded control response and were above the historical control range. Thyroid C-cell carcinomas were not seen in male mice, female mice or female rats exposed chronically to ELF magnetic fields in these oncogenicity bioassays.

A long-term oncogenicity bioassay of more limited design that was conducted to identify possible effects of exposure to ELF magnetic fields on the induction of leukaemia and lymphoma or of brain cancer in mice generated negative results.

Two multistage carcinogenesis studies combining exposure to N-methyl-N-nitroso-urea with exposure to static or 50-Hz magnetic fields were performed in the same laboratory using an uncharacterized outbred rat strain. The first study demonstrated an increase in mammary tumour incidence with exposure to the fields regardless of exposure to N-methyl-N-nitrosourea. The second study showed no effect at similar exposure levels.

Eleven multistage carcinogenesis studies combining exposure to 7,12-dimethyl-benz[a]anthracene with exposure to 50- or 60-Hz magnetic fields were performed in three different laboratories. One laboratory performed six 13-week studies and one 27-week study aimed at addressing exposure–response relationships for different magnitudes of exposure to magnetic fields. These studies reported significant increases in mammary tumour incidence at higher exposure levels. A pooled analysis of exposure–response from these studies yielded an average slope significantly different from zero. A second laboratory conducted three studies (two of which were considered inadequate to assess tumour incidence) to replicate these findings at the highest field strengths, but saw no enhancement of mammary tumorigenesis by exposure to ELF magnetic fields in one study, in which the sham control incidence was low enough to detect an increase. In the two other studies, high incidences of mammary tumours in sham controls limited comparisons to possible increases in tumour multiplicity; none were found. The third laboratory studied the impact of intermittent exposure to magnetic fields and saw no changes in tumour incidence or tumour multiplicity in either of two experiments.

Eight studies were performed in five different laboratories on promotion and/or co-promotion of skin tumorigenesis by 50- or 60-Hz magnetic fields using conventional mouse strains. The results of these studies were generally negative. However, a suggestion of accelerated progression to malignancy was observed in one study and a change in tumour multiplicity was observed in another. There was no consistent pattern of response in these studies, which were of effectively equivalent design. One study using a transgenic mouse model demonstrated an acceleration of skin tumorigenesis by ELF magnetic fields.

Three studies have been performed using the enzyme-altered liver foci model in rats or mice to determine tumour promoting and co-promoting effects of 50-Hz magnetic fields (0.5–500 µT). No enhancement of liver foci by magnetic field exposure was reported in two studies in rats. In the third study which used ionizing radiation with and without exposure to magnetic fields, the incidence of basophilic liver foci was significantly increased in exposed mice. This finding was not associated with a significant increase in liver cancer incidence.

Multistage studies have been carried out in both mice (conventional and transgenic strains) and rats to evaluate the effects of ELF magnetic fields on the development of leukaemia and lymphoma. In no study did exposure to ELF magnetic fields cause an increased incidence of leukaemia or lymphoma.

One study was performed to identify possible promoting effects of ELF magnetic field exposure on the induction of neurogenic tumours. The results of this study showed no enhancement of neurogenic tumour induction.

5.4   Other relevant data

Reproductive effects in humans and animals

Taken as a whole, the results of human studies do not establish an association of adverse reproductive outcomes with exposure to ELF electric and magnetic fields. Such adverse outcomes have been reported in a few studies, particularly at higher field intensities and in people exposed for longer durations. With exposures from video display terminals, a greater number of studies have been performed and these generally found no adverse reproductive effects.

Experiments with many different mammalian and non-mammalian experimental models consistently indicate lack of adverse effects on reproduction and development from exposure to strong static magnetic (0.25–1.0 T) and ELF electric (up to 150 kV/m) fields. Static magnetic fields with high spatial gradients and those mixed with alternating fields have been reported to affect embryonic development in frogs and mice, although the number of studies is small.

Prenatal exposure to ELF magnetic fields generally does not result in adverse effects on reproduction and development in mammals. When effects are observed, they usually consist of minor developmental anomalies. Non-mammalian classes of animals (fish, frogs, birds) show inconsistent effects of ELF electric and magnetic fields on development (including increased malformations).

Other effects in humans

Due to the small number of immunological and haematological studies in humans and very small sample sizes within the reported studies, no health-related conclusions can be drawn from the data on immunological and haematological effects after exposure to ELF electric and magnetic fields.

In humans, the principal element of neuroendocrine response to exposure to ELF electric and magnetic fields that has been investigated is the circadian production and release of melatonin. No effect on melatonin was seen following night-time exposure of human volunteers to 50 or 60-Hz magnetic fields under controlled laboratory conditions. In contrast, a small reduction in melatonin concentration has been observed in occupational and residential environments, but it is difficult to distinguish between effects of the magnetic field and those of other environmental factors.

Apart from established perceptual responses in humans to ELF electric fields at levels of tens of kilovolts per meter and the occurrence of magnetophosphenes (faint, flickering visual sensations) in response to exposure to relatively strong ELF magnetic fields (> 10 mT at 20 Hz), few behavioural effects of exposure to ELF electric and magnetic fields have been observed. Changes in electroencephalograms, cognition, mood, sleep electrophysiology and cardiac response tend to be few, subtle and transitory when they do occur during exposure. The evidence from epidemiological studies of residential and occupational exposure to ELF electric and magnetic fields in relation to the incidence of neurodegenerative disease, depression and suicide and cardiovascular disease is generally weak and inconsistent.

Other effects in animals

Studies to evaluate immune function and host resistance in animals have given negative effects for exposure to ELF electric and magnetic fields. In-vitro exposure of immune system cells generally did not cause changes in proliferation capacity.

Apart from occasional changes in some haematological parameters in one rat study, no consistent effects on blood formation were seen in experimental animals or their offspring exposed to either static magnetic fields or to 50- or 60-Hz electric and/or magnetic fields.

Most animal studies of endocrine function concern the pineal gland and melatonin, because of concerns related to cancer. Fewer studies have been carried out on the effects of exposure to ELF electric and magnetic fields on the pituitary hormones or those of other endocrine glands.

Some, but not all, studies of the effects of 50- or 60-Hz electric and magnetic fields in rodents show a reduction in pineal and/or serum melatonin concentrations. Differences in response have been reported for linearly polarized compared with circularly polarized magnetic fields. No convincing effect on melatonin concentrations has been seen in non-human primates chronically exposed to 50- or 60-Hz electric or magnetic fields.

With the possible exception of short-term stress (duration of minutes) following the onset of exposure to ELF electric fields at levels significantly above perception thresholds, no consistent effects have been seen in the stress-related hormones of the pituitary–adrenal axis in a variety of mammalian species.

Animals can perceive ELF electric fields (threshold 3–35 kV/m) and respond with activity changes or aversion. Such responses are generally not observed with magnetic fields.

Although exposure to magnetic fields has been reported to influence spatial learning and memory in rodents, it appears that no long-term behavioural deficits occur due to exposure to static or ELF electric and magnetic fields.

Genetic and related effects

A few studies on genetic effects have examined chromosomal aberrations and micronuclei in lymphocytes from workers exposed to ELF electric and magnetic fields. In these studies, confounding by genotoxic agents (tobacco, solvents) and comparability between the exposed and control groups are of concern. Thus, the studies reporting an increased frequency of chromosomal aberrations and micronuclei are difficult to interpret.

Many studies have been conducted to investigate the effects of ELF magnetic fields on various genetic end-points. Although increased DNA strand breaks have been reported in brain cells of exposed rodents, the results are inconclusive; most of the studies show no effects in mammalian cells exposed to magnetic fields alone at levels below 50 mT. However, extremely strong ELF magnetic fields have caused adverse genetic effects in some studies. In addition, several groups have reported that ELF magnetic fields enhance the effects of known DNA- and chromosome-damaging agents such as ionizing radiation.

The few animal studies on cancer-related non-genetic effects are inconclusive. Results on the effects on in-vitro cell proliferation and malignant transformation are inconsistent, but some studies suggest that ELF magnetic fields affect cell proliferation and modify cellular responses to other factors such as melatonin. An increase in apoptosis following exposure of various cell lines to ELF electric and magnetic fields has been reported in several studies with different exposure conditions. Numerous studies have investigated effects of ELF magnetic fields on cellular end-points associated with signal transduction, but the results are not consistent.

5.5   Evaluation

There is *limited evidence* in humans for the carcinogenicity of extremely low-frequency magnetic fields in relation to childhood leukaemia.

There is *inadequate evidence* in humans for the carcinogenicity of extremely low-frequency magnetic fields in relation to all other cancers.

There is *inadequate evidence* in humans for the carcinogenicity of static electric or magnetic fields and extremely low-frequency electric fields.

There is *inadequate evidence* in experimental animals for the carcinogenicity of extremely low-frequency magnetic fields.

No data relevant to the carcinogenicity of static electric or magnetic fields and extremely low-frequency electric fields in experimental animals were available.

Overall evaluation

Extremely low-frequency magnetic fields are *possibly carcinogenic to humans (Group 2B)*.

Static electric and magnetic fields and extremely low-frequency electric fields are *not classifiable as to their carcinogenicity to humans (Group 3)*.

6. References

Adair, R.K. (1991) Constraints on biological effects of weak extremely-low-frequency electromagnetic fields. *Phys. Rev. A*, **43**, 1039–1048

Adair, R.K. (1992) Criticism of Lednev's mechanism for the influence of weak magnetic fields on biological systems. *Bioelectromagnetics*, **13**, 231–235

Adair, R.K. (1993) Effect of ELF magnetic fields on biological magnetite. *Bioelectromagnetics*, **14**, 1–4

Adair, R.K. (1998) A physical analysis of the ion parametric resonance model. *Bioelectromagnetics*, **19**, 181–191

Adair, R.K. (1999) Effects of very weak magnetic fields on radical pair reformation. *Bioelectromagnetics*, **20**, 255–263

Agnew, D.A. (1992) Measurement of ELF fields. In: Greene, M.W., ed., *Non-Ionizing Radiation (Proceedings of the 2nd International Non-Ionizing Radiation Workshop, Vancouver, British Columbia, Canada, May 10–14, 1992)*, London, International Radiation Protection Association, pp. 368–382

Ahlbom, A., Day, N., Feychting, M., Roman, E., Skinner, J., Dockerty, J., Linet, M., McBride, M., Michaelis, J., Olsen, J.H., Tynes, T. & Verkasalo, P.K. (2000) A pooled analysis of magnetic fields and childhood leukaemia. *Br. J. Cancer*, **83**, 692–698

Åkerstedt, T., Arnetz, B., Ficca, G., Paulsson, L.-E. & Kallner, A. (1999) A 50-Hz electromagnetic field impairs sleep. *J. Sleep Res.*, **8**, 77–81

Al-Akhras, M.-A., Elbetieha, A., Hasan, M.-K., Al-Omari, I., Darmani, H. & Albiss, B. (2001) Effects of low-frequency magnetic field on fertility of adult male and female rats. *Bioelectromagnetics*, **22**, 340–344

Alfredsson, L., Hammar, N. & Karlehagen, S. (1996) Cancer incidence among male railway engine-drivers and conductors in Sweden, 1976–90. *Cancer Causes Control*, **7**, 377–381

Anderson, L.E., Boorman, G.A., Morris, J.E., Sasser, L.B., Mann, P.C., Grumbein, S.L., Hailey, J.R., McNally, A., Sills, R.C. & Haseman, J.K. (1999) Effect of 13 week magnetic field exposures on DMBA-initiated mammary gland carcinomas in female Sprague-Dawley rats. *Carcinogenesis*, **20**, 1615–1620

Andersson, B., Berg, M., Arnetz, B.B., Melin, L., Langlet, I. & Liden, S. (1996) A cognitive-behavioral treatment of patients suffering from 'electric hypersensitivity'. Subjective effects and reactions in a double-blind provocation study. *J. occup. environ. Med.*, **38**, 752–758

Anisimov, V.N., Zhukova, O.V., Beniashvili, D.S., Bilanishvili, V.G., Menabde, M.Z. & Gupta, D. (1996) [Effect of the light regime and electromagnetic fields on mammary carcinogenesis in female rats.] *Biofizika*, **41**, 807–814 (in Russian)

Antonopoulos, A., Yang, B., Stamm, A., Heller, W.-D. & Obe, G. (1995) Cytological effects of 50 Hz electromagnetic fields on human lymphocytes in vitro. *Mutat. Res.*, **346**, 151–157

Auvinen, A., Linet, M.S., Hatch, E.E., Kleinerman, R.A., Robison, L.L., Kaune, W.T., Misakian, M., Niwa, S., Wacholder, S. & Tarone, R.E. (2000) Extremely low-frequency magnetic fields and childhood acute lymphoblastic leukemia: An exploratory analysis of alternative exposure metrics. *Am. J. Epidemiol.*, **152**, 20–31

Azadniv, M., Klinge, C.M., Gelein, R., Carstensen, E.L., Cox, C. & Brayman, A.A. (1995) A test of the hypothesis that a 60-Hz magnetic field affects ornithine decarboxylase activity in mouse L929 cells *in vitro. Biochem. biophys. Res. Commun.*, **214**, 627–631

Babbitt, J.T., Kharazi, A.I., Taylor, J.M., Bonds, C.B., Mirell, S.G., Frumkin, E., Zhuang, D. & Hahn, T.J. (2000) Hematopoietic neoplasia in C57BL/6 mice exposed to split-dose ionizing radiation and circularly polarized 60 Hz magnetic fields. *Carcinogenesis*, **21**, 1379–1389

Bailey, W.H., Su, S.H., Bracken, T.D. & Kavet, R. (1997) Summary and evaluation of guidelines for occupational exposure to power frequency electric and magnetic fields. *Health Phys.*, **73**, 433–453

Bakos, J., Nagy, N., Thuróczy, G., Szabo, L.D. (1995) Sinusoidal 50 Hz, 500 microT magnetic field has no acute effect on urinary 6-sulphatoxymelatonin in Wistar rats. *Bioelectromagnetics*, **16**, 377–380

Bakos, J., Nagy, N., Thuróczy, G. & Szabo, L.D. (1997) Urinary 6-sulphatoxymelatonin excretion is increased in rats after 24 hours of exposure to vertical 50 Hz, 100 microT magnetic field. *Bioelectromagnetics*, **18**, 190–192

Bakos, J., Nagy, N. & Thuróczy, G. (1999) Urinary 6-sulphatoxymelatonin excretion of rats is not changed by 24 hours of exposure to a horizontal 50-Hz, 100-μT magnetic field. *Electro-Magnetobiol.*, **18**, 23–31

Balcer-Kubiczek, E.K., Zhang, X.-F., Harrison, G.H., McCready, W.A., Shi, Z.-M., Han, L.-H., Abraham, J.M., Ampey, L.L., III, Meltzer, S.J., Jacobs, M.C. & Davis, C.C. (1996) Rodent cell transformation and immediate early gene expression following 60-Hz magnetic field exposure. *Environ. Health Perspect.*, **104**, 1188–1198

Baraton, P. & Hutzler, B. (1995) *Magnetically-induced Currents in the Human Body* (Technical Report MISC TTA1), Geneva, International Electrotechnical Commission

Baraton, P., Cahouet, J. & Hutzler, B. (1993) *Three-dimensional Computation of the Electric Fields induced in a Human Body by Magnetic Fields* (Technical Report 93NV00013), Ecuelles, Electricité de France

Baris, D. & Armstrong, B. (1990) Suicide among electric utility workers in England and Wales. *Br. J. ind. Med.*, **47**, 788–789

Baris, D., Armstrong, B.G., Deadman, J. & Thériault, G. (1996a) A case cohort study of suicide in relation to exposure to electrical and magnetic fields among electrical utility workers. *Occup. environ. Med.*, **53**, 17–24

Baris, D., Armstrong, B.G., Deadman, J. & Thériault, G. (1996b) A mortality study of electrical utility workers in Quebec. *Occup. environ. Med.*, **53**, 25–31

Barker, A.T., Dixon, R.A., Sharrard, W.J. & Sutcliffe, M.L. (1984) Pulsed magnetic field therapy for tibial non-union. Interim results of a double-blind trial. *Lancet*, **i**, 994–996

Baroncelli, P., Battisti, S., Checcucci, A., Comba, P., Grandolfo, M., Serio, A. & Vecchia, P. (1986) A health examination of railway high-voltage substation workers exposed to ELF electromagnetic fields. *Am. J. ind. Med.*, **10**, 45–55

Barregård, L., Jarvholm, B. & Ungethüm, E. (1985) Cancer among workers exposed to strong static magnetic fields (Letter to the Editor). *Lancet*, **ii**, 892

Bassen, H., Litovitz, T., Penafiel, M. & Meister, R. (1992) ELF in vitro exposure systems for inducing uniform electric and magnetic fields in cell culture media. *Bioelectromagnetics*, **13**, 183–198

Basser, P.J. & Roth, B.J. (1991) Stimulation of a myelinated nerve axon by electromagnetic induction. *Med. Biol. Eng. Comput.*, **29**, 261–268

Bassett, C.A. (1989) Fundamental and practical aspects of therapeutic uses of pulsed electro-magnetic fields (PEMFs). *Crit. Rev. Biomed. Eng.*, **17**, 451–529

Bassett, C.A., Mitchell, S.N. & Gaston, S.R. (1981) Treatment of ununited tibial diaphyseal fractures with pulsing electromagnetic fields. *J. Bone Joint Surg. Am.*, **63**, 511–23

Bauchinger, M., Hauf, R., Schmid, E. & Dresp, J. (1981) Analysis of structural chromosome changes and SCE after occupational long-term exposure to electric and magnetic fields from 380 kV-systems. *Radiat. environ. Biophys.*, **19**, 235–238

Baum, A., Mevissen, M., Kamino, K., Mohr, U. & Löscher, W. (1995) A histopathological study on alterations in DMBA-induced mammary carcinogenesis in rats with 50 Hz, 100 µT magnetic field exposure. *Carcinogenesis*, **16**, 119–125

Bawin, S.M. & Adey, W.R. (1976) Sensitivity of calcium binding in cerebral tissue to weak environmental electric fields oscillating at low frequency. *Proc. natl Acad. Sci. USA*, **73**, 1999–2003

Belanger, K., Leaderer, B., Hellenbrand, K., Holford, T.R., McSharry, J., Power, M.E. & Bracken, M.B. (1998) Spontaneous abortion and exposure to electric blankets and heated water beds. *Epidemiology*, **9**, 36–42

Beniashvili, D.S., Bilanishvili, V.G. & Menabde, M.Z. (1991) Low-frequency electromagnetic radiation enhances the induction of rat mammary tumors by nitrosomethyl urea. *Cancer Lett.*, **61**, 75–79

Beniashvili, D.S., Bilanishvili, V.G., Menabde, M.Z., Gupta, D. & Anisimov, V.N. (1993) [Modifying effect of light and electromagnetic field on development of mammary tumours induced by *N*-methylnitroso-*N*-urea in female rats.] *Vopr. Onkol.*, **39**, 52–60 (in Russian)

Berman, E., Chacon, L., House, D., Koch, B.A., Koch, W.E., Leal, J., Løvtrup, S., Mantiply, E., Martin, A.H., Martucci, G.I., Mild, K.H., Monahan, J.C., Sandström, M., Shamsaifar, K., Tell, R., Trillo, M.A., Ubeda, A. & Wagner, P. (1990) Development of chicken embryos in a pulsed magnetic field. *Bioelectromagnetics*, **11**, 169–187

Bernhardt, J.H. (1988) The establishment of frequency dependent limits for electric and magnetic fields and evaluation of indirect effects. *Radiat. environ. Biophys.*, **27**, 1–27

Bianchi, N., Crosignani, P., Rovelli, A., Tittarelli, A., Carnelli, C.A., Rossitto, F., Vanelli, U., Porro, E. & Berrino, F. (2000) Overhead electricity power lines and childhood leukemia: A registry-based, case–control study. *Tumori*, **86**, 195–198

Binhi, V.N. (2000) Amplitude and frequency dissociation spectra of ion–protein complexes rotating in magnetic fields. *Bioelectromagnetics*, **21**, 34–45

Bjerkedal, T. & Egenaes, J. (1987) Video display terminals and birth defects: a study of pregnancy outcomes of employees of the Postal-Giro-Center, Oslo, Norway. In: Knave, B. & Wideback, P.G., eds, *Work with Display Units*. Amsterdam, Elsevier Science Publishers, pp. 111–114

Blackman, C.F., Benane, S.G., House, D.E. & Joines, W.T. (1985) Effects of ELF (1–120 Hz) and modulated (50 Hz) RF fields on the efflux of calcium ions from brain tissue in vitro. *Bioelectromagnetics*, **6**, 1–11

Blackman, C.F., Benane, S.G. & House, D.E. (1993) Evidence for direct effect of magnetic fields on neurite outgrowth. *FASEB J.*, **7**, 801–806

Blackman, C.F., Benane, S.G. & House, D.E. (1995) Frequency-dependent interference by magnetic fields of nerve growth factor-induced neurite outgrowth in PC-12 cells. *Bioelectromagnetics*, **16**, 387–395

Blackman, C.F., Blanchard, J.-P., Benane, S.G. & House, D.E. (1996) Effect of AC and DC magnetic field orientation on nerve cells. *Biochem. biophys. Res. Commun.*, **220**, 807–811

Blackman, C.F., Blanchard, J.-P., Benane, S.G. & House, D.E. (1999) Experimental determination of hydrogen bandwidth for the ion parametric resonance model. *Bioelectromagnetics*, **20**, 5–12

Blackman, C.F., Benane, S.G. & House, D.E. (2001) The influence of 1.2 μT, 60 Hz magnetic fields on melatonin- and tamoxifen-induced inhibition of MCF-7 cell growth. *Bioelectromagnetics*, **22**, 122–128

Blakemore, R. (1975) Magnetotactic bacteria. *Science*, **190**, 377–379

Blanchard, J.P. & Blackman, C.F. (1994) Clarification and application of an ion parametric resonance model for magnetic field interactions with biological systems. *Bioelectromagnetics*, **15**, 217–238

Blask, D.E. & Hill, S.M. (1986) Effects of melatonin on cancer: studies on *MCF-7* human breast cancer cells in culture. *J. neural. transm.*, **Suppl.**, 21433–21449

Blumenthal, N.C., Ricci, J., Breger, L., Zychlinsky, A., Solomon, H., Chen, G.G., Kuznetsov, D. & Dorfman, R. (1997) Effects of low-intensity AC and/or DC electromagnetic fields on cell attachment and induction of apoptosis. *Bioelectromagnetics*, **18**, 264–272

Bonhomme-Faivre, L., Marion, S., Bezie, Y., Auclair, H., Fredj, G. & Hommeau, C. (1998) Study of human neurovegetative and hematologic effects of environmental low-frequency (50-Hz) electromagnetic fields produced by transformers. *Arch. environ. Health*, **53**, 87–92

Boorman, G.A., Gauger, J.R., Johnson, T.R., Tomlinson, M.J., Findlay, J.C., Travlos, G.S. & McCormick, D.L. (1997) Eight-week toxicity study of 60 Hz magnetic fields in F344 rats and B6C3F$_1$ mice. *Fundam. appl. Toxicol.*, **35**, 55–63

Boorman, G.A., McCormick, D.L., Findlay, J.C., Hailey, J.R., Gauger, J.R., Johnson, T.R., Kovatch, R.M., Sills, R.C. & Haseman, J.K. (1999a) Chronic toxicity/oncogenicity evaluation of 60 Hz (power frequency) magnetic fields in F344/N rats. *Toxicol. Pathol.*, **27**, 267–278

Boorman, G.A., Anderson, L.E., Morris, J.E., Sasser, L.B., Mann, P.C., Grumbein, S.L., Hailey, J.R., McNally, A., Sills, R.C. & Haseman, J.K. (1999b) Effect of 26 week magnetic field exposures in a DMBA initiation-promotion mammary gland model in Sprague-Dawley rats. *Carcinogenesis*, **20**, 899–904

Borsalino, G., Bagnacani, M., Bettati, E., Fornaciari, F., Rocchi, R., Uluhogian, S., Ceccherelli, G., Cadossi, R. & Traina, G.C. (1988) Electrical stimulation of human femoral intertrochanteric osteotomies. Double-blind study. *Clin. Orthop.*, **237**, 256–263

Bowman, J.D. & Methner, M.M. (2000) Hazard surveillance for industrial magnetic fields: II. Field characteristics from waveform measurements. *Ann. occup. Hyg.*, **44**, 615–633

Bowman, J.D., Garabrant, D.H., Sobel, E. & Peters, J.M. (1988) Exposures to extremely low frequency (ELF) electromagnetic fields in occupations with elevated leukemia rates. *Appl. ind. Hyg.*, **3**, 189–194

Bracken, T.D. (1993) Exposure assessment for power frequency electric and magnetic fields. *Am. ind. Hyg. Assoc.*, **54**, 165–177

Bracken, M.B., Belanger, K., Hellenbrand, K., Dlugosz, L., Holford, T.R., McSharry, J.-E., Addesso, K. & Leaderer, B. (1995) Exposure to electromagnetic fields during pregnancy with emphasis on electrically heated beds: association with birthweight and intrauterine growth retardation. *Epidemiology*, **6**, 263–270

Bracken, T. D., Rankin, R.F., Wiley, J., Bittner, P.L., Patterson, R. & Bailey, W. (1997) *Recommendations for Guidelines for EMF Personal Exposure Measurements* (EMF RAPID Program Engineering Projects: Project 4), Oak Ridge, TN, Oak Ridge National Laboratory. Available at http://www.emf-data.org/related-projects.html

Brady, J.V. & Reiter, R.J. (1992) Neurobehavioural effects. In: *Health Effects of Low-Frequency Electric and Magnetic Fields. Prepared by an Oak Ridge Associated Universities Panel for the Committee on Interagency Radiation Research and Policy Coordination* (ORAU 92/F8, Chapter 7), Oak Ridge, TN, Oak Ridge Associated Universities, pp. vii-1–vii-56

Brent, R.L., Gordon, W.E., Bennett, W.R. & Beckman, D.A. (1993) Reproductive and teratologic effects of electromagnetic fields. *Reprod. Toxicol.*, **7**, 535–580

Brocklehurst, B. & McLauchlan, K.A. (1996) Free radical mechanism for the effects of environmental electromagnetic fields on biological systems. *Int. J. Radiat. Biol.*, **69**, 3–24

de Bruyn, L., de Jager, L. & Kuyl, J.M. (2001) The influence of long-term exposure of mice to randomly varied power frequency magnetic fields on their nocturnal melatonin secretion patterns. *Environ. Res.*, **85**, 115–121

Bryant, H.E. & Love, E.J. (1989) Video display terminal use and spontaneous abortion risk. *Int. J. Epidemiol.*, **18**, 132–138

Buiatti, E., Barchielli, A., Geddes, M., Nastasi, L., Kriebel, D., Franchini, M. & Scarselli, G. (1984) Risk factors in male infertility: a case–control study. *Arch. environ. Health*, **39**, 266–270

Bunin, G.R., Ward, E., Kramer, S., Rhee, C.A. & Meadows, A.T. (1990) Neuroblastoma and parental occupation. *Am. J. Epidemiol.*, **131**, 776–780

Burch, J.B., Reif, J.S., Yost, M.G., Keefe, T.J. & Pitrat, C.A. (1998) Nocturnal excretion of a urinary melatonin metabolite among electric utility workers. *Scand. J. Work Environ. Health*, **24**, 183–189

Burch, J.B., Reif, J.S., Yost, M.G., Keefe, T.J. & Pitrat, C.A. (1999) Reduced excretion of a melatonin metabolite in workers exposed to 60 Hz magnetic fields. *Am. J. Epidemiol.*, **150**, 27–36

Burch, J.B., Reif, J.S., Noonan, C.W. & Yost, M.G. (2000) Melatonin metabolite levels in workers exposed to 60-Hz magnetic fields: work in substations and with 3-phase conductors. *J. occup. environ. Med.*, **42**, 136–142

Byus, C.V., Pieper, S.E. & Adey, W.R. (1987) The effects of low-energy 60-Hz environmental electromagnetic fields upon the growth-related enzyme ornithine decarboxylase. *Carcinogenesis*, **8**, 1385–1389

Cain, C.D., Thomas, D.L. & Adey, W.R. (1993) 60 Hz magnetic field acts as co-promoter in focus formation of C3H10T1/2 cells. *Carcinogenesis*, **14**, 955–960

Calle, E.E. & Savitz, D.A. (1985) Leukemia in occupational groups with presumed exposure to electrical and magnetic fields (Letter to the Editor). *New Engl. J. Med.*, **313**, 1476–1477

Cameron, I.L., Hunter, K.E. & Winters, W.D. (1985) Retardation of embryogenesis by extremely low frequency 60 Hz electromagnetic fields. *Physiol. Chem. Phys. med. NMR*, **17**, 135–138

Cane, V., Botti, P. & Soana, S. (1993) Pulsed magnetic fields improve osteoblast activity during the repair of an experimental osseous defect. *J. orthopaed. Res.*, **11**, 664–670

Cantoni, O., Sestili, P., Fiorani, M. & Dachà, M. (1995) The effect of 50 Hz sinusoidal electric and/or magnetic fields on the rate of repair of DNA single/double strand breaks in oxidatively injured cells. *Biochem. mol. Biol. int.*, **37**, 681–689

Cantoni, O., Sestili, P., Fiorani, M. & Dachà, M. (1996) Effect of 50 Hz sinusoidal electric and/or magnetic fields on the rate of repair of DNA single strand breaks in cultured mammalian cells exposed to three different carcinogens: methylmethane sulphonate, chromate and 254 nm U.V. radiation. *Biochem. mol. Biol. int.*, **38**, 527–533

Cantor, K.P., Dosemeci, M., Brinton, L.A. & Stewart, P.A. (1995) Re: Breast cancer mortality among female electrical workers in the United States (Letter to the Editor). *J. natl Cancer Inst.*, **87**, 227–228

Caputa, K. & Stuchly, M.A. (1996) Computer controlled system for producing uniform magnetic fields and its application in biomedical research. *IEEE Trans. Instrum. Meas.*, **45**, 701–709

Chen, G., Upham, B.L., Sun, W., Chang, C.C., Rothwell, E.J., Chen, K.M., Yamasaki, H. & Trosko, J.E. (2000) Effect of electromagnetic field exposure on chemically induced differentiation of Friend erythroleukemia cells. *Environ. Health Perspect.*, **108**, 967–972

Chernoff, N., Rogers, J.M. & Kavet, R. (1992) A review of the literature on potential reproductive and developmental toxicity of electric and magnetic fields. *Toxicology*, **74**, 91–126

Ciccone, G., Mirabelli, D., Levis, A., Gavarotti, P., Rege-Cambrin, G., Davico, L. & Vineis, P. (1993) Myeloid leukemias and myelodysplastic syndromes: Chemical exposure, histologic subtype and cytogenetics in a case–control study. *Cancer Genet. Cytogenet.*, **68**, 135–139

Clairmont, B.A., Johnson, G.B., Zaffanella, L.E. & Zelingher, S. (1989) The effects of HVAC-HVDC line separation in a hybrid corridor. *IEEE Trans. Power Deliv.*, **4**, 1338–1350

Cocco, P., Dosemeci, M. & Heineman, E.F. (1998a) Occupational risk factors for cancer of the central nervous system: A case–control study on death certificates from 24 US states. *Am. J. ind. Med.*, **33**, 247–255

Cocco, P., Figgs, L., Dosemeci, M., Hayes, R., Linet, M.S. & Hsing, A.W. (1998b) Case–control study of occupational exposures and male breast cancer. *Occup. environ. Med.*, **55**, 599–604

Cocco, P., Heineman, E.F. & Dosemeci, M. (1999) Occupational risk factors for cancer of the central nervous system (CNS) among US women. *Am. J. ind. Med.*, **36**, 70–74

Coelho, A.M., Jr, Easley, S.P. & Rogers, W.R. (1991) Effects of exposure to 30 kV/m, 60 Hz electric fields on the social behavior of baboons. *Bioelectromagnetics*, **12**, 117–135

Coelho, A.M., Jr, Rogers, W.R. & Easley, S.P. (1995) Effects of concurrent exposure to 60 Hz electric and magnetic fields on the social behavior of baboons. *Bioelectromagnetics*, **Suppl. 3**, 71–92

Coghill, R.W., Steward, J. & Philips, A. (1996) Extra low frequency electric and magnetic fields in the bedplace of children diagnosed with leukaemia: A case–control study. *Eur. J. Cancer Prev.*, **5**, 153–158

Cohen, M., Lippman, M. & Chabner, B. (1978) Role of pineal gland in aetiology and treatment of breast cancer. *Lancet*, **ii**, 814–816

Cohen, M.M., Kunska A., Astemborski, J.A. & McCulloch, D. (1986a) The effect of low-level 60-Hz electromagnetic fields on human lymphoid cells. II. Sister-chromatid exchanges in peripheral lymphocytes and lymphoblastoid cell lines. *Mutat. Res.*, **172**, 177–184

Cohen, M.M., Kunska, A., Astemborski, J.A., McCulloch, D. & Paskewitz, D.A. (1986b) Effect of low-level, 60-Hz electromagnetic fields on human lymphoid cells: I. Mitotic rate and chromosome breakage in human peripheral lymphocytes. *Bioelectromagnetics*, **7**, 415–423

Coleman, M., Bell, J. & Skeet, R. (1983) Leukaemia incidence in electrical workers. *Lancet*, **i**, 982–983

Coleman, M.P., Bell, C.M.J., Taylor, H.-L. & Primic-Zakelj, M. (1989) Leukaemia and residence near electricity transmission equipment: A case–control study. *Br. J. Cancer*, **60**, 793–798

Conti, P., Gigante, G.E., Cifone, M.G., Alesse, E., Ianni, G., Reale, M. & Angeletti, P.U. (1983) Reduced mitogenic stimulation of human lymphocytes by extremely low frequency electromagnetic fields. *FEBS Lett.*, **162**, 156–160

Conti, P., Gigante, G.E., Alesse, E., Cifone, M.G., Fieschi, C., Reale, M. & Angeletti, P.U. (1985) A role for Ca^{2+} in the effect of very low frequency electromagnetic field on the blastogenesis of human lymphocytes. *FEBS Lett.*, **181**, 28–32

Coogan, P.F. & Aschengrau, A. (1998) Exposure to power frequency magnetic fields and risk of breast cancer in the Upper Cape Cod Cancer Incidence Study. *Arch. environ. Health*, **53**, 359–367

Coogan, P.F., Clapp, R.W., Newcomb, P.A., Wenzl, T.B., Bogdan, G., Mittendorf, R., Baron, J.A. & Longnecker, M.P. (1996) Occupational exposure to 60-hertz magnetic fields and risk of breast cancer in women. *Epidemiology*, **7**, 459–464

Cook, M.R., Graham, C., Cohen, H.D. & Gerkovich, M.M. (1992) A replication study of human exposure to 60-Hz fields: Effects on neurobehavioral measures. *Bioelectromagnetics*, **13**, 261–285

Cook, L.L., Persinger, M.A. & Koren, S.A. (2000) Differential effects of low frequency, low intensity (< 6 mG) nocturnal magnetic fields upon infiltration of mononuclear cells and numbers of mast cells in lewis rat brains. *Toxicol. Lett.*, **118**, 9–19

Cooper, M.S. (1984) Gap junctions increase the sensitivity of tissue cells to exogenous electric fields. *J. theor. Biol.*, **111**, 123–130

Cossarizza, A., Monti, D., Sola, P., Moschini, G., Cadossi, R., Bersani, F. & Franceschi, C. (1989a) DNA repair after γ irradiation in lymphocytes exposed to low-frequency pulsed electromagnetic fields. *Radiat. Res.*, **118**, 161–168

Cossarizza, A., Monti, D., Bersani, F., Cantini, M., Cadossi, R., Sacchi, A. & Franceschi, C. (1989b) Extremely low frequency pulsed electromagnetic fields increase cell proliferation in lymphocytes from young and aged subjects. *Biochem. biophys. Res. Commun.*, **160**, 692–698

Cossarizza, A., Monti, D., Bersani, F., Paganelli, R., Montagnani, G., Cadossi, R., Cantini, M. & Franceschi, C. (1989c) Extremely low frequency pulsed electromagnetic fields increase interleukin-2 (IL-2) utilization and IL-2 receptor expression in mitogen-stimulated human lymphocytes from old subjects. *FEBS Lett.*, **248**, 141–144

Cossarizza, A., Monti, D., Bersani, F., Scarfi, M.R., Zanotti, M., Cadossi, R. & Franceschi, C. (1991) Exposure to low-frequency pulsed electromagnetic fields increases mitogen-induced lymphocyte proliferation in Down's syndrome. *Aging*, **3**, 241–246

Coulton, L.A. & Barker, A.T. (1993) Magnetic fields and intracellular calcium: effects on lymphocytes exposed to conditions for 'cyclotron resonance'. *Phys. Med. Biol.*, **38**, 347–360

Cox, F.C., Brewer, L.J., Raeman, C.H., Schryver, C.A., Child, S.Z. & Carstensen, E.L. (1993) A test for teratological effects of power frequency magnetic fields on chick embryos. *IEEE Trans. biomed. Eng.*, **40**, 605–609

Crasson, M., Legros, J.-J., Scarpa, P. & Legros, W. (1999) 50 Hz magnetic field exposure influence on human performance: Two double-blind experimental studies. *Bioelectromagnetics*, **20**, 474–486

Creim, J.A., Lovely, R.H., Kaune, W.T. & Phillips, R.D. (1984) Attempts to produce taste-aversion learning in rats exposed to 60-Hz electric fields. *Bioelectromagnetics*, **5**, 271–282

Creim, J.A., Lovely, R.H., Weigel, R.J., Forsythe, W.C. & Anderson, L.E. (1993) Rats avoid exposure to HVdc electric fields: a dose response study. *Bioelectromagnetics*, **14**, 341–352

Creim, J.A., Lovely, R.H., Weigel, R.J., Forsythe, W.C. & Anderson, L.E. (1995) Failure to produce taste-aversion learning in rats exposed to HVdc electric fields. *Bioelectromagnetics*, **16**, 301–306

Cress, L.W., Owen, R.D. & Desta, A.B. (1999) Ornithine decarboxylase activity in L929 cells following exposure to 60 Hz magnetic fields. *Carcinogenesis*, **20**, 1025–1030

Cridland, N.A., Cragg, T.A., Haylock, R.G.E. & Saunders, R.D. (1996) Effects of 50 Hz magnetic field exposure on the rate of DNA synthesis by normal human fibroblasts. *Int. J. Radiat. Biol.*, **69**, 503–511

Davanipour, Z., Sobel, E., Bowman, J.D., Qian, Z. & Will, A.D. (1997) Amyotrophic lateral sclerosis and occupational exposure to electromagnetic fields. *Bioelectromagnetics*, **18**, 28–35

Davis, H.P., Mizumori, S.J.Y., Allen, H., Rosenzweig, M.R., Bennett, E.L. & Tenforde, T.S. (1984) Behavioral studies with mice exposed to DC and 60-Hz magnetic fields. *Bioelectromagnetics*, **5**, 147–164

Davis, S., Kaune, W.T., Mirick, D.K., Chu Chen, M.S. & Stevens, R.G. (2001) Residential magnetic fields, light-at-night, and nocturnal urinary 6-hydroxymelatonin in women. *Am. J. Epidemiol.*, **154**, 591–600

Dawson, T.W. & Stuchly, M.A. (1997) An analytic solution for verification of computer models for low-frequency magnetic induction. *Radio Sci.*, **32**, 343–367

Dawson, T.W. & Stuchly, M.A. (1998) High-resolution organ dosimetry for human exposure to low frequency magnetic fields. *IEEE Trans. Magn.*, **34**, 708–718

Dawson, T.W., Caputa, K. & Stuchly, M.A. (1997) Influence of human model resolution on computed currents induced in organs by 60 Hz-magnetic fields. *Bioelectromagnetics*, **18**, 478–490

Dawson, T.W., Caputa, K. & Stuchly, M.A. (1998) High-resolution organ dosimetry for human exposure to low-frequency electric fields. *IEEE Trans. Power Deliv.*, **13**, 366–372

Dawson, T.W., Caputa, K. & Stuchly, M.A. (1999a) Organ dosimetry for live-line occupational exposures to magnetic fields. *IEEE Trans. Power Deliv.*, **14**, 1234–1239

Dawson, T.W., Caputa, K. & Stuchly, M.A. (1999b) High-resolution magnetic field numerical dosimetry for live-line workers. *IEEE Trans. Magn.*, **35**, 1131–1134

Dawson, T.W., Caputa, K. & Stuchly, M.A. (1999c) Numerical evaluation of 60 Hz magnetic induction in the human body in complex occupational environments. *Phys. Med. Biol.*, **44**, 1025–1040

Dawson, T.W., Caputa, K., Stuchly, M.A. & Kavet, R. (2001) Electric fields in the human body resulting from 60-Hz contact currents. *IEEE Trans. Biomed. Eng.*, **48**, 1020–1026

Deadman, J.E., Camus, M., Armstrong, B.G., Héroux, P., Cyr, D., Plante, M. & Thériault, G. (1988) Occupational and residential 60-Hz electromagnetic fields and high-frequency electric transients: Exposure assessment using a new dosimeter. *Am. ind. Hyg. Assoc. J.*, **49**, 409–419

Deapen, D.M. & Henderson, B.E. (1986) A case–control study of amyotrophic lateral sclerosis. *Am. J. Epidemiol.*, **123**, 790–799

De Guire, L., Thériault, G., Iturra, H., Provencher, S., Cyr, D. & Case, B.W. (1988) Increased incidence of malignant melanoma of the skin in workers in a telecommunication industry. *Br. J. ind. Med.*, **45**, 824–828

Deibert, M.C., Mcleod, B.R., Smith, S.D. & Liboff, A.R. (1994) Ion resonance electromagnetic field stimulation of fracture healing in rabbits with a fibular ostectomy. *J. orthoped. Res.*, **12**, 878–885

Delgado, J.M.R., Leal, J., Monteagudo, J.L. & Gracia, M.G. (1982) Embryological changes induced by weak, extremely low frequency electromagnetic fields. *J. Anat.*, **134**, 533–551

Demers, P.A., Thomas, D.B., Rosenblatt, K.A., Jimenez, L.M., McTiernan, A., Stalsberg, H., Stemhagen, A., Thompson, W.D., McGrea Curnen, M.G., Satariano, W., Austin, D.F., Isacson, P., Greenberg, R.S., Key, C., Kolonel, L.N. & West, D.W. (1991) Occupational exposure to electromagnetic fields and breast cancer in men. *Am. J. Epidemiol.*, **134**, 340–347

Deno, D.W. & Zaffanella, L.E. (1975) Electrostatic effects of overhead transmission lines and stations. In: *Transmission Line Reference Books 345 kV and Above*, Palo Alto, CA, Electric Power Research Institute, pp. 248–279

Deno, D.W. & Zaffanella, L.E. (1982) Field effects of overhead transmission lines and stations. In: *Transmission Line Reference Books: 345 kV and Above*, 2nd Ed., Palo Alto, CA, Electric Power Research Institute, pp. 329–420

Devevey, L., Patinot, C., Debray, M., Thierry, D., Brugere, H., Lambrozo, J., Guillosson, J.-J. & Nafziger, J. (2000) Absence of the effects of 50 Hz magnetic fields on the progression of acute myeloid leukaemia in rats. *Int. J. Radiat. Biol.*, **76**, 853–862

De Vita, R., Cavallo, D., Raganella, L. Eleuteri, P., Grollino, M.G. & Calugi, A. (1995) Effects of 50 Hz magnetic fields on mouse spermatogenesis monitored by flow cytometric analysis. *Bioelectromagnetics*, **16**, 330–334

DiGiovanni, J., Johnston, D.A., Rupp, T., Sasser, L.B., Anderson, L.E., Morris, J.E., Miller, D.L., Kavet, R. & Walborg, E.F. (1999) Lack of effect of a 60 Hz magnetic field on bio-markers of tumor promotion in the skin of SENCAR mice. *Carcinogenesis*, **20**, 685–689

Dimbylow, P.J. (1997) FDTD calculations of the whole-body averaged SAR in an anatomically realistic voxel model of the human body from 1 MHz to 1 GHz. *Phys. Med. Biol.*, **42**, 479–490

Dimbylow, P.J. (1998) Induced current densities from low-frequency magnetic fields in a 2 mm resolution, anatomically realistic model of the body. *Phys. Med. Biol.*, **43**, 221–230

Dimbylow, P.J. (2000) Current densities in a 2 mm resolution anatomically realistic model of the body induced by low frequency electric fields. *Phys. Med. Biol.*, **45**, 1013–1022

Ding, G.-R., Wake, K., Taki, M. & Miyakoshi, J. (2001) Increase in hypoxanthine–guanine phosphoribosyl transferase gene mutations by exposure to electric field. *Life Sci.*, **68**, 1041–1046

Dlugosz, L., Vena, J., Byers, T., Sever, L., Bracken, M. & Marshall, E. (1992) Congenital defects and electric bed heating in New York State: A register-based case–control study. *Am. J. Epidemiol.*, **135**, 1000–1011

Dockerty, J.D., Elwood, J.M., Skegg, D.C.G. & Herbison, G.P. (1998) Electromagnetic field exposures and childhood cancers in New Zealand. *Cancer Causes Control*, **9**, 299–309

Dockerty, J.D., Elwood, J.M., Skegg, D.C.G. & Herbison, G.P. (1999) Electromagnetic field exposures and childhood leukaemia in New Zealand (Letter to the Editor). *Lancet*, **354**, 1967–1968

Dowson, D.I., Lewith, G.T., Campbell, M., Mullee, M.A. & Brewster, L.A. (1988) Overhead high-voltage cables and recurrent headache and depressions. *Practitioner*, **232**, 435–436

Dubrov, A.P. (1978) *The Geomagnetic Field and Life. Geomagnetobiology*, New York, Plenum Press

Easley, S.P., Coelho, A.M., Jr & Rogers, W.R. (1991) Effects of exposure to a 60-kV/m, 60-Hz electric field on the social behavior of baboons. *Bioelectromagnetics*, **12**, 361–375

Ebi, K.L., Zaffanella, L.E. & Greenland, S. (1999) Application of the case-specular method to two studies of wire codes and childhood cancers. *Epidemiology*, **10**, 398–404

Eichwald, C. & Walleczek, J. (1997) Low-frequency-dependent effects of oscillating magnetic fields on radical pair recombination in enzyme kinetics. *J. Chem. Phys.*, **107**, 4943–4950

Einhorn, J., Eklund, G. & Wiklund, K. (1980) [Leukaemia in telephone workers at the Tele-communication Office in Sweden.] *Läkartidningen*, **40**, 3519–3520 & 3526 (in Swedish)

Ekström, T., Mild, K.H. & Holmberg, B. (1998) Mammary tumours in Sprague-Dawley rats after initiation with DMBA followed by exposure to 50 Hz electromagnetic fields in a promotional scheme. *Cancer Lett.*, **123**, 107–111

Elias, Z., Mur, J.M., Pierre, F., Gilgenkrantz, S., Schneider, O., Baruthio, F., Daniere, M.C. & Fontana, J.M. (1989) Chromosome aberrations in peripheral blood lymphocytes of welders and characterization of their exposure by biological sample analysis. *J. occup. Med.*, **31**, 477–483

Ericson, A. & Källén, B. (1986a) An epidemiology study of work with video screens and pregnancy outcome: I. A registry study. *Am. J. ind. Med.*, **9**, 447–457

Ericson, A. & Källén, B. (1986b) An epidemiological study of work with video screens and pregnancy outcome: II. A case–control study. *Am. J. ind. Med.*, **9**, 459–475

Espinar, A., Piera, V., Carmona, A. & Guerrero, J.M. (1997) Histological changes during deve-lopment of the cerebellum in the chick embryo exposed to a static magnetic field. *Bio-electromagnetics*, **18**, 36–46

Eyres, K.S., Saleh, M., Kanis, J.A. (1996) Effect of pulsed electromagnetic fields on bone formation and bone loss during limb lengthening. *Bone.* **18**, 505–509

Fairbairn, D.W. & O'Neill, K.L. (1994) The effect of electromagnetic field exposure on the formation of DNA single strand breaks in human cells. *Cell. mol. Biol.*, **40**, 561–567

Fajardo-Gutiérrez, A., Velásquez-Pérez, L., Martínez-Méndez, J., Martínez-García, C. (1997) [Exposure to electromagnetic fields and its association with leukaemia in children living in the city of Mexico.] Mexico DF, Unidad de Investigación Médica en Epidemiología Clínica Hospital de Pediatria Centro Médico Nacional Siglo XXI

Fam, W.Z. & Mikhail, E.L. (1996) Lymphoma induced in mice chronically exposed to very strong low-frequency electromagnetic field. *Cancer Lett.*, **105**, 257–269

Fanelli, C., Coppola, S., Barone, R., Colussi, C., Gualandi, G., Volpe, P. & Ghibelli, L. (1999) Magnetic fields increase cell survival by inhibiting apoptosis via modulation of Ca2+ influx. *FASEB J.*, **13**, 95–102

Farrell, J.M., Litovitz, T.L., Penafiel, M., Montrose, C.J., Doinov, P., Barber, M., Brown, K.M. & Litovitz, T.A. (1997) The effect of pulsed and sinusoidal magnetic fields on the morphology of developing chick embryos. *Bioelectromagnetics*, **18**, 431–438

Fear, E.C. & Stuchly, M.A. (1998) Modeling assemblies of biological cells exposed to electric fields. *IEEE Trans. Biomed. Eng.*, **45**, 1259–1271

Fear, N.T., Roman, E., Carpenter, L.M., Newton, R. & Bull, D. (1996) Cancer in electrical workers: An analysis of cancer registrations in England 1981–87. *Br. J. Cancer*, **73**, 935–939

Fews, A.P., & Henshaw, D.L. (2000) Reply to letter from Swanson and Jeffers. *Int. J. Radiat. Biol.*, **76**, 1688–1691

Fews, A.P., Henshaw, D.L., Keitch, P.A., Close, J.J. & Wilding, R.J. (1999a) Increased exposure to pollutant aerosols under high voltage power lines. *Int. J. Radiat. Biol.*, **75**, 1505–1521

Fews, A.P., Henshaw, D.L., Wilding, R.J. & Keitch, P.A. (1999b) Corona ions from powerlines and increased exposure to pollutant aerosols. *Int. J. Radiat. Biol.*, **75**, 1523–1531

Feychting, M. & Ahlbom, A. (1992a) [Cancer and magnetic fields in persons living close to high voltage power lines in Sweden.] *Lakartidningen*, **89**, 4371–4374 (in Swedish)

Feychting, M. & Ahlbom, A. (1992b) *Magnetic Fields and Cancer in People Residing Near Swedish High Voltage Lines*, Stockholm, Institute for Environmental Medicine, Karolinska Institute

Feychting, M. & Ahlbom, A. (1993) Magnetic fields and cancer in children residing near Swedish high-voltage power lines. *Am. J. Epidemiol.*, **138**, 467–481

Feychting, M. & Ahlbom, A. (1994) Magnetic fields, leukemia, and central nervous system tumors in Swedish adults residing near high-voltage power lines. *Epidemiology*, **5**, 501–509

Feychting, M., Forssén, U. & Floderus, B. (1997) Occupational and residential magnetic field exposure and leukemia and central nervous system tumors. *Epidemiology*, **8**, 384–389

Feychting, M., Forssén, U., Rutqvist, L.E. & Ahlbom, A. (1998a) Magnetic fields and breast cancer in Swedish adults residing near high-voltage power lines. *Epidemiology*, **9**, 392–397

Feychting, M., Pedersen, N.L., Svedberg, P., Floderus, B. & Gatz, M. (1998b) Dementia and occupational exposure to magnetic fields. *Scand. J. Work Environ. Health*, **24**, 46–53

Feychting, M., Floderus, B. & Ahlbom, A. (2000) Parental occupational exposure to magnetic fields and childhood cancer (Sweden). *Cancer Causes Control*, **11**, 151–156

Fiorani, M., Cantoni, O., Sestili, P., Conti, R., Nicolini, P., Vetrano, F. & Dachà, M. (1992) Electric and/or magnetic field effects on DNA structure and function in cultured human cells. *Mutat. Res.*, **282**, 25–29

Fitzsimmons, R.J., Ryaby, J.T., Mohan, S., Magee, F.P. & Baylink, D.J. (1995) Combined magnetic fields increase insulin-like growth factor-II in TE-85 human osteosarcoma bone cell cultures. *Endocrinology*, **136**, 3100–3106

Flipo, D., Fournier, M., Benquet, C., Roux, P., Le Boulaire, C., Pinsky, C., LaBella, F.S. & Krzystyniak, K. (1998) Increased apoptosis, changes in intracellular Ca2+, and functional alterations in lymphocytes and macrophages after in vitro exposure to static magnetic field. *J. Toxicol. environ. Health*, **54A**, 63–76

Floderus, B., Persson, T., Stenlund, C., Wennberg, A., Öst, Å. & Knave, B. (1993) Occupational exposure to electromagnetic fields in relation to leukemia and brain tumors: A case–control study in Sweden. *Cancer Causes Control*, **4**, 465–476

Floderus, B., Törnqvist, S. & Stenlund, C. (1994) Incidence of selected cancers in Swedish railway workers, 1961–79. *Cancer Causes Control*, **5**, 189–194

Floderus, B., Persson, T. & Stenlund, C. (1996) Magnetic field exposures in the workplace: Reference distribution and exposures in occupational groups. *Int. J. occup. environ. Health*, **2**, 226–238

Floderus, B., Stenlund, C. & Persson, T. (1999) Occupational magnetic field exposure and site-specific cancer incidence: A Swedish cohort study. *Cancer Causes Control*, **10**, 323–332

Florig, H.K. & Hoburg, J.F. (1990) Power frequency magnetic fields from electric blankets. *Health Phys.*, **58**, 493–502

Foster, K.R. & Schwan, H.P. (1995) Dielectric properties of tissues. In: Polk, C. & Postow, E., eds, *Handbook of Biologic Effects of Electromagnetic Fields*, 2nd Ed., Boca Raton, CRC Press, pp. 25–102

Forssén, U.M., Feychting, M., Rutqvist, L.E., Floderus, B. & Ahlbom, A. (2000) Occupational and residential magnetic field exposure and breast cancer in females. *Epidemiology*, **11**, 24–29

Frazier, M.E., Reese, J.A., Morris, J.E., Jostes, R.F. & Miller, D.L. (1990) Exposure of mammalian cells to 60-Hz magnetic or electric fields: analysis of DNA repair of induced, single-strand breaks. *Bioelectromagnetics*, **11**, 229–234

Free, M.J., Kaune, W.T., Philips, R.D. & Chen H.-C. (1981) Endocrinological effects of strong 60-Hz electric fields on rats. *Bioelectromagnetics*, **2**, 105–121

Friedman, D.R., Hatch, E.E., Tarone, R., Kaune, W.T., Kleinerman, R.A., Wacholder, S., Boice, J.D., Jr & Linet, M.S. (1996) Childhood exposure to magnetic fields: Residential area measurements compared to personal dosimetry. *Epidemiology*, 7, 151–155

Fulton, J.P., Cobb, S., Preble, L., Leone, L. & Forman, E. (1980) Electrical wiring configurations and childhood leukemia in Rhode Island. *Am. J. Epidemiol.*, **111**, 292–296

Furse, C.M. & Gandhi, O.P. (1998) Calculation of electric fields and currents induced in a millimeter-resolution human model at 60 Hz using the FDTD method. *Bioelectromagnetics*, **19**, 293–299

Gabriel, S., Lau, R.W. & Gabriel, C. (1996) The dielectric properties of biological tissues: III. Parametric models for the dielectric spectrum of tissues. *Phys. Med. Biol.*, **41**, 2271–2293

Gallagher, R.P., McBride, M.L., Band, P.R., Spinelli, J.J., Threlfall, W.J. & Yang, P. (1990) Occupational electromagnetic field exposure, solvent exposure, and leukemia (Letter to the Editor). *J. occup. Med.*, **32**, 64–65

Gallagher, R.P., McBride, M.L., Band, P.R., Spinelli, J.J., Threlfall, W.J. & Tamaro, S. (1991) Brain cancer and exposure to electromagnetic fields (Letter to the Editor). *J. occup. Med.*, **33**, 944–945

Gamble, S.C., Wolff, H. & Arrand, J.E. (1999) Syrian hamster dermal cell immortalization is not enhanced by power line frequency electromagnetic field exposure. *Br. J. Cancer*, **81**, 377–380

Gammon, M.D., Schoenberg, J.B., Britton, J.A., Kelsey, J.L., Stanford, J.L., Malone, K.E., Coates, R.J., Brogan, D., Potischman, N., Swanson, C.A. & Brinton, L.A. (1998) Electric blanket use and breast cancer risk among younger women. *Am. J. Epidemiol.*, **148**, 556–563

Gandhi, O.P. (1995) Some numerical methods for dosimetry: extremely low frequencies to microwave frequencies. *Radio Sci.*, **30**, 161–177

Gandhi, O.P. & De Ford, J.F. (1988) Calculation of EM power deposition for operator exposure to RF induction heaters. *IEEE Trans. Electromagn. Compat.*, **30**, 63–68

Gandhi, O.P. & Chen, J.-Y. (1992) Numerical dosimetry at power line frequencies using anatomically based models. *Bioelectromagnetics*, **Suppl. 1**, 43–60

Gandhi, O.P., Kang, G., Wu, D. & Lazzi, G. (2001) Currents induced in anatomic models of the human for uniform and nonuniform power frequency magnetic fields. *Bioelectromagnetics*, **22**, 112–121

Garland, F.C., Shaw, E., Gorham, E.D., Garland, C.F., White, M.R. & Sinsheimer, P.J. (1990) Incidence of leukemia in occupations with potential electromagnetic field exposure in United States Navy personnel. *Am. J. Epidemiol.*, **132**, 293–303

Gauger, J.R. (1985) Household appliance magnetic field survey. *IEEE Trans. Power Appar. Syst.*, **104**, 2436–2444

Gauger, J.R., Johnson, T.R., Stangel, J.E., Patterson, R.C., Williams, D.A., Harder, J.B. & McCormick, D.L. (1999) Design, construction, and validation of a large capacity rodent magnetic field exposure laboratory. *Bioelectromagnetics*, **20**, 13–23

Gavalas, R.J., Walter, D.O., Hammer, J. & Adey, W.R. (1970) Effect of low-level, low-frequency electric fields on EEG and behavior in *Macaca nemestrina*. *Brain Res.*, **18**, 491–501

Gavalas-Medici, R. & Day-Magdaleno, S.R. (1976) Extremely low frequency, weak electric fields affect schedule-controlled behavior of monkeys. *Nature*, **261**, 256–259

Goldhaber, M.K., Polen, M.R. & Hiatt, R.A. (1988) The risk of miscarriage and birth defects among women who use visual display terminals during pregnancy. *Am. J. ind. Med.*, **13**, 695–706

Gona, A.G., Yu, M.C., Gona, O., Al-Rabiai, S., Von Hagen, S. & Cohen, E. (1993) Effects of 60 Hz electric and magnetic fields on the development of the rat cerebellum. *Bioelectromagnetics*, **14**, 433–447

Goodman, R. & Shirley-Henderson, A. (1991) Transcription and translation in cells exposed to extremely low frequency electromagnetic fields. *Bioelectrochem. Bioenerg.*, **25**, 335–355

Goodman, R., Blank, M., Lin, H., Dai, R., Khorkova, O., Soo, L., Weisbrot, D. & Henderson, A.H. (1994) Increased levels of hsp70 transcripts induced when cells are exposed to low frequency electromagnetic fields. *Bioelectrochem. Bioenerg.*, **33**, 115–120

Graham, C., Cohen, H.D., Cook, M.R., Phelps, J., Gerkovich, M. & Fotopoulos, S.S. (1987) A double-blind evaluation of 60-Hz field effects on human performance, physiology, and subjective state. In: Anderson, L.E., Weigel, R.J. & Kelman, B.J., eds, *Interaction of Biological Systems with Static and ELF Electric and Magnetic Fields* (DOE Symposium Series CONF-841041), Springfield, VA, National Technical Information Service, pp. 471–486

Graham, C., Cook, M.R., Cohen, H.D. & Gerkovich, M.M. (1994) Dose response study of human exposure to 60 Hz electric and magnetic fields. *Bioelectromagnetics*, **15**, 447–463

Graham, C., Cook, M.R., Riffle, D.W., Gerkovich, M.M. & Cohen, H.D. (1996) Nocturnal melatonin levels in human volunteers exposed to intermittent 60 Hz magnetic fields. *Bioelectromagnetics*, **17**, 263–273

Graham, C., Cook, M.R. & Riffle, D.W. (1997) Human melatonin during continuous magnetic field exposure. *Bioelectromagnetics*, **18**, 166–171

Graham, C., Cook, M.R., Cohen, H.D., Riffle, D.W., Hoffman, S. & Gerkovich, M.M. (1999) Human exposure to 60-Hz magnetic fields: neurophysiological effects. *Int. J. Psychophysiol.*, **33**, 169–175

Graham, C., Cook, M.R., Sastre, A., Riffle, D.W. & Gerkovich, M.M. (2000a) Multi-night exposure to 60 Hz magnetic fields: Effects on melatonin and its enzymatic metabolite. *J. Pineal Res.*, **28**, 1–8

Graham, C., Cook, M.R., Sastre, A., Gerkovich, M.M. & Kavet, R. (2000b) Cardiac autonomic control mechanisms in power-frequency magnetic fields: a multistudy analysis. *Environ. Health Perspect.*, **108**, 737–742

Graham, J.H., Fletcher, D., Tigue, J. & McDonald, M. (2000c) Growth and developmental stability of *Drosophila melanogaster* in low frequency magnetic fields. *Bioelectromagnetics*, **21**, 465–472

Graves, H.B. (1981) Detection of a 60-Hz electric field by pigeons. *Behav. neural Biol.*, **32**, 229–234

Graves H.B., Carter, J.H., Kellmel, D., Cooper, L., Poznaniak, D.T. & Bankoske, J.W. (1978) Perceptibility and electrophysiological response of small birds to intense 60-Hz electric fields. *IEEE Trans. Power Appar. Syst.*, **97**, 1070–1073

Grayson, J.K. (1996) Radiation exposure, socioeconomic status, and brain tumor risk in the US Air Force: A nested case–control study. *Am. J. Epidemiol.*, **143**, 480–486

Green, L.M., Miller, A.B., Agnew, D.A., Greenberg, M.L., Li, J., Villeneuve, P.J. & Tibshirani, R. (1999a) Childhood leukemia and personal monitoring of residential exposures to electric and magnetic fields in Ontario, Canada. *Cancer Causes Control*, **10**, 233–243

Green, L.M., Miller, A.B., Villeneuve, P.J., Agnew, D.A., Greenberg, M.L., Li, J. & Donnelly, K.E. (1999b) A case–control study of childhood leukemia in southern Ontario, Canada, and exposure to magnetic fields in residences. *Int. J. Cancer*, **82**, 161–170

Greene, J.J., Pearson, S.L., Skowronski, W.J., Nardone, R.M., Mullins, J.M. & Krause, D. (1993) Gene-specific modulation of RNA synthesis and degradation by extremely low frequency electromagnetic fields. *Cell. mol. Biol.*, **39**, 261–268

Greenland, S., Sheppard, A.R., Kaune, W.T., Poole, C. & Kelsh, M.A. for the Childhood Leukemia-EMF Study Group (2000) A pooled analysis of magnetic fields, wire codes, and childhood leukemia. *Epidemiology*, **11**, 624–634

Griffin, G.D., Williams, M.W. & Gailey, P.C. (2000) Cellular communication in clone 9 cells exposed to magnetic fields. *Radiat. Res.*, **153**, 690–698

Grota, L.J., Reiter, R.J., Keng, P. & Michaelson, S. (1994) Electric field exposure alters serum melatonin but not pineal melatonin synthesis in male rats. *Bioelectromagnetics*, **15**, 427–437

Grzesik, J., Bortel, M., Duda, D., Kuska, R., Ludyga, K., Michnik, J., Smolka, B., Sowa, B., Trzeciak, H. & Zielinski, G. (1988) Influence of a static magnetic field on the reproductive function of certain biochemical indices and behaviour of rats. *Pol. J. occup. Med.*, **1**, 329–339

Guénel, P., Nicolau, J., Imbernon, E., Warret, G. & Goldberg, M. (1993a) Design of a job exposure matrix on electric and magnetic fields: selection of an efficient job classification for workers in thermoelectric power production plants. *Int. J. Epidemiol.*, **22** (Suppl. 2), S16–S21

Guénel, P., Raskmark, P., Andersen, J.B. & Lynge, E. (1993b) Incidence of cancer in persons with occupational exposure to electromagnetic fields in Denmark. *Br. J. ind. Med.*, **50**, 758–764

Guénel, P., Nicolau, J., Imbernon, E., Chevalier, A. & Goldberg, M. (1996) Exposure to 50-Hz electric field and incidence of leukemia, brain tumors and other cancers among French electric utility workers. *Am. J. Epidemiol.*, **144**, 1107–1121

Gunnarsson, L.G., Bodin, L., Soderfeldt, B. & Axelson, O. (1992) A case–control study of motor neurone disease: Its relation to heritability, and occupational exposures, particularly to solvents. *Br. J. ind. Med.*, **49**, 791–798

Gurney, J.G., Mueller, B.A., Davis, S., Schwartz, S.M., Stevens, R.G. & Kopecky, K.J. (1996) Childhood brain tumor occurrence in relation to residential power line configurations, electric heating sources, and electric appliance use. *Am. J. Epidemiol.*, **143**, 120–128

Hackman, R.M. & Graves, H.B. (1981) Corticosterone levels in mice exposed to high intensity electric fields. *Behav. neural Biol.*, **32**, 201–213

Hallquist, A., Hardell, L., Degerman, A. & Boquist, L. (1993) Occupational exposures and thyroid cancer: Results of a case–control study. *Eur. J. Cancer Prev.*, **2**, 345–349

Hansen, N.H., Sobel, E., Davanipour, Z., Gillette, L.M., Niiranen, J. & Wilson, B.W. (2000) EMF exposure assessment in the Finnish garment industry: Evaluation of proposed EMF exposure metrics. *Bioelectromagnetics*, **21**, 57–67

Hansson Mild, K. (2000) Sources and levels of terrestrial electromagnetic field exposure. In: Matthes, R., Bernhardt, M.H. & Repacholi, M., eds, *Proceedings from a Joint Seminar 'International Seminar on Effects of Electromagnetic Fields on the Living Environment' of ICNIRP, WHO and BfS, ICNIRP 10/2000, Ismaning, Germany, October 4–5*, pp. 21–37

Hansson Mild, K., Sandström, M. & Løvtrup, S. (1981) Development of *Xenopus laevis* embryos in a static magnetic field. *Bioelectromagnetics*, **2**, 199–201

Hansson Mild, K., Lövdahl, L., Lövstrand, K.-G. & Løvtrup, S. (1982) Effect of high-voltage pulses on the viability of human leucocytes in vitro. *Bioelectromagnetics*, **3**, 213–218

Hansson Mild, K., Berglund, A. & Forsgren, P.G. (1991) [Reducing 50 Hz magnetic fields from a built-in transformer station.] (Undersökningsrapport 24), Umeå, *Arbetsmiljöinstitutet*

Hansson Mild, K., Sandström, M. & Johnsson, A. (1996) Measured 50 Hz electric and magnetic fields in Swedish and Norwegian residential buildings. *IEEE Trans. Instrument. Measure.*, **45**, 710–714

Harland, J.D. & Liburdy, R.P. (1997) Environmental magnetic fields inhibit the antiproliferative action of tamoxifen and melatonin in a human breast cancer cell line. *Bioelectromagnetics*, **18**, 555–562

Harland, J.D., Engström, S. & Liburdy, R. (1999) Evidence for a slow time-scale of interaction for magnetic fields inhibiting tamoxifen's antiproliferative action in human breast cancer cells. *Cell Biochem. Biophys.*, **31**, 295–306

Harrington, J.M., McBride, D.I., Sorahan, T., Paddle, G.M & van Tongeren, M. (1997) Occupational exposure to magnetic fields in relation to mortality from brain cancer among electricity generation and transmission workers. *Occup. environ. Med.*, **54**, 7–13

Harrington, J.M., Nichols, L., Sorahan, T. & van Tongeren, M. (2001) Leukaemia mortality in relation to magnetic field exposure: Findings from a study of United Kingdom electricity generation and transmission workers, 1973–97. *Occup. environ. Med.*, **58**, 307–314

Harris, A.W., Basten, A., Gebski, V., Noonan, D., Finnie, J., Bath, M.L., Bangay, M.J. & Repacholi, M.H. (1998) A test of lymphoma induction by long-term exposure of Eμ-*Pim1* transgenic mice to 50 Hz magnetic fields. *Radiat. Res.*, **149**, 300–307

Hatch, M. (1992) The epidemiology of electric and magnetic field exposures in the power frequency range and reproductive outcomes. *Paediatr. perinat. Epidemiol.*, **6**, 198–214

Hatch, E.E., Linet, M.S., Kleinerman, R.A., Tarone, R.E., Severson, R.K., Hartsock, C.T., Haines, C., Kaune, W.T., Friedman, D., Robison, L.L. & Wacholder, S. (1998) Association between childhood acute lymphoblastic leukemia and use of electrical appliances during pregnancy and childhood. *Epidemiology*, **9**, 234–245

Hatch, E.E., Kleinerman, R.A., Linet, M.S., Tarone, R.E., Kaune, W.T., Auvinen, A., Baris, D., Robison, L.L. & Wacholder, S. (2000) Do confounding or selection factors of residential wiring codes and magnetic fields distort findings of electromagnetic fields studies? *Epidemiology*, **11**, 189–198

Haussler, M., Thun-Battersby, S., Mevissen, M. & Löscher, W. (1999) Exposure of rats to a 50-Hz, 100 μT magnetic field does not affect the ex vivo production of interleukins by activated T or B lymphocytes. *Bioelectromagnetics*, **20**, 295–305

Heikkinen, P., Kumlin, T., Laitinen, J.T., Komulainen, H. & Juutilainen, J. (1999) Chronic exposure to 50 Hz magnetic fields or 900 MHz electromagnetic fields does not alter nocturnal 6-hydroxymelatonin sulphate secretion in CBA/S mice. *Electro- and Magnetobiology*, **18**, 33–42

Heikkinen, P., Kosma, V.M., Huuskonen, H., Komulainen, H., Kumlin, T., Penttila, I., Vaananen, A. & Juutilainen, J. (2001) Effects of 50 Hz magnetic fields on cancer induced by ionizing radiation in mice. *Int. J. Radiat. Biol.*, **77**, 483–495

Henshaw, D.L., Ross, A.N., Fews, A.P. & Preece, A.W. (1996) Enhanced deposition of radon daughter nuclei in the vicinity of power frequency electromagnetic fields. *Int. J. Radiat. Biol.*, **69**, 25–38

Héroux, P. (1991) A dosimeter for assessment of exposures to ELF fields. *Bioelectromagnetics*, **12**, 241–257

Hertz-Picciotto, I., Swan, S.H. & Neutra, R.R. (1992) Reporting bias and mode of interview in a study of adverse pregnancy outcomes and water consumption. *Epidemiology*, **3**, 104–112

Hietanen, M. & Jokela, K. (1990) Measurements of ELF and RF electromagnetic emissions from video display units. In: Berlinguet, L. & Berthelette, D., eds, *Work With Display Units 89*, Amsterdam, Elsevier Science Publishers, pp. 357–362

High, W.B., Sikora, J., Ugurbil, K. & Garwood, M. (2000) Subchronic in vivo effects of a high static magnetic field (9.4 T) in rats. *J. magn. Reson. Imaging*, **12**, 122–139

Hintenlang, D.E. (1993) Synergistic effects of ionizing radiation and 60 Hz magnetic fields. *Bioelectromagnetics*, **14**, 545–551

Hiraoka, M., Miyakoshi, J., Li, Y.P., Shung, B., Takebe, H. & Abe, M. (1992) Induction of c-*fos* gene expression by exposure to a static magnetic field in HeLaS3 cells. *Cancer Res.*, **52**, 6522–6524

Hirata, A., Caputa, K., Dawson, T.W. & Stuchly, M.A. (2001) Dosimetry in models of child and adult for low-frequency electric field. *IEEE Trans. Biomed. Eng.*, **48**, 1007–1012

Hisamitsu, T., Narita, K., Kasahara, T., Seto, A., Yu, Y. & Asano, K. (1997) Induction of apoptosis in human leukemic cells by magnetic fields. *Jpn J. Physiol.*, **47**, 307–310

Hjeresen D.L., Kaune, W.D., Decker, J.R. & Phillips, R.D. (1980) Effects of 60-Hz electric fields on avoidance behavior and activity of rats. *Bioelectromagnetics*, **1**, 299–312

Hjeresen, D.L., Miller, M.C., Kaune, W.T. & Phillips, R.D. (1982) A behavioral response of swine to 60-Hz electric field. *Bioelectromagnetics*, **2**, 443–451

Holder, J.W., Elmore, E. & Barrett, J.C. (1993) Gap junction function and cancer. *Cancer Res.*, **53**, 3475–3485

Holian, O., Astumian, R.D., Lee, R.C., Reyes, H.M., Attar, B.M. & Walter, R.J. (1996) Protein kinase C activity is altered in HL60 cells exposed to 60 Hz AC electric fields. *Bioelectromagnetics*, **17**, 504–509

Hong, C.Z., Huestis, P., Thompson, R. & Yu, J. (1988) Learning ability of young rats is unaffected by repeated exposure to a static electromagnetic field in early life. *Bioelectromagnetics*, **9**, 269–273

House, R.V., Ratajczak, H.V., Gauger, J.R., Johnson, T.R., Thomas, P.T. & McCormick, D.L. (1996) Immune function and host defense in rodents exposed to 60-Hz magnetic fields. *Fundam. appl. Toxicol.*, **34**, 228–239

Hutzler, B., Baraton, P., Vicente, J.L., Antoine, J.C., Roux, M. & Urbain, J.P. (1994) Exposure to 50 Hz magnetic fields during live work (36-106). *Proc. CIGRE*, 1–9, Paris, CIGRE

Huuskonen, H., Juutilainen, J. & Komulainen, H. (1993) Effects of low-frequency magnetic fields on fetal development in rats. *Bioelectromagnetics*, **14**, 205–213

Huuskonen, H., Lindbohm, M.-L. & Juutilainen, J. (1998a) Teratogenic and reproductive effects of low-frequency magnetic fields. *Mutat. Res.*, **410**, 167–183

Huuskonen, H., Juutilainen, J., Julkunen, A., Mäki-Paakkanen, J. & Komulainen, H. (1998b) Effects of low-frequency magnetic fields on fetal development in CBA/Ca mice. *Bioelectromagnetics*, **19**, 477–485

Huuskonen, H., Saastamoinen, V., Komulainen, H., Laitinen, J. & Juutilainen, J. (2001a) Effects of low-frequency magnetic fields on implantation in rats. *Reprod. Toxicol.*, **15**, 49–59

Huuskonen, H., Juutilainen, J. & Komulainen, H. (2001b) Development of preimplantation mouse embryos after exposure to a 50 Hz magnetic field *in vitro*. *Toxicol. Lett.*, **122**, 149–155

IARC (1984) *IARC Monographs on the Evaluation of the Carcinogenic Risk of Chemicals to Humans*, Vol. 34, *Polynuclear Aromatic Compounds, Part 3, Industrial Exposures in Aluminium Production, Coal Gasification, Coke Production, and Iron and Steel Founding*, Lyon, IARC*Press*, pp. 37–64

IARC (1985) *IARC Monographs on the Evaluation of the Carcinogenic Risk of Chemicals to Humans*, Vol. 35, *Polynuclear Aromatic Compounds, Part 4, Bitumens, Coal-tars and Derived Products, Shale-oils and Soots*, Lyon, IARC*Press*, pp. 83–159

IARC (1987a) *IARC Monographs on the Evaluation of Carcinogenic Risks to Humans*, Suppl. 7, *Overall Evaluations of Carcinogenicity: An Updating of* IARC Monographs *Volumes 1 to 42*, Lyon, IARCPress, pp. 89–91

IARC (1987b) *IARC Monographs on the Evaluation of Carcinogenic Risks to Humans*, Suppl. 7, *Overall Evaluations of Carcinogenicity: An Updating of* IARC Monographs *Volumes 1 to 42*, Lyon, IARCPress, pp. 174–175

IARC (1990) *IARC Monographs on the Evaluation of Carcinogenic Risks to Humans*, Vol. 49, *Chromium, Nickel and Welding*, Lyon, IARCPress, pp. 447–525

IEC (International Electrotechnical Commission) (1998) *Measurement of Low Frequency Magnetic and Electric Fields with Regard to Exposure of Human Beings — Special Requirements for Instruments and Guidance for Measurements* (IEC 61786 (1998-08)), Geneva

IEEE (Institute of Electrical and Electronics Engineers) (1995a) *Standard Procedures for Measurement of Power Frequency Electric and Magnetic Fields From AC Power Lines* (IEEE Standard 644-1994), New York

IEEE (Institute of Electrical and Electronics Engineers) (1995b) *Recommended Practice for Instrumentation: Specifications for Magnetic Flux Density and Electric Field Strength Meters- 10Hz to 3 kHz* (IEEE Standard 1308-1994), New York

Ikehata, M., Koana, T., Suzuki, Y., Shimizu, H. & Nakagawa, M. (1999) Mutagenicity and co-mutagenicity of static magnetic fields detected by bacterial mutation assay. *Mutat. Res.*, **427**, 147–156

International Commission on Non-Ionizing Radiation Protection (ICNIRP) (1998) Guidelines for limiting exposure to time-varying electric, magnetic and electromagnetic fields (up to 300 GHz). *Health Phys.*, **74**, 494–522

Irgens, A., Kruger, K., Skorve, A.H. & Irgens, L.M. (1997) Male proportion in offspring of parents exposed to strong static and extremely low-frequency electromagnetic fields in Norway. *Am. J. ind. Med.*, **32**, 557–561

Ismael, S.J., Callera, F., Garcia, A.B., Baffa, O. & Falcao R.P. (1998) Increased dexamethasone-induced apoptosis of thymocytes from mice exposed to long-term extremely low frequency magnetic fields. *Bioelectromagnetics*, **19**, 131–135

Jackson, J.D. (1992) Are the stray 60-Hz electromagnetic fields associated with the distribution and use of electric power a significant cause of cancer? *Proc. natl Acad. Sci. (USA)*, **89**, 3508–3510

Jacobson-Kram, D., Tepper, J., Kuo, P., San, R.H., Curry, P.T., Wagner, V.O. & Putman, D.L. (1997) Evaluation of potential genotoxicity of pulsed electric and electromagnetic fields used for bone growth stimulation. *Mutat. Res.*, **388**, 45–57

Jaffa K.C., Kim, H. & Aldrich T.E. (2000) The relative merits of contemporary measurements and historical calculated fields in the Swedish childhood cancer study. *Epidemiology*, **11**, 353–356

Jankovic, B.D., Maric, D., Ranin, J. & Veljic, J. (1991) Magnetic fields, brain and immunity: effect on humoral and cell-mediated immune responses. *Int. J. Neurosci.*, **59**, 25–43

Jankovic, B.D., Jovanova-Nesic, J. & Nikolic, V. (1993a) Locus ceruleus and immunity. III. Compromised immune function (antibody production, hypersensitivity skin reactions and experimental allergic encephalomyelitis) in rats with lesioned locus ceruleus is restored by magnetic fields applied to the brain. *Int. J. Neurosci.*, **69**, 251–269

Jankovic, B.D., Jovanova-Nesic, J., Nikolic, V. & Nikolic, P. (1993b) Brain-applied magnetic fields and immune response: role of the pineal gland. *Int. J. Neurosci.*, **70**, 127–134

Jankovic, B.D., Nikolic, P., Cupic, V. & Hladni, K. (1994) Potentiation of immune responsiveness in aging by static magnetic fields applied to the brain: role of the pineal gland. *Ann. N.Y. Acad. Sci.*, **719**, 410–418

Jeffers, D.E. (1996) Comment on the paper: Enhanced deposition of radon daughter nuclei in the vicinity of power frequency electromagnetic fields. *Int. J. Radiat. Biol.*, **69**, 651–652

Jeffers, D.E. (1999) Effects of wind and electric fields on ^{218}Po deposition from the atmosphere. *Int. J. Radiat. Biol.*, **75**, 1533–1539

Jenrow, K.A., Smith, C.H. & Liboff, A.R. (1995) Weak extremely low frequency magnetic fields and regeneration in the planarian Dugesia tigrina. *Bioelectromagnetics*, **16**, 106–112

Jenrow, K.A., Smith, C.H. & Liboff, A.R. (1996) Weak extremely-low-frequency magnetic field-induced regeneration anomalies in the planarian *Dugesia tigrina*. *Bioelectromagnetics*, **17**, 467–474

Johansen, C. & Olsen, J.H. (1998) Risk of cancer among Danish utility workers — A nationwide cohort study. *Am. J. Epidemiol.*, **147**, 548–555

John, T.M., Liu, G.Y. & Brown, G.M. (1998) 60 Hz magnetic field exposure and urinary 6-sulphatoxymelatonin levels in the rat. *Bioelectromagnetics*, **19**, 172–180

Johnson, G.B. (1998) Instrumentation and measurement technology. In: Bracken, T.D. & Montgomery, J.H., eds, *Proceedings of EMF Engineering Review Symposium, Status and Summary of EMF Engineering Research*, Oak Ridge, TN, Oak Ridge National Laboratory. Available at http://www.emf-data.org/symposium98.html

Johnson, C.C. & Spitz, M.R. (1989) Childhood nervous system tumours: An assessment of risk associated with paternal occupations involving use, repair or manufacture of electrical and electronic equipment. *Int. J. Epidemiol.*, **18**, 756–762

Juutilainen, J. (1986) Effects of low frequency magnetic fields on chick embryos. Dependence on incubation temperature and storage of the eggs. *Z. Naturforsch.*, **41C**, 1111–1115

Juutilainen, J. & Liimatainen, A. (1986) Mutation frequency in *Salmonella* exposed to weak 100-Hz magnetic fields. *Hereditas*, **104**, 145–147

Juutilainen, J. & Saali, K. (1986) Development of chick embryos in 1 Hz to 100 kHz magnetic fields. *Radiat. environ. Biophys.*, **25**, 135–140

Juutilainen, J., Harri, M., Saali, K. & Lahtinen, T. (1986) Effects of 100-Hz magnetic fields with various waveforms on the development of chick embryos. *Radiat. Environ. Biophys.*, **25**, 65–74

Juutilainen, J., Läära, E. & Saali, K. (1987) Relationship between field strength and abnormal development in chick embryos exposed to 50 Hz magnetic fields. *Int. J. Radiat. Biol.*, **52**, 787–793

Juutilainen, J., Läärä, E. & Pukkala, E. (1990) Incidence of leukaemia and brain tumours in Finnish workers exposed to ELF magnetic fields. *Int. Arch. occup. environ. Health*, **62**, 289–293

Juutilainen, J., Matilainen, P., Saarikoski, S., Laara, E. & Suonio, S. (1993) Early pregnancy loss and exposure to 50-Hz magnetic fields. *Bioelectromagnetics*, **14**, 229–236

Juutilainen, J., Stevens, R.G., Anderson, L.E., Hansen, N.H., Kilpeläinen, M., Kumlin, T., Laitinen, J.T., Sobel, E. & Wilson, B.W. (2000) Nocturnal 6-hydroxymelatonin sulfate excretion in female workers exposed to magnetic fields. *J. pineal Res.*, **28**, 97–104

Kang, K.-I., Bouhouche, I., Fortin, D., Baulieu, E.E. & Catelli, M.G. (1998) Luciferase activity and synthesis of Hsp70 and Hsp90 are insensitive to 50Hz electromagnetic fields. *Life Sci.*, **63**, 489–497

Kangarlu, A., Burgess, R.E., Zhu, H., Nakayama, T., Hamlin, R.L., Abduljalil, A.M. & Robataille, P.M.L. (1999) Cognitive, cardiac, and physiological safety studies in ultra high field magnetic resonance imaging. *Magn. Reson. Imaging*, **17**, 1407–1416

Kato, M., Honma, K., Shigemitsu, T. & Shiga, Y. (1993) Effects of exposure to a circularly polarized 50-Hz magnetic field on plasma and pineal melatonin levels in rats. *Bioelectromagnetics*, **14**, 97–106

Kato, M., Honma, K., Shigemitsu, T. & Shiga, Y. (1994a) Circularly polarized, sinusoidal, 50 Hz magnetic field exposure does not influence plasma testosterone levels of rats. *Bioelectromagnetics*, **15**, 513–518

Kato, M., Honma, K., Shigemitsu, T. & Shiga, Y. (1994b) Recovery of nocturnal melatonin concentration takes place within one week following cessation of 50 Hz circularly polarized magnetic field exposure for six weeks. *Bioelectromagnetics*, **15**, 489–492

Kato, M., Honma, K., Shigemitsu, T. & Shiga, Y. (1994c) Circularly polarized 50-Hz magnetic field exposure reduces pineal gland and blood melatonin concentrations of Long-Evans rats. *Neurosci. Lett.*, **166**, 59–62

Kato, M., Honma, K., Shigemitsu, T. & Shiga, Y. (1994d) Horizontal or vertical 50-Hz, 1-microT magnetic fields have no effect on pineal gland or plasma melatonin concentration of albino rats. *Neurosci. Lett.*, **168**, 205–208

Kaune, W.T. (1981a) Interactive effects in 60-Hz electric-field exposure systems. *Bioelectromagnetics*, **2**, 33–50

Kaune, W.T. (1981b) Power-frequency electric fields averaged over the body surfaces of grounded humans and animals. *Bioelectromagnetics*, **2**, 403–406

Kaune, W.T., Phillips, R.D., Hjeresen, D.L., Richardson, R.L. & Beamer, J.L. (1978) A method for the exposure of miniature swine to vertical 60-Hz electric fields. *IEEE Trans. biomed. Eng.*, **25**, 276–283

Kaune, W.T., Stevens, R.G., Callahan, N.J., Severson, R.K. & Thomas, D.B. (1987) Residential magnetic and electric fields. *Bioelectromagnetics*, **8**, 315–335

Kaune, W.T., Darby, S.D., Gardner, S.N., Hrubec, Z., Iriye, R.N. & Linet, M.S. (1994) Development of a protocol for assessing time-weighted average exposures of young children to power-frequency magnetic fields. *Bioelectromagnetics*, **15**, 33–51

Kaune, W.T., Bracken, T.D., Senior, R.S., Rankin, R.F., Niple, J.C. & Kavet, R. (2000) Rate of occurrence of transient magnetic field events in US residences. *Bioelectromagnetics*, **21**, 197–213

Kavet, R., Zaffanella, L.E., Daigle, J.P. & Ebi, K.L. (2000) The possible role of contact current in cancer risk associated with residential magnetic fields. *Bioelectromagnetics*, **21**, 538–553

Kavet, R., Stuchly, M.A., Bailey, W.H. & Bracken, T.D. (2001) Evaluation of biological effects, dosimetric models and exposure assessment related to ELF electric- and magnetic-field guidelines. *Appl. occup. environ. Hyg.*, **16**, 1118–1138

Kazantzis, N., Podd, J. & Whittington, C. (1996) Acute effects of 50 Hz, 100 µT magnetic field exposure on visual duration discrimination at two different times of the day. *Bioelectromagnetics*, **19**, 310–317

Kelsh, M.A., Kheifets, L. & Smith, R. (2000) The impact of work environment, utility, and sampling design on occupational magnetic field exposure summaries. *Am. ind. Hyg. Assoc. J.*, **61**, 174–182

Khalil, A.M. & Qassem, W. (1991) Cytogenetic effects of pulsing electromagnetic field on human lymphocytes *in vitro*: chromosome aberrations, sister-chromatid exchanges and cell kinetics. *Mutat. Res.*, **247**, 141–146

Kharazi, A.I., Babbitt, J.T. & Hahn, T.J. (1999) Primary brain tumor incidence in mice exposed to split-dose ionizing radiation and circularly polarized 60 Hz magnetic fields. *Cancer Lett.*, **147**, 149–156

Kheifets, L.I., Abdelmonem, A.A., Buffler, P.A. & Zhang, Z.W. (1995) Occupational electric and magnetic field exposure and brain cancer: A meta-analysis. *J. occup. environ. Med.*, **37**, 1327–1341

Kheifets, L.I., Kavet, R. & Sussman, S.S. (1997a) Wire codes, magnetic fields, and childhood cancer. *Bioelectromagnetics*, **18**, 99–110

Kheifets, L.I., London, S.J. & Peters, J.M. (1997b) Leukemia risk and occupational electric field exposure in Los Angeles County, California. *Am. J. Epidemiol.*, **146**, 87–90

Kheifets, L.I., Gilbert, E.S., Sussman, S.S., Guénel, P., Sahl, J.D., Savitz, D.A. & Thériault, G. (1999) Comparative analyses of the studies of magnetic fields and cancer in electric utility workers: Studies from France, Canada, and the United States. *Occup. environ. Med.*, **56**, 567–574

Kinouchi, Y., Yamaguchi, H. & Tenforde, T.S. (1996) Theoretical analysis of magnetic field interactions with aortic blood flow. *Bioelectromagnetics*, **17**, 21–32

Kirschvink, J.L. (1992) Uniform magnetic fields and double-wrapped coil systems: improved techniques for the design of bioelectromagnetic experiments. *Bioelectromagnetics*, **13**, 401–411

Kirschvink, J.L., Kobayashi-Kirschvink, A. & Woodford, B.J. (1992) Magnetite bio-mineralization in the human brain. *Proc. natl Acad. Sci. USA*, **89**, 7683–7687

Kirschvink, J.L., Padmanabha, S., Boyce, C.K. & Oglesby, J. (1997) Measurement of the threshold sensitivity of honeybees to weak, extremely low-frequency magnetic fields. *J. Exp. Biol.*, **200**, 1363–1368

Kleinerman, R.A., Linet, M.S., Hatch, E.E., Wacholder, S., Tarone, R.E., Severson, R.K., Kaune, W.T., Friedman, D.R., Haines, C.M., Muirhead, C.R., Boice, J.D., Jr & Robison, L.L. (1997) Magnetic field exposure assessment in a case–control study of childhood leukemia. *Epidemiology*, **8**, 575–583

Kleinerman, R.A., Kaune, W.T., Hatch, E.E., Wacholder, S., Linet, M.S., Robison, L.L., Niwa, S. & Tarone, R.E. (2000) Are children living near high-voltage power lines at increased risk of acute lymphoblastic leukemia? *Am. J. Epidemiol.*, **151**, 512–515

Kliukiene, J., Tynes, T., Martinsen, J.I., Blaasaas, K.G. & Andersen, A. (1999) Incidence of breast cancer in a Norwegian cohort of women with potential workplace exposure to 50 Hz magnetic fields. *Am. J. ind. Med.*, **36**, 147–154

Koana, T., Okada, M.O., Ikehata, M. & Nakagawa, M. (1997) Increase in the mitotic recom-bination frequency in *Drosophila melanogaster* by magnetic field exposure and its suppression by vitamin E supplement. *Mutat. Res.*, **373**, 55–60

Koana, T., Okada, M.-O., Takashima, Y., Ikehata, M. & Miyakoshi, J. (2001) Involvement of eddy currents in the mutagenicity of ELF magnetic fields. *Mutat. Res.*, **476**, 55–62

Koifman, S., Ferraz, I., Viana, T.S., Silveira, C.L., Carneiro, M.T.D., Koifman, R.J., Sarcinelli, P.N., de Cássia, C., Mattos, R., Lima, J.S., Silva, J.J.O., Moreira, J.C., de Fátima, A., Ferreira, M., Fernandes, C. & Bulcão, A.C. (1998) Cancer cluster among young Indian adults living near power transmission lines in Bom Jesus do Tocantins, Pará, Brazil. *Cad. Saude públ.*, **14** (Suppl. 3), 161–172

Konermann, G. & Mönig, H. (1986) [Studies on the influence of static magnetic fields on prenatal development of mice.] *Radiologie*, **26**, 490–497 (in German)

König, H.L., Krueger, A.P., Lang, S. & Sönning, W. (1981) *Biologic Effects of Environmental Electromagnetism*, New York, Springer-Verlag

Korneva, H.A., Grigoriev, V.A., Isaeva, E.N., Kaloshina, S.M. & Barnes, F.S. (1999) Effects of low-level 50 Hz magnetic fields on the level of host defense and on spleen colony formation. *Bioelectromagnetics*, **20**, 57–63

Korzh-Sleptsova, I.L., Lindström, E., Mild, K.H., Berglund, A. & Lundgren, E. (1995) Low frequency MFs increased inositol 1,4,5-triphosphate levels in the Jurkat cell line. *FEBS Lett.*, **359**, 151–154

Kowalczuk, C.I. & Saunders, R.D. (1990) Dominant lethal studies in male mice after exposure to a 50-Hz electric field. *Bioelectromagnetics*, **11**, 129–137

Kowalczuk, C.I., Sienkiewicz, Z.J. & Saunders, R.D. (1991) *Biological Effects of Exposure to Non-ionizing Electromagnetic Fields and Radiation. 1. Static Electric and Magnetic Fields* (NRPB-R238), Didcot, National Radiological Protection Board

Kowalczuk, C.I., Robbins, L., Thomas, J.M., Butland, B.K. & Saunders, R.D. (1994) Effects of prenatal exposure to 50 Hz magnetic fields on development in mice: I. Implantation rate and fetal development. *Bioelectromagnetics*, **15**, 349–361

Kowalczuk, C.I., Robbins, L., Thomas, J.M. & Saunders, R.D. (1995) Dominant lethal studies in male mice after exposure to a 50 Hz magnetic field. *Mutat. Res.*, **328**, 229–237

Kristupaitis, D., Dibirdik, I., Vassilev, A., Mahajan, S., Kurosaki, T., Chu, A., Tuel-Ahlgren, L., Tuong, D., Pond, D., Luben, R. & Uckun, F.M. (1998) Electromagnetic field-induced stimulation of Bruton's tyrosine kinase. *J. biol. Chem.*, **273**, 12397–12401

Kuijten, R.R., Bunin, G.R., Nass, C.C. & Meadows, A.T. (1992) Parental occupation and childhood astrocytoma: Results of a case–control study. *Cancer Res.*, **52**, 782–786

Kumlin, T., Kosma, V.M., Alhonen, L., Janne, J., Komulainen, H., Lang, S., Rytomaa, T., Servomaa, K. & Juutilainen, J. (1998a) Effects of 50 Hz magnetic fields on UV-induced skin tumorigenesis in ODC-transgenic and non-transgenic mice. *Int. J. Radiat. Biol.*, **73**, 113–121

Kumlin, T., Alhonen, L., Janne, J., Lang, S., Kosma, V.M. & Juutilainen, J. (1998b) Epidermal ornithine decarboxylase and polyamines in mice exposed to 50 Hz magnetic fields and UV radiation. *Bioelectromagnetics*, **19**, 388–391

Kwee, S. & Raskmark, P. (1995) Changes in cell proliferation due to environmental non-ionizing radiation. 1. ELF electromagnetic fields. *Bioelectrochem. Bioenerg.*, **36**, 109–114

Lacy-Hulbert, A., Wilkins, R.C., Hesketh, T.R. & Metcalfe, J.C. (1995) No effect of 60 Hz electromagnetic fields on *MYC* or β-actin expression in human leukemic cells. *Radiat. Res.*, **144**, 9–17

Laden, F., Neas, L.M., Tolbert, P.E., Holmes, M.D., Hankinson, S.E., Spiegelman, D., Speizer, F.E. & Hunter, D.J. (2000) Electric blanket use and breast cancer in the Nurses' Health Study. *Am. J. Epidemiol.*, **152**, 41–49

Lagroye, I. & Poncy, J.L. (1997) The effect of 50 Hz electromagnetic fields on the formation of micronuclei in rodent cell lines exposed to gamma radiation. *Int. J. Radiat. Biol.*, **72**, 249–254

Lagroye, I. & Poncy, J.L. (1998) Influence of 50-Hz magnetic fields and ionizing radiation on *c-jun* and *c-fos* oncoproteins. *Bioelectromagnetics*, **19**, 112–116

Lai, H. (1996) Spatial learning deficit in the rat after exposure to a 60 Hz magnetic field. *Bioelectromagnetics*, **17**, 494–496

Lai, H. & Singh, N.P. (1997a) Acute exposure to a 60 Hz magnetic field increases DNA strand breaks in rat brain cells. *Bioelectromagnetics*, **18**, 156–165

Lai, H. & Singh, N.P. (1997b) Melatonin and N-*t*-butyl-alpha-phenylnitrone block 60-Hz magnetic field-induced DNA single and double strand breaks in rat brain cells. *J. Pineal Res.*, **22**, 152–162

Lai, H., Carino, M.A. & Ushijima, I. (1998) Acute exposure to a 60 Hz magnetic field affects rats' water-maze performance. *Bioelectromagnetics*, **19**, 117–122

Leal, J., Shamsaifar, K., Trillo, M.A., Ubeda, A., Abraira, V. & Chacon, L. (1989) Embryonic development and weak changes of the geomagnetic field. *J. Bioelectr.*, **7**, 141–153

Lednev. V.V. (1991) Possible mechanism for the influence of weak magnetic fields on biological systems. *Bioelectromagnetics*, **12**, 71–75

Lee, J.M., Stormshak, F., Thompson, J.M., Thinesen, P., Painter, L.J., Olenchek, E.G., Hess, D.L., Forbes, R. & Foster, D.L. (1993) Melatonin secretion and puberty in female lambs exposed to environmental electric and magnetic fields. *Biol. Reprod.*, **49**, 857–864

Lee, J.M., Stormshak, F., Thompson, J.M., Hess, D.L. & Foster, D.L. (1995) Melatonin and puberty in female lambs exposed to EMF; a replicate study. *Bioelectromagnetics*, **16**, 119–123

Lee, G.M., Neutra, R.R., Hristova, L., Yost, M. & Hiatt, R.A. (2000) The use of electric bed heaters and the risk of clinically recognized spontaneous abortion. *Epidemiology*, **11**, 406–415

Lerchl, A., Reiter, R.J., Howes, K.A., Nonaka, K.O. & Stokkan, K.-A. (1991) Evidence that extremely low frequency Ca2+-cyclotron resonance depresses pineal melatonin synthesis *in vitro*. *Neurosci. Lett.*, **124**, 213–215

Levin, M. & Ernst, S.G. (1997) Applied DC magnetic fields cause alterations in the time of cell divisions and developmental abnormalities in early sea urchin embryos. *Bioelectromagnetics*, **18**, 255–263

Li, C.-Y., Thériault, G. & Lin, R.S. (1997) Residential exposure to 60-Hertz magnetic fields and adult cancers in Taiwan. *Epidemiology*, **8**, 25–30

Li, C.-Y., Lee, W.-C. & Lin, R.S. (1998) Risk of leukemia in children living near high-voltage transmission lines. *J. occup. environ. Med.*, **40**, 144–147

Li, C.M., Chiang, H., Fu, Y.D., Shao, B.J., Shi, J.R. & Yao, G.D. (1999) Effects of 50 Hz magnetic fields on gap junctional intercellular communication. *Bioelectromagnetics*, **20**, 290–294

Li, D.-K., Checkoway, H. & Mueller, B.A. (1995) Electric blanket use during pregnancy in relation to the risk of congenital urinary tract anomalies among women with a history of subfertility. *Epidemiology*, **6**, 485–489

Li, D.K., Odouli, R., Wi, S., Janevic, T., Golditch, I., Bracken, T.D., Senior, R., Rankin, R. & Iriye, R. (2002) A population-based prospective cohort study of personal exposure to magnetic fields during pregnancy and the risk of miscarriage. *Epidemiology*, **13**, 9–20

Liboff, A.R. (1985) Geomagnetic cyclotron resonance in living cells. *J. biol. Phys.*, **13**, 99–102

Liboff, A.R. & Parkinson, W.C. (1991) Search for ion-cyclotron resonance in a Na⁺-transport system. *Bioelectromagnetics*, **12**, 77–83

Liboff, A.R., Williams, T., Jr, Strong, D.M. & Wistar, R., Jr (1984) Time-varying magnetic fields: Effect on DNA synthesis. *Science*, **223**, 818–820

Liboff, A.R., Thomas, J.R. & Schrot, J. (1989) Intensity threshold for 60-Hz magnetically induced behavioral changes in rats. *Bioelectromagnetics*, **10**, 111–113

Liboff, A.R. Jenrow, K.A. & McLeod, B.R. (1993) ELF-induced proliferation at 511 mG in HSB-2 cell culture as a function of 60 Hz field intensity. In: Blank, M., ed., *Electricity and Magnetism in Biology and Medicine*, San Francisco, CA, San Francisco Press Inc., pp. 623–627

Liburdy, R.P., Sloma, T.R., Sokolic, R.R. & Yaswen, P. (1993) ELF magnetic fields, breast cancer, and melatonin: 60 Hz fields block melatonin's oncostatic action on ER⁺ breast cancer cell proliferation. *J. pineal Res.*, **14**, 89–97

Lin, R.S. & Lee, W.C. (1994) Risk of childhood leukemia in areas passed by high power lines. *Rev. environ. Health*, **10**, 97–103

Lin, R.S., Dischinger, P.C., Conde, J. & Farrell, K.P. (1985) Occupational exposure to electromagnetic fields and the occurrence of brain tumors. An analysis of possible associations. *J. occup. Med.*, **27**, 413–419

Lin, H., Goodman, R. & Shirley-Henderson, A. (1994) Specific region of the c-*myc* promoter is responsive to electric and magnetic fields. *J. cell. Biochem.*, **54**, 281–288

Lin, H., Han, L., Blank, M., Head, M. & Goodman, R. (1998a) Magnetic field activation of protein–DNA binding. *J. cell. Biochem.*, **70**, 297–303

Lin, H., Head, M., Blank, M., Han, L., Jin, M. & Goodman, R. (1998b) *Myc*-mediated transactivation of HSP70 expression following exposure to magnetic fields. *J. cell. Biochem.*, **69**, 181–188

Lindbohm, M.-L., Hietanen, M., Kyyronen, P., Sallmen, M., von Nandelstadh, P., Taskinen, H., Pekkarinen, M., Ylikoski, M. & Hemminki, K. (1992) Magnetic fields of video display terminals and spontaneous abortion. *Am. J. Epidemiol.*, **136**, 1041–1051

Lindström, E., Lindström, P., Berglund, A., Hansson Mild, K. & Lundgren, E. (1993) Intracellular calcium oscillations induced in a T-cell line by a weak 50 Hz magnetic field. *J. cell. Physiol.*, **156**, 395–398

Lindström, E., Lindström, P., Berglund, A., Lundgren, E. & Hansson Mild, K. (1995a) Intracellular calcium oscillations in a T-cell line after exposure to extremely-low-frequency magnetic fields with variable frequencies and flux densities. *Bioelectromagnetics*, **16**, 41–47

Lindström, E., Berglund, A., Mild, K.H., Lindström, P. & Lundgren, E. (1995b) CD45 phosphatase in Jurkat cells is necessary for response to applied ELF magnetic fields. *FEBS Lett.*, **370**, 118–122

Lindström, E., Lindström, P., Berglund, A., Mild, K.H. & Lundgren, E. (1996) Intracellular calcium oscillations induced in a T-cell line by a weak 50 Hz magnetic field. *J. cell. Physiol.*, **156**, 395–398

Linet, M.S., Hatch, E.E., Kleinerman, R.A., Robison, L.L., Kaune, W.T., Friedman, D.R., Severson, R.K., Haines, C.M., Hartsock, C.T., Niwa, S., Wacholder, S. & Tarone, R.E. (1997) Residential exposure to magnetic fields and acute lymphoblastic leukemia in children. *New Engl. J. Med.*, **337**, 1–7

Litovitz, T.A., Krause, D. & Mullins, J.M. (1991) Effect of coherence time of the applied magnetic field on ornithine decarboxylase activity. *Biochem. biophys. Res. Commun.*, **178**, 862–865

Livingston, G.K., Witt, K.L., Gandhi, O.P., Chatterjee, I. & Roti Roti, J.L. (1991) Reproductive integrity of mammalian cells exposed to power frequency electromagnetic fields. *Environ. mol. Mutag.*, **17**, 49–58

London, S.J., Thomas, D.C., Bowman, J.D., Sobel, E., Cheng, T.-C. & Peters, J.M. (1991) Exposure to residential electric and magnetic fields and risk of childhood leukemia. *Am. J. Epidemiol.*, **134**, 923–937

London, S.J., Thomas, D.C., Bowman, J.D., Sobel, E., Cheng, T.-C. & Peters, J.M. (1993) Exposure to residential electric and magnetic fields and risk of childhood leukemia. *Am. J. Epidemiol.*, **137**, 381 (Erratum)

London, S.J., Bowman, J.D., Sobel, E., Thomas, D.C., Garabrant, D.H., Pearce, N., Bernstein, L. & Peters, J.M. (1994) Exposure to magnetic fields among electrical workers in relation to leukemia risk in Los Angeles County. *Am. J. ind. Med.*, **26**, 47–60

Lonne-Rahm, S., Andersson, B., Melin, L., Schultzberg, M., Arnetz, B. & Berg, M. (2000) Provocation with stress and electricity of patients with 'sensitivity to electricity'. *J. occup. environ. Med.*, **42**, 512–516

Loomis, D.P. (1992) Cancer of breast among men in electrical occupations (Letter to the Editor). *Lancet*, **339**, 1482–1483

Loomis, D.P. & Savitz, D.A. (1990) Mortality from brain cancer and leukaemia among electrical workers. *Br. J. ind. Med.*, **47**, 633–638

Loomis, D.P., Peipins, L.A., Browning, S.R., Howard, R.L., Kromhout, H. & Savitz, D.A. (1994a) Organization and classification of work history data in industry-wide studies: An application to the electric power industry. *Am. J. ind. Med.*, **26**, 413–425

Loomis, D.P., Savitz, D.A. & Ananth, C.V. (1994b) Breast cancer mortality among female electrical workers in the United States. *J. natl Cancer Inst.*, **86**, 921–925

Loomis, A., Kromhout, H., Kleckner, R.C. & Savitz, D.A. (1998) Effects of the analytical treatment of exposure data on associations of cancer and occupational magnetic field exposure. *Am. J. ind. Med.*, **34**, 49–56

Lorimore, S.A., Kowalczuk, C.I., Saunders, R.D. & Wright, E.G. (1990) Lack of acute effects of 20 mT, 50 Hz magnetic fields on murine haemopoiesis. *Int. J. Radiat. Biol.*, **58**, 713–723

Löscher, W. & Mevissen, M. (1995) Linear relationship between flux density and tumor co-promoting effect of prolonged magnetic field exposure in a breast cancer model. *Cancer Lett.*, **96**, 175–180

Löscher, W., Mevissen, M., Lehmacher, W. & Stamm, A. (1993) Tumor promotion in a breast cancer model by exposure to a weak alternating magnetic field. *Cancer Lett.*, **71**, 75–81

Löscher, W., Wahnschaffe, U., Mevissen, M., Lerchl, A. & Stamm, A. (1994) Effects of weak alternating magnetic fields on nocturnal melatonin production and mammary carcinogenesis in rats. *Oncology*, **51**, 288–295

Löscher, W., Mevissen, M. & Lerchl, A. (1998) Exposure of female rats to a 100-microT 50 Hz magnetic field does not induce consistent changes in nocturnal levels of melatonin. *Radiat. Res.*, **150**, 557–567

Lovely, R.H. (1988) Recent Studies in the behavioral toxicology of ELF electric and magnetic fields. In: O'Connor, M.E. & Lovely, R.H., eds, *Electromagnetic Fields and Neurobehavioral Function*, New York, A.R. Liss, pp. 327–347

Lovely, R.H., Creim, J.A., Kaune, W.T., Miller, M.C., Phillips, R.D. & Anderson, L.E. (1992) Rats are not aversive when exposed to 60-Hz magnetic fields at 3.03 mT. *Bioelectromagnetics*, **13**, 351–362

Lovely, R.H., Buschbom, R.L., Slavich, A.L., Anderson, L.E., Hansen, N.H. & Wilson, B.W. (1994) Adult leukemia risk and personal appliance use: A preliminary study. *Am. J. Epidemiol.*, **140**, 510–517

Lövsund, P., Öberg, P.Å. & Nilsson, S.E.G. (1979) Influence on vision of extremely low frequency electromagnetic fields. Industrial measurements, magnetophosphene studies in volunteers and intraretinal studies in animals. *Acta Ophthalmol.*, **57**, 812–821

Lövsund, P., Öberg, P.Å. & Nilsson, S.E.G. (1980a) Magneto- and electrophosphenes: a comparative study. *Med. Biol. Eng. Comput.*, **18**, 758–764

Lövsund, P., Öberg, P.Å., Nilsson, S.E.G. & Reuter, T. (1980b) Magnetophosphenes: a qualitative analysis of thresholds. *Med. Biol. Eng. Comput.*, **18**, 326–334

Lövsund, P., Öberg, P.Å. & Nilsson, S.E.G. (1982) ELF magnetic fields in electrosteel and welding industries. *Radio Sci.*, **17**, 35S–38S

Lundsberg., L.S., Bracken, M.B. & Belanger, K. (1995) Occupationally related magnetic field exposure and male subfertility. *Fertil. Steril.*, **63**, 384–391

Lyle, D.B., Fuchs, T.A., Casamento, J.P., Davis, C.C. & Swicord, M.L. (1997) Intracellular calcium signaling by Jurkat T-lymphocytes exposed to a 60 Hz magnetic field. *Bioelectromagnetics*, **18**, 439–445

Lyle, D.B., Ayotte, R.D., Sheppard, A.R. & Adey, W.R. (1998) Suppression of T-lymphocyte cytotoxicity following exposure to 60-Hz sinusoidal electric fields. *Bioelectromagnetics*, **9**, 303–313

Lymangrover, J.R., Keku, E. & Seto, Y.J. (1983) 60-Hz electric field alters the steroidogenic response of rat adrenal tissue *in vitro*. *Life Sci.*, **32**, 691–696

Lyskov, E.B., Juutilainen, J., Jousmäki, V., Partanen, J., Medvedev, S. & Hänninen, O. (1993a) Effects of 45-Hz magnetic fields on the functional state of the human brain. *Bioelectromagnetics*, **14**, 87–95

Lyskov, E., Juutilainen, J., Jousmäki, V., Hänninen, O., Medvedev, S. & Partanen, J. (1993b) Influence of short-term exposure of magnetic field on the bioelectrical processes of the brain and performance. *Int. J. Psychophysiol.*, **14**, 227–231

Maddock, B.J. (1992) Overhead line design in relation to electric and magnetic field limits. *Power Engin. J.*, **September**, 217–224

Maes, A., Collier, M., Vandoninck, S., Scarpa, P. & Verschaeve, L. (2000) Cytogenetic effects of 50 Hz magnetic fields of different magnetic flux densities. *Bioelectromagnetics*, **21**, 589–596

Maestroni, G.J., Conti, A. & Pierpaoli, W. (1986) Role of the pineal gland in immunity. Circadian synthesis and release of melatonin modulates the antibody response and antagonizes the immunosuppressive effect of corticosterone. *J. Neuroimmunol.*, **13**, 19–30

Maffeo, S., Miller, M.W. & Carstensen, E.L. (1984) Lack of effect of weak low frequency electromagnetic fields on chick embryogenesis. *J. Anat.*, **139**, 613–618

Maffeo, S., Brayman, A.A., Miller, M.W., Carstensen, E.L., Ciaravino, V. & Cox, C. (1988) Weak low frequency electromagnetic fields and chick embryogenesis: failure to reproduce findings. *J. Anat.*, **157**, 101–104

Male, J.C., Norris, W.T. & Watts, M.W. (1987) Exposure of people to power-frequency electric and magnetic fields. In: Anderson, L.E., Kelman, B.J. & Weigel, R.J., eds, *Interaction of Biological Systems with Static and ELF Electric and Magnetic Fields (Twenty-third Hanford Life Sciences Symposium, October 2–4, 1984, Richland, WA)*, Richland, WA, Pacific Northwest Laboratory, pp. 407–418

Malmivuo, J. & Plonsey, R. (1995) *Bioelectromagnetism*, New York, Oxford Press

Mammi, G.I., Rocchi, R., Cadossi, R., Massari, L. & Traina, G.C. (1993) The electrical stimulation of tibial osteotomies. Double-blind study. *Clin. Orthop. rel. Res.*, **288**, 246–253

Mandeville, R., Franco, E., Sidrac-Ghali, S., Paris-Nadon, L., Rocheleau, N. Mercier, G., Désy, M. & Gaboury, L. (1997) Evaluation of the potential carcinogenicity of 60 Hz linear sinusoidal continuous-wave magnetic fields in Fischer F344 rats. *FASEB J.*, **11**, 1127–1136

Mandeville, R., Franco, E., Sidrac-Ghali, S., Paris-Nadon, L., Rocheleau, N., Mercier, G., Désy, M., Devaux, C. & Gaboury, L. (2000) Evaluation of the potential promoting effect of 60 Hz magnetic fields on N-ethyl-N-nitrosourea induced neurogenic tumors in female F344 rats. *Bioelectromagnetics*. **21**, 84–93

Margonato, V., Veicsteinas, A., Conti, R., Nicolini, P. & Cerretelli, P. (1993) Biologic effects of prolonged exposure to ELF electromagnetic fields in rats. I. 50 Hz electric fields. *Bioelectromagnetics*, **14**, 479–493

Margonato, V., Nicolini, P., Conti, R., Zecca, L., Veicsteinas, A. & Cerretelli, P. (1995) Biologic effects of prolonged exposure to ELF electromagnetic fields in rats: II. 50 Hz magnetic fields. *Bioelectromagnetics*, **16**, 343–355

Marino, A.A., Wolcott, R.M., Chervenak, R., Jourd'Heuil, F., Nilsen, E. & Frilot, C., II (2000) Nonlinear response of the immune system to power-frequency magnetic fields. *Am. J. Physiol. Regul. Integr. Comp. Physiol.*, **279**, R761–R768

Markkanen, A., Juutilainen, J., Lang, S., Pelkonen, J., Rytömaa, T. & Naarala, J. (2001) Effects of 50 Hz magnetic field on cell cycle kinetics and the colony forming ability of budding yeast exposed to ultraviolet radiation. *Bioelectromagnetics*, **22**, 345–350

Martinez-Soriano, F., Gimenez-Gonzalez, M., Armanazas, E. & Ruiz-Torner, A. (1992) Pineal 'synaptic ribbons' and serum melatonin levels in the rat following the pulse action of 52-Gs (50-Hz) magnetic fields: An evolutive analysis over 21 days. *Acta Anat.*, **143**, 289–293

Maruvada, P.S., Jutras, P. & Plante, M. (2000) An investigation to identify possible sources of electromagnetic field transients responsible for exposures reported in recent epidemiological studies. *IEEE Trans. Power Del.*, **15**, 266–271

Massot, O., Grimaldi, B., Bailly, J.-M., Kochanek, M., Deschamps, R., Lambrozo, J. & Fillion, G. (2000) Magnetic field desensitizes 5-HT1B receptor in brain: pharmacological and functional studies. *Brain Res.*, **858**, 143–150

Matanoski, G.M., Breysse, P.N. & Elliott, E.A. (1991) Electromagnetic field exposure and male breast cancer (Letter to the Editor). *Lancet*, **337**, 737

Matanoski, G.M., Elliott, E.A., Breysse, P.N. & Lynberg, M.C. (1993) Leukemia in telephone linemen. *Am. J. Epidemiol.*, **137**, 609–619

Matthes, R. (1998) Response to questions and comments on ICNIRP. *Health Phys.*, **75**, 438–439

Mattos, I.E. & Koifman, S. (1996) [Cancer mortality among electricity utility workers in the state of São Paulo, Brazil.] *Rev. Saúde públ.*, **30**, 564–575 (in Portuguese)

Mattsson, M.-O. & Rehnholm, U. (1993) Gene expression in tumor cell lines after exposure to a 50Hz sinusoidal magnetic field. In: Blank, M., ed., *Electricity and Magnetism in Biology and Medicine*, San Francisco, San Francisco Press, pp. 500–502

McBride, M.L., Gallagher, R.P., Thériault, G., Armstrong, B.G., Tamaro, S., Spinelli, J.J., Deadman, J.E., Fincham, S., Robson, D. & Choi, W. (1999) Power-frequency electric and magnetic fields and risk of childhood leukemia in Canada. *Am. J. Epidemiol.*, **149**, 831–842 (Erratum: *Am. J. Epidemiol.*, 1999, **150**, 223)

McCormick, D.L., Ryan, B.M., Findlay, J.C., Gauger, J.R., Johnson, T.R., Morrissey, R.L. & Boorman, G.A. (1998) Exposure to 60 Hz magnetic fields and risk of lymphoma in PIM transgenic and TSG-*p53* (*p53* knockout) mice. *Carcinogenesis*, **19**, 1649–1653

McCormick, D.L., Boorman, G.A., Findlay, J.C., Hailey, J.R., Johnson, T.R., Gauger, J.R., Pletcher, J.M., Sills, R.C. & Haseman, J.K. (1999) Chronic toxicity/oncogenicity evaluation of 60 Hz (power frequency) magnetic fields in $B6C3F_1$ mice. *Toxicol. Pathol.*, **27**, 279–285

McCredie, M., Maisonneuve, P. & Boyle, P. (1994) Perinatal and early postnatal risk factors for malignant brain tumours in New South Wales children. *Int. J. Cancer*, **56**, 11–15

McDonald, A.D., Cherry, N.M., Delorme, C. & McDonald, J.C. (1986) Visual display units and pregnancy: evidence from the Montreal survey. *J. occup. Med.*, **28**, 1226–1231

McDowall, M.E. (1983) Leukaemia mortality in electrical workers in England and Wales (Letter to the Editor). *Lancet*, **i**, 246

McDowall, M.E. (1986) Mortality of persons resident in the vicinity of electricity transmission facilities. *Br. J. Cancer*, **53**, 271–279

McFarlane, E.H., Dawe, G.S., Marks, M. & Campbell, I.C. (2000) Changes in neurite outgrowth but not in cell division induced by low EMF exposure: Influence of field strength and culture conditions on responses in rat PC12 pheochromocytoma cells. *Bioelectrochemistry*, **52**, 23–28

McGivern, R.F., Sokol, R.Z. & Adey, W.R. (1990) Prenatal exposure to a low-frequency electromagnetic field demasculinizes adult scent marking behavior and increases accessory sex organ weights in rats. *Teratology*, **41**, 1–8

McKay, B.E. & Persinger, M.A. (1999) Geophysical variables and behavior: LXXXVII. Effects of synthetic and natural geomagnetic patterns on maze learning. *Percept. Mot. Skills*, **89**, 1023–1024

McLean, J.R., Stuchly, M.A., Mitchel, R.E., Wilkinson, D., Yang, H., Goddard, M., Lecuyer, D., Schunk, M., Callary, E. & Morrison, D. (1991) Cancer promotion in a mouse-skin model by a 60-Hz magnetic field: II. Tumor development and immune response. *Bioelectromagnetics*, **12**, 273–287

McLean, J., Thansandote, A., Lecuyer, D., Goddard, M., Tryphonas, L., Scaiano, J.C. & Johnson, F. (1995) A 60-Hz magnetic field increases the incidence of squamous cell carcinomas in mice previously exposed to chemical carcinogens. *Cancer Lett.*, **92**, 121–125

McLean, J.R.N., Thansandote, A., Lecuyer, D. & Goddard, M. (1997) The effect of 60-Hz magnetic fields on co-promotion of chemically induced skin tumors on SENCAR mice: A discussion of three studies. *Environ. Health Perspect.*, **105**, 94–96

McLeod, K.J., Lee, R.C. & Ehrlich, H.P. (1987) Frequency dependence of electric field modulation of fibroblast protein synthesis. *Science*, **236**, 1465–1469

McMahan, S., Ericson, J. & Meyer, J. (1994) Depressive symptomatology in women and residential proximity to high-voltage transmission lines. *Am. J. Epidemiol.*, **39**, 58–63

Merchant, C.J., Renew, D.C. & Swanson, J. (1994) Occupational exposure to power-frequency magnetic fields in the electricity supply industry. *J. Radiol. Protect.*, **14**, 155–164

Merritt, R., Purcell, C. & Stroink, G. (1983) Uniform magnetic field produced by three, four and five square coils. *Rev. Sci. Instr.*, **54**, 879–882

Mevissen, M., Stamm, A., Buntenkötter, S., Zwingelberg, R., Wahnschaffe, U. & Löscher, W. (1993) Effects of magnetic fields on mammary tumor development induced by 7,12-dimethylbenz(a)anthracene in rats. *Bioelectromagnetics*, **14**, 131–143

Mevissen, M., Buntenkötter, S. & Löscher, W. (1994) Effects of static and time-varying (50 Hz) magnetic fields on reproduction and fetal development in rats. *Teratology*, **50**, 229–237

Mevissen, M., Kietzmann, M. & Loscher, W. (1995) In vivo exposure of rats to a weak alternating magnetic field increases ornithine decarboxylase activity in the mammary gland by a similar extent as the carcinogen DMSA. *Cancer Lett.*, **90**, 207–214

Mevissen, M., Lerchl, A. & Löscher, W. (1996a) Study of pineal function and DMBA-induced breast cancer formation in rats during exposure to a 100-mG, 50-Hz magnetic field. *J. Toxicol. environ. Health*, **48**, 169–185

Mevissen, M., Lerchl, A., Szamel, M. & Löscher, W. (1996b) Exposure of DMBA-treated female rats in a 50-Hz, 50 μTesla magnetic field: effects on mammary tumor growth, melatonin levels, and T lymphocyte activation. *Carcinogenesis*, **17**, 903–910

Mevissen, M., Haussler, M., Lerchl, A. & Löscher, W. (1998a) Acceleration of mammary tumorigenesis by exposure of 7,12-dimethylbenz[a]anthracene-treated female rats in a 50-Hz, 100-microT magnetic field: replication study. *J. Toxicol. environ. Health*, **53**, 401–418

Mevissen, M., Haussler, M., Szamel, M., Emmendorffer, A., Thun-Battersby, S. & Löscher, W. (1998b) Complex effects of long-term 50 Hz magnetic field exposure in vivo on immune functions in female Sprague-Dawley rats depend on duration of exposure. *Bioelectromagnetics*, **19**, 259–270

Mezei, G., Kheifets, L.I., Nelson, L.M., Mills, K.M., Iriye, R. & Kelsey, J.L. (2001) Household appliance use and residential exposure to 60-Hz magnetic fields. *J. Expos. Anal. environ. Epidemiol.*, **11**, 41–49

Michaelis, J., Schüz, J., Meinert, R., Menger, M., Grigat, J.-P., Kaatsch, P., Kaletsch, U., Miesner, A., Stamm, A., Brinkmann, K. & Kärner, H. (1997) Childhood leukemia and electromagnetic fields: Results of a population-based case–control study in Germany. *Cancer Causes Control*, **8**, 167–174

Michaelis, J., Schüz, J., Meinert, R., Zemann, E., Grigat, J.-P., Kaatsch, P., Kaletsch, U., Miesner, A., Brinkmann, K., Kalkner, W. & Kärner, H. (1998) Combined risk estimates for two German population-based case–control studies on residential magnetic fields and childhood acute leukemia. *Epidemiology*, **9**, 92–94

Milham, S., Jr (1982) Mortality from leukemia in workers exposed to electrical and magnetic fields (Letter to the Editor). *New Engl. J. Med.*, **307**, 249

Milham, S., Jr (1985a) Silent keys: Leukaemia mortality in amateur radio operators (Letter to the Editor). *Lancet*, **i**, 812

Milham, S., Jr (1985b) Mortality in workers exposed to electromagnetic fields. *Environ. Health Perspectives*, **62**, 297–300

Milham, S. & Ossiander, E.M. (2001) Historical evidence that residential electrification caused the emergence of the childhood leukemia peak. *Med. Hypotheses*, **56**, 290–295

Miller, S.C. & Furniss, M.J. (1998) Bruton's tyrosine kinase activity and inositol 1,4,5-tris-phosphate production are not altered in DT40 lymphoma B cells exposed to power line frequency magnetic fields. *J. biol. Chem.*, **273**, 32618–32626

Miller, A.B., To, T., Agnew, D.A., Wall, C. & Green, L.M. (1996) Leukemia following occupational exposure to 60-Hz electric and magnetic fields among Ontario electricity utility workers. *Am. J. Epidemiol.*, **144**, 150–160

Mills, K.M., Kheifets, L.I., Nelson, L.M., Bloch, D.A., Takemoto-Hambleton, R. & Kelsey, J.L. (2000) Reliability of proxy-reported and self-reported household appliance use. *Epidemiology*, **11**, 581–588

Milunsky, A., Ulcickas, M., Rothman, K.J., Willett, W., Jick, S.S. & Jick, H. (1992) Maternal heat exposure and neural tube defects. *J. Am. med. Assoc.*, **268**, 882–885

Minder, C.E. & Pfluger, D.H. (2001) Leukemia, brain tumors, and exposure to extremely low frequency electromagnetic fields in Swiss railway employees. *Am. J. Epidemiol.*, **153**, 825–835

Misakian, M., Sheppard, A. R., Krause, D., Frazier, M. E., & Miller, D. L. (1993) Biological, physical, and electrical parameters for in vitro studies with ELF magnetic and electric fields: a primer. *Bioelectromagnetics*, **Suppl. 2**, S1–S73

Miyakoshi, J., Ohtsu, S., Tatsumi-Miyajima, J. & Takebe, H. (1994) A newly designed experimental system for exposure of mammalian cells to extremely low frequency magnetic fields. *J. Radiat. Res.*, **36**, 26–34

Miyakoshi, J., Yamagishi, N., Ohtsu, S., Mohri, K. & Takebe, H. (1996a) Increase in hypoxanthine–guanine phosphoribosyl transferase gene mutations by exposure to high-density 50-Hz magnetic fields. *Mutat. Res.*, **349**, 109–114

Miyakoshi, J., Ohtsu, S., Shibata, T. & Takebe, H. (1996b) Exposure to magnetic field (5 mT at 60 Hz) does not affect cell growth and *c-myc* gene expression. *J. Radiat. Res.*, **37**, 185–191

Miyakoshi, J., Kitagawa, K. & Takebe, H. (1997) Mutation induction by high-density, 50-Hz magnetic fields in human MeWo cells exposed in the DNA synthesis phase. *Int. J. Radiat. Biol.*, **71**, 75–79

Miyakoshi, J., Mori, Y., Yamagishi, N., Yagi, K. & Takebe, H. (1998a) Suppression of high-density magnetic field (400 mT at 50 Hz)-induced mutations by wild-type p53 expression in human osteosarcoma cells. *Biochem. biophys. Res. Commun.*, **243**, 579–584

Miyakoshi, J., Tsukada, T., Tachiiri, S., Bandoh, S., Yamaguchi, K. & Takebe, H. (1998b) Enhanced NOR-1 gene expression by exposure of Chinese hamster cells to high-density 50 Hz magnetic fields. *Mol. cell. Biochem.*, **181**, 191–195

Miyakoshi, J., Koji, Y., Wakasa, T. & Takebe, H. (1999) Long-term exposure to a magnetic field (5 mT at 60 Hz) increases X-ray-induced mutations. *J. Radiat. Res.*, **40**, 13–21

Miyakoshi, J., Yoshida, M., Shibuya, K. & Hiraoka, M. (2000a) Exposure to strong magnetic fields at power frequency potentiates X-ray-induced DNA strand breaks. *J. Radiat. Res.*, **41**, 293–302

Miyakoshi, J., Mori, Y., Yaguchi, H., Ding, G. & Fujimori, A. (2000b) Suppression of heat-induced hsp-70 by simultaneous exposure to 50 mT magnetic field. *Life Sci.*, **66**, 1187–1196

Miyakoshi, J., Yoshida, M., Yaguchi, H. & Ding, G.-R. (2000c) Exposure to extremely low frequency magnetic fields suppresses X-ray-induced transformation in mouse C3H10T1/2 cells. *Biochem. biophys. Res. Commun.*, **271**, 323–327

Monson, R. (1990) *Occupational Epidemiology*, 2nd Ed., Boca Raton, FL, CRC Press

Mooney, V. (1990) A randomized double-blind prospective study of the efficacy of pulsed electromagnetic fields for interbody lumbar fusions. *Spine.* **15**, 708–712

Mooney, N.A., Smith, R.E. & Watson, B.W. (1986) Effect of extremely-low-frequency pulsed magnetic fields on the mitogenic response of peripheral blood mononuclear cells. *Bioelectromagnetics*, **7**, 387–394

Morandi, M.A., Pak, C.M., Caren, R.P. & Caren, L.D. (1996) Lack of an EMF-induced genotoxic effect in the Ames assay. *Life Sci.*, **59**, 263–271

Morgan, M.G. & Nair, I. (1992) Alternative functional relationships between ELF field exposure and possible health effects: report on an expert workshop. *Bioelectromagnetics*, **13**, 335–350

Morris, J.E. & Ragan, H.A. (1979) Immunological studies of 60-Hz electric fields. In: Phillips, X., Gillis, X., Kaune, X. & Mahlum, X., eds, *Biological Effects of Extremely Low Frequency Electromagnetic Fields* (Hanford Life Sciences Symposium 18; DOE Symposium Series 50; CONF-781016), Washington DC, Richland, pp. 326–334

Morris, J.E., Sasser, L.B., Miller, D.L., Dagle, G.E., Rafferty, C.N., Ebi, K.L. & Anderson, L.E. (1999) Clinical progression of transplanted large granular lymphocytic leukemia in Fischer 344 rats exposed to 60 Hz magnetic fields. *Bioelectromagnetics*, **20**, 48–56

Mullins, R.D., Sisken, J.E., Hejase, H. & Sisken, B.F. (1993) Design and characterization of a system for exposure of cultured cells to extremely low frequency electric and magnetic fields over a wide range of field strengths. *Bioelectromagnetics*, **14**, 173–186

Mullins, J.M., Penafiel, L.M., Juutilainen, J. & Litovitz, T.A. (1999) Dose–response of electromagnetic field-enhanced ornithine decarboxylase activity. *Bioelectrochem. Bioenerg.*, **48**, 193–199

Murthy, K.K., Rogers, W.R. & Smith, H.D. (1995) Initial studies on the effects of combined 60 Hz electric and magnetic field exposure on the immune system of nonhuman primates. *Bioelectromagnetics*, **3**, 93–102

Mutnick, A. & Muscat, J.E. (1997) Primary brain cancer in adults and the use of common household appliances: A case–control study. *Rev. environ. Health*, **12**, 59–62 (erratum: *Rev. environ. Health*, **12**, 131)

Myers, A., Clayden, A.D., Cartwright, R.A. & Cartwright, S.C. (1990) Childhood cancer and overhead powerlines: A case–control study. *Br. J. Cancer*, **62**, 1008–1014

Nafziger, J., Desjobert, H., Benamar, B., Guillosson, J.J. & Adolphe, M. (1993) DNA mutations and 50 Hz electromagnetic fields. *Bioelectrochem. Bioenerg.*, **30**, 133–141

Nakagawa, M. & Matsuda, Y. (1988) A strong static-magnetic field alters operant responding by rats. *Bioelectromagnetics*, **9**, 25–37

Narita, K., Hanakawa, K., Kasahara, T., Hisamitsu, T. & Asano, K. (1997) Induction of apoptotic cell death in human leukemic cell line, HL-60, by extremely low frequency electric magnetic fields: analysis of the possible mechanisms *in vitro*. *In Vivo*, **11**, 329–335

Nasca, P.C., Baptiste, M.S., MacCubbin, P.A., Metzger, B.B., Carlton, K., Greenwald, P., Armbrustmacher, V.W., Earle, K.M. & Waldman, J. (1988) An epidemiologic case–control study of central nervous system tumors in children and parental occupational exposures. *Am. J. Epidemiol.*, **128**, 1256–1265

National Geophysical Data Center (United States Department of Commerce, National Oceanic and Atmospheric Administration, National Environment Satellite, Data and Information Service), http://www.ngdc.noaa.gov

National Institute of Environmental Health Sciences (NIEHS) (1999) *Health Effects from Exposure to Power-line Frequency Electric and Magnetic Fields* (NIH Publication No. 99-4493), Cincinnati, OH

National Radiological Protection Board (NRPB) (1992) *Electromagnetic Fields and the Risk of Cancer, Report of an Advisory Group on Non-ionising Radiation* (Doc NRPB 3), Chilton, UK

National Radiological Protection Board (NRPB) (2001) *ELF Electromagnetic Fields and the Risk of Cancer, Report of an Advisory Group on Non-ionizing Radiation* (Doc NRPB 12), Chilton, UK

National Research Council (1997) *Possible Health Effects to Residential Electric and Magnetic Fields*, Washington DC, National Academy Press

National Toxicology Program (NTP) (1996) *Toxicity, Reproductive and Developmental Studies of 60-Hz Magnetic Fields Administered by Whole-body Exposure to F344/N rats, Sprague-Dawley rats and B6C3F$_1$ mice (Toxicity Report Series No. 58)*, NIH Publication No. 96-3939), US Department of Health and Human Services, Public Health Service, Research Triangle Park, NC

National Toxicology Program (1999a) *Toxicology and Carcinogenesis Studies of 60-Hz Magnetic Fields in F344/N Rats and B6C3F$_1$ Mice (Whole-Body Exposure Studies)* (NTP Technical Report 488), Research Triangle Park, NC

National Toxicology Program (1999b) *Toxicology and Carcinogenesis Studies of Magnetic Field Promotion (DMBA Initiation) in Female Sprague-Dawley Rats (Whole-Body Exposure, Gavage Studies)* (NTP Technical Report 489), Research Triangle Park, NC

Neurath, P.W. (1968) High gradient magnetic field inhibits embryonic development of frogs. *Nature*, **219**, 1358–1359

Neutra, R.R. & DelPizzo, V. (1996) When 'wire codes' predict cancer better than spot measurements of magnetic fields (Editorial). *Epidemiology*, **7**, 217–218

Neutra, R.R., DelPizzo, V., Lee, G., Leonard, A., Stevenson, M. & Hristova, L. (1996) Progress Report from the California EMF Program. In: The Annual review of Research on Biological Effects of Electric and Magnetic Fields from the Generation, Delivery and Use of Electricity, P-78, DOE, NIEH, EPRI, San Antonio, Texas

Nguyen, P., Bournias-Vardiabasis, N., Haggren, W., Adey, W.R. & Phillips, J.L. (1995) Exposure of *Drosophila melanogaster* embryonic cell cultures to 60-Hz sinusoidal magnetic fields: assessment of potential teratogenic effects. *Teratology*, **51**, 273–277

Niehaus, M., Bruggemeyer, H., Behre, H.M. & Lerchl, A. (1997) Growth retardation, testicular stimulation, and increased melatonin synthesis by weak magnetic fields (50 Hz) in Djungarian hamsters, *Phodopus sungorus*. *Biochem. biophys. Res. Commun.*, **234**, 707–711

Nielsen, C.V. & Brandt, L.P. (1990) Spontaneous abortion among women using video display terminals. *Scand. J. Work environ. Health*, **16**, 323–328

Nishimura, I., Yamazaki, K., Shigemitsu, T., Negishi, T. & Sasano, T. (1999) Linearly and circularly polarized, 50 Hz magnetic fields did not alter intracellular calcium in rat immune cells. *Ind. Health*, **37**, 289–299

Nolte, C.M., Pittman, D.W., Kalevitch, B., Henderson, R. & Smith, J.C. (1998) Magnetic field conditioned taste aversion in rats. *Physiol. Behav.*, **63**, 683–688

Nordenson, I., Hansson-Mild, K., Nordstrom, S., Sweins, A. & Birke, E. (1984) Clastogenic effects in human lymphocytes of power frequency electric fields: in vivo and in vitro studies. *Radiat. environ. Biophys.*, **23**, 191–201

Nordenson, I., Mild, K.H., Östman, U. & Ljungberg, H. (1988) Chromosomal effects in lymphocytes of 400 kV-substation workers. *Radiat. environ. Biophys.*, **27**, 39–47

Nordenson, I., Mild, K.H., Andersson, G. & Sandstrom, M. (1994) Chromosomal aberrations in human amniotic cells after intermittent exposure to fifty Hertz magnetic fields. *Bioelectromagnetics*, **15**, 293–301

Nordenson, I., Hansson Mild, K., Järventaus, H., Hirvonen, A., Sandström, M., Wilén, J., Blix, N. & Norppa, H. (2001) Chromosomal aberrations in peripheral lymphocytes of train engine drivers. *Bioelectromagnetics*, **22**, 306–315

Nordström, S., Birke, E. & Gustavsson, L. (1983) Reproductive hazards among workers at high voltage substations. *Bioelectromagnetics*, **4**, 91–101

Ohtsu, S., Miyakoshi, J., Tsukada, T., Hiraoka, M., Abe, M. & Takebe, H. (1995) Enhancement of β-galactosidase gene expression in rat pheochromocytoma cells by exposure to extremely low frequency magnetic fields. *Biochem. biophys. Res. Commun.*, **212**, 104–109

Okonogi, H., Nakagawa, M. & Tsuji, Y. (1996) The effects of a 4.7 tesla static magnetic field on the frequency of micronucleated cells induced by mitomycin C. *Tohoku J. exp. Med.*, **180**, 209–215

Olin, R., Vågerö, D. & Ahlbom, A. (1985) Mortality experience of electrical engineers. *Br. J. ind. Med.*, **42**, 211–212

Olsen, R.G. (1994) Power-transmission electromagnetics. *IEEE Antennas Propagation Mag.*, **36**, 7–16

Olsen, J.H., Nielsen, A. & Schulgen, G. (1993) Residence near high voltage facilities and risk of cancer in children. *Br. med. J.*, **307**, 891–895

Orr, J.L., Rogers, W.R. & Smith, H.D. (1995a) Exposure of baboons to combined 60 Hz electric and magnetic fields does not produce work stoppage or affect operant performance on a match to sample task. *Bioelectromagnetics*, **Suppl. 3**, 61–70

Orr, J.L., Rogers, W.R. & Smith, H.D. (1995b) Detection thresholds for 60 Hz electric fields by nonhuman primates. *Bioelectromagnetics*, **Suppl. 3**, 23–34

Pacini, S., Aterini, S., Pacini, P., Ruggiero, C., Gulisano, M. & Ruggiero, M. (1999a) Influence of static magnetic field on the antiproliferative effects of vitamin D on human breast cancer cells. *Oncol. Res.*, **11**, 265–271

Pacini, S., Vannelli, G.B., Barni, T., Ruggiero, M., Sardi, I., Pacini, P. & Gulisano, M. (1999b) Effect of 0.2 T static magnetic field on human neurons: remodeling and inhibition of signal transduction without genome instability. *Neurosci. Lett.*, **267**, 185–188

Pafková, H. & Jerábek, J. (1994) Interaction of MF 50 Hz, 10 mT with high dose of X-rays: Evaluation of embryotoxicity in chick embryos. *Rev. environ. Health*, **10**, 235–241

Pafková, H., Jerábek, J., Tejnorová, I. & Bednár, V. (1996) Developmental effects of magnetic field (50 Hz) in combination with ionizing radiation and chemical teratogens. *Toxicol. Lett.*, **88**, 313–316

Paile, W., Jokela, K., Koivistoinen, A. & Salomaa, S. (1995) Effects of 50 Hz sinusoidal magnetic fields and spark discharges on human lymphocytes in vitro. *Bioelectrochem. Bioenerg.*, **36**, 15–22

Parkinson, W.C. & Hanks, C.T. (1989) Search for cyclotron resonance in cells *in vitro*. *Bioelectromagnetics*, **10**, 129–145

Parkinson, W.C. & Sulik, G.L. (1992) Diatom response to extremely low-frequency magnetic fields. *Radiat. Res*, **130**, 319–330

Pearce, N.E., Sheppard, R.A., Howard, J.K., Fraser, J. & Lilley, B.M. (1985) Leukaemia in electrical workers in New Zealand (Letter to the Editor). *Lancet*, **i**, 811–812

Pearce, N.E., Reif, J. & Fraser, J. (1989) Case–control studies of cancer in New Zealand electrical workers. *Int. J. Epidemiol.*, **18**, 55–59

Percy, C., Stenek, E. III. & Gloeckler, L. (1981) Accuracy of cancer death certificates and its effect on cancer mortality statistics. *Am. J. Public Health*, **71**, 242–250

Perry, F.S., Reichmanis, M., Marino, A.A. & Becker, R.O. (1981) Environmental power-frequency magnetic fields and suicide. *Health Phys.*, **41**, 267–277

Persinger, M.A. & Pear, J.J. (1972) Prenatal exposure to an ELF-rotating magnetic field and subsequent increase in conditioned suppression. *Dev. Psychobiol.*, **5**, 269–274

Pessina, G.P. & Aldinucci, C. (1997) Short cycles of both static and pulsed electromagnetic fields have no effect on the induction of cytokines by peripheral blood mononuclear cells. *Bioelectromagnetics*, **18**, 548–554

Pessina, G.P. & Aldinucci, C. (1998) Pulsed electromagnetic fields enhance the induction of cytokines by peripheral blood mononuclear cells challenged with phytohemagglutinin. *Bioelectromagnetics*, **19**, 445–451

Peto, R. (1974) Guidelines on the analysis of tumour rates and death rates in experimental animals. *Br. J. Cancer*, **29**, 101–105

Petralia, S.A., Chow, W.-H., McLaughlin, J., Jin, F., Gao, Y.-T. & Dosemeci, M. (1998) Occupational risk factors for breast cancer among women in Shanghai. *Am. J. ind. Med.*, **34**, 477–483

Petridou, E., Hsieh, C.C., Skalkidis, Y., Toupadaki, N. & Athanassopoulos, Y. (1993) Suggestion of concomitant changes of electric power consumption and childhood leukemia in Greece. *Scand. J. Soc. Med.*, **21**, 281–285

Petridou, E., Trichopoulos, D., Kravaritis, A., Pourtsidis, A., Dessypris, N., Skalkidis, Y., Kogevinas, M., Kalmanti, K., Koliouskas, D., Kosmidis, H., Panagiotou, J.P., Piperopoulou, F., Tzortzatou, F. & Kalapothaki, V. (1997) Electrical power lines and childhood leukemia: A study from Greece. *Int. J. Cancer*, **73**, 345–348

Pfluger, D.H. & Minder, C.E. (1996) Effects of exposure to 16.7 Hz magnetic fields on urinary 6-hydroxymelatonin sulfate excretion of Swiss railway workers. *J. pineal Res.*, **21**, 91–100

Phillips, J.L., Haggren, W., Thomas, W.J., Ishida-Jones, T. & Adey, W.R. (1992) Magnetic field-induced changes in specific gene transcription. *Biochim. biophys. Acta*, **1132**, 140–144

Philips, K.L., Morandi, M.T., Oehme, D. & Cloutier, P.A. (1995) Occupational exposure to low frequency magnetic fields in health care facilities. *Am. ind. Hyg. Assoc. J.*, **56**, 677–685

Picazo, M.L., deMiguel, M.P., Leyton, V., Franco, P., Varela, L., Paniagua, R. & Bardasano, J.L. (1995) Long-term effects of ELF magnetic fields on the mouse testis and serum testosterone levels. *Electro- & Magnetobiol.*, **14**, 127–134

Picazo, M.L., Catalá, M.D., Romo, M.A. & Bardasano, J.L. (1998) Inhibition of melatonin in the plasma of third-generation male mice under the action of ELF magnetic fields. *Electro- & Magnetobiol.*, **17**, 75–85

Pira, E., Turbiglio, M., Maroni, M., Carrer, P., La Vecchia, C., Negri, E. & Lachetta, R. (1999) Mortality among workers in the geothermal power plants at Larderello, Italy. *Am. J. ind. Med.*, **35**, 536–539

Plonsey, R. & Barr, R.C. (1988) *Bioelectricity: A quantitative approach*, New York, Plenum Press

Podd, J.V., Whittington, C.J., Barnes, G.R.G., Page, W.H. & Rapley, B.I. (1995) Do ELF magnetic fields affect human reaction time? *Bioelectromagnetics*, **16**, 317–323

Polk, C. (1992) Dosimetry of extremely-low-frequency magnetic fields. *Bioelectromagnetics*, **Suppl. 1**, 209–235

Polk, C. (1993) Physical mechanisms for biological effects of ELF low-intensity electric and magnetic fields: thermal noise limit and counterion polarization. In: Blank, M., ed., *Electricity and Magnetism in Biology and Medicine*, San Francisco, CA, San Francisco Press, Inc., pp. 543–546

Polk, C. (1994) Effects of extremely-low frequency magnetic fields on biological magnetite. *Bioelectromagnetics*, **15**, 261–270

Polk, C. (1995) Electric and magnetic fields for bone and soft tissue repair. In: Polk, C. & Postow, E., eds, *Handbook of Biologic Effects of Electromagnetic fields*, 2nd Ed., Boca Raton, CRC Press, pp. 231–246

Poole, C., Kavet, R., Funch, D.P., Donelan, K., Charry, J.M. & Dreyer, N.A. (1993) Depressive symptoms and headaches in relation to proximity of residence to an alternating-current transmission line right-of-way. *Am. J. Epidemiol.*, **137**, 318–330

Portet, R. & Cabanes, J. (1988) Development of young rats and rabbits exposed to a strong electric field. *Bioelectromagnetics*, **9**, 95–104

Portier, C. (1986) Type 1 error and power of the linear trend test in proportion under the National Toxicology Program's modified pathology protocol. *Fund. appl. Toxicol.*, **6**, 515–519

Portier, C.J. & Wolfe, M.D., eds (1998) *Assessment of Health Effects from Exposure to Power-line Frequency Electric and Magnetic Fields*, NIEHS Working Group Report (NIH Publication No. 98-3981), Research Triangle Park, NC, National Institute of Environmental Health Sciences

Potter, M.E., Okoniewski, M. & Stuchly, M.A. (2000) Low frequency finite difference time domain (FDTD) for modeling of induced fields in humans close to line sources. *J. Comput. Phys.*, **162**, 82–103

Prasad, N., Wright, D.A. & Forster, J.D. (1982) Effect of nuclear magnetic resonance on early stages of amphibian development. *Magn. Reson. Imaging*, **1**, 35–38

Preece, A.W., Kaune, W., Grainger, P., Preece, S. & Golding, J. (1997) Magnetic fields from domestic appliances in the UK. *Phys. med. Biol.*, **42**, 67–76

Preece, A.W., Wesnes, K.A. & Iwi, G.R. (1998) The effect of a 50 Hz magnetic field on cognitive function in humans. *Int. J. Radiat. Biol.*, **74**, 463–470

Preston-Martin, S., Peters, J.M., Yu, M.C., Garabrant, D.H. & Bowman, J.D. (1988) Myelogenous leukemia and electric blanket use. *Bioelectromagnetics*, **9**, 207–213

Preston-Martin, S., Mack, W. & Henderson, B.E. (1989) Risk factors for gliomas and meningiomas in males in Los Angeles county. *Cancer Res.*, **49**, 6137–6143

Preston-Martin, S., Navidi, W., Thomas, D., Lee, P.-J., Bowman, J. & Pogoda, J. (1996a) Los
Angeles study of residential magnetic fields and childhood brain tumors. *Am. J. Epidemiol.*,
143, 105–119

Preston-Martin, S., Gurney, J.G., Pogoda, J.M., Holly, E.A. & Mueller, B.A. (1996b) Brain
tumor risk in children in relation to use of electric blankets and water bed heaters. *Am. J.
Epidemiol.*, **143**, 1116–1122

Price, J.A. & Strattan, R.D. (1998) Analysis of the effect of a 60 Hz AC field on histamine
release by rat peritoneal mast cells. *Bioelectromagnetics*, **19**, 192–198

Pulsoni, A., Stazi, A., Cotichini, R., Allione, B., Cerri, R., di Bona, E., Nosari, A.M., Pagano,
L., Recchia, A., Ribersani, M., Rocchi, L., Veneri, D., Visani, G., Mandelli, F. & Mele, A.
(1998) Acute promyelocytic leukaemia: Epidemiology and risk factors. A report of the
GIMEMA Italian archive of adult acute leukaemia. *Eur. J. Haematol.*, **61**, 327–332

Quittan, M., Schuhfried, O., Wiesinger, G.F. & Fialka-Moser, V. (2000) [Clinical effectiveness
of magnetic field therapy — a review of the literature.] *Acta Med. Austriaca*, **27**, 61–68
(in German)

Ragan, H.A., Buschbom, R.L., Pipes, M.J., Phillips, R.D. & Kaune, W.T. (1983) Hematologic
and serum chemistry studies in rats exposed to 60-Hz electric fields. *Bioelectromagnetics*,
4, 79–90

Ramirez, E., Monteagudo, J.L., Garcia-Garcia, M. & Delgado, J.M.R. (1983) Oviposition and
development of Drosophila modified by magnetic fields. *Bioelectromagnetics*, **4**, 315–326

Ramoni, C., Dupuis, M.L., Vecchia, P., Polichetti, A., Petrini, C., Bersani, F., Capri, M.,
Cossarizza, A., Franceschi, C. & Grandolfo, M. (1995) Human natural killer cytotoxic
activity is not affected by in vitro exposure to 50-Hz sinusoidal magnetic fields. *Int. J.
Radiat. Biol.*, **68**, 693–705

Rannug, A., Holmberg, B., Ekström, T. & Mild, K.H. (1993a) Rat liver foci study on coexposure
with 50 Hz magnetic fields and known carcinogens. *Bioelectromagnetics*, **14**, 17–27

Rannug, A., Ekström, T., Mild, K.H., Holmberg, B., Gimenez-Conti, I. & Slaga, T.J. (1993b)
A study on skin tumour formation in mice with 50 Hz magnetic field exposure. *Carcinogenesis*, **14**, 573–578

Rannug, A., Holmberg, B. & Mild, K.H. (1993c) A rat liver foci promotion study with 50-Hz
magnetic fields. *Environ. Res.*, **62**, 223–229

Rannug, A., Holmberg, B., Ekström, T., Mild, K.H., Gimenez-Conti, I. & Slaga, T.J. (1994)
Intermittent 50 Hz magnetic field and skin tumour promotion in SENCAR mice. *Carcinogenesis*, **15**, 153–157

Rao, S. & Henderson, A.S. (1996) Regulation of c-*fos* is affected by electromagnetic fields.
J. cell. Biochem., **63**, 358–365

Reese, J.A., Jostes, R.F. & Frazier, M.E. (1988) Exposure of mammalian cells to 60-Hz
magnetic or electric fields: analysis for DNA single-strand breaks. *Bioelectromagnetics*, **9**,
237–247

Reichmanis, M., Perry, F.S., Marino, A.A. & Becker, R.O. (1979) Relation between suicide and
the electromagnetic field of overhead power lines. *Physiol. Chem. Physics*, **11**, 395–403

Reilly, J.P. (1992) *Electrical Stimulation and Electropathology*, New York, Cambridge University Press

Reilly, J.P. (1998) Stimulation via electric and magnetic fields. *Applied Bioelectricity: From Electrical Stimulation to Electropathology*, New York, Springer, pp. 341–411

Reipert, B.M., Allan, D., Reipert, S. & Dexter, T.M. (1997) Apoptosis in haematopoietic progenitor cells exposed to extremely low-frequency magnetic fields. *Life Sci.*, **61**, 1571–1582

Reiter, R.J., Anderson, L.E., Buschbom, R.L. & Wilson, B.W. (1988) Reduction of the nocturnal rise in pineal melatonin levels in rats exposed to 60-Hz electric fields in utero and for 23 days after birth. *Life Sci.*, **42**, 2203–2206

Reiter, R.J., Melchiorri, D., Sewerynek, E., Poeggeler, B., Barlow-Walden, L., Chuang, J., Ortiz, G.G. & Acuna-Castroviejo, D. (1995) A review of the evidence supporting melatonin's role as an antioxidant. *J. pineal Res.*, **18**, 1–11

Repacholi, M.H. & Greenebaum, B. (1999) Interaction of static and extremely low frequency electric and magnetic fields with living systems: Health effects and research needs. *Bioelectromagnetics*, **20**, 133–160

Richardson, S., Zittoun, R., Bastuji-Garin, S., Lasserre, V., Guihenneuc, C., Cadiou, M., Viguie, F. & Laffont-Faust, I. (1992) Occupational risk factors for acute leukaemia: A case–control study. *Int. J. Epidemiol.*, **21**, 1063–1073

Rivas, L., Rius, C., Tello, I. & Oroza, M.A. (1985) Effects of chronic exposure to weak electromagnetic fields in mice. *IRCS med. Sci.*, **13**, 661–662

Rivas, L., Oroza, M.A. & Delgado, J.M.R. (1987) Influence of electromagnetic fields on body weight and serum chemistry in second generation mice. *Med. Sci. Res.*, **15**, 1041–1042

Robinson, C.F., Lalich, N.R., Burnett, C.A., Sestito, J.P., Frazier, T.M. & Fine, L.J. (1991) Electromagnetic field exposure and leukemia mortality in the United States. *J. occup. Med.*, **33**, 160–162

Rockette, H.E. & Arena, V.C. (1983) Mortality studies of aluminum reduction plant workers: Potroom and carbon department. *J. occup. Med.*, **25**, 549–557

Rodegerdts, E.A., Grönewäller, E.F., Kehlback, R., Roth, P., Wiskirchen, J., Gebert, R., Claussen, C.D. & Duda, S.H. (2000) In vitro evaluation of teratogenic effects by time-varying MR gradient fields on fetal human fibroblasts. *J. magn. Reson. Imaging*, **12**, 150–156

Rodvall, Y., Ahlbom, A, Stenlund, C., Preston-Martin, S., Lindh, T & Spännare, B. (1998) Occupational exposure to magnetic fields and brain tumours in Central Sweden. *Eur. J. Epidemiol.*, **14**, 563–569

Rogers, W.R., Reiter, R.J., Barlow-Walden, L., Smith, H.D. & Orr, J.L. (1995a) Regularly scheduled, day-time, slow-onset 60 Hz electric and magnetic field exposure does not depress serum melatonin concentration in nonhuman primates. *Bioelectromagnetics*, **Suppl.**, 3111–3118

Rogers, W.R., Reiter, R.J., Smith, H.D. & Barlow-Walden, L. (1995b) Rapid-onset/offset, variably scheduled 60 Hz electric and magnetic field exposure reduces nocturnal serum melatonin concentration in nonhuman primates. *Bioelectromagnetics*, **Suppl.**, 3119–3122

Rogers, W.R., Orr, J.L. & Smith, H.D. (1995c) Initial exposure to 30 kV/m or 60 kV/m 60 Hz electric fields produces temporary cessation of operant behavior of nonhuman primates. *Bioelectromagnetics*, **3**, 35–47

Rogers, W.R., Orr, J.L. & Smith, H.D. (1995d) Nonhuman primates will not respond to turn off strong 60 Hz electric fields. *Bioelectromagnetics*, **3**, 48–60

Rommereim, D.N., Kaune, W.T., Buschbom, R.L., Phillips, R.D. & Sikov, M.R. (1987) Reproduction and development in rats chronologically exposed to 60-Hz electric fields. *Bioelectromagnetics*, **8**, 243–258

Rommereim, D.N., Kaune, W.T., Anderson, L.E. & Sikov, M.R. (1989) Rats reproduce and rear
litters during chronic exposure to 150-kV/m, 60-Hz electric fields. *Bioelectromagnetics*,
10, 385–389

Rommereim, D.N., Rommereim, R.L., Sikov, M.R., Buschbom, R.L. & Anderson, L.E. (1990)
Reproduction, growth, and development of rats during chronic exposure to multiple field
strengths of 60-Hz electric fields. *Fundam. appl. Toxicol.*, **14**, 608–621

Rommereim, D.N., Rommereim, R.L., Miller, D.L., Buschbom, R.L. & Anderson, L.E. (1996)
Developmental toxicology evaluation of 60-Hz horizontal magnetic fields in rats. *Appl.
occup. environ. Hyg.*, **11**, 307–312

Rønneberg, A., Haldorsen, T., Romundstad, P. & Andersen, A. (1999) Occupational exposure
and cancer incidence among workers from an aluminum smelter in western Norway.
Scand. J. Work Environ. Health, **25**, 207–214

Rosenbaum, P.F., Vena, J.E., Zielezny, M.A. & Michalek, A.M. (1994) Occupational exposures
associated with male breast cancer. *Am. J. Epidemiol.*, **139**, 30–36

Rosenberg, R.S., Duffy, P.H. & Sacher, G.A. (1981) Effects of intermittent 60-Hz high voltage
electric fields on metabolism, activity, and temperature in mice. *Bioelectromagnetics*, **2**,
291–303

Rosenberg, R.S., Duffy, P.H., Sacher, G.A. & Ehret, C.F. (1983) Relationship between field
strength and arousal response in mice exposed to 60-Hz electric fields. *Bioelectromagnetics*,
4, 181–191

Rosenthal, M. & Obe, G. (1989) Effects of 50-Hertz electromagnetic fields on proliferation and
on chromosomal alterations in human peripheral lymphocytes untreated or pretreated with
chemical mutagens. *Mutat. Res.*, **210**, 329–335

Ross, S.M. (1990) Combined DC and ELF magnetic fields can alter cell proliferation. *Bioelectromagnetics*, 11, 27–36

Rozek, R.J., Sherman, M.L., Liboff, A.R., McLeod, B.R. & Smith, S.D. (1987) Nifedipine is an
antagonist to cyclotron resonance enhancement of ^{45}Ca incorporation in human
lymphocytes. *Cell Calcium*, 8, 413–427

Ryan, P., Lee, M.W., North, J.B. & McMichael, A.J. (1992) Risk factors for tumors of the brain
and meninges: Results from the Adelaide adult brain tumor study. *Int. J. Cancer*, **51**, 20–27

Ryan, B.M., Mallett, E., Jr, Johnson, T.R., Gauger, J.R. & McCormick, D.L. (1996) Developmental toxicity study of 60 Hz (power frequency) magnetic fields in rats. *Teratology*, **54**,
73–83

Ryan, B.M., Symanski, R.R., Pomeranz, L.A., Johnson, T.R., Gauger, J.R. & McCormick, D.L.
(1999) Multigeneration reproductive toxicity assessment of 60-Hz magnetic fields using a
continuous breeding protocol in rats. *Teratology*, **59**, 156–162

Ryan, B.M., Polen, M., Gauger, J.R., Mallett, E., Jr, Kearns, M.B., Bryan, L. & McCormick,
D.L. (2000) Evaluation of the developmental toxicity of 60 Hz magnetic fields and
harmonic frequencies in Sprague-Dawley rats. *Radiat. Res.*, **153**, 637–641

Saalman, E., Önfelt, A. & Gillstedt-Hedman, B. (1991) Lack of c-mitotic effects in V79 Chinese
hamster cells exposed to 50 Hz magnetic fields. *Bioelectrochem. Bioenerg.*, **26**, 335–338

Saffer, J.D. & Thurston, S.J. (1995) Short exposures to 60 Hz magnetic fields do not alter *MYC* expression in HL60 or Daudi cells. *Radiat. Res.*, **144**, 18–25

Saffer, J.D., Chen, G., Colburn, N.H. & Thurston, S.J. (1997) Power frequency magnetic fields do not contribute to transformation of JB6 cells. *Carcinogenesis*, **18**, 1365–1370

Sahl, J.D., Kelsh, M.A. & Greenland, S. (1993) Cohort and nested case–control studies of hematopoietic cancers and brain cancer among electric utility workers. *Epidemiology*, **4**, 104–114

Sakamoto, S., Hagino, N. & Winters, W.D. (1993) In vivo studies of the effect of magnetic field exposure on ontogeny of choline acetyltransferase in the rat brain. *Bioelectromagnetics*, **14**, 373–381

Salerno, S., Lo Casto, A., Caccamo, N., D'Anna, C., De Maria, M., Lagalla, R., Scola, L. & Cardinale, A.E. (1999) Static magnetic fields generated by 0.5 T MRI unit affects in vitro expression of activation markers and interleukin release in human peripheral blood mononuclear cells (PBMC). *Int. J. Radiat. Biol.*, **75**, 457–463

Salzinger, K., Freimark, S., McCullough, M., Phillips, D. & Birenbaum, L. (1990) Altered operant behavior of adult rats after perinatal exposure to a 60-Hz electromagnetic field. *Bioelectromagnetics*, **11**, 105–116

Sandström, M., Hansson Mild, K., Stenberg, B. & Wall, S. (1993) A survey of electric and magnetic fields among VDT operators in offices. *IEEE Trans. EMC*, **35**, 394–397

Sasser, L.B., Morris, J.E., Miller, D.L., Rafferty, C.N., Ebi, K.L. & Anderson, L.E. (1996) Exposure to 60 Hz magnetic fields does not alter clinical progression of LGL leukemia in Fischer rats. *Carcinogenesis*, **17**, 2681–2687

Sasser, L.B., Anderson, L.E., Morris, J.E., Miller, D.L., Walborg, E.F., Kavet, R., Johnston, D.A. & DiGiovanni, J. (1998) Lack of a co-promoting effect of a 60 Hz magnetic field on skin tumorigenesis in SENCAR mice. *Carcinogenesis*, **19**, 1617–1621

Sastre, A., Cook, M.R. & Graham, C. (1998) Nocturnal exposure to intermittent 60 Hz magnetic field alters human cardiac rhythm. *Bioelectromagnetics*, **19**, 96–106

Satter-Syed, A., Islam, M.S., Rabbani, K.S. & Talukder, M.S. (1999) Pulsed electromagnetic fields for the treatment of bone fractures. *Bangladesh Med. Res. Counc. Bull.*, **25**, 6–10

Savitz, D.A. & Ananth, C.V. (1994) Residential magnetic fields, wire codes, and pregnancy outcome. *Bioelectromagnetics*, **15**, 271–273

Savitz, D.A. & Loomis, D.P. (1995) Magnetic field exposure in relation to leukemia and brain cancer mortality among electric utility workers. *Am. J. Epidemiol.*, **141**, 123–134

Savitz, D.A., Wachtel, H., Barnes, F.A., John, E.M. & Tvrdik, J.G. (1988) Case–control study of childhood cancer and exposure to 60-Hz magnetic fields. *Am. J. Epidemiol.*, **128**, 21–38

Savitz, D.A., John, E.M. & Kleckner, R.C. (1990) Magnetic field exposure from electric appliances and childhood cancer. *Am. J. Epidemiol.*, **131**, 763–773

Savitz, D.A., Boyle, C.A. & Holmgreen, P. (1994) Prevalence of depression among electrical workers. *Am. J. ind. Med.*, **25**, 165–176

Savitz, D.A., Dufort, V., Armstrong, B. & Thériault, G. (1997) Lung cancer in relation to employment in the electrical utility industry and exposure to magnetic fields. *Occup. environ. Med.*, **54**, 396–402

Savitz, D., Checkoway, H. & Loomis, D. (1998a) Magnetic field exposure and neurodegenerative disease mortality among electric utility workers. *Epidemiology*, **9**, 398–404

Savitz, D., Loomis, D. & Chiu-Kit, T. (1998b) Electrical occupations and neurodegenerative disease: Analysis of U.S. mortality data. *Arch. environ. Health*, **53**, 1–5

Savitz, D.A., Liao, D., Sastre, A., Kleckner, R.C. & Kavet, R. (1999) Magnetic field exposure and cardiovascular disease mortality among electric utility workers. *Am. J. Epidemiol.*, **149**, 135–142

Savitz, D.A., Cai, J., van Wijngaarden, E., Loomis, D., Mihlan, G., Dufort, V., Kleckner, R.C., Nylander-French, L., Kromhout, H. & Zhou, H. (2000) Case–cohort analysis of brain cancer and leukemia in electric utility workers using a refined magnetic field job–exposure matrix. *Am. J. ind. Med.*, **38**, 417–425

Scarfi, M.R., Bersani, F., Cossarizza, A., Monti, D., Castellani, G., Cadossi, R., Franceschetti, G. & Franceschi, C. (1991) Spontaneous and mitomycin-C-induced micronuclei in human lymphocytes exposed to extremely low frequency pulsed magnetic fields. *Biochem. biophys. Res. Commun.*, **176**, 194–200

Scarfi, M.R., Bersani, F., Cossarizza, A., Monti, D., Zeni, O., Lioi, M.B., Franceschetti, G., Capri, M. & Franceschi, C. (1993) 50 Hz AC sinusoidal electric fields do not exert genotoxic effects (micronucleus formation) in human lymphocytes. *Radiat. Res.*, **135**, 64–68

Scarfi, M.R., Lioi, M.B., Zeni, O., Franceschetti, G., Franceschi, C. & Bersani, F. (1994) Lack of chromosomal aberration and micronucleus induction in human lymphocytes exposed to pulsed magnetic fields. *Mutat. Res.*, **306**, 129–133

Scarfi, M.R., Prisco, F., Bersani, F., Lioi, M.B., Zeni, O., Di-Pietro, R., Franceschi, C., Motta, M., Iafusco, D. & Stoppoloni, G. (1996) Spontaneous and mitomycin-C-induced micronuclei in lymphocytes from subjects affected by Turner's syndrome. *Mutat. Res.*, **357**, 183–190

Scarfi, M.R., Prisco, F., Lioi, M.B., Zeni, O., Delle Noce, M., DiPietro, R., Franceschi, C., Iafusco, D., Motta, M. & Bersani, F. (1997) Cytogenetic effects induced by extremely low frequency pulsed magnetic fields in lymphocytes from Turner's syndrome subjects. *Bioelectrochem. Bioenerg.*, **43**, 221–226

Scarfi, M.R., Lioi, M.B., Zeni, O., Della Noce, M., Franceschi, C. & Bersani, F. (1999) Micronucleus frequency and cell proliferation in human lymphocytes exposed to 50 Hz sinusoidal magnetic fields. *Health Phys.*, **76**, 244–250

Schenck, J.F., Dumoulin, C.L., Redington, R.W., Kressel, H.Y., Elliott, R.T. & McDougall, I.L. (1992) Human exposure to 4.0-T magnetic fields in a whole-body scanner. *Med. Phys.*, **19**, 1089–1098

Schimmelpfeng, J. & Dertinger, H. (1997) Action of a 50 Hz magnetic field on proliferation of cells in culture. *Bioelectromagnetics*, **18**, 177–183

Schnorr, T.M., Grajewski, B.A., Hornung, R.W., Thun, M.J., Egeland, G.M., Murray, W.E., Conover, D.L. & Halperin, W.E. (1991) Video display terminals and the risk of spontaneous abortion. *New Engl. J. Med.*, **324**, 727–733

Schreiber, G.H., Swaen, G.M.H., Meijers, J.M.M., Slangen, J.J.M. & Sturmans, F. (1993) Cancer mortality and residence near electricity transmission equipment: A retrospective cohort study. *Int. J. Epidemiol.*, **22**, 9–15

Schroeder, J.C. & Savitz, D.A. (1997) Lymphoma and multiple myeloma mortality in relation to magnetic field exposure among electric utility workers. *Am. J. ind. Med.*, **32**, 392–402

Schüz, J., Grigat, J.-P., Störmer, B., Rippin, G., Brinkmann, K. & Michaelis, J. (2000) Extremely low frequency magnetic fields in residences in Germany. Distribution of measurements, comparison of two methods for assessing exposure, and predictors for the occurrence of magnetic fields above background level. *Radiat. environ. Biophys.*, **39**, 233–240

Schüz, J., Grigat, J.-P., Brinkmann, K. & Michaelis, J. (2001a) Residential magnetic fields as a risk factor for childhood acute leukaemia: Results from a German population-based case–control study. *Int. J. Cancer*, **91**, 728–735

Schüz, J., Kaletsch, U., Kaatsch, P., Meinert, R. & Michaelis, J. (2001b) Risk factors for pediatric tumours of the central nervous system: Results from a German population-based case–control study. *Med. pediatr. Oncol.*, **36**, 274–282

Schüz, J., Grigat, J.P., Brinkmann, K. & Michaelis, J. (2001c) Childhood acute leukaemia and residential 16.7 Hz magnetic fields in Germany. *Br. J. Cancer*, **84**, 697–699

Selmaoui, B. & Touitou, Y. (1995) Sinusoidal 50-Hz magnetic fields depress rat pineal NAT activity and serum melatonin. Role of duration and intensity of exposure. *Life Sci.*, **57**, 1351–1358

Selmaoui, B., Bogdan, A., Auzeby, A., Lambrozo, J. & Touitou, Y. (1996a) Acute exposure to 50 Hz magnetic field does not affect hematological or immunologic functions in healthy young men: a circadian study. *Bioelectromagnetics*, **17**, 364–372

Selmaoui, B., Lambrozo, J. & Touitou, Y. (1996b) Magnetic fields and pineal function in humans: evaluation of nocturnal acute exposure to extremely low frequency magnetic fields on serum melatonin and urinary 6-sulfatoxymelatonin circadian rhythms. *Life Sci.*, **58**, 1539–1549

Selmaoui, B., Lambrozo, J. & Touitou, Y. (1997) Endocrine functions in young men exposed for one night to a 50-Hz magnetic field. A circadian study of pituitary, thyroid and adreno-cortical hormones. *Life Sci.*, **61**, 473–486

Seto, Y.J., Majeau-Chargois, D., Lymangrover, J.R., Dunlap, W.P., Walker, C.F. & Hsieh, S.T. (1984) Investigation of fertility and *In Utero* effects in rats chronically exposed to a high-intensity 60-Hz electric field. *IEEE Trans. biomed. Eng.*, **31**, 693–702

Severson, R.K., Stevens, R.G., Kaune, W.T., Thomas, D.B., Heuser, L., Davis, S. & Sever, L.E. (1988) Acute nonlymphocytic leukemia and residential exposure to power frequency magnetic fields. *Am. J. Epidemiol.*, **128**, 10–20

Sharrard, W.J. (1990) A double-blind trial of pulsed electromagnetic fields for delayed union of tibial fractures. *J. Bone Joint Surg. Br.*, **72**, 347–355

Shaw, G.M. (2001) Adverse human reproductive outcomes and electromagnetic fields: A brief summary of the epidemiologic literature. *Bioelectromagnetics*, **Suppl.** **5**, S5–S18

Shaw, G.M. & Croen, L.A. (1993) Human adverse reproductive outcomes and electromagnetic field exposures: review of epidemiologic studies. *Environ. Health Perspect.*, **101** (Suppl. 4), 107–119

Shen, Y.H., Shao, B.J., Chiang, H., Fu, Y.D. & Yu, M. (1997) The effects of 50 Hz magnetic field exposure on dimethylbenz(*a*)anthracene induced thymic lymphoma/leukemia in mice. *Bioelectromagnetics*, **18**, 360–364

Sienkiewicz, Z.J., Robbins, L., Haylock, R.G.E. & Saunders, R.D. (1994) Effects of prenatal exposure to 50 Hz magnetic fields on development in mice: II. Postnatal development and behavior. *Bioelectromagnetics*, **15**, 363–375

Sienkiewicz, Z.J., Larder, S. & Saunders, R.D. (1996) Prenatal exposure to a 50 Hz magnetic field has no effect on spatial learning in adult mice. *Bioelectromagnetics*, **17**, 249–252

Sienkiewicz, Z.J., Haylock, R.G.E. & Saunders, R.D. (1998) Deficits in spatial learning after exposure of mice to 50 Hz magnetic field. *Bioelectromagnetics*, **19**, 79–84

Sikov, M.R., Montgomery, L.D., Smith, L.G. & Phillips, R.D. (1984) Studies on prenatal and postnatal development in rats exposed to 60-Hz electric fields. *Bioelectromagnetics*, **5**, 101–112

Sikov, M.R., Rommereim, D.N., Beamer, J.L., Buschbom, R.L., Kaune, W.T. & Phillips, R.D. (1987) Developmental studies of Hanford miniature swine exposed to 60-Hz electric fields. *Bioelectromagnetics*, **8**, 229–242

Silny, J. (1984) Changes in VEP caused by strong magnetic fields. In: Nodar, R.H. & Barber, C., eds, *Visual Evoked Potentials II*, Boston, Butterworth Publishers, pp. 272–279

Silny, J. (1985) Effects of low-frequency, high intensity magnetic field on the organism. In: *IEE International Conference on Electric and Magnetic Fields in Medicine and Biology, London, December 1985*, London, Institute of Electric Engineers, pp. 103–107

Silny, J. (1986) The influence threshold of the time-varying magnetic field in the human organism. In: Bernhardt, J.H., ed., *Biological Effects of Static and Extremely Low Frequency Magnetic Fields*, Munich, MMV Medizin Verlag, pp. 105–112

Simkó, M., Kriehuber, R., Weiss, D.G. & Luben, R.A. (1998a) Effects of 50 Hz EMF exposure on micronucleus formation and apoptosis in transformed and nontransformed human cell lines. *Bioelectromagnetics*, **19**, 85–91

Simkó, M., Kriehuber, R. & Lange, S. (1998b) Micronucleus formation in human amnion cells after exposure to 50 Hz MF applied horizontally and vertically. *Mutat. Res.*, **418**, 101–111

Singh, N. & Lai, H. (1998) 60 Hz magnetic field exposure induces DNA crosslinks in rat brain cells. *Mutat. Res.*, **400**, 313–320

Skauli, K.S., Reitan, J.B. & Walther, B.T. (2000) Hatching in zebrafish (*Danio rerio*) embryos exposed to a 50 Hz magnetic field. *Bioelectromagnetics*, **21**, 407–410

Skyberg, K., Hansteen, I.L. & Vistnes, A.I. (1993) Chromosome aberrations in lymphocytes of high-voltage laboratory cable splicers exposed to electromagnetic fields. *Scand. J. Work Environ. Health*, **19**, 29–34

Skyberg, K., Hansteen, I.L. & Vistnes, A.I. (2001) Chromosomal aberrations in lymphocytes of employees in transformer and generator production exposed to electromagnetic fields and mineral oil. *Bioelectromagnetics*, **22**, 150–160

Smith, S.D., McLeod, B.R., Liboff, A.R. & Cooksey, K.E. (1987) Calcium cyclotron resonance and diatom   motility. *Stud. biophys.*, **119**, 131–136

Smith, S.D., McLeod, B.R. & Liboff, A.R. (1993) Effects of CR-tuned 60 Hz magnetic fields on sprouting and early growth of *Raphanus sativus*. *Bioelectrochem. Bioenerg.*, **32**, 67–76

Smith, R.F., Clarke, R.L. & Justesen, D.R. (1994) Behavioral sensitivity of rats to extremely-low-frequency magnetic fields. *Bioelectromagnetics*, **15**, 411–426

Sobel, E., Davanipour, Z., Sulkava, R., Erkinjuntti, T., Wikstrom, J., Henderson, V.W., Buckwalter, G., Bowman, J.D. & Lee, P.J. (1995) Occupations with exposure to electro-magnetic fields: A possible risk factor for Alzheimer's disease. *Am. J. Epidemiol.*, **142**, 515–524

Sobel, E., Dunn, M., Davanipour, Z., Qian, Z. & Chui, H.C. (1996) Elevated risk of Alzheimer's disease among workers with likely electromagnetic field exposure. *Neurology*, **47**, 1477–1481

Sorahan, T., Nichols, L., van Tongeren, M. & Harrington, J.M. (2001) Occupational exposure to magnetic fields relative to mortality from brain tumours: Updated and revised findings from a study of UK electricity generation and transmission workers, 1973–97. *Occup. environ. Med.*, **58**, 626–630

Speers, M.A., Dobbins, J.G. & Miller, V.S. (1988) Occupational exposures and brain cancer mortality: A preliminary study of East Texas residents. *Am. J. ind. Med.*, **13**, 629–638

Spinelli, J.J., Band, P.R., Svirchev, L.M. & Gallagher, R.P. (1991) Mortality and cancer incidence in aluminum reduction plant workers. *J. occup. Med.*, **33**, 1150–1155

Spitz, M.R. & Johnson, C.C. (1985) Neuroblastoma and paternal occupation. A case–control analysis. *Am. J. Epidemiol.*, **121**, 924–929

Stather, J.W., Bailey, M.R., Birchall, A. & Miles, J.C.H. (1996) Comment on the paper: enhanced deposition of radon daughter nuclei in the vicinity of power frequency electromagnetic fields. *Int. J. Radiat. Biol.*, **69**, 645–649

Stenlund, C. & Floderus, B. (1997) Occupational exposure to magnetic fields in relation to male breast cancer and testicular cancer: A Swedish case–control study. *Cancer Causes Control*, **8**, 184–191

Stern, R.M. (1987) Cancer incidence among welders: Possible effects of exposure to extremely low frequency electromagnetic radiation (ELF) and to welding fumes. *Environ. Health Perspect.*, **76**, 221–229

Stern, S. & Laties, V.G. (1985) Behavioral detection of 60-Hz electric fields by rats. *Bioelectromagnetics*, **6**, 99–103

Stern, S., Laties, V.G., Stancampiano, C.V., Cox, C. & de Lorge, J.O. (1983) Behavioral detection of 60-Hz electric fields by rats. *Bioelectromagnetics*, **4**, 215–247

Stern, S., Laties, V.G., Nguyen, Q.A. & Cox, C. (1996) Exposure to combined static and 60 Hz magnetic fields: failure to replicate a reported behavioral effect. *Bioelectromagnetics*, **17**, 279–292

Stevens, R.G. (1987) Electric power use and breast cancer: a hypothesis. *Am. J. Epidemiol.*, **125**, 556–561

Stevens, R.G. (1993) Biologically based epidemiological studies of electric power and cancer. *Environ. Health Perspect.*, **101**(Suppl. 4), 93–100

Stevens, R.G. (1995) Risk of premenopausal breast cancer and use of electric blankets (Letter). *Am. J. Epidemiol.*, **142**, 446

Stevens, R.G. & Davis, S. (1996) The melatonin hypothesis: Electric power and breast cancer. *Environ. Health Perspect.*, **104** (Suppl.), 1135–1140

Stollery, B.T. (1985) Human exposure to 50-Hz electric currents. In: Grandolfo, M., Michaelson, S.M. & Rindi, A., eds, *Biological Effects and Dosimetry of Static and ELF Electromagnetic Fields*, New York, NY, Plenum Press, pp. 445–454

Strahler, A.N. (1963) *The Earth Sciences*, New York, NY, Harper and Row, pp. 153–155

Strand, J.A., Abernethy, C.S., Skalski, J.R. & Genoway, R.G. (1983) Effects of magnetic field exposure on fertilization success in rainbow trout, *Salmo gairdneri*. *Bioelectromagnetics*, **4**, 295–301

Stuchly, M.A. & Lecuyer, D.W. (1989) Exposure to electromagnetic fields in arc welding. *Health Phys.*, **56**, 297–302

Stuchly, M.A. & Zhao, S. (1996) Magnetic field-induced currents in the human body in proximity of power lines. *IEEE Trans. Power Deliv.*, **11**, 102–109

Stuchly, M.A. & Dawson, T.W. (2000) Interaction of low frequency electric and magnetic fields with the human body. *Proc. IEEE*, **88**, 643–664

Stuchly, M.A. & Gandhi, O.P. (2000) Inter-laboratory comparison of numerical dosimetry for human exposure to 60 Hz electric and magnetic fields. *Bioelectromagnetics*, **21**, 167–174

Stuchly, M.A., Lecuyer, D.W. & McLean, J. (1991) Cancer promotion in a mouse-skin model by a 60-Hz magnetic field: 1. Experimental design and exposure system. *Bioelectromagnetics*, **12**, 261–271

Stuchly, M.A., McLean, J.R.N., Burnett, R., Goddard, M., Lecuyer, D.W. & Mitchel, R.E.J. (1992) Modification of tumor promotion in the mouse skin by exposure to an alternating magnetic field. *Cancer Lett.*, **65**, 1–7

Sun, W.Q., Heroux, P., Clifford, T., Sadilek, V. & Hamade, F. (1995) Characterization of the 60-Hz magnetic fields in schools of the Carleton Board of Education. *Am. ind. Hyg. Assoc. J.*, **56**, 1215–1224

Suri, A., deBoer, J., Kusser, W. & Glickman, B.W. (1996) A 3 milliTesla 60 Hz magnetic field is neither mutagenic nor co-mutagenic in the presence of menadione and MNU in a transgenic rat cell line. *Mutat. Res.*, **372**, 23–31

Sussman, S.S. & Kheifets, L.I. (1996) Re: Adult leukemia risk and personal appliance use: A preliminary study (Letter to the Editor). *Am. J. Epidemiol.*, **143**, 743–744

Svedenstål, B.-M. & Holmberg, B. (1993) Lymphoma development among mice exposed to X-rays and pulsed magnetic fields. *Int. J. Radiat. Biol.*, **64**, 119–125

Svedenstål, B.-M., Johanson, K.J. & Mild, K.H. (1999a) DNA damage induced in brain cells of CBA mice exposed to magnetic fields. *In Vivo*, **13**, 551–552

Svedenstål, B.-M., Johanson, K.J., Mattsson, M.O. & Paulsson, L.E. (1999b) DNA damage, cell kinetics and ODC activities studied in CBA mice exposed to electromagnetic fields generated by transmission lines. *In Vivo*, **13**, 507–513

Swan, S.H., Beaumont, J.J., Hammond, S.K., VonBehren, J., Green, R.S., Hallock, M.F., Woskie, S.R., Hines, C.J. & Schenker, M.B. (1995) Historical cohort study of spontaneous abortion among fabrication workers in the Semiconductor Health Study: agent-level analysis. *Am. J. ind. Med.*, **28**, 751–769

Swanson, J. (1996) Long-term variations in the exposure of the population of England and Wales to power-frequency magnetic fields. *J. Radiol. Prot.*, **16**, 287–301

Swanson, J. (1999) Residential power-frequency electric and magnetic fields: Sources and exposures. *Radiat. Protect. Dosim.*, **83**, 9–14

Swanson, J. & Jeffers, D. (1999) Possible mechanisms by which electric fields from power lines might affect airborne particles harmful to health. *J. radiol. Prot.*, **19**, 213–229

Swanson, J. & Kaune, W.T. (1999) Comparison of residential power-frequency magnetic fields away from appliances in different countries. *Bioelectromagnetics*, **20**, 244–254

Swanson, J. & Jeffers, D.E. (2000) Comment on the papers: increased exposure to pollutant aerosols under high voltage power lines; and Corona ions from powerlines and increased exposure to pollutant aerosols [Letter to the Editor and Reply from Fews & Henshaw]. *Int. J. Radiat. Biol.*, **76**, 1685–1691

Sweetland, J., Kertesatz, A., Prato, F.S. & Nantau, K. (1987) The effect of magnetic resonance imaging on human cognition. *Magn. Reson. Imaging*, **5**, 129–135

Tablado, L., Soler, C., Nunez, M., Nunez, J. & Pérez-Sánchez, F. (2000) Development of mouse testis and epididymis following intrauterine exposure to a static magnetic field. *Bioelectromagnetics*, **21**, 19–24

Tachiiri, S., Takebe, H., Hiraoka, M. & Miyakoshi, J. (1999) Magnetic fields (60Hz, 5mT) do not influence MCF-7 growth in melatonin insensitive cells. In: Bersani, F., ed., *Electricity and Magnetism in Biology and Medicine*, New York, NY, Kluwer Academic/Plenum Publishers, pp. 841–843

Takahashi, K., Kaneko, I., Date, M. & Fukada, E. (1986) Effect of pulsing electromagnetic fields on DNA synthesis in mammalian cells in culture. *Experientia*, **42**, 185–186

Takahashi, K., Kaneko, I., Date, M. & Fukada, E. (1987) Influence of pulsing electromagnetic field on the frequency of sister-chromatid exchanges in cultured mammalian cells. *Experientia*, **43**, 331–332

Tarone, R.E., Kaune, W.T., Linet, M.S., Hatch, E.E., Kleinerman, R.A., Robison, L.L., Boice, J.D., Jr & Wacholder, S. (1998) Residential wire codes: Reproducibility and relation with measured magnetic fields. *Occup. environ. Med.*, **55**, 333–339

Tenforde, T.S. (1992) Interaction mechanisms and biological effects of static magnetic fields. *Automedica*, **14**, 271–293

Tenforde, T.S. (1993) Cellular and molecular pathways of extremely low frequency electromagnetic field interactions with living systems. In: Blank, M., ed., *Electricity and Magnetism in Biology and Medicine*, San Francisco, CA, San Francisco Press, pp. 1–8

Tenforde, T.S. & Shifrine, M. (1984) Assessment of the immune responsiveness of mice exposed to 1.5-T stationary magnetic field. *Bioelectromagnetics*, **5**, 443–446

Terol, F.F. & Panchon, A. (1995) Exposure of domestic quail embryos to extremely low frequency magnetic fields. *Int. J. Radiat. Biol.*, **68**, 321–330

Thériault, G., Goldberg, M., Miller, A.B., Armstrong, B., Guénel, P., Deadman, J., Imbernon, E., To, T., Chevalier, A., Cyr, D. & Wall, C. (1994) Cancer risks associated with occupational exposure to magnetic fields among electric utility workers in Ontario and Quebec, Canada and France: 1970–1989. *Am. J. Epidemiol.*, **139**, 550–572

Thomas, J.R., Schrot, J. & Liboff, A.R. (1986) Low-intensity magnetic fields alter operant behavior in rats. *Bioelectromagnetics*, **7**, 349–357

Thomas, T.L., Stolley, P.D., Stemhagen, A., Fontham, E.T.H., Bleecker, M.L., Stewart, P.A. & Hoover, R.N. (1987) Brain tumor mortality risk among men with electrical and electronics jobs: A case–control study. *J. natl Cancer Inst.*, **79**, 233–238

Thommesen, G. & Bjølseth, P. (1992) [*Static and Low Frequency Magnetic Fields in Norwegian Alloy and Electrolysis Plants*], Oslo, National Institute of Radiation Hygiene (in Norwegian)

Thomson, R.A.E., Michaelson, S.M. & Nguyen, Q.A. (1988) Influence of 60-Hertz fields on leukemia. *Bioelectromagnetics*, **9**, 149–158

Thun-Battersby, S., Mevissen, M. & Löscher, W. (1999) Exposure of Sprague-Dawley rats to a 50-Hertz, 100-µTesla magnetic field for 27 weeks facilitates mammary tumorigenesis in the 7,12-dimethylbenz[a]anthracene model of breast cancer. *Cancer Res.*, **59**, 3627–3633

Tobey, R.A., Price, H.J., Scott, L.D. & Ley, K.D. (1981) *Lack of Effect of 60-Hz Fields on Growth of Cultured Mammalian Cells*, Report No. LA-8831-MS, Los Alamos, NM, Los Alamos Scientific Laboratory

Tomenius, L. (1986) 50-Hz electromagnetic environment and the incidence of childhood tumors in Stockholm county. *Bioelectromagnetics*, **7**, 191–207

Törnqvist, S. (1998) Paternal work in the power industry: effects on children at delivery. *J. occup. environ. Med.*, **40**, 111–117

Törnqvist, S., Norell, S., Ahlbom, A. & Knave, B. (1986) Cancer in the electric power industry. *Br. J. ind. Med.*, **43**, 212–213

Törnqvist, S., Knave, B., Ahlbom, A. & Persson, T. (1991) Incidence of leukaemia and brain tumours in some 'electrical occupations'. *Br. J. ind. Med.*, **48**, 597–603

Tremblay, L., Houde, M., Mercier, G., Gagnon, J. & Mandeville, R. (1996) Differential modulation of natural and adaptor immunity in Fischer rats exposed for 6 weeks to 60 Hz linear sinusoidal continuous-wave magnetic fields. *Bioelectromagnetics*, **17**, 373–383

Trillo, M.A., Ubeda, A., Blanchard, J.-P., House, D.E. & Blackman, C.F. (1996) Magnetic fields at resonant conditions for the hydrogen ion affect neurite outgrowth in PC-12 cells: a test of the ion parametric resonance model. *Bioelectromagnetics*, **17**, 10–20

Truong, H. & Yellon, S.M. (1997) Effect of various acute 60 Hz magnetic field exposures on the nocturnal melatonin rise in the adult Djungarian hamster. *J. pineal Res.*, **22**, 177–183

Truong, H., Smith, J.C. & Yellon, S.M. (1996) Photoperiod control of the melatonin rhythm and reproductive maturation in the juvenile Djungarian hamster: 60-Hz magnetic field exposure effects. *Biol. Reprod.*, **55**, 455–460

Trzeciak, H.I., Grzesik, J., Bortel, M., Kuska, R., Duda, D., Michnik, J. & Malecki, A. (1993) Behavioral effects of long-term exposure to magnetic fields in rats. *Bioelectromagnetics*, **14**, 287–297

Tuschl, H., Neubauer, G., Schmid, G., Weber, E. & Winker, N. (2000) Occupational exposure to static, ELF, VF and VLF magnetic fields and immune parameters. *Int. J. occup. Med. environ. Health*, **13**, 39–50

Tyndall, D.A. (1993) MRI effects on cranofacial size and crown–rump length in C57BL/6J mice in 1.5 T fields. *Oral Surg. oral Med. oral Pathol.*, **76**, 655–660

Tyndall, D.A. & Sulik, K.K. (1991) Effects of magnetic resonance imaging on eye development in the C57BL/6J mouse. *Teratology*, **43**, 263–275

Tynes, T. & Andersen, A. (1990) Electromagnetic fields and male breast cancer (Letter to the Editor). *Lancet*, **336**, 1596

Tynes, T. & Haldorsen, T. (1997) Electromagnetic fields and cancer in children residing near Norwegian high-voltage power lines. *Am. J. Epidemiol.*, **145**, 219–226

Tynes, T., Andersen, A. & Langmark, F. (1992) Incidence of cancer in Norwegian workers potentially exposed to electromagnetic fields. *Am. J. Epidemiol.*, **136**, 81–88

Tynes, T., Reitan, J.B. & Andersen, A. (1994a) Incidence of cancer among workers in Norwegian hydroelectric power companies. *Scand. J. Work Environ. Health*, **20**, 339–344

Tynes, T., Jynge, H. & Vistness, A.I. (1994b) Leukemia and brain tumors in Norwegian railway workers, a nested case–control study. *Am. J. Epidemiol.*, **139**, 645–653

Ubeda, A., Leal, J., Trillo, M.A., Jiménez, M.A. & Delgado, J.M.R. (1983) Pulse shape of magnetic fields influences chick embryogenesis. *J. Anat.*, **137**, 513–536

Ubeda, A., Leal, J., Trillo, M.A., Jiménez, M.A. & Delgado, J.M.R. (1985) Authors' correction to data. *J. Anat.*, **140**, 721

Ubeda, A., Trillo, M.A. & Leal, J. (1987) Magnetic field effects on embryonic development: influence of the organism orientation. *Med. Sci. Res.*, **15**, 531–532

Ubeda, A., Trillo, M.A., Chacon, L., Blanco, M.J. & Leal, J. (1994) Chick embryo development can be irreversibly altered by early exposure to weak extremely-low-frequency magnetic fields. *Bioelectromagnetics*, **15**, 385–398

Ubeda, A., Trillo, M.A., House, D.E. & Blackman, C.F. (1995) A 50 Hz magnetic field blocks melatonin-induced enhancement of junctional transfer in normal C3H/10T1/2 cells. *Carcinogenesis*, **16**, 2945–2949

Uckun, F.M., Kurosaki, T., Jin, J., Jun, X., Morgan, A., Takata, M., Bolen, J. & Luben, R. (1995) Exposure of B-lineage lymphoid cells to low energy electromagnetic fields stimulates Lyn kinase. *J. biol. Chem.*, **270**, 27666–27670

Ueno, S., Harada, K. & Shiokawa, K. (1984) The embryonic development of frogs under strong dc magnetic fields. *IEEE Trans. Magnet.*, **20**, 1663–1665

UK Childhood Cancer Study Investigators (UKCCSI) (1999) Exposure to power-frequency magnetic fields and the risk of childhood cancer. *Lancet*, **354**, 1925–1931

UK Childhood Cancer Study Investigators (UKCCSI) (2000a) Childhood cancer and residential proximity to power lines. *Br. J. Cancer*, **83**, 1573–1580

UK Childhood Cancer Study Investigators (UKCCSI) (2000b) The United Kingdom childhood cancer study: Objectives, materials and methods. *Br. J. Cancer*, **82**, 1073–1102

Vågerö, D. & Olin, R. (1983) Incidence of cancer in the electronics industry: Using the new Swedish Cancer Environment Registry as a screening instrument. *Br. J. ind. Med.*, **40**, 188–192

Vågerö, D., Ahlbom, A., Olin, R. & Sahlsten, S. (1985) Cancer morbidity among workers in the telecommunications industry. *Br. J. ind. Med.*, **42**, 191–195

Valberg, P.A. (1995) Designing EMF experiments: what is required to characterize 'exposure'? *Bioelectromagnetics*, **16**, 396–401, 406

Valjus, J., Norppa, H., Jarventaus, H., Sorsa, M., Nykyri, E., Salomaa, S., Jarvinen, P. & Kajander, J. (1993) Analysis of chromosomal aberrations, sister chromatid exchanges and micronuclei among power linesmen with long-term exposure to 50-Hz electromagnetic fields. *Radiat. environ. Biophys.*, **32**, 325–336

Valtersson, U., Mild, K.H. & Mattsson, M.-O. (1997) Ornithine decarboxylase activity and polyamine levels are different in Jurkat and CEM-CM3 cells after exposure to a 50 Hz magnetic field. *Bioelectrochem. Bioenerg.*, **43**, 169–172

van Wijngaarden, E., Savitz, D.A., Kleckner, R.C., Cai, J. & Loomis, D. (2000) Exposure to electromagnetic fields and suicide among electric utility workers: A nested case–control study. *West. J. Med.*, **173**, 94–100

Vena, J.E., Graham, S., Hellmann, R., Swanson, M. & Brasure, J. (1991) Use of electric blankets and risk of postmenopausal breast cancer. *Am. J. Epidemiol.*, **134**, 180–185

Vena, J.E., Freudenheim, J.L., Marshall, J.R., Laughlin, R., Swanson, M. & Graham, S. (1994) Risk of premenopausal breast cancer and the use of electric blankets. *Am. J. Epidemiol.*, **140**, 974–979

Vena, J.E., Marshall, J.R., Freudenheim, J.L., Swanson, M. & Graham, S. (1995) Re: 'Risk of premenopausal breast cancer and use of electric blankets' — The author's reply. *Am. J. Epidemiol.*, **142**, 446–447

Verkasalo, P.K. (1996) Magnetic fields and leukemia — Risk for adults living close to power lines. *Scand. J. Work Environ. Health*, **22** (Suppl. 2), 1–56

Verkasalo, P.K., Pukkala, E., Hongisto, M.Y., Valjus, J.E., Järvinen, P.J., Keikkilä, K.V. & Koskenvuo, M. (1993) Risk of cancer in Finnish children living close to power lines. *Br. med. J.*, **307**, 895–899

Verkasalo, P.K., Pukkala, E., Kaprio, J., Heikkilä, K.V. & Koskenvuo, M. (1996) Magnetic fields of high voltage power lines and risk of cancer in Finnish adults: Nationwide cohort study. *Br. med. J.*, **313**, 1047–1051

Verkasalo, P.K., Kaprio, J., Varjonen, J., Romanov, K., Heikkila, K. & Koskenvuo, M. (1997) Magnetic fields of transmission lines and depression. *Am. J. Epidemiol.*, **146**, 1037–1045

Villeneuve, P.J., Agnew, D.A., Miller, A.B., Corey, P.N. & Purdham, J.T. (2000) Leukemia in electric utility workers: The evaluation of alternative indices of exposure to 60 Hz electric and magnetic fields. *Am. J. ind. Med.*, **37**, 607–617

Wake, K., Tanaka, T., Kawasumi, M. & Taki, M. (1998) Induced current density distribution in a human related to magnetophosphenes. *Trans. IEE Japan*, **118-A**, 806–811

Walleczek, J. (1995) Magnetokinetic effects on radical pairs: a paradigm for magnetic field interactions with biological systems at lower than thermal energy. In: Blank, M., ed., *Electromagnetic Fields: Biological Interactions and Mechanisms* (Advances in Chemistry 250), Washington DC, American Chemical Society, pp. 395–420

Walleczek, J. & Liburdy, R.P. (1990) Nonthermal 60 Hz sinusoidal magnetic-field exposure enhances $^{45}Ca^{2+}$ uptake in rat thymocytes: dependence on mitogen activation. *FEBS Lett.*, **271**, 157–160

Walleczek, J., Shiu, E.C. & Hahn, G.M. (1999) Increase in radiation-induced *HPRT* gene mutation frequency after nonthermal exposure to nonionizing 60 Hz electromagnetic fields. *Radiat. Res.*, **151**, 489–497

Walters, E. & Carstensen, E.L. (1987) Test for the effects of 60-Hz magnetic fields on fecundity and development in *Drosophila*. *Bioelectromagnetics*, **8**, 351–354

Weaver, J.C. & Astumian, R.D. (1990) The response of living cells to very weak electric fields: the thermal noise limit. *Science*, **247**, 459–462

Weigel, R.J. & Lundstrom, D.L. (1987) Effect of relative humidity on the movement of rat vibrissae in a 60-Hz electric field. *Bioelectromagnetics*, **8**, 107–110

Weigel, R.J., Jaffe, R.A., Lundstrom, D.L., Forsythe, W.C. & Anderson, L.E. (1987) Stimulation of cutaneous mechanoreceptors by 60-Hz electric fields. *Bioelectromagnetics*, **8**, 337–350

Wertheimer, N. & Leeper, E. (1979) Electrical wiring configurations and childhood cancer. *Am. J. Epidemiol.*, **109**, 273–284

Wertheimer, N. & Leeper, E. (1982) Adult cancer related to electrical wires near the home. *Int. J. Epidemiol.*, **11**, 345–355

Wertheimer, N. & Leeper, E. (1986) Possible effects of electric blankets and heated waterbeds on fetal development. *Bioelectromagnetics*, **7**, 13–22

Wertheimer, N. & Leeper, E. (1987) Magnetic field exposure related to cancer subtypes. *Ann. NY Acad. Sci.*, **502**, 43–53

Wertheimer, N. & Leeper, E. (1989) Fetal loss associated with two seasonal sources of electro-magnetic field exposure. *Am. J. Epidemiol.*, **129**, 220–224

Wertheimer, N., Savitz, D.A. & Leeper, E. (1995) Childhood cancer in relation to indicators of magnetic fields from ground current sources. *Bioelectromagnetics*, **16**, 86–96

West, R.W., Hinson, W.G. & Swicord, M.L. (1996) Anchorage-independent growth with JB6 cells exposed to 60 Hz magnetic fields at several flux densities. *Bioelectrochem. Bioenerg.*, **39**, 175–179

Westerholm, P., & Ericson, A. (1987) *Pregnancy Outcome and VDU-work in a Cohort of Insurance Clerks, Proceedings of the International Scientific Conference: Work with Display Units, 1986*, Stockholm, pp. 104–107

Wey, H.E., Conover, D.P., Mathias, P., Toraason, M. & Lotz, W.G. (2000) 50-Hertz magnetic field and calcium transients in Jurkat cells: results of a research and public information dissemination (RAPID) program study. *Environ. Health Perspect.*, **108**, 135–140

Whittington, C.J., Podd, J.V. & Rapley, B.R. (1996) Acute effects of 50 Hz magnetic field exposure on human visual task and cardiovascular performance. *Bioelectromagnetics*, **17**, 131–137

WHO (1984) *Extremely Low Frequency (ELF) Fields* (Environmental Health Criteria Report 35), Geneva, World Health Organization

WHO (1987) *Magnetic Fields* (Environmental Health Criteria Report 69), Geneva, World Health Organization

Wiklund, K., Einhorn, J. & Eklund, G. (1981) An application of the Swedish Cancer-Environment Registry. Leukaemia among telephone operators at the telecommunications administration in Sweden. *Int. J. Epidemiol.*, **10**, 373–376

Wilkins, J.R., III & Koutras, R.A. (1988) Paternal occupation and brain cancer in offspring: A mortality-based case–control study. *Am. J. ind. Med.*, **14**, 299–318

Wilkins, J.R., III & Hundley, V.D. (1990) Paternal occupational exposure to electromagnetic fields and neuroblastoma in offspring. *Am. J. Epidemiol.*, **131**, 995–1008

Wilkins, J.R., III & Wellage, L.C. (1996) Brain tumor risk in offspring of men occupationally exposed to electric and magnetic fields. *Scand. J. Work Environ. Health*, **22**, 339–345

Wilson, B.W., Anderson, L.E., Hilton, D.I. & Phillips, R.D. (1981) Chronic exposure to 60-Hz electric fields: Effects on pineal function in the rat. *Bioelectromagnetics*, **2**, 371–380

Wilson, B.W., Anderson, L.E., Hilton, D.I. & Phillips, R.D. (1983) Chronic exposure to 60-Hz electric fields: Effects on pineal function in the rat. *Bioelectromagnetics*, **4**, 293

Wilson, B.W., Chess, E.K. & Anderson, L.E. (1986) 60-Hz electric-field effects on pineal melatonin rhythms: Time course for onset and recovery. *Bioelectromagnetics*, **7**, 239–242

Wilson, B.W., Stevens, R.G. & Anderson, L.E. (1989) Neuroendocrine mediated effects of electromagnetic-field exposure: possible role of the pineal gland. *Life Sci.*, **45**, 1319–1332

Wilson, B.W., Wright, C.W., Morris, J.E., Buschbom, R.L., Brown, D.P., Miller, D.L., Sommers-Flannigan, R. & Anderson, L.E. (1990) Evidence for an effect of ELF electromagnetic fields on human pineal gland function. *J. pineal Res.*, **9**, 259–269

Wilson, B.W., Caputa, K., Stuchly, M.A., Saffer, J.D., Davis, K.C., Washam, C.E., Washam, L.G., Washam, G.R. & Wilson, M.A. (1994) Design and fabrication of well confined uniform magnetic field exposure systems. *Bioelectromagnetics*, **15**, 563–577

Wilson, B.W., Lee, G.M., Yost, M.G., Davis, K.C., Heimbigner, T. & Buschbom, R.L. (1996) Magnetic field characteristics of electric bed-heating devices. *Bioelectromagnetics*, **17**, 174–179

Wilson, B.W., Matt, K.S., Morris, J.E., Sasser, L.B., Miller, D.L. & Anderson, L.E. (1999) Effects of 60 Hz magnetic field exposure on the pineal and hypothalamic-pituitary-gonadal axis in the Siberian hamster (*Phodopus sungorus*). *Bioelectromagnetics*, **20**, 224–232

Windham, G.C., Fenster, L., Swan, S.H. & Neutra, R.R. (1990) Use of video display terminals during pregnancy and risk of spontaneous abortion, low birthweight, or intrauterine growth retardation. *Am. J. ind. Med.*, **18**, 675–688

Wiskirchen, J., Groenewaeller, E.F., Kehlbach, R., Heinzelmann, F., Wittau, M., Rodemann, H.P., Claussen, C.D. & Duda, S.H. (1999) Long-term effects of repetitive exposure to a static magnetic field (1.5 T) on proliferation of human fetal lung fibroblasts. *Magn. Reson. Med.*, **41**, 464–468

Wiskirchen, J., Grönewäller, E.F., Heinzelmann, F., KehlbacH, R., Rodegerdts, E., Wittau, M., Rodemann, H.P., Claussen, C.D. & Duda, S.H. (2000) Human fetal lung fibroblasts: in vitro study of repetitive magnetic field exposure at 0.2, 1.0, and 1.5 T. *Radiology*, **215**, 858–862

Withers, H.R., Mason, K.A. & Davis, C.A. (1985) MR effect on murine spermatogenesis. *Radiology*, **156**, 741–742

Wolpaw, J.R., Seegal, R.F. & Dowman, R. (1989) Chronic exposure of primates to 60-Hz electric and magnetic fields: I. Exposure system and measurements of general health and performance. *Bioelectromagnetics*, **10**, 277–288

Wood, A.W., Armstrong, S.M., Sait, M.L., Devine, L. & Martin, M.J. (1998) Changes in human plasma melatonin profiles in response to 50 Hz magnetic field exposure. *J. pineal Res.*, **25**, 116–127

Woods, M., Bobanovic, F., Brown, D. & Alexander, D.R. (2000) Lyn and Syk tyrosine kinases are not activated in B-lineage lymphoid cells exposed to low-energy electromagnetic fields. *FASEB J.*, **14**, 2284–2290

Wrensch, M., Yost, M., Miike, R., Lee, G. & Touchstone, J. (1999) Adult glioma in relation to residential power frequency electromagnetic field exposures in the San Francisco Bay area. *Epidemiology*, **10**, 523–527

Wright, W.E., Peters, J.M. & Mack, T.M. (1982) Leukaemia in workers exposed to electrical and magnetic fields (Letter to the Editor). *Lancet*, **ii**, 1160–1161

Yaguchi, H., Yoshida, M., Ejima, Y. & Miyakoshi, J. (1999) Effect of high-density extremely low frequency magnetic field on sister chromatid exchanges in mouse m5S cells. *Mutat. Res.*, **440**, 189–194

Yaguchi, H., Yoshida, M., Ding, G.-R., Shingu, K. & Miyakoshi, J. (2000) Increased chromatid-type chromosomal aberrations in mouse m5S cells exposed to power-line frequency magnetic fields. *Int. J. Radiat. Biol.*, **76**, 1677–1684

Yasui, M., Kikuchi, T., Ogawa, M., Otaka, Y., Tsuchitani, M. & Iwata, H. (1997) Carcinogenicity test of 50 Hz sinusoidal magnetic fields in rats. *Bioelectromagnetics*, **18**, 531–540

Yellon, S.M. (1994) Acute 60 Hz magnetic field exposure effects on the melatonin rhythm in the pineal gland and circulation of the adult Djungarian hamster. *J. pineal Res.*, **16**, 136–144

Yellon, S.M. (1996a) 60-Hz magnetic field exposure effects on the melatonin rhythm and photoperiod control of reproduction. *Am. J. Physiol.*, **270**, E816–E821

Yellon, S.M. (1996b) Daily melatonin treatments regulate the circadian melatonin in the adult Djungarian hamster. *J. Biol. Rhythms*, **11**, 4–13

Yellon, S.M. & Truong, H.N. (1998) Melatonin rhythm onset in the adult siberian hamster: Influence of photoperiod but not 60-Hz magnetic field exposure on melatonin content in the pineal gland and in circulation. *J. Biol. Rhythms*, **13**, 52–59

Yip, Y.P., Capriotti, C., Norbash, S.G., Talagala, S.L. & Yip, J.W. (1994) Effects of MR exposure on cell proliferation and migration of chick motoneurons. *J. magn. Reson. Imaging*, **4**, 799–804

Youngson, J.H.A.M., Clayden, A.D., Myers, A. & Cartwright, R.A. (1991) A case/control study of adult haematological malignancies in relation to overhead powerlines. *Br. J. Cancer*, **63**, 977–985

Yu, M.C., Gona, A.G., Gona, O., Al-Rabiai, S., Von Hagen, S. & Cohen, E. (1993) Effects of 60 Hz electric and magnetic fields on maturation of the rat neopallium. *Bioelectromagnetics*, **14**, 449–458

Zaffanella, L. (1996) *Environmental Surveys* (EMF RAPID Engineering Project 3), Washington DC, US Department of Energy, Report available at ftp://ftp.emf-data.org/pub/emf-data/datasets/004/rapid3.pdf

Zaffanella, L. & Kalton, G.W. (1998) *Survey of Personal Magnetic Field Exposure. Phase II: 1000-Person Survey* (EMF RAPID Program Engineering Project #6), Oak Ridge, TN, Lockheed Martin Energy Systems, Report available at http://www.emf-data.org

Zaffanella, L.E., Kavet, R., Pappa, J.R. & Sullivan, T.P. (1997) Modeling magnetic fields in residences: Validation of the RESICALC [computer] program. *J. Expo. anal. Environ. Epidemiol.*, **7**, 241–259

Zecca, L., Ferrario, P. & Dal Conte, G. (1985) 765-Toxicological and teratological studies in rats after exposure to pulsed magnetic fields. *Bioelectrochem. Bioenerget.*, **14**, 63–69

Zecca, L., Mantegazza, C., Margonato, V., Cerretelli, P., Caniatti, M., Piva, F., Dondi, D. & Hagino, N. (1998) Biological effects of prolonged exposure to ELF electromagnetic fields in rats: III. 50 Hz electromagnetic fields. *Bioelectromagnetics*, **19**, 57–66

Zhadin, M.N. (1998) Combined action of static and alternating magnetic fields on ion motion in a macromolecule: theoretical aspects. *Bioelectromagnetics*, **19**, 279–292

Zhadin, M.N., Deryugina, O.N. & Pisachenko, T.M. (1999) Influence of combined DC and AC magnetic fields on rat behavior. *Bioelectromagnetics*, **20**, 378–386

Zheng, T., Holford, T.R., Mayne, S.T., Owens, P.H., Zhang, B., Boyle, P., Carter, D., Ward, B., Zhang, Y. & Hoar Zahm, S. (2000) Exposure to electromagnetic fields from use of electric blankets and other in-home electrical appliances and breast cancer risk. *Am. J. Epidemiol.*, **151**, 1103–1111

Zhu, K., Weiss, N.S., Stanford, J.L., Daling, J.R., Stergachis, A., McKnight, B., Brawer, M.K. & Levine, R.S. (1999) Prostate cancer in relation to the use of electric blanket or heated water bed. *Epidemiology*, **10**, 83–85

Zimmermann, B. & Hentschel, D. (1987) [Effect of a static magnetic field (3.5 T) on the reproductive behavior of mice, on the embryo and fetal development and on selected hematologic parameters]. *Digit. Bilddiagn.*, **7**, 155–161 (in German)

Zimmerman, S., Zimmerman, A.M., Winters, W.D. & Cameron, I.L. (1990) Influence of 60 Hz magnetic fields on sea urchin development. *Bioelectromagnetics*, **11**, 37–45

Zmyslony, M., Palus, J., Jajte, J., Dziubaltowska, E. & Rajkowska, E. (2000) DNA damage in rat lymphocytes treated in vitro with iron cations and exposed to 7 mT magnetic fields (static or 50 Hz). *Mutat. Res.*, **453**, 89–96

Zubal, I.G., Harrell, C.R., Smith, E.O., Rattner, Z., Gindi, G. & Hoffer, P.H. (1994) Computerized three-dimensional segmented human anatomy. *Med. Phys.*, **21**, 299–302

Zusman, I., Yaffe, P., Pinus, H. & Ornoy, A. (1990) Effects of pulsing electromagnetic fields on the prenatal and postnatal development in mice and rats: In Vivo and In Vitro studies. *Teratology*, **42**, 157–170

Zwingelberg, R., Obe, G., Rosenthal, M., Mevissen, M., Buntenkötter, S. & Löscher, W. (1993) Exposure of rats to a 50-Hz, 30-mT magnetic field influences neither the frequencies of sister-chromatid exchanges nor proliferation characteristics of cultured peripheral lymphocytes. *Mutat. Res.*, **302**, 39–44

LIST OF ABBREVIATIONS

AC: alternating current
ALL: acute lymphoblastic leukaemia
AMEX: average magnetic exposure (a personal exposure meter)
AML: acute myeloid leukaemia
ANLL: acute non-lymphocytic leukaemia
bw: body weight
CI: confidence interval
CLL: chronic lymphocytic leukaemia
CML: chronic myeloid leukaemia
DC: direct current
DMBA: 7,12-dimethylbenz[*a*]anthracene
EEG: electroencephalogram
ELF: extremely low frequency (the frequency range from 3–3000 Hz)
EMDEX: electric and magnetic field digital exposure system (a personal exposure meter)
ENU: *N*-ethyl-*N*-nitrosourea
Fe$_3$O$_4$: magnetite
γGT: γ-glutamyltranspeptidase
GST-P: glutathione-*S*-transferase, placental form
HRV: heart rate variability
ICD: International Classification of Diseases
IEC: International Electrotechnical Commission
IEEE: Institute of Electrical and Electronics Engineers, USA
JEM: job–exposure matrix
MDS: myelodysplastic syndrome
MNU: *N*-methyl-*N*-nitrosourea
MRI: magnetic resonance imaging
NDEA: *N*-nitrosodiethylamine
NRPB: National Radiological Protection Board
ODC: ornithine decarboxylase
OHCC: ordinary high current configuration

OLCC: ordinary low current configuration
OR: odds ratio
PEM: personal exposure meter
PEMF: pulsed electromagnetic fields
PIR: proportionate incidence ratio
PMR: proportionate mortality ratio
PRR: proportionate registration ratio
rms: root-mean-square (see Glossary)
SIR: standardized incidence ratio
SMR: standardized mortality ratio
TPA: 12-*O*-tetradecanoylphorbol 13-acetate
TWA: time-weighted average (see Glossary)
UG: underground (buried cables)
VDT: video (visual) display terminal
VDU: video display unit
VHCC: very high current configuration
VLCC: very low current configuration
VLF: very low frequency

GLOSSARY

Atmospherics: the electromagnetic processes associated with lightning discharges (also called 'sferics')

Busbars: electrical connections between the transformer and other parts of an electricity substation.

Characteristics: detailed physical properties of electric or magnetic fields, such as the magnitude, frequency spectrum, polarization, etc.

Counterion polarization: the physical phenomenon responsible for the dispersion at low frequencies.

Dosimeter: an instrument that can be worn on the body of a person for measuring exposure over time.

Electric field: a vector field **E** measured in volts per metre.

Electromagnetic fields: the combination of electric and magnetic fields in the environment. This term is often confused with 'electromagnetic radiation' and can therefore be misleading when used with extremely low frequencies for which the radiation is barely detectable. For this reason the term 'electric and magnetic fields' is used throughout this Monograph.

Electrostatic fields: static fields produced by fixed potential differences.

Exposure: the amount of a chemical or physical agent in the environment that a person comes into contact with over a period of time.

Exposure assessment: the evaluation of a person's exposure by measurements, modelling, information about sources or other means.

Exposure metric: a single number that summarizes exposure to an electric and/or magnetic field. The metric is usually determined by a combination of the instrument's signal processing and the data analysis performed after the measurement.

Frequency response: the output of an instrument as a function of frequency relative to the magnitude of the input signal. The specification of the frequency response of an instrument includes the type of filter and its bandwidth.

Gap junction: an aqueous pore or channel through which neighbouring cell membranes are connected.

Geomagnetic field: magnetic field originating from the earth (including the atmosphere). Predominantly a static magnetic field, but includes some oscillating components and transients.

Harmonic (frequency): frequencies that are integral multiples of the power frequency or some other reference frequency.

High-voltage power lines: usually taken to mean power lines operating at 100 kV or 132 kV (also referred to as transmission lines).

Intermittent fields: fields with a root-mean-square vector magnitude that changes rapidly. In contrast to transients, intermittent fields may reach high levels for longer times and are generally in the ELF frequency range.

Magnetic field: In studies at extremely low frequency, this term is generally used for the magnetic flux density (B field).

Magnetic field strength: a vector field **H** with units of ampere per metre.

Magnetic flux density: a vector field **B** with units of tesla.

Magnetostatic fields: static fields established by permanent magnets and by steady currents.

Phosphenes: weak visual sensations that occur in response to magnetic fields (threshold, 20 Hz, 8 mT) or by direct electrostimulation. The effect is believed to result from the interaction of the induced current with electrically excitable cells in the retina.

Power frequency: the frequency at which AC electricity is generated. For electric utilities, the power frequency is 60 Hz in North America, Brazil and parts of Japan, and 50 Hz in much of the rest of the world.

Right-of-way: corridor of defined width within which the power line runs.

Root-mean-square (rms): the most versatile mathematical function for averaging the magnitude of time-varying electric and magnetic fields.

Spot measurement: an instantaneous measurement at a designated location.

Static field: a field vector that does not vary with time. In most environments, electric and magnetic fields change with time, but their frequency spectrum has a component at 0 Hz. This 'quasi-static' component of the field can be measured by averaging the oscillating signal over the sample time.

Time-weighted average (TWA): a weighted average of exposure measurements taken over a period of time with the weighting factor equal to the time interval between measurements. When the measurements are made with a monitor with a fixed sampling rate, the TWA is equal to the arithmetic mean of the measurements.

Transients: brief bursts of high-frequency fields, usually resulting from mechanical switching of AC electricity.

Transmission lines: *see* high-voltage power lines.

Transposed phasing: arrangement in which the wires or bundles of wire — phases — in the circuit on one side of the tower have the opposite order to those on the other side. This arrangement results in fields that decrease more rapidly with distance from the lines than other configurations.

Voxels: cubic cells with sides of 1–10 mm used to represent animal and human tissues in dosimetry models.

Waveform: a single component of the field measured as a function of time by an instrument with a response time much faster than the field's frequency of oscillation. The term also refers to the shape of the wave as displayed on a graph or oscilloscope trace.

Wire coding: a non-intrusive method for classifying homes on the basis of their distance from visible electrical installations and the characteristics of these installations.

CUMULATIVE CROSS INDEX TO *IARC MONOGRAPHS ON THE EVALUATION OF CARCINOGENIC RISKS TO HUMANS*

The volume, page and year of publication are given. References to corrigenda are given in parentheses.

A

A-α-C	*40*, 245 (1986); *Suppl. 7*, 56 (1987)
Acetaldehyde	*36*, 101 (1985) (*corr. 42*, 263); *Suppl. 7*, 77 (1987); *71*, 319 (1999)
Acetaldehyde formylmethylhydrazone (*see* Gyromitrin)	
Acetamide	*7*, 197 (1974); *Suppl. 7*, 56, 389 (1987); *71*, 1211 (1999)
Acetaminophen (*see* Paracetamol)	
Aciclovir	*76*, 47 (2000)
Acridine orange	*16*, 145 (1978); *Suppl. 7*, 56 (1987)
Acriflavinium chloride	*13*, 31 (1977); *Suppl. 7*, 56 (1987)
Acrolein	*19*, 479 (1979); *36*, 133 (1985); *Suppl. 7*, 78 (1987); *63*, 337 (1995) (*corr. 65*, 549)
Acrylamide	*39*, 41 (1986); *Suppl. 7*, 56 (1987); *60*, 389 (1994)
Acrylic acid	*19*, 47 (1979); *Suppl. 7*, 56 (1987); *71*, 1223 (1999)
Acrylic fibres	*19*, 86 (1979); *Suppl. 7*, 56 (1987)
Acrylonitrile	*19*, 73 (1979); *Suppl. 7*, 79 (1987); *71*, 43 (1999)
Acrylonitrile-butadiene-styrene copolymers	*19*, 91 (1979); *Suppl. 7*, 56 (1987)
Actinolite (*see* Asbestos)	
Actinomycin D (*see also* Actinomycins)	*Suppl. 7*, 80 (1987)
Actinomycins	*10*, 29 (1976) (*corr. 42*, 255)
Adriamycin	*10*, 43 (1976); *Suppl. 7*, 82 (1987)
AF-2	*31*, 47 (1983); *Suppl. 7*, 56 (1987)
Aflatoxins	*1*, 145 (1972) (*corr. 42*, 251); *10*, 51 (1976); *Suppl. 7*, 83 (1987); *56*, 245 (1993)
Aflatoxin B₁ (*see* Aflatoxins)	
Aflatoxin B₂ (*see* Aflatoxins)	
Aflatoxin G₁ (*see* Aflatoxins)	
Aflatoxin G₂ (*see* Aflatoxins)	
Aflatoxin M₁ (*see* Aflatoxins)	
Agaritine	*31*, 63 (1983); *Suppl. 7*, 56 (1987)
Alcohol drinking	*44* (1988)
Aldicarb	*53*, 93 (1991)
Aldrin	*5*, 25 (1974); *Suppl. 7*, 88 (1987)

Allyl chloride   *36*, 39 (1985); *Suppl. 7*, 56 (1987);
                                                                                        71, 1231 (1999)

Allyl isothiocyanate   *36*, 55 (1985); *Suppl. 7*, 56 (1987);
                                                                                        73, 37 (1999)

Allyl isovalerate   *36*, 69 (1985); *Suppl. 7*, 56 (1987);
                                                                                        71, 1241 (1999)

Aluminium production   *34*, 37 (1984); *Suppl. 7*, 89 (1987)
Amaranth   *8*, 41 (1975); *Suppl. 7*, 56 (1987)
5-Aminoacenaphthene   *16*, 243 (1978); *Suppl. 7*, 56 (1987)
2-Aminoanthraquinone   *27*, 191 (1982); *Suppl. 7*, 56 (1987)
para-Aminoazobenzene   *8*, 53 (1975); *Suppl. 7*, 56, 390
                                                                                        (1987)

ortho-Aminoazotoluene   *8*, 61 (1975) (*corr. 42*, 254);
                                                                                        Suppl. 7, 56 (1987)

para-Aminobenzoic acid   *16*, 249 (1978); *Suppl. 7*, 56 (1987)
4-Aminobiphenyl   *1*, 74 (1972) (*corr. 42*, 251);
                                                                                        Suppl. 7, 91 (1987)

2-Amino-3,4-dimethylimidazo[4,5-*f*]quinoline (*see* MeIQ)
2-Amino-3,8-dimethylimidazo[4,5-*f*]quinoxaline (*see* MeIQx)
3-Amino-1,4-dimethyl-5*H*-pyrido[4,3-*b*]indole (*see* Trp-P-1)
2-Aminodipyrido[1,2-*a*:3',2'-*d*]imidazole (*see* Glu-P-2)
1-Amino-2-methylanthraquinone   *27*, 199 (1982); *Suppl. 7*, 57 (1987)
2-Amino-3-methylimidazo[4,5-*f*]quinoline (*see* IQ)
2-Amino-6-methyldipyrido[1,2-*a*:3',2'-*d*]imidazole (*see* Glu-P-1)
2-Amino-1-methyl-6-phenylimidazo[4,5-*b*]pyridine (*see* PhIP)
2-Amino-3-methyl-9*H*-pyrido[2,3-*b*]indole (*see* MeA-α-C)
3-Amino-1-methyl-5*H*-pyrido[4,3-*b*]indole (*see* Trp-P-2)
2-Amino-5-(5-nitro-2-furyl)-1,3,4-thiadiazole   *7*, 143 (1974); *Suppl. 7*, 57 (1987)
2-Amino-4-nitrophenol   *57*, 167 (1993)
2-Amino-5-nitrophenol   *57*, 177 (1993)
4-Amino-2-nitrophenol   *16*, 43 (1978); *Suppl. 7*, 57 (1987)
2-Amino-5-nitrothiazole   *31*, 71 (1983); *Suppl. 7*, 57 (1987)
2-Amino-9*H*-pyrido[2,3-*b*]indole (*see* A-α-C)
11-Aminoundecanoic acid   *39*, 239 (1986); *Suppl. 7*, 57 (1987)
Amitrole   *7*, 31 (1974); *41*, 293 (1986) (*corr.*
                                                                                        52, 513; *Suppl. 7*, 92 (1987);
                                                                                        79, 381 (2001)

Ammonium potassium selenide (*see* Selenium and selenium compounds)
Amorphous silica (*see also* Silica)   *42*, 39 (1987); *Suppl. 7*, 341 (1987);
                                                                                        68, 41 (1997)

Amosite (*see* Asbestos)
Ampicillin   *50*, 153 (1990)
Amsacrine   *76*, 317 (2000)
Anabolic steroids (*see* Androgenic (anabolic) steroids)
Anaesthetics, volatile   *11*, 285 (1976); *Suppl. 7*, 93 (1987)
Analgesic mixtures containing phenacetin (*see also* Phenacetin)   *Suppl. 7*, 310 (1987)
Androgenic (anabolic) steroids   *Suppl. 7*, 96 (1987)
Angelicin and some synthetic derivatives (*see also* Angelicins)   *40*, 291 (1986)
Angelicin plus ultraviolet radiation (*see also* Angelicin and some   *Suppl. 7*, 57 (1987)
    synthetic derivatives)
Angelicins   *Suppl. 7*, 57 (1987)
Aniline   *4*, 27 (1974) (*corr. 42*, 252);
                                                                                        27, 39 (1982); *Suppl. 7*, 99 (1987)

B

Benz[*a*]anthracene	*3*, 45 (1973); *32*, 135 (1983); *Suppl. 7*, 58 (1987)
Benzene	*7*, 203 (1974) (*corr. 42*, 254); *29*, 93, 391 (1982); *Suppl. 7*, 120 (1987)
Benzidine	*1*, 80 (1972); *29*, 149, 391 (1982); *Suppl. 7*, 123 (1987)
Benzidine-based dyes	*Suppl. 7*, 125 (1987)
Benzo[*b*]fluoranthene	*3*, 69 (1973); *32*, 147 (1983); *Suppl. 7*, 58 (1987)
Benzo[*j*]fluoranthene	*3*, 82 (1973); *32*, 155 (1983); *Suppl. 7*, 58 (1987)
Benzo[*k*]fluoranthene	*32*, 163 (1983); *Suppl. 7*, 58 (1987)
Benzo[*ghi*]fluoranthene	*32*, 171 (1983); *Suppl. 7*, 58 (1987)
Benzo[*a*]fluorene	*32*, 177 (1983); *Suppl. 7*, 58 (1987)
Benzo[*b*]fluorene	*32*, 183 (1983); *Suppl. 7*, 58 (1987)
Benzo[*c*]fluorene	*32*, 189 (1983); *Suppl. 7*, 58 (1987)
Benzofuran	*63*, 431 (1995)
Benzo[*ghi*]perylene	*32*, 195 (1983); *Suppl. 7*, 58 (1987)
Benzo[*c*]phenanthrene	*32*, 205 (1983); *Suppl. 7*, 58 (1987)
Benzo[*a*]pyrene	*3*, 91 (1973); *32*, 211 (1983) (*corr. 68*, 477); *Suppl. 7*, 58 (1987)
Benzo[*e*]pyrene	*3*, 137 (1973); *32*, 225 (1983); *Suppl. 7*, 58 (1987)
1,4-Benzoquinone (see *para*-Quinone)	
1,4-Benzoquinone dioxime	*29*, 185 (1982); *Suppl. 7*, 58 (1987); *71*, 1251 (1999)
Benzotrichloride (*see also* α-Chlorinated toluenes and benzoyl chloride)	*29*, 73 (1982); *Suppl. 7*, 148 (1987); *71*, 453 (1999)
Benzoyl chloride (*see also* α-Chlorinated toluenes and benzoyl chloride)	*29*, 83 (1982) (*corr. 42*, 261); *Suppl. 7*, 126 (1987); *71*, 453 (1999)
Benzoyl peroxide	*36*, 267 (1985); *Suppl. 7*, 58 (1987); *71*, 345 (1999)
Benzyl acetate	*40*, 109 (1986); *Suppl. 7*, 58 (1987); *71*, 1255 (1999)
Benzyl chloride (*see also* α-Chlorinated toluenes and benzoyl chloride)	*11*, 217 (1976) (*corr. 42*, 256); *29*, 49 (1982); *Suppl. 7*, 148 (1987); *71*, 453 (1999)
Benzyl violet 4B	*16*, 153 (1978); *Suppl. 7*, 58 (1987)
Bertrandite (*see* Beryllium and beryllium compounds)	
Beryllium and beryllium compounds	*1*, 17 (1972); *23*, 143 (1980) (*corr. 42*, 260); *Suppl. 7*, 127 (1987); *58*, 41 (1993)

Beryllium acetate (*see* Beryllium and beryllium compounds)
Beryllium acetate, basic (*see* Beryllium and beryllium compounds)
Beryllium-aluminium alloy (*see* Beryllium and beryllium compounds)
Beryllium carbonate (*see* Beryllium and beryllium compounds)
Beryllium chloride (*see* Beryllium and beryllium compounds)
Beryllium-copper alloy (*see* Beryllium and beryllium compounds)
Beryllium-copper-cobalt alloy (*see* Beryllium and beryllium compounds)
Beryllium fluoride (*see* Beryllium and beryllium compounds)
Beryllium hydroxide (*see* Beryllium and beryllium compounds)
Beryllium-nickel alloy (*see* Beryllium and beryllium compounds)
Beryllium oxide (*see* Beryllium and beryllium compounds)

C

Cabinet-making (*see* Furniture and cabinet-making)
Cadmium acetate (*see* Cadmium and cadmium compounds)
Cadmium and cadmium compounds

    2, 74 (1973); *11*, 39 (1976)
    (*corr. 42*, 255); *Suppl. 7*, 139
    (1987); *58*, 119 (1993)

Cadmium chloride (*see* Cadmium and cadmium compounds)
Cadmium oxide (*see* Cadmium and cadmium compounds)
Cadmium sulfate (*see* Cadmium and cadmium compounds)
Cadmium sulfide (*see* Cadmium and cadmium compounds)
Caffeic acid   *56*, 115 (1993)
Caffeine   *51*, 291 (1991)
Calcium arsenate (*see* Arsenic and arsenic compounds)
Calcium chromate (see Chromium and chromium compounds)
Calcium cyclamate (*see* Cyclamates)
Calcium saccharin (*see* Saccharin)
Cantharidin   *10*, 79 (1976); *Suppl. 7*, 59 (1987)
Caprolactam   *19*, 115 (1979) (*corr. 42*, 258);
    39, 247 (1986) (*corr. 42*, 264);
    Suppl. 7, 59, 390 (1987); *71*, 383
    (1999)
Captafol   *53*, 353 (1991)
Captan   *30*, 295 (1983); *Suppl. 7*, 59 (1987)
Carbaryl   *12*, 37 (1976); *Suppl. 7*, 59 (1987)
Carbazole   *32*, 239 (1983); *Suppl. 7*, 59
    (1987); *71*, 1319 (1999)
3-Carbethoxypsoralen   *40*, 317 (1986); *Suppl. 7*, 59 (1987)
Carbon black   *3*, 22 (1973); *33*, 35 (1984);
    Suppl. 7, 142 (1987); *65*, 149
    (1996)
Carbon tetrachloride   *1*, 53 (1972); *20*, 371 (1979);
    Suppl. 7, 143 (1987); *71*, 401
    (1999)
Carmoisine   *8*, 83 (1975); *Suppl. 7*, 59 (1987)
Carpentry and joinery   *25*, 139 (1981); *Suppl. 7*, 378
    (1987)
Carrageenan   *10*, 181 (1976) (*corr. 42*, 255); *31*,
    79 (1983); *Suppl. 7*, 59 (1987)
Catechol   *15*, 155 (1977); *Suppl. 7*, 59
    (1987); *71*, 433 (1999)
CCNU (*see* 1-(2-Chloroethyl)-3-cyclohexyl-1-nitrosourea)
Ceramic fibres (see Man-made mineral fibres)
Chemotherapy, combined, including alkylating agents (*see* MOPP and
    other combined chemotherapy including alkylating agents)
Chloral   *63*, 245 (1995)
Chloral hydrate   *63*, 245 (1995)
Chlorambucil   *9*, 125 (1975); *26*, 115 (1981);
    Suppl. 7, 144 (1987)
Chloramphenicol   *10*, 85 (1976); *Suppl. 7*, 145
    (1987); *50*, 169 (1990)
Chlordane (*see also* Chlordane/Heptachlor)   *20*, 45 (1979) (*corr. 42*, 258)
Chlordane and Heptachlor   *Suppl. 7*, 146 (1987); *53*, 115
    (1991); *79*, 411 (2001)

Coal-tars	*35*, 83 (1985); *Suppl. 7*, 175 (1987)
Cobalt[III] acetate (*see* Cobalt and cobalt compounds)	
Cobalt-aluminium-chromium spinel (*see* Cobalt and cobalt compounds)	
Cobalt and cobalt compounds (*see also* Implants, surgical)	*52*, 363 (1991)
Cobalt[II] chloride (*see* Cobalt and cobalt compounds)	
Cobalt-chromium alloy (*see* Chromium and chromium compounds)	
Cobalt-chromium-molybdenum alloys (*see* Cobalt and cobalt compounds)	
Cobalt metal powder (*see* Cobalt and cobalt compounds)	
Cobalt naphthenate (*see* Cobalt and cobalt compounds)	
Cobalt[II] oxide (*see* Cobalt and cobalt compounds)	
Cobalt[II,III] oxide (*see* Cobalt and cobalt compounds)	
Cobalt[II] sulfide (*see* Cobalt and cobalt compounds)	
Coffee	*51*, 41 (1991) (*corr. 52*, 513)
Coke production	*34*, 101 (1984); *Suppl. 7*, 176 (1987)
Combined oral contraceptives (*see* Oral contraceptives, combined)	
Conjugated equine oestrogens	*72*, 399 (1999)
Conjugated oestrogens (*see also* Steroidal oestrogens)	*21*, 147 (1979); *Suppl. 7*, 283 (1987)
Contraceptives, oral (*see* Oral contraceptives, combined; Sequential oral contraceptives)	
Copper 8-hydroxyquinoline	*15*, 103 (1977); *Suppl. 7*, 61 (1987)
Coronene	*32*, 263 (1983); *Suppl. 7*, 61 (1987)
Coumarin	*10*, 113 (1976); *Suppl. 7*, 61 (1987); *77*, 193 (2000)
Creosotes (*see also* Coal-tars)	*35*, 83 (1985); *Suppl. 7*, 177 (1987)
meta-Cresidine	*27*, 91 (1982); *Suppl. 7*, 61 (1987)
para-Cresidine	*27*, 92 (1982); *Suppl. 7*, 61 (1987)
Cristobalite (*see* Crystalline silica)	
Crocidolite (*see* Asbestos)	
Crotonaldehyde	*63*, 373 (1995) (*corr. 65*, 549)
Crude oil	*45*, 119 (1989)
Crystalline silica (*see also* Silica)	*42*, 39 (1987); *Suppl. 7*, 341 (1987); *68*, 41 (1997)
Cycasin (*see also* Methylazoxymethanol)	*1*, 157 (1972) (*corr. 42*, 251); *10*, 121 (1976); *Suppl. 7*, 61 (1987)
Cyclamates	*22*, 55 (1980); *Suppl. 7*, 178 (1987); *73*, 195 (1999)
Cyclamic acid (*see* Cyclamates)	
Cyclochlorotine	*10*, 139 (1976); *Suppl. 7*, 61 (1987)
Cyclohexanone	*47*, 157 (1989); *71*, 1359 (1999)
Cyclohexylamine (*see* Cyclamates)	
Cyclopenta[*cd*]pyrene	*32*, 269 (1983); *Suppl. 7*, 61 (1987)
Cyclopropane (*see* Anaesthetics, volatile)	
Cyclophosphamide	*9*, 135 (1975); *26*, 165 (1981); *Suppl. 7*, 182 (1987)
Cyproterone acetate	*72*, 49 (1999)

D

2,4-D (*see also* Chlorophenoxy herbicides; Chlorophenoxy herbicides, occupational exposures to)	*15*, 111 (1977)

Dacarbazine	*26*, 203 (1981); *Suppl. 7*, 184 (1987)
Dantron	*50*, 265 (1990) (*corr. 59*, 257)
D&C Red No. 9	*8*, 107 (1975); *Suppl. 7*, 61 (1987); *57*, 203 (1993)
Dapsone	*24*, 59 (1980); *Suppl. 7*, 185 (1987)
Daunomycin	*10*, 145 (1976); *Suppl. 7*, 61 (1987)
DDD (*see* DDT)	
DDE (*see* DDT)	
DDT	*5*, 83 (1974) (*corr. 42*, 253); *Suppl. 7*, 186 (1987); *53*, 179 (1991)
Decabromodiphenyl oxide	*48*, 73 (1990); *71*, 1365 (1999)
Deltamethrin	*53*, 251 (1991)
Deoxynivalenol (*see* Toxins derived from *Fusarium graminearum, F. culmorum* and *F. crookwellense*)	
Diacetylaminoazotoluene	*8*, 113 (1975); *Suppl. 7*, 61 (1987)
N,N'-Diacetylbenzidine	*16*, 293 (1978); *Suppl. 7*, 61 (1987)
Diallate	*12*, 69 (1976); *30*, 235 (1983); *Suppl. 7*, 61 (1987)
2,4-Diaminoanisole and its salts	*16*, 51 (1978); *27*, 103 (1982); *Suppl. 7*, 61 (1987); *79*, 619 (2001)
4,4'-Diaminodiphenyl ether	*16*, 301 (1978); *29*, 203 (1982); *Suppl. 7*, 61 (1987)
1,2-Diamino-4-nitrobenzene	*16*, 63 (1978); *Suppl. 7*, 61 (1987)
1,4-Diamino-2-nitrobenzene	*16*, 73 (1978); *Suppl. 7*, 61 (1987); *57*, 185 (1993)
2,6-Diamino-3-(phenylazo)pyridine (*see* Phenazopyridine hydrochloride)	
2,4-Diaminotoluene (*see also* Toluene diisocyanates)	*16*, 83 (1978); *Suppl. 7*, 61 (1987)
2,5-Diaminotoluene (*see also* Toluene diisocyanates)	*16*, 97 (1978); *Suppl. 7*, 61 (1987)
ortho-Dianisidine (*see* 3,3'-Dimethoxybenzidine)	
Diatomaceous earth, uncalcined (*see* Amorphous silica)	
Diazepam	*13*, 57 (1977*); Suppl. 7*, 189 (1987); *66*, 37 (1996)
Diazomethane	*7*, 223 (1974); *Suppl. 7*, 61 (1987)
Dibenz[*a,h*]acridine	*3*, 247 (1973); *32*, 277 (1983); *Suppl. 7*, 61 (1987)
Dibenz[*a,j*]acridine	*3*, 254 (1973); *32*, 283 (1983); *Suppl. 7*, 61 (1987)
Dibenz[*a,c*]anthracene	*32*, 289 (1983) (*corr. 42*, 262); *Suppl. 7*, 61 (1987)
Dibenz[*a,h*]anthracene	*3*, 178 (1973) (*corr. 43*, 261); *32*, 299 (1983); *Suppl. 7*, 61 (1987)
Dibenz[*a,j*]anthracene	*32*, 309 (1983); *Suppl. 7*, 61 (1987)
7*H*-Dibenzo[*c,g*]carbazole	*3*, 260 (1973); *32*, 315 (1983); *Suppl. 7*, 61 (1987)
Dibenzodioxins, chlorinated (other than TCDD) (*see* Chlorinated dibenzodioxins (other than TCDD))	
Dibenzo[*a,e*]fluoranthene	*32*, 321 (1983); *Suppl. 7*, 61 (1987)
Dibenzo[*h,rst*]pentaphene	*3*, 197 (1973); *Suppl. 7*, 62 (1987)
Dibenzo[*a,e*]pyrene	*3*, 201 (1973); *32*, 327 (1983); *Suppl. 7*, 62 (1987)
Dibenzo[*a,h*]pyrene	*3*, 207 (1973); *32*, 331 (1983); *Suppl. 7*, 62 (1987)

Di(2-ethylhexyl) phthalate	*29*, 269 (1982) (*corr. 42*, 261); *Suppl. 7*, 62 (1987); *77*, 41 (2000)
1,2-Diethylhydrazine	*4*, 153 (1974); *Suppl. 7*, 62 (1987); *71*, 1401 (1999)
Diethylstilboestrol	*6*, 55 (1974); *21*, 173 (1979) (*corr. 42*, 259); *Suppl. 7*, 273 (1987)
Diethylstilboestrol dipropionate (*see* Diethylstilboestrol)	
Diethyl sulfate	*4*, 277 (1974); *Suppl. 7*, 198 (1987); *54*, 213 (1992); *71*, 1405 (1999)
N,N'-Diethylthiourea	*79*, 649 (2001)
Diglycidyl resorcinol ether	*11*, 125 (1976); *36*, 181 (1985); *Suppl. 7*, 62 (1987); *71*, 1417 (1999)
Dihydrosafrole	*1*, 170 (1972); *10*, 233 (1976) *Suppl. 7*, 62 (1987)
1,8-Dihydroxyanthraquinone (*see* Dantron)	
Dihydroxybenzenes (*see* Catechol; Hydroquinone; Resorcinol)	
Dihydroxymethylfuratrizine	*24*, 77 (1980); *Suppl. 7*, 62 (1987)
Diisopropyl sulfate	*54*, 229 (1992); *71*, 1421 (1999)
Dimethisterone (*see also* Progestins; Sequential oral contraceptives)	*6*, 167 (1974); *21*, 377 (1979))
Dimethoxane	*15*, 177 (1977); *Suppl. 7*, 62 (1987)
3,3'-Dimethoxybenzidine	*4*, 41 (1974); *Suppl. 7*, 198 (1987)
3,3'-Dimethoxybenzidine-4,4'-diisocyanate	*39*, 279 (1986); *Suppl. 7*, 62 (1987)
para-Dimethylaminoazobenzene	*8*, 125 (1975); *Suppl. 7*, 62 (1987)
para-Dimethylaminoazobenzenediazo sodium sulfonate	*8*, 147 (1975); *Suppl. 7*, 62 (1987)
trans-2-[(Dimethylamino)methylimino]-5-[2-(5-nitro-2-furyl)-vinyl]-1,3,4-oxadiazole	*7*, 147 (1974) (*corr. 42*, 253); *Suppl. 7*, 62 (1987)
4,4'-Dimethylangelicin plus ultraviolet radiation (*see also* Angelicin and some synthetic derivatives)	*Suppl. 7*, 57 (1987)
4,5'-Dimethylangelicin plus ultraviolet radiation (*see also* Angelicin and some synthetic derivatives)	*Suppl. 7*, 57 (1987)
2,6-Dimethylaniline	*57*, 323 (1993)
N,N-Dimethylaniline	*57*, 337 (1993)
Dimethylarsinic acid (*see* Arsenic and arsenic compounds)	
3,3'-Dimethylbenzidine	*1*, 87 (1972); *Suppl. 7*, 62 (1987)
Dimethylcarbamoyl chloride	*12*, 77 (1976); *Suppl. 7*, 199 (1987); *71*, 531 (1999)
Dimethylformamide	*47*, 171 (1989); *71*, 545 (1999)
1,1-Dimethylhydrazine	*4*, 137 (1974); *Suppl. 7*, 62 (1987); *71*, 1425 (1999)
1,2-Dimethylhydrazine	*4*, 145 (1974) (*corr. 42*, 253); *Suppl. 7*, 62 (1987); *71*, 947 (1999)
Dimethyl hydrogen phosphite	*48*, 85 (1990); *71*, 1437 (1999)
1,4-Dimethylphenanthrene	*32*, 349 (1983); *Suppl. 7*, 62 (1987)
Dimethyl sulfate	*4*, 271 (1974); *Suppl. 7*, 200 (1987); *71*, 575 (1999)
3,7-Dinitrofluoranthene	*46*, 189 (1989); *65*, 297 (1996)
3,9-Dinitrofluoranthene	*46*, 195 (1989); *65*, 297 (1996)
1,3-Dinitropyrene	*46*, 201 (1989)
1,6-Dinitropyrene	*46*, 215 (1989)
1,8-Dinitropyrene	*33*, 171 (1984); *Suppl. 7*, 63 (1987); *46*, 231 (1989)

Lead carbonate (*see* Lead and lead compounds)
Lead chloride (*see* Lead and lead compounds)
Lead chromate (*see* Chromium and chromium compounds)
Lead chromate oxide (*see* Chromium and chromium compounds)
Lead naphthenate (*see* Lead and lead compounds)
Lead nitrate (*see* Lead and lead compounds)
Lead oxide (*see* Lead and lead compounds)
Lead phosphate (*see* Lead and lead compounds)
Lead subacetate (*see* Lead and lead compounds)
Lead tetroxide (*see* Lead and lead compounds)

Leather goods manufacture	*25*, 279 (1981); *Suppl. 7*, 235 (1987)
Leather industries	*25*, 199 (1981); *Suppl. 7*, 232 (1987)
Leather tanning and processing	*25*, 201 (1981); *Suppl. 7*, 236 (1987)
Ledate (*see also* Lead and lead compounds)	*12*, 131 (1976)
Levonorgestrel	*72*, 49 (1999)
Light Green SF	*16*, 209 (1978); *Suppl. 7*, 65 (1987)
d-Limonene	*56*, 135 (1993); *73*, 307 (1999)

Lindane (*see* Hexachlorocyclohexanes)
Liver flukes (*see Clonorchis sinensis, Opisthorchis felineus* and
    Opisthorchis viverrini)

Lumber and sawmill industries (including logging)	*25*, 49 (1981); *Suppl. 7*, 383 (1987)
Luteoskyrin	*10*, 163 (1976); *Suppl. 7*, 65 (1987)
Lynoestrenol	*21*, 407 (1979); *Suppl. 7*, 293 (1987); *72*, 49 (1999)

M

Magenta	*4*, 57 (1974) (*corr. 42*, 252); *Suppl. 7*, 238 (1987); *57*, 215 (1993)
Magenta, manufacture of (*see also* Magenta)	*Suppl. 7*, 238 (1987); *57*, 215 (1993)
Malathion	*30*, 103 (1983); *Suppl. 7*, 65 (1987)
Maleic hydrazide	*4*, 173 (1974) (*corr. 42*, 253); *Suppl. 7*, 65 (1987)
Malonaldehyde	*36*, 163 (1985); *Suppl. 7*, 65 (1987); *71*, 1037 (1999)
Malondialdehyde (*see* Malonaldehyde)	
Maneb	*12*, 137 (1976); *Suppl. 7*, 65 (1987)
Man-made mineral fibres	*43*, 39 (1988)
Mannomustine	*9*, 157 (1975); *Suppl. 7*, 65 (1987)
Mate	*51*, 273 (1991)
MCPA (*see also* Chlorophenoxy herbicides; Chlorophenoxy herbicides, occupational exposures to)	*30*, 255 (1983)
MeA-α-C	*40*, 253 (1986); *Suppl. 7*, 65 (1987)
Medphalan	*9*, 168 (1975); *Suppl. 7*, 65 (1987)
Medroxyprogesterone acetate	*6*, 157 (1974); *21*, 417 (1979) (*corr. 42*, 259); *Suppl. 7*, 289 (1987); *72*, 339 (1999)
Megestrol acetate	*Suppl. 7*, 293 (1987); *72*, 49 (1999)

4,4'-Methylene bis(2-chloroaniline)	*4*, 65 (1974) (*corr. 42*, 252); *Suppl. 7*, 246 (1987); *57*, 271 (1993)
4,4'-Methylene bis(*N,N*-dimethyl)benzenamine	*27*, 119 (1982); *Suppl. 7*, 66 (1987)
4,4'-Methylene bis(2-methylaniline)	*4*, 73 (1974); *Suppl. 7*, 248 (1987)
4,4'-Methylenedianiline	*4*, 79 (1974) (*corr. 42*, 252); *39*, 347 (1986); *Suppl. 7*, 66 (1987)
4,4'-Methylenediphenyl diisocyanate	*19*, 314 (1979); *Suppl. 7*, 66 (1987); *71*, 1049 (1999)
2-Methylfluoranthene	*32*, 399 (1983); *Suppl. 7*, 66 (1987)
3-Methylfluoranthene	*32*, 399 (1983); *Suppl. 7*, 66 (1987)
Methylglyoxal	*51*, 443 (1991)
Methyl iodide	*15*, 245 (1977); *41*, 213 (1986); *Suppl. 7*, 66 (1987); *71*, 1503 (1999)
Methylmercury chloride (*see* Mercury and mercury compounds)	
Methylmercury compounds (*see* Mercury and mercury compounds)	
Methyl methacrylate	*19*, 187 (1979); *Suppl. 7*, 66 (1987); *60*, 445 (1994)
Methyl methanesulfonate	*7*, 253 (1974); *Suppl. 7*, 66 (1987); *71*, 1059 (1999)
2-Methyl-1-nitroanthraquinone	*27*, 205 (1982); *Suppl. 7*, 66 (1987)
N-Methyl-*N'*-nitro-*N*-nitrosoguanidine	*4*, 183 (1974); *Suppl. 7*, 248 (1987)
3-Methylnitrosaminopropionaldehyde [*see* 3-(*N*-Nitrosomethylamino)-propionaldehyde]	
3-Methylnitrosaminopropionitrile [*see* 3-(*N*-Nitrosomethylamino)-propionitrile]	
4-(Methylnitrosamino)-4-(3-pyridyl)-1-butanal [*see* 4-(*N*-Nitrosomethyl-amino)-4-(3-pyridyl)-1-butanal]	
4-(Methylnitrosamino)-1-(3-pyridyl)-1-butanone [*see* 4-(-Nitrosomethyl-amino)-1-(3-pyridyl)-1-butanone]	
N-Methyl-*N*-nitrosourea	*1*, 125 (1972); *17*, 227 (1978); *Suppl. 7*, 66 (1987)
N-Methyl-*N*-nitrosourethane	*4*, 211 (1974); *Suppl. 7*, 66 (1987)
N-Methylolacrylamide	*60*, 435 (1994)
Methyl parathion	*30*, 131 (1983); *Suppl. 7*, 66, 392 (1987)
1-Methylphenanthrene	*32*, 405 (1983); *Suppl. 7*, 66 (1987)
7-Methylpyrido[3,4-*c*]psoralen	*40*, 349 (1986); *Suppl. 7*, 71 (1987)
Methyl red	*8*, 161 (1975); *Suppl. 7*, 66 (1987)
Methyl selenac (*see also* Selenium and selenium compounds)	*12*, 161 (1976); *Suppl. 7*, 66 (1987)
Methylthiouracil	*7*, 53 (1974); *Suppl. 7*, 66 (1987); *79*, 75 (2001)
Metronidazole	*13*, 113 (1977); *Suppl. 7*, 250 (1987)
Mineral oils	*3*, 30 (1973); *33*, 87 (1984) (*corr. 42*, 262); *Suppl. 7*, 252 (1987)
Mirex	*5*, 203 (1974); *20*, 283 (1979) (*corr. 42*, 258); *Suppl. 7*, 66 (1987)
Mists and vapours from sulfuric acid and other strong inorganic acids	*54*, 41 (1992)
Mitomycin C	*10*, 171 (1976); *Suppl. 7*, 67 (1987)
Mitoxantrone	*76*, 289 (2000)
MNNG (*see N*-Methyl-*N'*-nitro-*N*-nitrosoguanidine)	

6-Nitrobenzo[*a*]pyrene	*33*, 187 (1984); *Suppl. 7*, 67 (1987); *46*, 255 (1989)
4-Nitrobiphenyl	*4*, 113 (1974); *Suppl. 7*, 67 (1987)
6-Nitrochrysene	*33*, 195 (1984); *Suppl. 7*, 67 (1987); *46*, 267 (1989)
Nitrofen (technical-grade)	*30*, 271 (1983); *Suppl. 7*, 67 (1987)
3-Nitrofluoranthene	*33*, 201 (1984); *Suppl. 7*, 67 (1987)
2-Nitrofluorene	*46*, 277 (1989)
Nitrofural	*7*, 171 (1974); *Suppl. 7*, 67 (1987); *50*, 195 (1990)
5-Nitro-2-furaldehyde semicarbazone (*see* Nitrofural)	
Nitrofurantoin	*50*, 211 (1990)
Nitrofurazone (*see* Nitrofural)	
1-[(5-Nitrofurfurylidene)amino]-2-imidazolidinone	*7*, 181 (1974); *Suppl. 7*, 67 (1987)
N-[4-(5-Nitro-2-furyl)-2-thiazolyl]acetamide	*1*, 181 (1972); *7*, 185 (1974); *Suppl. 7*, 67 (1987)
Nitrogen mustard	*9*, 193 (1975); *Suppl. 7*, 269 (1987)
Nitrogen mustard *N*-oxide	*9*, 209 (1975); *Suppl. 7*, 67 (1987)
Nitromethane	*77*, 487 (2000)
1-Nitronaphthalene	*46*, 291 (1989)
2-Nitronaphthalene	*46*, 303 (1989)
3-Nitroperylene	*46*, 313 (1989)
2-Nitro-*para*-phenylenediamine (*see* 1,4-Diamino-2-nitrobenzene)	
2-Nitropropane	*29*, 331 (1982); *Suppl. 7*, 67 (1987); *71*, 1079 (1999)
1-Nitropyrene	*33*, 209 (1984); *Suppl. 7*, 67 (1987); *46*, 321 (1989)
2-Nitropyrene	*46*, 359 (1989)
4-Nitropyrene	*46*, 367 (1989)
N-Nitrosatable drugs	*24*, 297 (1980) (*corr. 42*, 260)
N-Nitrosatable pesticides	*30*, 359 (1983)
N'-Nitrosoanabasine	*37*, 225 (1985); *Suppl. 7*, 67 (1987)
N'-Nitrosoanatabine	*37*, 233 (1985); *Suppl. 7*, 67 (1987)
N-Nitrosodi-*n*-butylamine	*4*, 197 (1974); *17*, 51 (1978); *Suppl. 7*, 67 (1987)
N-Nitrosodiethanolamine	*17*, 77 (1978); *Suppl. 7*, 67 (1987); *77*, 403 (2000)
N-Nitrosodiethylamine	*1*, 107 (1972) (*corr. 42*, 251); *17*, 83 (1978) (*corr. 42*, 257); *Suppl. 7*, 67 (1987)
N-Nitrosodimethylamine	*1*, 95 (1972); *17*, 125 (1978) (*corr. 42*, 257); *Suppl. 7*, 67 (1987)
N-Nitrosodiphenylamine	*27*, 213 (1982); *Suppl. 7*, 67 (1987)
para-Nitrosodiphenylamine	*27*, 227 (1982) (*corr. 42*, 261); *Suppl. 7*, 68 (1987)
N-Nitrosodi-*n*-propylamine	*17*, 177 (1978); *Suppl. 7*, 68 (1987)
N-Nitroso-*N*-ethylurea (*see* *N*-Ethyl-*N*-nitrosourea)	
N-Nitrosofolic acid	*17*, 217 (1978); *Suppl. 7*, 68 (1987)
N-Nitrosoguvacine	*37*, 263 (1985); *Suppl. 7*, 68 (1987)
N-Nitrosoguvacoline	*37*, 263 (1985); *Suppl. 7*, 68 (1987)
N-Nitrosohydroxyproline	*17*, 304 (1978); *Suppl. 7*, 68 (1987)
3-(*N*-Nitrosomethylamino)propionaldehyde	*37*, 263 (1985); *Suppl. 7*, 68 (1987)
3-(*N*-Nitrosomethylamino)propionitrile	*37*, 263 (1985); *Suppl. 7*, 68 (1987)
4-(*N*-Nitrosomethylamino)-4-(3-pyridyl)-1-butanal	*37*, 205 (1985); *Suppl. 7*, 68 (1987)

Oestrogen-progestin replacement therapy (*see* Post-menopausal
 oestrogen-progestogen therapy)
Oestrogen replacement therapy (*see* Post-menopausal oestrogen
 therapy)
Oestrogens (*see* Oestrogens, progestins and combinations)
Oestrogens, conjugated (*see* Conjugated oestrogens)
Oestrogens, nonsteroidal (*see* Nonsteroidal oestrogens)
Oestrogens, progestins (progestogens) and combinations   *6* (1974); *21* (1979); *Suppl. 7*, 272
                                                                (1987); *72*, 49, 339, 399, 531
                                                                (1999)

Oestrogens, steroidal (*see* Steroidal oestrogens)
Oestrone   *6*, 123 (1974); *21*, 343 (1979)
                                                                (*corr. 42*, 259); *Suppl. 7*, 286
                                                                (1987); *72*, 399 (1999)

Oestrone benzoate (*see* Oestrone)
Oil Orange SS   *8*, 165 (1975); *Suppl. 7*, 69 (1987)
Opisthorchis felineus (infection with)   *61*, 121 (1994)
Opisthorchis viverrini (infection with)   *61*, 121 (1994)
Oral contraceptives, combined   *Suppl. 7*, 297 (1987); *72*, 49 (1999)
Oral contraceptives, sequential (*see* Sequential oral contraceptives)
Orange I   *8*, 173 (1975); *Suppl. 7*, 69 (1987)
Orange G   *8*, 181 (1975); *Suppl. 7*, 69 (1987)
Organolead compounds (*see also* Lead and lead compounds)   *Suppl. 7*, 230 (1987)
Oxazepam   *13*, 58 (1977); *Suppl. 7*, 69 (1987);
                                                                66, 115 (1996)
Oxymetholone (*see also* Androgenic (anabolic) steroids)   *13*, 131 (1977)
Oxyphenbutazone   *13*, 185 (1977); *Suppl. 7*, 69 (1987)

P

Paint manufacture and painting (occupational exposures in)   *47*, 329 (1989)
Palygorskite   *42*, 159 (1987); *Suppl. 7*, 117
                                                                (1987); *68*, 245 (1997)

Panfuran S (*see also* Dihydroxymethylfuratrizine)   *24*, 77 (1980); *Suppl. 7*, 69 (1987)
Paper manufacture (*see* Pulp and paper manufacture)
Paracetamol   *50*, 307 (1990); *73*, 401 (1999)
Parasorbic acid   *10*, 199 (1976) (*corr. 42*, 255);
                                                                Suppl. 7, 69 (1987)
Parathion   *30*, 153 (1983); *Suppl. 7*, 69 (1987)
Patulin   *10*, 205 (1976); *40*, 83 (1986);
                                                                Suppl. 7, 69 (1987)
Penicillic acid   *10*, 211 (1976); *Suppl. 7*, 69 (1987)
Pentachloroethane   *41*, 99 (1986); *Suppl. 7*, 69 (1987);
                                                                71, 1519 (1999)

Pentachloronitrobenzene (see Quintozene)
Pentachlorophenol (*see also* Chlorophenols; Chlorophenols,   *20*, 303 (1979); *53*, 371 (1991)
 occupational exposures to; Polychlorophenols and their sodium salts)
Permethrin   *53*, 329 (1991)
Perylene   *32*, 411 (1983); *Suppl. 7*, 69 (1987)
Petasitenine   *31*, 207 (1983); *Suppl. 7*, 69 (1987)
Petasites japonicus (*see also* Pyrrolizidine alkaloids)   *10*, 333 (1976)
Petroleum refining (occupational exposures in)   *45*, 39 (1989)
Petroleum solvents   *47*, 43 (1989)

Polystyrene (*see also* Implants, surgical)	*19*, 245 (1979); *Suppl. 7*, 70 (1987)
Polytetrafluoroethylene (*see also* Implants, surgical)	*19*, 288 (1979); *Suppl. 7*, 70 (1987)
Polyurethane foams (*see also* Implants, surgical)	*19*, 320 (1979); *Suppl. 7*, 70 (1987)
Polyvinyl acetate (*see also* Implants, surgical)	*19*, 346 (1979); *Suppl. 7*, 70 (1987)
Polyvinyl alcohol (*see also* Implants, surgical)	*19*, 351 (1979); *Suppl. 7*, 70 (1987)
Polyvinyl chloride (*see also* Implants, surgical)	*7*, 306 (1974); *19*, 402 (1979); *Suppl. 7*, 70 (1987)
Polyvinyl pyrrolidone	*19*, 463 (1979); *Suppl. 7*, 70 (1987); *71*, 1181 (1999)
Ponceau MX	*8*, 189 (1975); *Suppl. 7*, 70 (1987)
Ponceau 3R	*8*, 199 (1975); *Suppl. 7*, 70 (1987)
Ponceau SX	*8*, 207 (1975); *Suppl. 7*, 70 (1987)
Post-menopausal oestrogen therapy	*Suppl. 7*, 280 (1987); *72*, 399 (1999)
Post-menopausal oestrogen-progestogen therapy	*Suppl. 7*, 308 (1987); *72*, 531 (1999)
Potassium arsenate (*see* Arsenic and arsenic compounds)	
Potassium arsenite (*see* Arsenic and arsenic compounds)	
Potassium bis(2-hydroxyethyl)dithiocarbamate	*12*, 183 (1976); *Suppl. 7*, 70 (1987)
Potassium bromate	*40*, 207 (1986); *Suppl. 7*, 70 (1987); *73*, 481 (1999)
Potassium chromate (*see* Chromium and chromium compounds)	
Potassium dichromate (*see* Chromium and chromium compounds)	
Prazepam	*66*, 143 (1996)
Prednimustine	*50*, 115 (1990)
Prednisone	*26*, 293 (1981); *Suppl. 7*, 326 (1987)
Printing processes and printing inks	*65*, 33 (1996)
Procarbazine hydrochloride	*26*, 311 (1981); *Suppl. 7*, 327 (1987)
Proflavine salts	*24*, 195 (1980); *Suppl. 7*, 70 (1987)
Progesterone (*see also* Progestins; Combined oral contraceptives)	*6*, 135 (1974); *21*, 491 (1979) (*corr. 42*, 259)
Progestins (*see* Progestogens)	
Progestogens	*Suppl. 7*, 289 (1987); *72*, 49, 339, 531 (1999)
Pronetalol hydrochloride	*13*, 227 (1977) (*corr. 42*, 256); *Suppl. 7*, 70 (1987)
1,3-Propane sultone	*4*, 253 (1974) (*corr. 42*, 253); *Suppl. 7*, 70 (1987); *71*, 1095 (1999)
Propham	*12*, 189 (1976); *Suppl. 7*, 70 (1987)
β-Propiolactone	*4*, 259 (1974) (*corr. 42*, 253); *Suppl. 7*, 70 (1987); *71*, 1103 (1999)
n-Propyl carbamate	*12*, 201 (1976); *Suppl. 7*, 70 (1987)
Propylene	*19*, 213 (1979); *Suppl. 7*, 71 (1987); *60*, 161 (1994)
Propyleneimine (*see* 2-Methylaziridine)	
Propylene oxide	*11*, 191 (1976); *36*, 227 (1985) (*corr. 42*, 263); *Suppl. 7*, 328 (1987); *60*, 181 (1994)
Propylthiouracil	*7*, 67 (1974); *Suppl. 7*, 329 (1987); *79*, 91 (2001)

Sawmill industry (including logging) (*see* Lumber and
    sawmill industry (including logging))

Scarlet Red	*8*, 217 (1975); *Suppl. 7*, 71 (1987)
Schistosoma haematobium (infection with)	*61*, 45 (1994)
Schistosoma japonicum (infection with)	*61*, 45 (1994)
Schistosoma mansoni (infection with)	*61*, 45 (1994)
Selenium and selenium compounds	*9*, 245 (1975) (*corr. 42*, 255); *Suppl. 7*, 71 (1987)

Selenium dioxide (*see* Selenium and selenium compounds)
Selenium oxide (*see* Selenium and selenium compounds)

Semicarbazide hydrochloride	*12*, 209 (1976) (*corr. 42*, 256); *Suppl. 7*, 71 (1987)
Senecio jacobaea L. (*see also* Pyrrolizidine alkaloids)	*10*, 333 (1976)
Senecio longilobus (*see also* Pyrrolizidine alkaloids)	*10*, 334 (1976)
Seneciphylline	*10*, 319, 335 (1976); *Suppl. 7*, 71 (1987)
Senkirkine	*10*, 327 (1976); *31*, 231 (1983); *Suppl. 7*, 71 (1987)
Sepiolite	*42*, 175 (1987); *Suppl. 7*, 71 (1987); *68*, 267 (1997)
Sequential oral contraceptives (*see also* Oestrogens, progestins and combinations)	*Suppl. 7*, 296 (1987)
Shale-oils	*35*, 161 (1985); *Suppl. 7*, 339 (1987)
Shikimic acid (*see also* Bracken fern)	*40*, 55 (1986); *Suppl. 7*, 71 (1987)

Shoe manufacture and repair (*see* Boot and shoe manufacture
    and repair)

Silica (*see also* Amorphous silica; Crystalline silica)	*42*, 39 (1987)

Silicone (*see* Implants, surgical)

Simazine	*53*, 495 (1991); *73*, 625 (1999)

Slagwool (*see* Man-made mineral fibres)
Sodium arsenate (*see* Arsenic and arsenic compounds)
Sodium arsenite (*see* Arsenic and arsenic compounds)
Sodium cacodylate (*see* Arsenic and arsenic compounds)

Sodium chlorite	*52*, 145 (1991)

Sodium chromate (*see* Chromium and chromium compounds)
Sodium cyclamate (*see* Cyclamates)
Sodium dichromate (*see* Chromium and chromium compounds)

Sodium diethyldithiocarbamate	*12*, 217 (1976); *Suppl. 7*, 71 (1987)

Sodium equilin sulfate (*see* Conjugated oestrogens)
Sodium fluoride (*see* Fluorides)
Sodium monofluorophosphate (*see* Fluorides)
Sodium oestrone sulfate (*see* Conjugated oestrogens)

Sodium *ortho*-phenylphenate (*see also* *ortho*-Phenylphenol)	*30*, 329 (1983); *Suppl. 7*, 71, 392 (1987); *73*, 451 (1999)

Sodium saccharin (*see* Saccharin)
Sodium selenate (*see* Selenium and selenium compounds)
Sodium selenite (*see* Selenium and selenium compounds)
Sodium silicofluoride (*see* Fluorides)

Solar radiation	*55* (1992)
Soots	*3*, 22 (1973); *35*, 219 (1985); *Suppl. 7*, 343 (1987)
Spironolactone	*24*, 259 (1980); *Suppl. 7*, 344 (1987); *79*, 317 (2001)

Tannic acid   *10*, 253 (1976) (*corr. 42*, 255);
                                                             Suppl. 7, 72 (1987)
Tannins (*see* also Tannic acid)   *10*, 254 (1976); *Suppl. 7*, 72 (1987)
TCDD (*see* 2,3,7,8-Tetrachlorodibenzo-*para*-dioxin)
TDE (*see* DDT)
Tea   *51*, 207 (1991)
Temazepam   *66*, 161 (1996)
Teniposide   *76*, 259 (2000)
Terpene polychlorinates   *5*, 219 (1974); *Suppl. 7*, 72 (1987)
Testosterone (*see also* Androgenic (anabolic) steroids)   *6*, 209 (1974); *21*, 519 (1979)
Testosterone oenanthate (*see* Testosterone)
Testosterone propionate (*see* Testosterone)
2,2′,5,5′-Tetrachlorobenzidine   *27*, 141 (1982); *Suppl. 7*, 72 (1987)
2,3,7,8-Tetrachlorodibenzo-*para*-dioxin   *15*, 41 (1977); *Suppl. 7*, 350
                                                             (1987); *69*, 33 (1997)
1,1,1,2-Tetrachloroethane   *41*, 87 (1986); *Suppl. 7*, 72 (1987);
                                                             71, 1133 (1999)
1,1,2,2-Tetrachloroethane   *20*, 477 (1979); *Suppl. 7*, 354
                                                             (1987); *71*, 817 (1999)
Tetrachloroethylene   *20*, 491 (1979); *Suppl. 7*, 355
                                                             (1987); *63*, 159 (1995) (*corr. 65*,
                                                             549)
2,3,4,6-Tetrachlorophenol (*see* Chlorophenols; Chlorophenols,
       occupational exposures to; Polychlorophenols and their sodium salts)
Tetrachlorvinphos   *30*, 197 (1983); *Suppl. 7*, 72 (1987)
Tetraethyllead (*see* Lead and lead compounds)
Tetrafluoroethylene   *19*, 285 (1979); *Suppl. 7*, 72
                                                             (1987); *71*, 1143 (1999)
Tetrakis(hydroxymethyl)phosphonium salts   *48*, 95 (1990); *71*, 1529 (1999)
Tetramethyllead (*see* Lead and lead compounds)
Tetranitromethane   *65*, 437 (1996)
Textile manufacturing industry, exposures in   *48*, 215 (1990) (*corr. 51*, 483)
Theobromine   *51*, 421 (1991)
Theophylline   *51*, 391 (1991)
Thioacetamide   *7*, 77 (1974); *Suppl. 7*, 72 (1987)
4,4′-Thiodianiline   *16*, 343 (1978); *27*, 147 (1982);
                                                             Suppl. 7, 72 (1987)
Thiotepa   *9*, 85 (1975); *Suppl. 7*, 368 (1987);
                                                             50, 123 (1990)
Thiouracil   *7*, 85 (1974); *Suppl. 7*, 72 (1987);
                                                             79, 127 (2001)
Thiourea   *7*, 95 (1974); *Suppl. 7*, 72 (1987);
                                                             79, 703 (2001)
Thiram   *12*, 225 (1976); *Suppl. 7*, 72
                                                             (1987); *53*, 403 (1991)
Titanium (*see* Implants, surgical)
Titanium dioxide   *47*, 307 (1989)
Tobacco habits other than smoking (*see* Tobacco products, smokeless)
Tobacco products, smokeless   *37* (1985) (*corr. 42*, 263; *52*, 513);
                                                             Suppl. 7, 357 (1987)
Tobacco smoke   *38* (1986) (*corr. 42*, 263); *Suppl. 7*,
                                                             359 (1987)
Tobacco smoking (*see* Tobacco smoke)
ortho-Tolidine (*see* 3,3′-Dimethylbenzidine)

List of IARC Monographs on the Evaluation of Carcinogenic Risks to Humans*

---

*Certain older volumes, marked out-of-print, are still available directly from IARCPress. Further, high-quality photocopies of all out-of-print volumes may be purchased from University Microfilms International, 300 North Zeeb Road, Ann Arbor, MI 48106-1346, USA (Tel.: 313-761-4700, 800-521-0600).